DIGITAL SATELLITE COMMUNICATIONS

The Macmillan Database/Data Communications Series

Digital Satellite Communications

Tri T. Ha

Virginia Polytechnic Institute
and State University

Macmillan Publishing Company
NEW YORK

Collier Macmillan Publishers
LONDON

Macmillan Publishing Company
866 Third Avenue, New York, NY 10022

Collier Macmillan Canada, Inc.

Printed in the United States of America

printing number   year
1   2   3   4   5   6   7   8   9   10   6   7   8   9   0   1   2   3   4   5

Library of Congress Cataloging-in-Publication Data

Ha, Tri T., 1949-
  Digital satellite communications.

  Includes bibliographies and index.
  1. Artificial satellites in telecommunication.
2. Digital communications. I. Title.
TK5104.H3   1986   621.38'0422   85-18873
ISBN 0-02-948860-5

To Minh-Hien

for love and support

in difficult times

Contents

Preface

This book addresses the fundamental principles of satellite communications. Although most materials deal with digital systems, analog systems are also treated in detail. The book is written at a level suitable for use by seniors and first-year graduate students in electrical engineering. Examples are used throughout the text to help students develop a good understanding of the subject. Because the material in each chapter is independent or nearly independent of each other, the book provides full flexibility for the instructor. For a senior course I suggest Chapters 1–5 (one quarter) or Chapters 1–6 with selected materials in Chapter 7 (one semester). For a first-year graduate course, Chapters 6–9 are appropriate for one quarter, and Chapters 6–11 for one semester. The end-of-chapter problems are an important and integral part of the book. They are intended not only to help students test their comprehension of the subject but also to extend the discussion sometimes made brief because of the page constraint. A list of symbols and a list of acronyms are provided to make reading more pleasant.

Because satellite technology is a big picture, it is impossible to cover all topics in a book of this length. Therefore, I have focused only on fundamental principles, and I have tried to keep the mathematics to a minimum level. Advanced concepts of satellite communications are referred to in appropriate references. I hope that this book on satellite communications will serve students and satellite engineers well.

It would not have been possible for me to complete this book without the direct and indirect help of many persons, whom I wish to acknowledge. Professor D. B. Hodge (EE Department Head) provided a light teaching load in my first year at VPI&SU. Professor C. W. Bostian (Clayton Ayre Professor and Head of Satellite Communication Group) has given me the opportunity to work in satellite projects. Professor S. Haykin (McMaster University) and Professor L. Couch II (University of Florida) reviewed the book and offered many valuable comments. T. Battle, A. Bennani, N. Miles, and W. Wattaul provided supporting materials; Tuan Ha helped on TDMA. I also wish to thank Professor R. W. Newcomb for being an academic mentor during my graduate career at the University of Maryland. Finally, I wish to thank my wife Hien, to whom this book is dedicated, for typing and retyping the manuscript. Her support and encouragement during the past three years made this project possible.

List of Acronyms

ACU:	antenna control unit
ARFA:	assisted receive frame acquisition
AWGN:	additive white Gaussian noise
BPF:	bandpass filter
CCIR:	International Radio Consultative Committee
CCITT:	International Telegraph and Telephone Consultative Committee
CCR:	carrier and clock recovery
CDMA:	code division multiple access
CONUS:	Continental United States
CSC:	common signaling channel
DA-FDMA:	demand assignment–frequency division multiple access
DA-TDMA:	demand assignment–time division multiple access
DAMA:	demand assignment multiple access
DAU:	data acquisition unit
DBS:	direct broadcasting satellite
DC:	downconverter
DS:	direct sequence
DS-CDMA:	direct sequence–code division multiple access
DSI:	digital speech interpolation
EIRP:	effective isotropic radiated power

EQL:	equalizer
FCC:	Federal Communication Commission
FDM:	frequency division multiplexing
FDMA:	frequency division multiple access
FH:	frequency hop
FH-CDMA:	frequency hop–code division multiple access
FM:	frequency modulation
FSK:	frequency-shift keying
HPA:	high-power amplifier
IF:	intermediate frequency
INTELSAT:	International Telecommunications Satellite Organi-zation
ITU:	International Telecommunication Union
LNA:	low-noise amplifier
LO:	local oscillator
LPF:	low-pass filter
M&C:	monitoring and control
M-ary FSK:	M-ary frequency-shift keying
M-ary PSK:	M-ary phase-shift keying
MSK:	minimum shift keying
MTTF:	mean time to failure
MTTR:	mean time to repair
NCC:	network control center
OMT:	orthogonal mode transducer
OQPSK:	off-set quaternary phase shift keying
PCM:	pulse code modulation
PLL:	phase-locked loop
PN:	pseudo-noise
PRB:	primary reference burst
PRS:	primary reference station
PSK:	phase-shift keying
QPSK:	quaternary phase-shift keying
RBT:	receive burst timing
RF:	radio frequency
RFA:	receive frame acquisition
RFS:	receive frame synchronization
RFT:	receive frame timing
SB:	short burst
SCPB:	single channel per burst
SCPB-DAMA:	single channel per burst–demand assignment mul-tiple access
SCPC:	single channel per carrier
SCPC-DAMA:	single channel per carrier–demand assignment mul-tiple access

SRB:	secondary reference burst
SRS:	secondary reference station
SS-TDMA:	satellite-switched TDMA
SSB:	superframe short burst
SSB:	single sideband
SSB-AM-FDMA:	single sideband-amplitude modulation-frequency division multiple access
TA:	transmit acquisition
TB:	traffic burst
TBT:	transmit burst timing
TDM:	time division multiplexing
TDMA:	time division multiple access
TFA:	transmit frame acquisition
.TFS:	transmit frame synchronization
TFT:	transmit frame timing
TIM:	terrestrial interface module
TRT:	timing and reference transponder
TS:	transmit synchronization
TV:	television
TWTA:	traveling wave tube amplifier
UC:	upconverter
VCO:	voltage-controlled oscillator
WARC:	World Administrative Radio Conference

List of Symbols

a:	orbital radius (42,164.2 km), traffic intensity
A:	amplitude, area, azimuth angle
$b(t)$:	baseband signal
B:	bandwidth
$B(n,a)$:	blocking probability
B_3:	3-dB bandwidth
B_{rms}:	rms bandwidth (10.95)
BO_i:	input back-off
BO_o:	output back-off
c:	light velocity (2.997925×10^5 km/s)
C:	carrier power
$\mathscr{C}$:	channel capacity
d:	Hamming distance
d_d:	downlink slant range
d_N:	distance from satellite to station N
d_u:	uplink slant range
D:	antenna diameter
D_N:	transmit frame delay
E:	elevation angle
E_b:	energy per bit
E_c:	energy per coded bit
E_s:	energy per symbol

$E\{x\}$:	expected value of x
Erfc(x):	complementary error function (9.31)
f:	frequency
F:	force, noise figure
g:	gravitational constant, rectangular pulse
G:	antenna gain, channel traffic
$h(t)$:	impulse response
H:	orbital altitude (35786.045 km), transfer function
i:	inclination angle
I:	interference power
$I_n(x)$:	modified Bessel function of order n of the first kind
j:	imaginary number, summation index, product index
J:	jamming power
J_0:	jamming density
$J_n(x)$:	Bessel function of order n of the first kind
k:	Boltzmann constant (1.38×10^{-23} J/K), information block length in a codeword, number of users in DS-CDMA
K:	degree Kelvin, randomized interval
ln:	natural logarithm
log:	logarithm of base 10
$\log_2$:	logarithm of base 2
L:	length, power loss
L_a:	atmospheric attenuation
L_d:	downlink free-space attenuation
L_r:	rain attenuation
L_u:	uplink free space attenuation
m:	mass
M:	$M = 2^k$ in M-ary signaling
n:	length of a code word
N:	noise power, processing gain
$\mathcal{N}$:	effective noise power
$N_0/2$:	noise power spectral density
$\mathcal{N}_0/2$:	effective noise power spectral density
p:	transition probability
$p(x)$:	probability density function of x
P:	power
P_A:	availability probability
P_b:	probability of bit error
P_d:	detection probability
P_F:	false alarm probability
P_s:	probability of symbol error
Pr$\{x\}$:	probability of x
q:	successful probability of a packet
$Q(x)$:	Gaussian integral (9.29)

r:	distance
R:	reliability, resistance, channel capacity
R_b:	bit rate
R_e:	earth radius (6378.155 km)
R_p:	chip rate
R_s:	symbol rate
$s(t)$:	signal
S:	channel throughput
$S(f)$:	power spectral density
t:	error-correcting capability, time
T:	noise temperature
T_b:	bit duration
T_0:	ambient temperature (290 K)
T_e:	equivalent noise temperature
T_f:	frame length
T_R:	satellite propagation delay
T_s:	symbol duration, system noise temperature
$u(t)$:	unit step function
v:	true anomaly
V:	orbital velocity, voltage
W:	aperture window
X:	cross-polarization discrimination, x coordinate
$y(t)$:	envelope of a burst
Y:	y coordinate
z:	a random variable
Z:	z coordinate
α:	coupling coefficient
β:	coupling coefficient
β_{max}:	maximum root mean square of partial cross-correlations
γ:	partial cross-correlation
Γ:	reflection coefficient
δ:	duty cycle
ΔA:	differential attenuation
$\Delta\psi$:	differential phase
ϵ:	rms surface error of antenna, threshold of unique word detection
η:	antenna efficiency
θ:	carrier phase, antenna off-axis angle
θ_1:	earth station latitude
θ_L:	earth station longitude
λ:	wavelength, a constant in Chernoff bound, latitude, arrival rate
μ:	mean value, mean hang-up rate
π:	3.1416

ρ:	failure rate, traffic intensity
$\rho(\tau)$:	autocorrelation
ρ_{nm}:	cross-correlation
σ^2:	variance
τ:	time delay
ϕ:	carrier phase
ω:	angular frequency
Ω:	carrier power flux density
Ω_{sat}:	saturation carrier flux density
x^*:	conjugate of x
$\lvert x \rvert$:	absolute value of x
$\overline{f(t)}$:	time average of $f(t)$
$f(t) \circledast g(t)$:	convolution of $f(t)$ and $g(t)$
$\mathrm{Re}(x)$:	real part of x

DIGITAL SATELLITE COMMUNICATIONS

chapter *1*

Elements of Satellite Communication

The unique feature of communications satellites is their ability to simultaneously link all users on the earth's surface, thereby providing distance-insensitive point-to-multipoint communications. This capability applies to fixed terminals on earth and to mobile terminals on land, in the air, and at sea. Also, with satellites, capacity can be dynamically allocated to users who need it. These features make satellite communications systems unique in design. This chapter serves as an overview of satellite communication and prepares the reader for more elaborate study in the rest of the book.

Arthur C. Clarke, author of many famous books on exploration, wrote in *Wireless World* in 1945 [1] that a satellite with a circular equatorial orbit at a correct altitude of 35,786 km would make one revolution every 24 h; that is, it would rotate at the same angular velocity as the earth. An observer looking at such a geostationary satellite would see it hanging at a fixed spot in the sky. Clarke showed that three geostationary satellites powered by solar energy could provide worldwide communications for all possible types of services. Clarke's vision became a reality 20 years later when the *International Telecommunications Satellite Organization* (INTELSAT), established in 1964, launched the *Early Bird (INTELSAT I)* in April 1965. By mid-1983 a total of 33 INTELSAT satellites had been launched, ranging from instruments with a small capacity (240 voice circuits or one television channel) to those with a

1

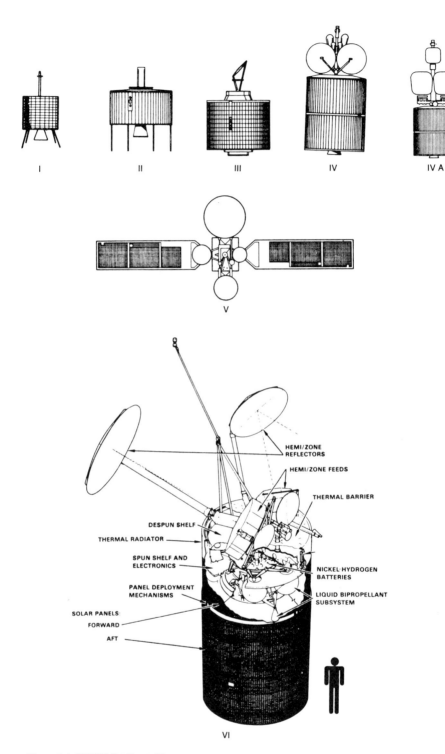

Figure 1.1 INTELSAT satellites.

Table 1.1 Electromagnetic frequency spectrum

Frequency	Wavelength (m)	Designation
3 Hz–30 kHz	10^8–10^4	Very low frequency (VLF)
30–300 kHz	10^4–10^3	Low frequency (LF)
300 kHz–3 MHz	10^3–10^2	Medium frequency (MF)
3–30 MHz	10^2–10	High frequency (HF)
30–300 MHz	10–1	Very high frequency (VHF)
300 MHz–3 GHz	1–10^{-1}	Ultrahigh frequency (UHF)
3–30 GHz	10^{-1}–10^{-2}	Superhigh frequency (SHF)
30–300 GHz	10^{-2}–10^{-3}	Extremely high frequency (EHF)
10^3–10^7 GHz	3×10^{-5}–3×10^{-9}	Infrared, visible light, ultraviolet

huge capacity (12,500 voice circuits and two television channels) and covering three regions—the Atlantic, Pacific, and Indian oceans (Fig. 1.1). By 1984 hundreds of geostationary satellites were in service, as summarized in [2].

1.1 SATELLITE FREQUENCY BANDS

Communications systems employ the electromagnetic frequency spectrum shown in Table 1.1. The frequencies used for satellite communications are allocated in superhigh-frequency (SHF) and extremely high-frequency (EHF) bands which are broken down into subbands as summarized in Table 1.2. Spectrum management is an important activity that facilitates the orderly use of the electromagnetic frequency spectrum not only for satellite communications but for other telecommunications applications as well. This is done under the auspices of the

Table 1.2 Satellite frequency spectrum

Frequency band	Range (GHz)
L	1–2
S	2–4
C	4–8
X	8–12
Ku	12–18
K	18–27
Ka	27–40
Millimeter	40–300

International Telecommunication Union (ITU) which is a specialized agency of the *United Nations* (UN). It predates the UN, having come into existence in 1932 as a result of the merging of the *International Telegraph Union* (1865–1932) and the *Radio Telegraph Union* (1903–1932). There are four permanent organs of the ITU: (1) the *General Secretariat,* headquartered in *Geneva* and responsible for executive management and technical cooperation; (2) the *International Frequency Registration Board* (IFRB), responsible for recording frequencies and orbital positions and for advising member countries on operation of the maximum practical number of radio channels in portions of the spectrum where harmful interference may occur; (3) the *International Radio Consultative Committee* (CCIR, from the initial letters in French), responsible for studying technical and operational questions relating to radiocommunications which results in reports, recommendations, resolutions, and decisions published as a group in the *Green Books* every 4 yr following CCIR plenary assemblies; and (4) the *International Telegraph* and *Telephone Consultative Committee* (CCITT), responsible for studying technical, operational, and tarriff questions relating to telegraphy and telephony and for adopting reports and recommendations.

The ITU has developed rules and guidelines called radio regulations at a series of international radio conferences held since 1903. The 1979 *World Administrative Radio Conference* (WARC-79) was the most recent in this long series. The frequency bands allocated by WARC-79 for satellite communications involve 17 service categories (although some of them represent special subcategories), as listed in Table 1.3, and three geographic regions: region 1 which includes Europe, Africa, the USSR, and Mongolia; region 2 which includes North and South America and Greenland; and region 3 which includes Asia (except the USSR and Mongolia), Australia, and the Southwest Pacific. Tables 1.4 and 1.5 show the WARC-79 frequency allocations for *fixed satellite service* (FSS) and *broadcasting satellite service* (BSS).

Table 1.3 Satellite services

Fixed	Meteorological
Intersatellite	Space operation
Mobile	Amateur
Land mobile	Radiodetermination
Maritime mobile	Radionavigation
Aeronautical mobile	Aeronautical radionavigation
Broadcasting	Maritime radionavigation
Earth exploration	Standard frequency and time signal
Space research	

Table 1.4 Frequency allocations for fixed satellite service

Frequency range (GHz)	Restrictions[a]	Frequency range (GHz)	Restrictions[a]
2.5–2.535	1n, 2d, 3d	18.1–21.2	d
2.535–2.655	1n, 2b, 3n	27–27.5	1n, 2u[c], 3u[c]
2.655–2.690	1n, 2b, 3u	27.5–31	u
3.4–4.2	d	37.5–40.5	d
4.5–4.8	d	42.5–43.5	u
5.725–5.85	1u, 2n, 3n	47.2–49.2	u[c]
5.85–7.075	u	49.2–50.2	u
7.25–7.75	d	50.4–51.4	u
7.9–8.4	u	71–74	u
10.7–11.7	1b[b], 2d, 3d	74–75.5	u
11.7–12.3	1n, 2d, 3n	81–84	d
12.5–12.7	1b, 2n, 3d	92–95	u
12.7–12.75	1b, 2u, 3d	102–105	d
12.75–13.25	u	149–164	d
14–14.5	u	202–217	u
14.5–14.8	u[b]	231–241	d
17.3–17.7	u[b]	265–275	u
17.7–18.1	b[b]		

[a]1, Region 1; 2, region 2; 3, region 3; u, uplink (earth to space); d, downlink (space to earth); n, not allocated; b, bidirectional.

[b]uplink limited to BSS feeder links.

[c]Intended for but not limited to BSS feeder links.

Table 1.5 Frequency allocations for broadcasting satellite service

Frequency range (GHz)	Restriction[a]
0.62–0.79	t
2.5–2.69	c
11.7–12.1	1, 3 only
12.1–12.2	
12.2–12.5	1, 2 only
12.5–12.7	2, 3c only
12.7–12.75	3c only
22.5–23	2, 3 only
40.5–42.5	
84–86	

[a]t, Television only; c, community reception only; 1, region 1; 2, region 2; 3, region 3.

1.2 SATELLITE SYSTEMS

A satellite system consists basically of a satellite in space which links many earth stations on the ground, as shown schematically in Fig. 1.2. The user generates the baseband signal which is routed to the earth station through the terrestrial network. The terrestrial network can be a telephone switch or a dedicated link to the earth station. At the earth station the baseband signal is processed and transmitted by a modulated radio frequency (RF) carrier to the satellite. The satellite can be thought of as a large repeater in space. It receives the modulated RF carriers in its uplink (earth-to-space) frequency spectrum from all the earth stations in the network, amplifies these carriers, and retransmits them back to earth in the downlink (space-to-earth) frequency spectrum which is different from the uplink frequency spectrum in order to avoid interference. The receiving earth station processes the modulated RF carrier down to the baseband signal which is sent through the terrestrial network to the user.

Most commercial communications satellites today utilize a 500-MHz bandwidth on the uplink and a 500-MHz bandwidth on the downlink. The most widely used frequency spectrum is the 6/4-GHz band, with an uplink of 5.725 to 7.075 GHz and a downlink of 3.4 to 4.8 GHz. The 6/4-GHz band for geostationary satellites is becoming overcrowded because it is also used by common carriers for terrestrial microwave links. Satellites are now being operated in the 14/12-GHz band using an uplink of

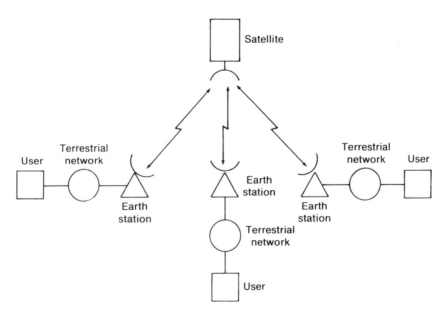

Figure 1.2 A basic satellite system.

12.75 to 14.8 GHz and a downlink of either 10.7 to 12.3 GHz or 12.5 to 12.7 GHz. The 14/12-GHz band will be used extensively in the future and is not yet congested, but one problem exists—rain, which attenuates 14/12-GHz signals much more than it does those at 6/4 GHz. The frequency spectrum in the 30/20-GHz band has also been set aside for commercial satellite communications, with a downlink of 18.1 to 21.2 GHz and an uplink of 27.5 to 31 GHz. Equipment for the 30/20-GHz band is still in the experimental stage and is expensive.

The 500-MHz satellite bandwidth at the 6/4 and 14/12-GHz bands can be segmented into many satellite transponder bandwidths. For example, eight transponders can be provided, each with a nominal bandwidth of 54 MHz and a center-to-center frequency spacing of 61 MHz. Modern communications satellites also employ frequency reuse to increase the number of transponders in the 500 MHz allocated to them. Frequency reuse can be accomplished through orthogonal polarizations where one transponder operates in one polarization (e.g., vertical polarization) and a cross-polarized transponder operates in the orthogonal polarization (e.g., horizontal polarization). Isolation of the two polarizations can be maintained at 30 dB or more by staggering the center frequencies of the cross-polarized transponders so that only sideband energy of the RF carriers overlaps, as shown in Fig. 1.3. With orthogonal polarizations a satellite can double the number of transponders in the available 500-MHz bandwidth, hence double its capacity. A review of orthogonal polarizations will be presented in Sec. 1.6.

With this brief discussion of a general satellite system we will now take a look at an earth station that transmits information to and receives information from a satellite. Figure 1.4 shows the functional elements of a digital earth station. Digital information in the form of binary digits from the terrestrial network enters the transmit side of the earth station and is then processed (buffered, multiplexed, formatted, etc.) by the baseband equipment so that these forms of information can be sent to the appropriate destinations. The presence of noise and the nonideal nature of any communication channel introduce errors in the information being sent and thus limit the rate at which it can be transmitted between the source and the destination. Users generally establish an error rate above which the received information is not usable. If the received information does not meet the error rate requirement, error-correction coding performed by the encoder can often be used to reduce the error rate to the acceptable level by inserting extra digits into the digital stream from the output of the baseband equipment. These extra digits carry no information but are used to accentuate the uniqueness of each information message. They are always chosen so as to make it unlikely that the channel disturbance will corrupt enough digits in a message to destroy its uniqueness.

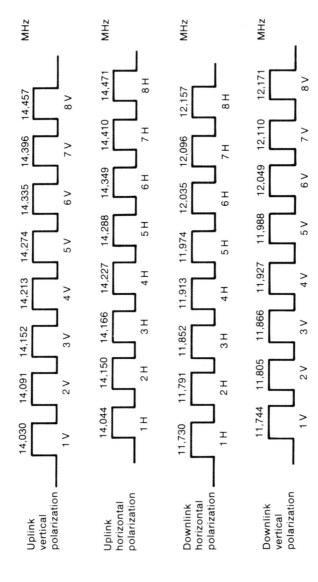

Figure 1.3 Staggering frequency reuse Ku-band transponders.

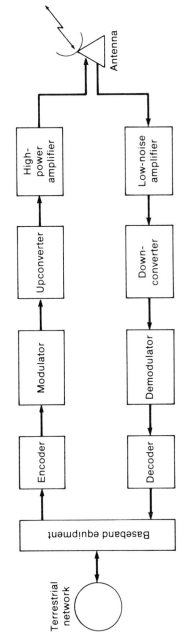

Figure 1.4 Functional block diagram of a digital earth station.

In order to transmit the baseband digital information over a satellite channel that is a bandpass channel, it is necessary to transfer the digital information to a carrier wave at the appropriate bandpass channel frequency. This technique is called *digital carrier modulation*. The function of the modulator is to accept the symbol stream from the encoder and use it to modulate an intermediate frequency (IF) carrier. In satellite communications, the IF carrier frequency is chosen at 70 MHz for a communication channel using a 36-MHz transponder bandwidth and at 140 MHz for a channel using a transponder bandwidth of 54 or 72 MHz. A carrier wave at an intermediate frequency rather than at the satellite RF uplink frequency is chosen because it is difficult to design a modulator that works at the uplink frequency spectrum (6 or 14 GHz, as discussed previously). For binary modulation schemes, each output digit from the encoder is used to select one of two possible waveforms. For M-ary modulation schemes, the output of the encoder is segmented into sets of k digits, where $M = 2^k$ and each k-digit set or symbol is used to select one of the M waveforms. For example, in one particular binary modulation scheme called *phase-shift keying* (PSK), the digit 1 is represented by the waveform $s_1(t) = A \cos \omega_0 t$ and the digit 0 is represented by the waveform $s_0(t) = -A \cos \omega_0 t$, where ω_0 is the intermediate frequency. (In this book the letter symbols ω and f will be used to denote angular frequency and frequency, respectively, and we will refer to both of them as "frequency.")

The modulated IF carrier from the modulator is fed to the upconverter, where its intermediate frequency ω_0 is translated to the uplink RF frequency ω_u in the uplink frequency spectrum of the satellite. This modulated RF carrier is then amplified by the high-power amplifier (HPA) to a suitable level for transmission to the satellite by the antenna.

On the receive side the earth station antenna receives the low-level modulated RF carrier in the downlink frequency spectrum of the satellite. A low-noise amplifier (LNA) is used to amplify this low-level RF carrier to keep the carrier-to-noise ratio at a level necessary to meet the error rate requirement. The downconverter accepts the amplified RF carrier from the output of the low-noise amplifier and translates the downlink frequency ω_d to the intermediate frequency ω_0. The reason for downconverting the RF frequency of the received carrier wave to the intermediate frequency is that it is much easier to design the demodulator to work at 70 or 140 MHz than at a downlink frequency of 4 or 12 GHz. The modulated IF carrier is fed to the demodulator, where the information is extracted. The demodulator estimates which of the possible symbols was transmitted based on observation of the received IF carrier. The probability that a symbol will be correctly detected depends on the carrier-to-noise ratio of the modulated carrier, the characteristics of the satellite channel, and the detection scheme employed. The decoder performs a function opposite that of the encoder. Because the sequence of symbols recovered by the demodulator may contain errors, the decoder must use the uniqueness of the redundant digits introduced by the encoder to correct the errors and recover information-bearing digits. The information stream is fed to the baseband equipment for processing for delivery to the terrestrial network.

In the United States the *Federal Communications Commission* (FCC) assigns orbital positions for all communications satellites to avoid interference between adjacent satellite systems operating at the same frequency. Before 1983 the spacing was established at 4° of the equatorial arc, and the smallest earth station antenna for a simultaneous transmit-receive operation allowed by the FCC is 5 m in diameter.

In 1983, the FCC ruled that fixed service communications satellites in the geostationary orbit should be spaced every 2° along the equatorial arc instead of 4°. This closer spacing allows twice as many satellites to occupy the same orbital arc.

The FCC ruling poses a major challenge to antenna engineers to design a directional feed for controlling the amount of energy received off-axis by the antenna feed, thus reducing interference from an adjacent satellite. This challenge is especially great because the trend in earth stations is toward smaller antennas, but smaller antennas have a wider beamwidth and thus look at a wider angle in the sky.

The FCC ruling specified that, as of July 1, 1984, all new satellite earth station antennas had to be manufactured to accommodate the spacing of 2° and that, as of January 1, 1987, all existing antennas must be modified to conform to the new standards.

1.3 TRANSMISSION AND MULTIPLEXING

In the above section we took a look at a simplified satellite communications system where digital information (a sequence of symbols instead of continuous signals) is carried between terrestrial networks. Historically, analog transmission has dominated satellite communications since its inception. Even today many satellite systems still transmit telephone and television signals using frequency modulation (FM), and this trend will continue for some time to come because of the large investment in existing earth stations. With the advent of digital electronics and computers, many earth stations have begun to use digital transmission to improve satellite capacity over analog transmission. These digital earth stations can interconnect digital terrestrial networks or analog terrestrial networks with appropriate analog-to-digital (A/D) conversion equipment. A clear advantage of digital transmission is that it permits integration of information in various forms. Such analog information as speech and visu-

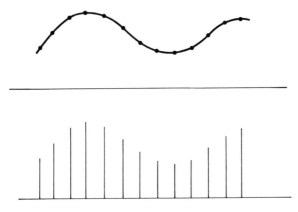

Figure 1.5 Sampling of an analog signal.

al signals can be converted to digital form and thereby combined with data for transmission, switching, processing, and retrieval.

1.3.1 Pulse Code Modulation

One commonly used technique for converting an analog signal to digital form is *pulse code modulation* (PCM) which requires three operations: sampling, quantizing, and coding. *Sampling* converts the continuous analog signal into a set of periodic pulses, the amplitudes of which represent the instantaneous amplitudes of the analog signal at the sampling instant, as shown in Fig. 1.5. A question naturally arises: What sampling rate is required in order to reconstruct the signal completely from these samples? Nyquist [3,4] proved that, if the analog signal is band-limited to a bandwidth of B hertz, the signal can be completely reconstructed if the sampling rate is at least the Nyquist rate which is $2B$. For example, telephone speech is band-limited to 4 kHz and thus requires 8000 samples per second. Since analog signals have a continuous amplitude range, the samples are also continuous in amplitude. When the continuous-amplitude samples are transmitted over a noisy channel, the receiver cannot discern the exact sequences of transmitted values. This effect of noise in the system can be minimized by breaking the sample amplitude into discrete levels and transmitting these levels using a binary scheme. The process of representing the continuous amplitude of the samples by a finite set of levels is called *quantizing*. If V quantized levels are employed to represent the amplitude range, it will take $\log_2 V$ bits to code each sample. In telephone transmission 256 quantized levels are employed, hence each sample is coded using $\log_2 256 = 8$ bits, and thus the digital bit rate is $8000 \times 8 = 64,000$ bits per second (bps).

1.3.2 Delta Modulation

It has been found that analog signals such as speech and video signals generally have a considerable amount of redundancy; that is, there is a significant correlation between successive samples when these signals are sampled at a rate slightly higher than the Nyquist rate. For example, the frequency spectrum of the human voice is 300 to 3400 Hz but it is sampled at 8000 samples per second in a PCM system. When these correlated samples are coded as in a PCM system, the resulting digital stream contains redundant information. The redundancy in these analog signals makes it possible to predict a sample value from the preceding sample values and to transmit the difference between the actual sample value and the predicted sample value estimated from past samples. This results in a technique called *difference encoding*. One of the simplest forms of difference encoding is *delta modulation* which provides a staircase approximation of the sampled version of the analog input signal as shown in Fig. 1.6. The difference between the input and the approximation is quantized into two levels, $+\Delta$ and $-\Delta$, corresponding to a positive and a negative difference, respectively. Thus at any sampling instant the approximation is increased by Δ or decreased by Δ depending on whether it is below or above the analog input signal. A digital output of 1 or 0 can be generated according to whether the difference is $+\Delta$ or $-\Delta$. In delta modulation, overloading can occur if the amplitude of the analog input signal changes too fast for the encoding to keep up. Increasing the step size Δ will result in poor resolution, and increasing the sampling rate will lead to a higher digital bit rate. A better scheme for avoiding overloading is to detect the overloading condition and to adjust the step size Δ to a larger value. This is called *adaptive delta modulation*. Delta modulation has been used to encode speech with good quality at 32,000 bps. Another important approach in digital encoding of analog signals is

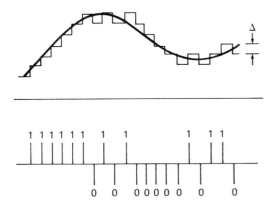

Figure 1.6 Delta modulation.

differential pulse code modulation. This is basically a modification of delta modulation where the difference between the analog input signal and its approximation at the sampling instant is quantized into V levels and the output of the encoder is coded into $\log_2 V$ bits. Differential PCM combines the simplicity of delta modulation and the multilevel quantizing feature of PCM, and in many applications it can provide good reproduction of analog signals comparable to PCM with a considerable reduction in the digital bit rate.

1.3.3 Time Division Multiplexing—Pulse Code Modulation

Information traffic between a terrestrial network and an earth station involves much more than a PCM channel at 64,000 bps. In order to carry many more channels simultaneously over a single transmission facility such as a wire pair or coaxial cable, multiplexing must be employed. One of the most widely used multiplexing techniques for telephone speech signals is *time division multiplexing—pulse code modulation* (TDM-PCM) as shown in Fig. 1.7. Here 24 speech signals are fed to 24 contacts of a pair of synchronized electronic switches at the transmit and receive ends. The continuous amplitude of the speech signals is repeatedly sampled as the switch rotates. Each of the 24 speech signals is sampled every 125 μs and interleaved to form a time division-multiplexed signal. Each sample of the time division-multiplexed signal is quantized and converted to a 8-bit PCM codeword. The 8-bit PCM codeword forms a time slot corresponding to a sample from one of the 24 speech signals. Twenty-four time slots form a 125-μs frame which consists of 192 bits and an additional 193rd bit at the end of the frame that is used for establishing and maintaining frame timing. Normally the receiver checks the 193rd bit every frame to make sure that it has not lost synchronization. If synchronization has been lost, the receiver can scan for the framing pattern and be resynchronized. Since there are 193 bits/125 μs, the total bit rate is 1.544 Mbps.

In addition to the voice signal, the frame also carries signaling information needed to transmit telephone dial tones as well as on-hook and off-hook signals. Every sixth frame, the least significant bit (the eighth bit) of each voice channel is deleted and a signaling bit is inserted in its place. This type of TDM-PCM bit stream is employed in the Bell System's T1 carrier which is used in North America [5].

An international standard also exists for PCM transmission. The CCITT has a recommendation for a PCM carrier at 2.048 Mbps. In this carrier, there are 32 8-bit time slots in each 125-μs frame. Thirty of these time slots are used for speech at a bit rate of 64 kbps, one for synchronization and one for signaling. The 2.048-Mbps PCM carriers are used outside North America and Japan.

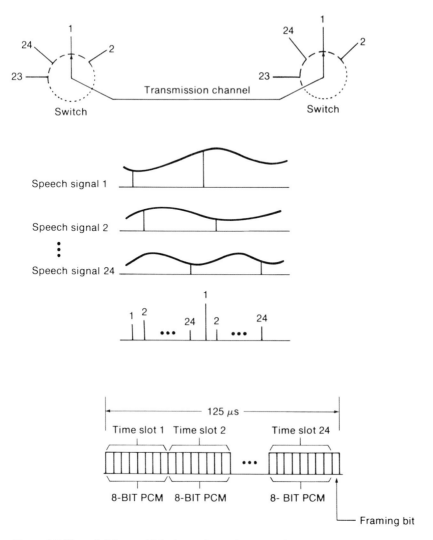

Figure 1.7 Time division multiplexing–pulse code modulation.

1.3.4 Digital Hierarchy

To transmit digitized analog signals such as telephone speech and visual signals having different bit rates, and data with a diversified bit rate over the same transmission channel, higher-order digital multiplexing or a digital hierarchy must be used. Figure 1.8 illustrates the Bell System digital hierarchy which consists of four levels. The respective data signals with bit rates of 1.544 Mbps (T1), 6.312 Mbps (T2), 44.736 Mpbs (T3), and

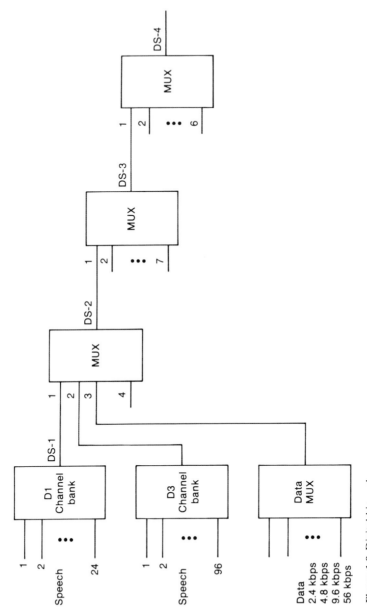

Figure 1.8 Digital hierarchy.

274.176 Mbps (T4) correspond to levels 1, 2, 3, and 4. Level 1 is the output of a D1 channel bank which time division-multiplexes and PCM-encodes 24 speech signals, or one of the four outputs of a D3 channel bank which multiplexes and encodes 96 speech channels, or the output of a data multiplexer which multiplexes data with bit rates of 2.4, 4.8, 9.6, and 56 kbps. The DS-1 data signal is carried by the T1 carrier system over a wire pair. Level 2 is formed by multiplexing four DS-1 data signals and is carried by a T2 carrier system over a wire pair. Level 3 is formed by multiplexing seven DS-2 data signals and is carried by a T3 carrier system over coaxial cable. Level 4 is formed by multiplexing six DS-3 data signals and is carried by a T4 carrier system over coaxial cable.

1.3.5 Frequency Division Multiplexing

Another form of multiplexing that characterizes analog communications is *frequency division multiplexing* (FDM), as shown in Fig. 1.9. Twelve speech signals, each of which occupies a bandwidth from 300 to 3400 Hz, are used to modulate 12 separate carriers each 4 kHz apart. The output of the modulator, which is the product of the speech signal and the carrier, consists of a lower sideband and an upper sideband centered around the carrier frequency. The signals are then passed through 4-kHz bandpass filters that reject the upper sideband and pass only the lower sideband. This technique is called *single-sideband suppressed carrier* (SSBSC) generation. Twelve lower sidebands are then combined to form a group that occupies the frequency band from 60 to 108 kHz. Five groups can be multiplexed in a similar fashion to form a supergroup of 60 speech signals that occupies the band from 312 to 552 kHz, and five supergroups form a master group of 300 speech channels that occupies the band from 812 to 2044 kHz.

1.3.6 Transmultiplexing

Despite the explosive progress in digital telecommunications technology, a major portion of terrestrial transmission facilities still uses a frequency division multiplexing hierarchy and will do so in the future because of the large investment in existing systems. For a digital earth station to interface with such an analog terrestrial network, some means for conversion between a FDM hierarchy and a digital hierarchy is needed. This can be achieved by a FDM-TDM converter, for example between two supergroups (120 speech channels) and five T1 carriers (120 PCM channels) as shown in Fig. 1.10. It consists of a FDM multiplexer or demultiplexer and PCM channel banks connected back to back. On the transmit side the FDM supergroups are demultiplexed to 120 individual speech channels

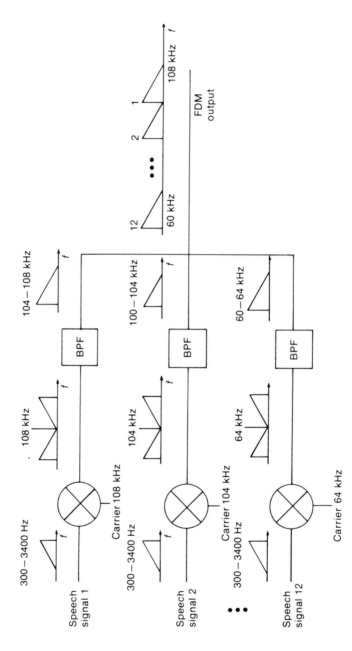

Figure 1.9 Frequency division multiplexing.

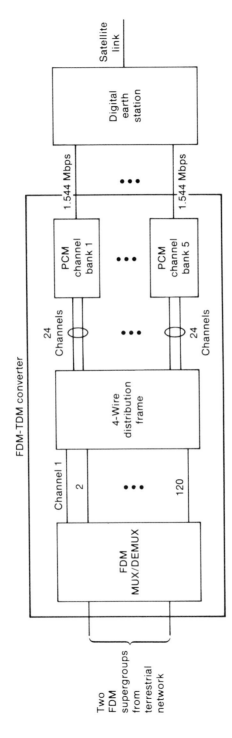

Figure 1.10 FDM-TDM converter.

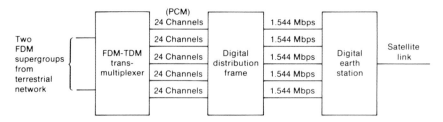

Figure 1.11 FDM-TDM transmultiplexer.

which are then time division-multiplexed and PCM-encoded by the channel banks into five separate 1.544-Mbps T1 carriers. On the receive side, the five separate 1.544-Mbps T1 carriers are PCM-decoded and time division-demultiplexed into 120 individual speech channels which are then frequency division-multiplexed into two supergroups. A four-wire distribution frame is used for the purpose of individual channel manipulation such as reordering, adding, deleting, and testing.

A special type of FDM-TDM converter called a *transmultiplexer* can interconvert FDM speech signals and TDM-PCM signals at the multiplex level and thus avoid breaking the signals down into individual speech channels, as shown in Fig. 1.11 where two supergroups are converted to five T1 carriers at 1.544 Mbps each. With the use of a distribution frame, individual channels can be reordered, added, deleted, and tested. The fundamental element of a transmultiplexer is shown in Fig. 1.12. FDM-TDM conversion is accomplished by removing any unwanted out-of-band components from the FDM signal with an analog bandpass filter (312 to 552 kHz for two supergroups). The filtered signal is passed through an A/D converter to produce a digital stream. The individual channel in this digital stream is processed by the digital signal processor via a real-time signal-processing algorithm. Transmultiplexer performance can be superior to that achievable with the conventional FDM-TDM conversion equipment shown in Fig. 1.10, with approximately 2:1 and 5:1 reduction advantages in cost per voice channel and in size, respectively.

Figure 1.12 Fundamental elements of a transmultiplexer.

1.4 MODULATION

As mentioned in Sec. 1.2, modulation must be employed to transmit baseband information over a bandpass channel. In *analog modulation* such as frequency modulation, which is extremely popular in satellite communications, the signal-to-noise ratio at the output of the FM demodulator is an intuitive measure of how well the FM demodulator can recover the analog information signal from the received modulated carrier in the presence of additive white Gaussian noise (AWGN). The *output signal-to-noise ratio* is defined as the ratio of the average power of the analog information signal to the average power of the noise at the output of the demodulator. In *digital modulation,* the performance of the demodulator is measured in terms of the average probability of bit error, or the *bit error rate* as it is often called. The binary information, which consists of sequences of 1 and 0 digits, can be used to modulate the phase, frequency, or amplitude of a carrier. Consider the carrier $A \cos(\omega_c t + \phi)$, where A is the carrier amplitude, ω_c is the carrier frequency, and ϕ is the carrier phase. To transmit the binary digit or bit 1, ϕ is set to 0 rad, and to transmit the bit 0, ϕ is set to π radians. Thus 1 is represented by the waveform $A \cos \omega_c t$, and 0 is represented by the waveform $A \cos(\omega_c t + \pi) = -A \cos \omega_c t$. This type of discrete phase modulation is called phase-shift keying (PSK). Similarly, 1 can be transmitted by using the waveform $A \cos \omega_1 t$ and 0 transmitted by using the waveform $A \cos \omega_2 t$, where $\omega_1 \neq \omega_2$. This type of digital modulation is called *frequency-shift keying* (FSK), where two waveforms at different carrier frequencies ω_1 and ω_2 are used to convey the binary information. The problem with digital modulation is that sometimes the binary digit 1 is transmitted but the demodulator decodes it as a 0, or vice versa, because of perturbation of the carrier by noise; this results in bit errors in the demodulation of the binary information. The average probability of bit error P_b is a convenient measure of the performance of the demodulator and is a function of the ratio of the energy per bit to the noise density, E_b/N_0, where the energy per bit E_b is the energy of the carrier during a signaling interval or bit duration T_b and $N_0/2$ is the noise power spectral density. When the baseband information is transmitted at a rate of R bits per second, the bit duration is simply $T_b = 1/R$ seconds, and this is also the signaling interval of the waveform that represents a particular bit. For example, in PSK modulation,

$$s_1(t) = A \cos \omega_c t \qquad 0 \leq t \leq T_b$$

$$s_2(t) = -A \cos \omega_c t \qquad 0 \leq t \leq T_b$$

where $s_1(t)$ represents 1 and $s_2(t)$ represents 0. By definition we have

$$E_b = \int_0^{T_b} s_1^2(t) \; dt = \int_0^{T_b} s_2^2(t) \; dt = \int_0^{T_b} A^2 \cos^2 \omega_c t \; dt \qquad (1.1)$$

Note that $E_b \approx A^2 T_b/2$ when $\omega_c > 2\pi/T_b$. The quantity E_b/N_0 can be related to the average carrier power C, and the noise power N measured within the receiver noise bandwidth B. By definition, the average carrier power is

$$C = \frac{1}{T_b} \int_0^{T_b} E[s^2(t)] \; dt \qquad (1.2)$$

where $s(t)$ is the carrier waveform during the signaling interval T_b and $E[\cdot]$ is the expected value. If all the carrier waveforms have identical energy E_b during any signaling interval, then

$$C = \frac{E_b}{T_b} \qquad (1.3)$$

Recall that the power spectral density of noise is $N_0/2$ and that the noise bandwidth is B. Hence the noise power measured within the noise bandwidth for both positive and negative frequencies is

$$N = N_0 B \qquad (1.4)$$

Therefore it is seen that the ratio of the energy per bit to the noise density can be expressed as

$$\frac{E_b}{N_0} = \frac{CT_b}{N/B} = T_b B \left(\frac{C}{N} \right) \qquad (1.5)$$

where C/N is the average carrier-to-noise ratio. In satellite communications, it is the quantity C/N that is directly evaluated, as we will discuss in Chap. 4. Once the C/N is known and the bandwidth of the receiver is selected, E_b/N_0 can be calculated, as well as the average probability of bit error P_b which is a function of E_b/N_0.

1.5 MULTIPLE ACCESS

One advantage of communications satellites over other transmission media is their ability to link all earth stations together, thereby providing point-to-multipoint communications. A satellite transponder can be accessed by many earth stations, and therefore it is necessary to have techniques for allocating transponder capacity to each of them. For example, consider a transponder with a bandwidth of 72 MHz. Assume that the bit duration–bandwidth product $T_b B$ in (1.5) is chosen to be 0.6; that is, every 0.6 Hz of the transponder bandwidth can be used to transmit 1 bps. (There are digital modulation schemes that can easily achieve $T_b B = 0.6$, which we will discuss in Chap. 9.) Then the transponder capacity is 120 Mbps, which can handle about 3562 voice channels at 32 kbps, assuming the transponder efficiency is 95%. It is unlikely that a single earth station would have this much traffic, therefore the transponder capacity must be wisely allocated to other earth station. Furthermore, to avoid chaos, we want the earth stations to gain access to the transponder capacity allocated to them in an orderly fashion. This is called *multiple access*. The two most commonly used multiple access schemes are *frequency division multiple access* (FDMA) and *time division multiple access* (TDMA).

FDMA has been used since the inception of satellite communication. Here each earth station in the community of earth stations that share the transponder capacity transmits one or more carriers to the satellite transponder at different center frequencies. Each carrier is assigned a frequency band in the transponder bandwidth, along with a small guard band to avoid interference between adjacent carriers. The satellite transponder receives all the carriers in its bandwidth, amplifies them, and retransmits them back to earth. The earth station in the satellite antenna beam served by the transponder can select the carrier that contains the messages intended for it. FDMA is illustrated in Fig. 1.13. The carrier modulation used in FDMA is FM or PSK.

In TDMA the earth stations that share the satellite transponder use a carrier at the same center frequency for transmission on a time division basis. Earth stations are allowed to transmit traffic bursts in a periodic time frame called the *TDMA frame*. During the burst, an earth station has the entire transponder bandwidth available to it for transmission. The transmit timing of the bursts is carefully synchronized so that all the bursts arriving at the satellite transponder are closely spaced in time but

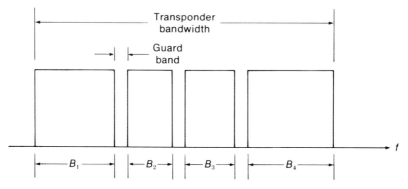

Figure 1.13 Concept of FDMA. B_m, Bandwidth reserved explicitly for station m, $m = 1, 2, 3, 4$.

do not overlap. The satellite transponder receives one burst at a time, amplifies it, and retransmits it back to earth. Thus every earth station in the satellite beam served by the transponder can receive the entire burst stream and extract the bursts intended for it. TDMA is illustrated in Fig. 1.14. The carrier modulation used in TDMA is always a digital modulation scheme. TDMA possesses many advantages over FDMA, especially in medium to heavy traffic networks, because there are a number of efficient techniques such as demand assignment and digital speech interpolation (Chap. 7) that are inherently suitable for TDMA and can maximize the amount of terrestrial traffic that can be handled by a satellite transponder. For example, a 72-MHz transponder can handle about 1781 satellite PCM voice channels. With a digital speech interpolation technique it can handle about twice this number, 3562 terrestrial PCM voice channels. In many TDMA networks employing demand assignment the amount of terrestrial traffic handled by the transponder can be increased up to 10 times. Of course these efficient techniques depend on the terrestrial traffic distribution in the network and must be used in situations that are suited to the characteristics of the technique. Although TDMA has many advantages, this does not mean that FDMA has no advantages over TDMA. Indeed, in networks with many links of low traffic, FDMA with demand assignment is overwhelmingly preferred to TDMA because of the low cost of equipment, as we will discuss in Chap. 7.

Besides FDMA and TDMA, a satellite system may also employ *random multiple access* schemes to serve a large population of users with bursty (low duty cycle) traffic. Here each user transmits at will and, if a collision (two users transmitting at the same time, causing severe interference that destroys their data) occurs, retransmits at a randomly selected

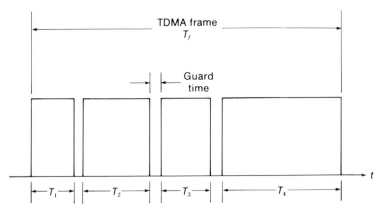

Figure 1.14 Concept of TDMA. T_m, Time slot reserved explicitly for station m, $m = 1, 2, 3, 4$, every frame. The TDMA frame length T_f is also the frame period.

time to avoid repeated collisions. Another type of multiple access scheme is *code division multiple access,* where each user employs a particular code address to spread the carrier bandwidth over a much larger bandwidth so that the earth station community can transmit simultaneously without frequency or time separation and with low interference.

1.6 FREQUENCY REUSE BY ORTHOGONAL POLARIZATIONS

One method of obtaining frequency reuse is to transmit two signals on the same frequency band (cochannel) by placing each on orthogonal polarizations; thereby doubling the information capacity carried by the satellite. A fundamental requirement of dual-polarized transmission is to maintain a good level of isolation between two polarizations so that the cochannel interference is acceptable.

The polarization of a radiated electromagnetic wave is the curve traced by the end point of the instantaneous electric field vector as observed along the direction of propagation. Polarization may be classified as linear, circular, or elliptical (Fig. 1.15). If the vector oscillates along a line, the field is linearly polarized. If the vector remains constant in length but traces a circle, the field is circularly polarized. When the sense of rotation of the vector is counterclockwise (the wave direction is off the page), the field is right-hand circularly polarized. If the rotation is clockwise, the field is left-hand circularly polarized. In general, the field may be elliptically polarized with either a right- or a left-hand sense of rotation.

A polarization ellipse is shown in Fig. 1.16. The wave associated with it travels in the $+z$ direction (off the page). The instantaneous electric field can be described as follows:

$$\mathscr{E} = \mathscr{E}_x \hat{\mathbf{x}} + \mathscr{E}_y \hat{\mathbf{y}}$$

$$= E_1 \cos \omega t \hat{\mathbf{x}} + E_2 \cos(\omega t + \theta)\hat{\mathbf{y}}$$

where E_1, E_2, θ, $\hat{\mathbf{x}}$, and $\hat{\mathbf{y}}$ are the peak value, phase difference, x unit vector, and y unit vector of the x and y components of ϵ, respectively. The tilt angle of the ellipse τ is the angle between the x axis and the major axis of the ellipse. The axial ratio r of the ellipse is defined as the ratio of the electric field component along the major axis to that along the minor axis. When these components are in phase, $\theta = 0°$, the wave is linearly polarized, and the orientation depends on the relative values of E_1 and E_2. For example, if $E_2 = 0$, then $\mathscr{E} = E_1 \cos \omega t \,\hat{\mathbf{x}}$ and the wave is horizontally polarized. When $E_1 = 0$, then $\mathscr{E} = E_2 \cos(\omega t + \theta)\hat{\mathbf{y}}$ and the wave is vertically polarized. Horizontal and vertical polarizations are two orthogonal linear polarizations. When $E_1 = E_2$ and $\theta = \pm90°$, the polarization is circular, with the right-hand rotation corresponding to $-90°$ and the left-

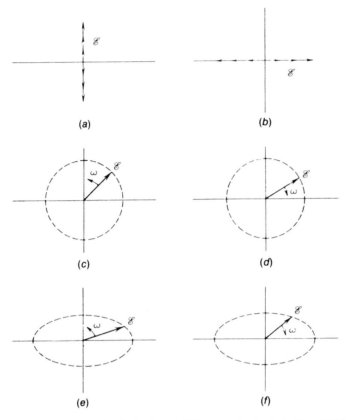

Figure 1.15 Types of polarizations. (*a*) Linear vertical polarization. (*b*) Linear horizontal polarization. (*c*) Right-hand circular polarization. (*d*) Left-hand circular polarization. (*e*) Right-hand elliptical polarization. (*f*) Left-hand elliptical polarization.

hand rotation corresponding to 90°. Right- and left-hand circular polarizations are two orthogonal circular polarizations.

In theory there is infinite isolation between orthogonal polarizations; that is, two signals in the same frequency band but each on orthogonal polarizations do not interfere with each other. A right-hand circularly polarized antenna absorbs the maximum amount of power from a right-hand circularly polarized wave and absorbs no power from a left-hand circularly polarized wave. A polarization mismatch factor p is normally employed to describe the coupling between polarizations. It varies between 1 and 0 and is equal to 1 when the two polarizations are the same or the antenna and the wave have the same polarization state. It is 0 when the two polarizations are orthogonal. For two general polarization ellipses with axial ratios r_1 and r_2, p is given by [6]

$$p(\pm) = \frac{1}{2} + \frac{\pm 4r_1 r_2 + (1 - r_1^2)(1 - r_2^2) \cos 2(\tau_1 - \tau_2)}{2(1 + r_1^2)(1 + r_2^2)}$$

$$= \frac{(r_1 \pm r_2)^2 + (1 - r_1^2)(1 - r_2^2) \cos^2(\tau_1 - \tau_2)}{(1 + r_1^2)(1 + r_2^2)} \tag{1.6}$$

The $+$ sign indicates the same sense of rotation, and the $-$ sign indicates the opposite sense of rotation.

As an example, consider right-hand and left-hand circular polarizations; thus $r_1 = r_2 = 1$ and $\tau_1 = \tau_2$.

$$p(-) = \frac{1}{2} + \frac{-4(1)(1) + (1 - 1)(1 - 1)(1)}{2(1 + 1)(1 + 1)} = 0$$

$$p(+) = 1$$

Now consider horizontal and vertical linear polarizations; $r_1 = r_2 = \infty$ and $\tau_1 - \tau_2 = \pm 90°$. Dividing the numerator and denominator by $r_1^2 r_2^2$ and taking the limit gives

$$p = \tfrac{1}{2} + \tfrac{1}{2} \cos 2(\tau_1 - \tau_2) = \cos^2(\tau_1 - \tau_2) = 0$$

For two aligned linear polarizations, $\tau_1 = \tau_2$, and thus $p = \cos^2 0 = 1$.

In a satellite system using dual orthogonal polarizations as a means of signal discrimination, an important parameter in determination of the quality of the system is the *cross-polarization discrimination ratio*. It is defined as the ratio of the polarization mismatch factors of an incident

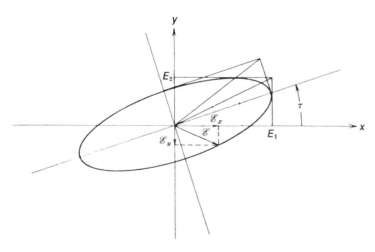

Figure 1.16 Polarization ellipse.

wave, when measured at the antenna port having the same polarization state (copolarized), to that measured at the port having the orthogonal polarization (cross-polarized). Thus

$$X = \frac{(r_1 + r_2)^2 + (1 - r_1^2)(1 - r_2^2) \cos^2(\tau_1 - \tau_2)}{(r_1 - r_2)^2 + (1 - r_1^2)(1 - r_2^2) \cos^2(\tau_1 - \tau_2)} \qquad (1.7)$$

As an example consider the case where the wave is left-hand circularly polarized with an axial ratio of $r_2 = 1$ and the antenna with an axial ratio of $r_1 \geq 1$. The copolarized port is designated as left-hand circular polarization, and $\tau_2 = \tau_1$. Then,

$$X = \frac{(r_1 + 1)^2}{(r_1 - 1)^2} = 10 \log \frac{(r_1 + 1)^2}{(r_1 - 1)^2} \text{ dB}$$

For an ideal antenna $r_1 = 1$ and $X = \infty$. For an antenna with $r_1 = 1.1$, $X = 26.4$ dB.

1.7 ADVENT OF DIGITAL SATELLITE COMMUNICATION

The future trend in satellite communications is toward digital techniques. Frequency division multiplexing–frequency modulation–frequency division multiple access (FDM-FM-FDMA) (Sec. 5.1) has been the most popular analog technique used in commercial satellite systems because it has been field-proven and makes it easy to provide quality satellite links at a low cost. As the number of earth stations increases, the transponder capacity decreases markedly in a FDM-FM-FDMA system. In addition, FDM-FM-FDMA is inflexible in responding to traffic changes. On the other hand, a digital satellite system such as quaternary phase shift keying–time division multiple access (QPSK-TDMA) (Chap. 6) can accommodate a large number of earth stations with only a small loss in transponder capacity. Furthermore, it can quickly respond to traffic variations. Also associated with digital satellite communications are techniques such as demand assignment and digital speech interpolation (Chap. 7) to further increase the efficiency. With advanced satellite systems with on-board switching and processing, multiple spot beams, and beam hopping (Sec. 6.11), a digital system can serve a mixture of large, medium, and small earth stations with high efficiency. Unlike an analog satellite system, a digital satellite system can employ error-correction coding (Sec. 10.10) to trade bandwidth for power. Finally, the use of code-division multiple access (CDMA) (Chap. 11) for low data rate applications enables users to employ micro earth stations (0.5-m antenna) at an extremely low cost ($3000) to obtain premium quality services. The flexibility of digital satellite systems will make them even more promising when integrated digital networks become fully implemented.

REFERENCES

1. A. C. Clarke, "Extraterrestrial Relays," *Wireless World,* Vol. 51, No. 10, Oct. 1945, pp. 305–308.
2. W. L. Morgan, "Satellite Locations—1984," *Proc. IEEE,* Vol. 72, No. 11, Nov. 1984, pp. 1434–1444.
3. M. Schwartz, *Information Transmission, Modulation, and Noise,* 3d ed. New York: McGraw-Hill, 1980.
4. S. Haykin, *Communication Systems,* 2d ed. New York: Wiley, 1983.
5. H. H. Henning and J. W. Pan, "D2 Channel Bank: System Aspects," *Bell Syst. Tech. J.,* Vol. 51, Oct. 1972, pp. 1641–1657.
6. M. L. Kales, "Elliptically Polarized Waves and Antennas," *Proc. IRE,* Vol. 39, No. 5, May 1951, pp. 544–549.
7. P. L. Bargellini, "Commerical U.S. Satellites," *IEEE Spectrum,* Vol. 16, No. 10, Oct. 1979, pp. 30–37.
8. W. L. Pritchard, "Satellite Communications—An Overview of the Problems and Programs," *Proc. IEEE,* Vol. 65, No. 3, Mar. 1977, pp. 294–307.
9. H. L. Van Trees et al., "Communications Satellites: Looking to the 1980's," *IEEE Spectrum,* Vol. 14, No. 12, Dec. 1977, pp. 42–51.
10. Special issues on satellite networks. *Proc. IEEE,* Vol. 72, No. 11, Nov. 1984.

PROBLEMS

1.1 A noiseless 10-kHz channel is sampled every 1 ms. What is the maximum data rate?

1.2 A noiseless channel is 3 MHz wide. How many bits per second can be sent if an eight-level digital signal is used?

1.3 For a noisy channel, Shannon's theorem states that the maximum data rate C of any channel whose bandwidth is B hertz and whose signal-to-noise ratio is S/N is given by

$$C = B \log_2 \left(1 + \frac{S}{N} \right) \tag{1.8}$$

What is the signal-to-noise ratio needed to transmit data at a rate of 10 kbps over a channel with a bandwidth of $B = 3000$ Hz?

1.4 A computer terminal is connected to a computer through a voice-grade telephone line having a bandwidth of 3000 Hz and an output signal-to-noise ratio of 10 dB. Assume that the terminal has 128 characters and uses a binary scheme. Find the channel capacity in characters per second using Shannon's equation (1.8).

1.5 Shannon's theorem indicates that a noiseless channel, that is, $S/N = \infty$, has an infinite capacity.

    (*a*) Is it true that the channel capacity approaches infinity as the bandwidth approaches infinity? Explain.

    (*b*) If the noise power spectral density is $N_0/2$, find the channel maximum data rate as $B \rightarrow \infty$.

1.6 Find the efficiency of a T1 carrier.

1.7 The data rate of a digital television channel is 44.736 Mbps. How many additional PCM voice channels can one put on a T4 carrier already carrying four television channels?

1.8 Find the signal-to-noise ratio needed to put a T2 carrier on a 200-kHz channel.

1.9 What is the polarization factor for a linearly polarized wave incident on a circularly polarized antenna?

1.10 The magnitude of the axial ratio r of a polarization ellipse is normally given in decibels, where $r \, (\mathrm{dB}) = 20 \log r$. For a right-hand elliptically polarized wave with $r = 2$ dB and a tilt angle of $\tau = 45°$. Find the polarization mismatch factor for the following receive antennas:

(a) Horizontal linear

(b) Right-hand elliptical with $r = 2.2$ dB and $\tau = 50°$. Find the cross-polarization discrimination ratio for the wave and the antenna in (b).

chapter 2

Communications Satellite: Orbit and Description

This chapter addresses the orbital mechanics of communications satellites, together with their construction, especially in relation to a geostationary satellite that appears to an observer on earth to be hanging perfectly still at one spot in the sky. But this is all relative—an observer in space sees a geostationary satellite orbiting the earth at a speed of 11,068.8 km/h. At this velocity the satellite makes one revolution around the earth in exactly the same amount of time its takes the earth to rotate once on its axis. Since the only great circle that is moving exactly parallel to the direction of the earth's rotation is the equator, the geostationary orbit lies in the equatorial plane at a distance of approximately 42,164 km from the earth's center. A satellite that has a 24-h nonequatorial orbit is called a *synchronous satellite*. (The term *geosynchronous satellite* is also used to refer to a synchronous satellite, but it is also used in the literature to refer to a geostationary satellite. To avoid confusion the term *geostationary satellite* is employed throughout this book.)

Why use a geostationary satellite? Because it is stationary relative to a point on earth, there is no requirement for a tracking antenna and the cost of the space and earth segments is much less than for lower-altitude satellite systems. This is the principal advantage.

2.1 ORBITAL PERIOD AND VELOCITY

The motion of a satellite orbiting the earth can be described by Newton's laws of motion and the law of gravitation. Consider the earth as having a mass of m_1 and the satellite a mass of m_2 at distances r_1 and r_2 from some inertial origin as shown in Fig. 2.1. From Newton's second law of motion, which says that a force acting on a body is equal to the product of its mass and its acceleration, the forces $\mathbf{F}_1$ on the earth and $\mathbf{F}_2$ on the satellite are given by

$$\mathbf{F}_1 = m_1 \frac{d^2\mathbf{r}_1}{dt^2}$$

$$\mathbf{F}_2 = m_2 \frac{d^2\mathbf{r}_2}{dt^2}$$

Also according to Newton's law of gravitation, the attractive force be-tween any two bodies is directly proportional to the product of their masses and inversely proportional to the square of the distance between them. Thus

$$\mathbf{F}_1 = -\mathbf{F}_2 = g\frac{m_1m_2}{r^2}\left(\frac{\mathbf{r}}{r}\right)$$

where g is the earth's gravitational constant. From the above three equa-tions we deduce that

$$\frac{d^2\mathbf{r}_1}{dt^2} = g\,\frac{m_2}{r^2}\left(\frac{\mathbf{r}}{r}\right)$$

$$\frac{d^2\mathbf{r}_2}{dt^2} = -g\frac{m_1}{r^2}\left(\frac{\mathbf{r}}{r}\right)$$

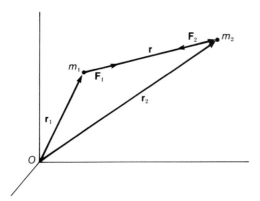

Figure 2.1 Satellite-earth coordinates.

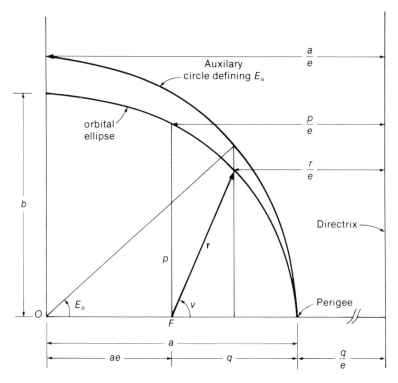

Figure 2.2 Satellite orbit.

Substituting $\mathbf{r} = \mathbf{r}_2 - \mathbf{r}_1$ gives

$$\frac{d^2\mathbf{r}}{dt^2} = -g(m_1 + m_2)\frac{\mathbf{r}}{r^3}$$

$$= -\mu\frac{\mathbf{r}}{r^3} \qquad (2.1)$$

where $\mu = g(m_1 + m_2) \approx gm_1$, since the mass of the satellite is negligible compared to that of the earth. The value gm_1 is given in [1] as $\mu \approx gm_1 = 3.986013 \times 10^5$ km^3/s^2. Equation (2.1) is known as the two-body equation of motion in relative form. It describes the motion of a satellite orbiting the earth.

A satellite orbit is either elliptical or circular, as shown in Fig. 2.2, and its characteristics are governed by Kepler's laws:

First law. The orbit of a satellite is an ellipse with the center of the earth at one focus.

Second law. The line joining the center of the earth and the satellite sweeps over equal areas in equal time intervals.

Third law. The squares of the orbital periods of two satellites have the same ratio as the cubes of their mean distances from the center of the earth.

In Fig. 2.2 the following notation is used:

r = distance of satellite from primary focus F which is the center of the earth

v = true anomaly, measured from primary focus F in the direction of motion from the perigee to the satellite position vector **r**

a = semimajor axis of ellipse

b = semiminor axis of ellipse

e = eccentricity

E_a = eccentric anomaly defined by an auxiliary circle of radius a having the center O of the ellipse as origin

p = semiparameter

q = perigee distance, the point on the orbit closest to focus F

Q = apogee distance, the point on the orbit farthest from the focus F (not shown in Fig. 2.2).

The first law is stated as the polar equation of the ellipse with origin at the primary focus:

$$p = r(1 + e \cos v) \tag{2.2}$$

The second law, the law of areas, can be derived by finding the cross-product of the position vector **r** and the acceleration vector $d^2\mathbf{r}/dt^2$ given by Newton's laws in (2.1):

$$\mathbf{r} \times \frac{d^2\mathbf{r}}{dt^2} = \mathbf{r} \times \left(-\mu \frac{\mathbf{r}}{r^3}\right) \tag{2.3}$$

With the help of the first law the integral of the above cross-product yields

$$\left|\mathbf{r} \times \frac{d\mathbf{r}}{dt}\right| = r^2 \frac{dv}{dt} = \sqrt{\mu\, p} \tag{2.4}$$

which means that the area swept out by the radial vector **r** in an infinitesimal time is constant. Rewriting the above equation and using the fact that $p = a(1 - e^2)$, as seen in Fig. 2.2, we have

$$\begin{aligned} r^2\, dv &= \sqrt{\mu\, p}\; dt \\ &= \sqrt{\mu\, a(1 - e^2)}\; dt \end{aligned} \tag{2.5}$$

By using the relations

$$\cos v = \frac{\cos E_a - e}{1 - e \cos E_a} \tag{2.6a}$$

and

$$\sin v = \frac{\sqrt{1 - e^2} \sin E_a}{1 - e \cos E_a} \tag{2.6b}$$

we obtain, by differentiating (2.6a),

$$-\sin v \, dv = \frac{dE_a}{(1 - e \cos E_a)^2} \left[-\sin E_a (1 - e \cos E_a) \right.$$
$$\left. - e \sin E_a (\cos E_a - e) \right]$$

and, with the substitution of (2.6b),

$$\frac{\sqrt{1 - e^2} \sin E_a \, dv}{1 - e \cos E_a} = \frac{dE_a}{(1 - e \cos E_a)^2} (1 - e^2) \sin E_a$$

$$dv = \frac{\sqrt{1 - e^2}}{1 - e \cos E_a} dE_a$$

By using the relation $r = a(1 - e \cos E_a)$ and dv in the expression of $r^2 \, dv$ in (2.5), we obtain

$$a^2(1 - e \cos E_a) \sqrt{1 - e^2} \, dE_a = \sqrt{\mu} \, a(1 - e^2) \, dt$$

or

$$(1 - e \cos E_a) \, dE_a = \frac{\sqrt{\mu}}{a^{3/2}} \, dt$$

The integral of this equation is called *Kepler's equation*:

$$M = E_a - e \sin E_a = \frac{\sqrt{\mu}}{a^{3/2}} (t - t_0) \tag{2.7}$$

M is called the *mean anomaly* and increases at a steady rate n, known as the *mean angular motion*:

$$n = \frac{\sqrt{\mu}}{a^{3/2}} \tag{2.8}$$

To obtain the third law, the orbital period law, set $E_a = 2\pi$ and $T = t - t_0$ for the satellite period:

$$2\pi = \frac{\sqrt{\mu}}{a^{3/2}} T$$

or

$$T = 2\pi \frac{a^{3/2}}{\sqrt{\mu}} \tag{2.9}$$

Note that the circular orbit is just a special case of the elliptical orbit where $a = b = r$.

To derive the orbital velocity of the satellite, we find the scalar product of the acceleration $d^2\mathbf{r}/dt^2$ in (2.1) and $d\mathbf{r}/dt$, obtaining

$$\frac{d\mathbf{r}}{dt} \cdot \frac{d^2\mathbf{r}}{dt^2} = -\frac{\mu}{r^3}\left(\frac{d\mathbf{r}}{dt} \cdot \mathbf{r}\right) = -\frac{\mu}{r^2}\left(\frac{dr}{dt}\right)$$

The integral of this equation is

$$\frac{1}{2}\left(\frac{d\mathbf{r}}{dt} \cdot \frac{d\mathbf{r}}{dt}\right) = \frac{1}{2}V^2 = \frac{\mu}{r} + C = \frac{1}{2}\left[\left(\frac{dr}{dt}\right)^2 + \left(r\frac{dv}{dt}\right)^2\right]$$

where V is the velocity. At the perigee $dr/dt = 0$ and $r = q$, hence from (2.4)

$$r\frac{dv}{dt} = \frac{\sqrt{\mu p}}{r} = \frac{\sqrt{\mu p}}{q}$$

and

$$\frac{\mu}{q} + C = \frac{\mu p}{2q^2}$$

$$C = \frac{\mu}{q}\left(\frac{p}{2q} - 1\right) = \frac{\mu}{a(1-e)}\left[\frac{a(1-e^2)}{2a(1-e)} - 1\right]$$

$$= -\frac{\mu}{2a}$$

Hence the orbital velocity is given by

$$V = \sqrt{\mu\left(\frac{2}{r} - \frac{1}{a}\right)} \tag{2.10}$$

As derived before, the orbital period of the satellite is expressed in terms of mean solar time; it is not as accessible to measurement as another kind of time which is determined from the culminations of stars—sidereal time. A *sidereal day* is defined as the time required for the earth to rotate once on its axis relative to the stars. Observations of other stars can be made more precisely than observations of the sun because of their greater distance from the earth. Stars appear as point sources in telescopes, and their culminations can be accurately timed. A sidereal day is measured as 23 h, 56 min, and 4.09 s of mean solar time. A satellite with a circular orbital period of one sidereal day is called a *synchronous satellite* and has an orbit radius of

$$a = \left(\frac{T\sqrt{\mu}}{2\pi}\right)^{2/3} = 42,164.2 \text{ km}$$

If the synchronous orbit is over the equator and the satellite travels in the same direction as the earth's surface, the satellite will appear to be stationary over one point on earth. This type of orbit is called a *geostationary orbit*. By taking the mean equatorial radius of the earth to be 6378.155 km [1], the distance from the satellite to the subsatellite point is found to be $42,164.2 - 6378.155 = 35,786.045$ km for a geostationary orbit. (The subsatellite point is the point where the equator meets the line joining the center of the earth and the satellite.)

The geostationary orbit is now employed for most commercial satellites because of the following advantages:

1. The satellite remains stationary with respect to one point on earth; therefore the earth station antenna is not required to track the satellite periodically. Instead, the earth station antenna beam can be accurately aimed toward the satellite by using the elevation angle and the azimuth angle (Fig. 2.6). This reduces the station's cost considerably.
2. With a 5° minimum elevation angle of the earth station antenna, the geostationary satellite can cover almost 38% of the surface of the earth.
3. Three geostationary satellites (120° apart) can cover the entire surface of the earth with some overlapping, except for the polar regions above latitudes 76°N and 76°S, assuming a 5° minimum elevation angle.
4. The Doppler shift caused by a satellite drifting in orbit (because of the gravitational attraction of the moon and the sun) is small for all the earth stations within the geostationary satellite coverage. This is desirable for many synchronous digital systems.

To cover the polar regions and to provide higher elevation angles for earth stations at high northern and southern latitudes, inclined orbits such as the one in Fig. 2.3 can be used. The disadvantages of an inclined orbit are that the earth station antenna must acquire and track the satellite and the necessity for switching from a setting satellite to a rising satellite. This handover problem can be minimized by designing the orbit so that the satellite is over a certain region for a relatively long part of its period. The Russian *Molniya* satellite has a highly inclined elliptical orbit with a 63° inclination angle and an orbital period of 12 h. The apogee is above the Northern Hemisphere. Communications are established when the satellite is in the apogee region where the orbital period is small and antenna tracking is slow. The satellite visibility for a station above 60° latitude with an antenna elevation greater than 20° is between 4.5 and 10.5 h.

Although a geostationary satellite appears to be stationary in its orbit, the gravitational attraction of the moon and to a lesser extent that of the sun cause it to drift from its stationary position and the satellite orbit

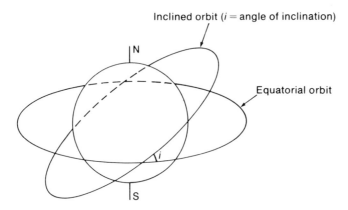

Figure 2.3 Inclined orbit.

tends to become inclined at a rate of about 1°/year. Also, the nonuniformity of the earth's gravitational field and the radiation pressure of the sun cause the satellite to drift in longitude. But this drift is several orders of magnitude smaller than that resulting from the attraction of the moon and the sun. Stationkeeping is therefore required to maintain the position of a satellite accurately so that satellites in a geostationary orbit do not drift close together and cause adjacent satellite interference. North-south stationkeeping is required to prevent a drift in latitude, and east-west stationkeeping is needed to prevent a drift in longitude.

2.2 EFFECTS OF ORBITAL INCLINATION

The maximum drift in both latitude and longitude due to *orbit inclination* can be determined by considering Fig. 2.4. Let R be the satellite's instantaneous projection on the earth's surface (the intersection of the earth's surface and the line joining the satellite and the earth's center) and let λ and ψ denote the instantaneous latitude and relative longitude (with respect to the ascending node P) of R. Considering the spherical triangle PRQ and a nonrotating earth we obtain

$$\sin \psi = \frac{\tan \lambda}{\tan i}$$

The arc PR is the trace of the orbit and subtends an angle u given by

$$\sin u = \frac{\sin \lambda}{\sin i}$$

Now consider a rotating earth. Let t_P be the time at which the satellite passes the ascending node P; then, when the satellite reaches the latitude

λ in time t, the earth has rotated eastward through an angle u. Therefore, if the relative longitude of the satellite projection R is ψ, it becomes $\psi - u$. Thus the relative longitude of the satellite for any given lattitude λ is equal to

$$\psi = \sin^{-1}\left(\frac{\tan \lambda}{\tan i}\right) - \sin^{-1}\left(\frac{\sin \lambda}{\sin i}\right) \tag{2.11}$$

The trace of a synchronous satellite (the orbital period is the same as the sidereal period of the earth) with inclination i is plotted in Fig. 2.5. It is seen that this inclination in effect gives the satellite an apparent movement in the form of a *"figure eight."* The maximum latitude deviation from the equator is given by

$$\lambda_{\max} = i \tag{2.12}$$

and the maximum longitude deviation from the ascending node when i is small ($i < 5°$) can be approximated by

$$\psi_{\max} = \frac{i^2}{228} \tag{2.13}$$

where i is in degrees. From (2.12) and (2.13) it is seen that the displacement in latitude is more pronounced than the displacement in longitude for a synchronous satellite with a small inclination. In this case the dis-

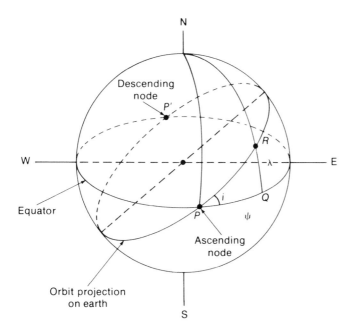

Figure 2.4 Longitude and latitude of a satellite in an inclined and synchronous orbit.

placement D_λ (corresponding to λ_{max}) and D_ψ (corresponding to ψ_{max}) can be calculated as follows:

$$\frac{D_\lambda}{R_e\,\lambda_{max}} = \frac{a}{R_e}$$

$$\frac{D_\psi}{D_\lambda} = \frac{\psi_{max}}{\lambda_{max}} = \frac{i}{228}$$

where a is the orbital radius ($a = 42{,}164.2$ km) and R_e is the earth's radius. Therefore,

$$D_\lambda = a\lambda_{max}$$

$$= \frac{42{,}164.2i}{360/2\,\pi} = 735.9i \text{ (km)} \tag{2.14}$$

$$D_\psi = \frac{iD_\lambda}{228}$$

$$= 3.23i^2 \text{ (km)} \tag{2.15}$$

where i in (2.14) and (2.15) is in degrees. As an example, consider the case when $i = 1°$; then $D_\lambda = 735.9$ km and $D_\psi = 3.23$ km.

To correct the orbital inclination, it is necessary to apply a velocity

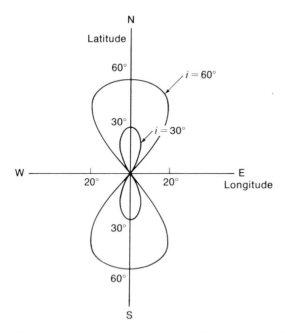

Figure 2.5 Apparent movement of a satellite in an inclined and synchronous orbit with respect to the ascending node.

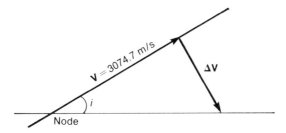

Figure 2.6 Correction of the inclination of a synchronous orbit.

impulse perpendicular to the orbital plane when the satellite passes through the nodes (see Fig. 2.4), as indicated in Fig. 2.6. For a given i, the impulse amplitude is given by

$$\Delta V = V \tan i$$

$$= \sqrt{\frac{\mu}{a}} \tan i \qquad (2.16)$$

where $V = \sqrt{\mu/a} = 3074.7$ m/s and is the orbital velocity. For $i = 1°$, $\Delta V = 3074.7 \tan 1° = 53.7$ m/s.

2.3 AZIMUTH AND ELEVATION

As previously mentioned, a satellite in a geostationary orbit appears to be stationary with respect to a point on the earth. Therefore, if an earth station is within the coverage of the satellite, it can communicate with the satellite by simply pointing its antenna toward it. Such a positioning of the earth station antenna can be accomplished using the azimuth angle A and the elevation angle E based on a knowledge of the earth station latitude θ_1 and longitude θ_L and the satellite longitude θ_S as shown in Fig. 2.7a. The *azimuth angle* is defined as the angle produced by intersection of the local horizontal plane TMP and the plane TSO (passing through the earth station, the satellite, and the earth's center) with the true north. Depending on the location of the earth station with respect to the subsatellite point, the azimuth angle A is given by:

1. Northern Hemisphere
   Earth station west of satellite: $A = 180° - A'$
   Earth station east of satellite: $A = 180° + A'$
2. Southern hemisphere
   Earth station west of satellite: $A = A'$
   Earth station east of satellite: $A = 360° - A'$

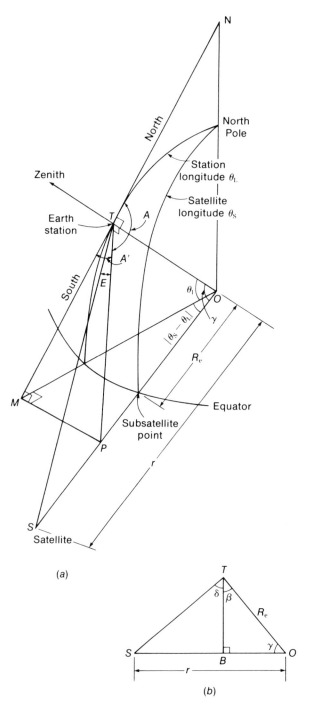

Figure 2.7 (*a*) Azimuth and elevation. (*b*) Triangle to calculate elevation.

where A' is defined in Fig. 2.7a. The *elevation angle E* is defined as the angle produced by the intersection of the local horizontal plane TMP and the plane TSO with the line of sight between the earth station and the satellite. Of course we assume that the earth is a perfect sphere with radius R_e. From Fig. 2.7a we have

$$A' = \tan^{-1}\left(\frac{MP}{MT}\right)$$

$$= \tan^{-1}\left(\frac{MO \tan |\theta_S - \theta_L|}{R_e \tan \theta_1}\right)$$

$$= \tan^{-1}\left[\frac{(R_e/\cos \theta_1) \tan |\theta_S - \theta_L|}{R_e \tan \theta_1}\right]$$

$$= \tan^{-1}\left(\frac{\tan |\theta_S - \theta_L|}{\sin \theta_1}\right) \qquad (2.17)$$

To calculate the elevation angle E, consider the triangle TSO shown in Fig. 2.7a and redrawn in Fig. 2.7b. We have

$$E = \beta + \delta - 90°$$

$$= (90° - \gamma) + \delta - 90°$$

$$= \delta - \gamma$$

The angle γ can be evaluated from the triangle TPO as follows:

$$\gamma = \cos^{-1}\left(\frac{R_e}{OP}\right)$$

Since $OP = MO/\cos|\theta_S - \theta_L| = R_e/\cos \theta_1 \cos|\theta_S - \theta_L|$ as seen from the triangles MPO and TMO, we have

$$\gamma = \cos^{-1}(\cos \theta_1 \cos|\theta_S - \theta_L|)$$

To evaluate the angle δ in Fig. 2.7b, we note that

$$\delta = \tan^{-1}\left(\frac{SB}{TB}\right)$$

$$= \tan^{-1}\left(\frac{r - R_e \cos \gamma}{R_e \sin \gamma}\right)$$

$$= \tan^{-1}\left(\frac{r - R_e \cos \theta_1 \cos|\theta_S - \theta_L|}{R_e \sin [\cos^{-1}(\cos \theta_1 \cos|\theta_S - \theta_L|)]}\right)$$

Thus the elevation angle E can be expressed by

$$E = \tan^{-1}\left(\frac{r - R_e \cos \theta_1 \cos|\theta_S - \theta_L|}{R_e \sin [\cos^{-1}(\cos \theta_1 \cos|\theta_S - \theta_L|)]}\right)$$

$$- \cos^{-1}(\cos \theta_1 \cos|\theta_S - \theta_L|) \qquad (2.18)$$

where r is the geostationary orbital radius and is equal to 42,164.2 km.

Example 2.1 Consider an earth station located at longitude $\theta_L = 80°$W and latitude $\theta_1 = 40°$N and a geostationary satellite at longitude $\theta_S = 120°$W.

Because the earth station is in the Northern Hemisphere and east of the satellite, its azimuth angle A is

$$A = 180° + A' = 180° + 52.5° = 232.5°$$

The elevation angle E is given by (2.18) as follows:

$$E = 28.3°$$

2.4 COVERAGE ANGLE AND SLANT RANGE

A satellite is capable of communicating with an earth station using a global coverage antenna if the station is in the footprint of the satellite, which is a function of time except for a geostationary satellite. Consider Fig. 2.8 where the earth coverage angle $2\alpha_{max}$ is the total angle subtended by the earth as seen from the satellite. This angle is important in the

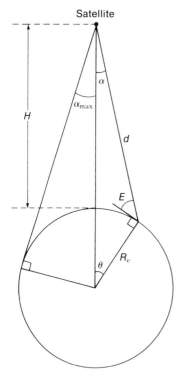

Figure 2.8 Coverage angle and slant range.

design of a global coverage antenna and depends on the satellite altitude. The communication coverage angle 2α is similarly defined, except that the minimum elevation angle E_{min} of the earth station antenna must be taken into account. For an elevation angle E of the earth station antenna, the communication coverage angle 2α is given by the relation

$$\frac{\sin \alpha}{R_e} = \frac{\sin (90° + E)}{R_e + H} = \frac{\cos E}{R_e + H}$$

where a spherical earth with radius R_e is assumed and H is the altitude of the satellite orbit and is a function of time except for a geostationary satellite where $H = 35{,}786$ km. Thus,

$$2\alpha = 2 \sin^{-1} \left(\frac{R_e}{R_e + H} \cos E \right) \tag{2.19}$$

The earth coverage angle is calculated simply by setting $E = 0°$:

$$2\alpha_{max} = 2 \sin^{-1} \left(\frac{R_e}{R_e + H} \right) \tag{2.20}$$

For a geostationary orbit where R_e is assumed to be about 6378 km, the earth coverage angle is $2\alpha_{max} = 17.4°$. The central angle θ, which is the angular radius of the satellite footprint, is

$$\theta = 180° - (90° + E + \alpha) = 90° - E - \alpha \tag{2.21}$$

For a geostationary orbit, the central angle θ corresponding to the earth coverage angle α_{max} is obtained by setting $\alpha = \alpha_{max}$ and $E = 0°$, which yields $\theta = 81.3°$. If a minimum elevation angle of $5°$ is required for the earth station antenna, then $\theta = 76.3°$. Thus it is seen that the polar regions above these northern and southern latitudes of $76.3°$ will not be covered by the footprint of the satellite.

Besides the coverage angle, it is important to know the slant range from the earth station to the satellite, because this range determines the satellite roundtrip delay to the earth station. From Fig. 2.8 the slant range d can be determined as

$$d^2 = (R_e + H)^2 + R_e^2 - 2R_e(R_e + H) \cos \theta$$

$$= (R_e + H)^2 + R_e^2 - 2R_e(R_e + H) \times$$

$$\sin \left[E + \sin^{-1} \left(\frac{R_e}{R_e + H} \cos E \right) \right] \tag{2.22}$$

For a geostationary orbit and a minimum elevation angle of $E_{min} = 5°$, the maximum slant range is $d = 41{,}127$ km, yielding a satellite roundtrip delay of $2d/c = 0.274$ s, where $c = 2.997925 \times 10^5$ km/s and is the speed of light.

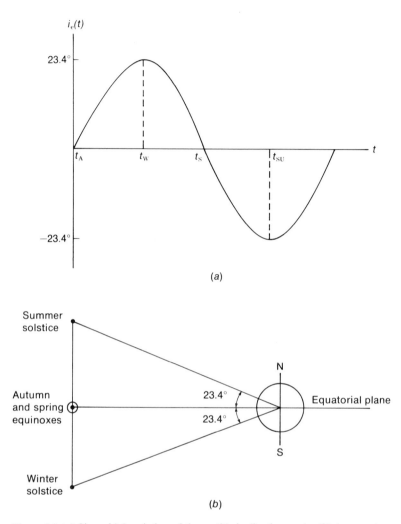

Figure 2.9 (*a*) Sinusoidal variation of the earth's inclination angle. (*b*) Apparent movement of the sun.

2.5 ECLIPSE

For a geostationary satellite that utilizes solar energy, the duration and periodicity of solar eclipses are important because no solar energy is available during eclipses. The earth's equatorial plane is inclined at an angle $i_e(t)$ with respect to the direction of the sun. This annual sinusoidal variation is given in degrees (Fig. 2.9) by

$$i_e(t) = 23.4 \sin \frac{2\pi t}{T} \qquad (2.23)$$

where the annual period $T = 365$ days and the maximum inclination is $i_{e,max} = 23.4°$. The time t_A and t_S when the inclination angle i_e is zero are called the *autumn equinox* and the *spring equinox* and occur about September 21 and March 21, respectively. The times t_W and t_{SU} when the inclination angle i_e is at its maximum are called the *winter solstice* and the *summer soltice* and occur about December 21 and June 21; respectively.

To find the eclipse duration consider Fig. 2.10 where the finite diameter of the sun is ignored (the sun is assumed to be at infinity with respect to the earth), hence the earth's shadow is considered to be a cylinder of constant diameter. The maximum shadow angle occurs at the equinoxes and is given by

$$\phi_{max} = 180° - 2 \cos^{-1}\left(\frac{R_e}{a}\right)$$

$$= 180° - 2 \cos^{-1}\left(\frac{6378.155}{42,164.2}\right)$$

$$= 17.4° \qquad (2.24)$$

Because a geostationary satellite period is 24 h, this maximum shadow angle is equivalent to a maximum daily eclipse duration:

$$\tau_{max} = \frac{17.4°}{360°} \times 24 = 1.16 \text{ h} \qquad (2.25)$$

The first day of eclipse before an equinox and the last day of eclipse after an equinox correspond to the relative position of the sun such that the sun rays tangent to the earth pass through the satellite orbit. Thus the inclination angle of the equatorial plane with respect to the direction of

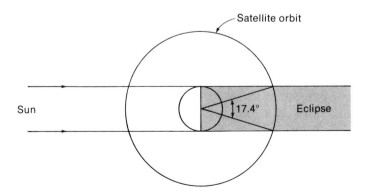

Figure 2.10 Eclipse when the sun is at equinox.

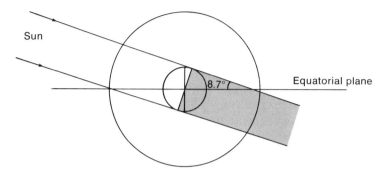

Figure 2.11 Earth inclination at first day of eclipse before equinox.

the sun in these cases (Fig. 2.11) is

$$i_e = \tfrac{1}{2}\, \phi_{max} = 8.7° \qquad (2.26)$$

Substituting $i_e = 8.7°$ into (2.23) yields the time from the first day of eclipse to the equinox and also the time from the equinox to the last day of eclipse:

$$t = \frac{365}{2\pi} \sin^{-1} \left(\frac{8.7}{23.4} \right)$$

$$= 22.13 \text{ days} \qquad (2.27)$$

where the angle is in radians.

2.6 PLACEMENT OF A SATELLITE IN A GEOSTATIONARY ORBIT

The placement of a satellite in a geostationary orbit involves many complex sequences and is shown schematically in Fig. 2.12. First, the launch vehicle (a rocket or a space shuttle) places the satellite in an elliptical transfer orbit whose apogee distance is equal to the radius of the geosynchronous orbit (42,164.2 km). The perigee distance of the elliptical transfer orbit is in general about 6678.2 km (about 300 km above the earth's surface). The satellite is then spin-stabilized in the transfer orbit so that the ground control can communicate with its telemetry system. When the orbit and attitude of the satellite have been determined exactly and when the satellite is at the apogee of the transfer orbit, the apogee kick motor is fired to circularize the orbit. This circular orbit, with a radius of 42,164.2 km, is a geostationary orbit if the launch is carried out at 0° latitude (i.e., at the equator). What happens if the satellite is launched from, say, Cape Kennedy at 28°N latitude? Then the orbit will be a synchro-

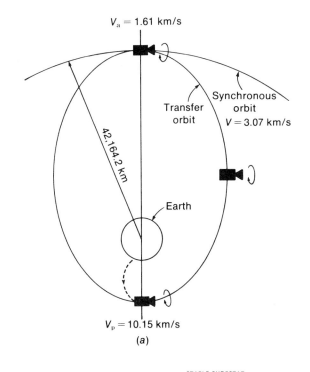

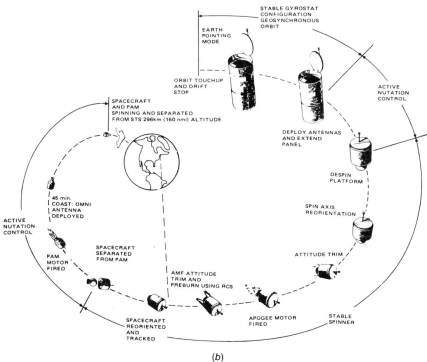

Figure 2.12 Placement of a satellite in a geostationary orbit. (*a*) Concept. (*b*) Actual launch (*Courtesy of Hughes Aircraft Co.*).

nous orbit with an inclination i greater than or equal to the latitude θ_1 of the launch point. The inclination i is equal to θ_1 when the injection at the perigee is horizontal.

The velocity at the perigee and apogee of the transfer orbit can be calculated from (2.10):

$$V = \sqrt{\mu \left(\frac{2}{r} - \frac{1}{a} \right)}$$

At the perigee $r = 6678.2$ km, $a = (6678.2 + 42{,}164.2)/2 = 24{,}421.2$ km, and the velocity is

$$V_p = 10.15 \text{ km/s}$$

At the apogee $r = 42{,}164.2$ km, hence the velocity is

$$V_a = 1.61 \text{ km/s}$$

Since the velocity in a synchronous orbit ($r = a = 42{,}164.2$ km) is

$$V_c = 3.07 \text{ km/s}$$

the incremental velocity required to circularize the orbit at the apogee of the transfer orbit must be

$$\Delta V_c = V_c - V_a$$
$$= 3.07 - 1.61 = 1.46 \text{ km/s} \tag{2.28}$$

Since the plane of the transfer orbit is formed by the position vector $\mathbf{r}$ (Fig. 2.2) and the velocity vector $\mathbf{V}$ of the satellite at a given instant in time, the inclination correction can be made at the ascending or descending node where the orbit intersects the equatorial plane at an incremental velocity vector $\Delta\mathbf{V}$ is such a way that the sum of the node velocity vector $\mathbf{V}_n$ and the incremental velocity vector $\Delta\mathbf{V}$ is a vector $\mathbf{V}$ in the equatorial plane. The inclination correction is shown schematically in Fig. 2.13 where the magntiude of $\Delta\mathbf{V}$ required to correct the inclination is

$$\Delta V = \sqrt{V_n^2 + V_c^2 - 2V_n V_c \cos i} \tag{2.29}$$

If the line connecting the apogee and perigee is the node line, and the inclination correction is made at the apogee in conjunction with orbit circularization, then

$$\Delta V = \sqrt{V_a^2 + V_c^2 - 2V_a V_c \cos i} \tag{2.30}$$

For $i = 28°$, $V_a = 1.61$ km/s, and $V_c = 3.07$ km/s, the incremental velocity required to correct the orbit inclination and to achieve orbit circularization is

$$\Delta V = 1.81 \text{ km/s}$$

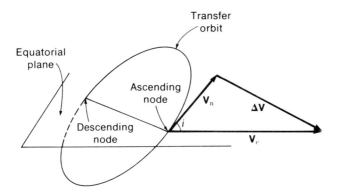

Figure 2.13 Simultaneous orbit circularization and inclination correction.

Satellites are now being placed in geostationary orbits by two major organizations: the *National Aeronautics and Space Administration* (NASA) and the *European Space Agency* (ESA). NASA uses a *space transportation system* (STS) or *space shuttle* to take the satellite to a circular parking orbit between 300 and 500 km with an inclination of 28°. The propulsion requirements for establishing the final goestationary orbit are satisfied by two impulsive maneuvers. The first maneuver imparts a velocity increment of approximately 2.42 km/s for a 300-km parking orbit (in this orbit the satellite velocity is about 7.73 km/s, while the velocity at the perigee of the transfer orbit is 10.15 km/s) at the first equatorial cross-ing of the parking orbit. This establishes the elliptical transfer orbit with the perigee at the equatorial crossing where the maneuver is executed ($r = 6678.2$ km) and the apogee at the geostationary radius ($r = 42,164.2$ km). The second maneuver has been described previously. ESA uses the *Ariane* rocket to carry the satellite directly to the elliptical transfer orbit. Since the transfer orbit established from the *Ariane* launch site in French Guiana is inclined only 5.3° to the geostationary orbit, less fuel is required in the second maneuver for the *Ariane* launch. The *Ariane* is also capable of placing a satellite directly in a geostationary orbit.

2.7 SATELLITE DESCRIPTION

This section provides a description of communications satellites in a geo-stationary orbit. To keep the satellite antennas pointing at a selected por-tion of the earth, the satellite must be stabilized in orbit. There are two classes of satellites, namely, spin-stabilized satellites and three-axis body-stabilized satellites. *Spin-stabilized satellites* have gyroscopic stiffness as a result of rotating about their axis of maximum moment of inertia. They are normally dual-spin satellites with a spinning section and a despun section which is kept stationary by counterrotation so that the antennas mounted on it are constantly pointing toward the earth. A typical dual-spin satellite is shown in Fig. 2.14. *Three-axis body-stabilized* satellites are attitude-controlled about their three axes, namely, the yaw axis, the pitch axis and the roll axis, as shown in Fig. 2.15, to provide gyroscopic stiffness. The satellite body is aligned with the local vertical and the orbit normal. Antennas are mounted on the surface facing the earth. A typical three-axis body-stabilized satellite is shown in Fig. 2.16.

Although their design can differ depending on the mission, communications satellites generally consist of the following subsystems: a communications subsystem, a telemetry, command, and ranging subsystem, an attitude control subsystem, an electrical power subsystem, a reaction control subsystem, and an apogee kick motor.

Figure 2.14 Dual-spin stabilized satellite. (*Courtesy of Hughes Aircraft Co.*).

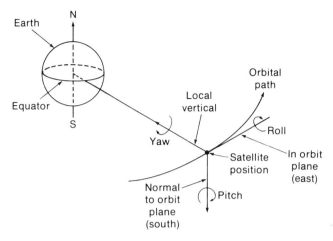

Figure 2.15 Yaw, pitch, and roll.

2.7.1 Communications Subsystem

The communications subsystem provides the receive and transmit coverage for the satellite. It consists of a communications antenna and a communications repeater. The communications antenna serves as an interface between the earth stations on the ground and various satellite subsystems during operation. The main function of the antenna is to provide shaped downlink and uplink beams for transmission and reception of communications signals in the operating frequency bands (e.g., the C or Ku

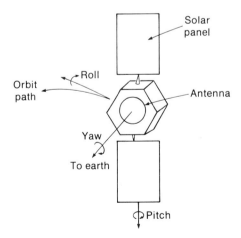

Figure 2.16 Three-axis body-stabilized satellite.

band). In addition the antenna may be used to provide a signal link for the satellite telemetry, command, and ranging subsystem which in conjunction with the attitude control subsystem provides beacon tracking signals for precise pointing of the antenna toward the earth coverage areas, as shown in Fig. 2.17. To increase the communications capacity of the satellite, *frequency reuse* is often employed through the use of orthogonally polarized beams; hence the communication antenna requires two reflectors, one for each polarization, that can be incorporated into a single physical structure such as that shown in Fig. 2.18. The single structure consists of two polarization-selective gridded offset parabolic reflectors stacked one behind the other to facilitate aperture sharing. With this geometry, the two reflectors are fed by two independent multihorn feed arrays which do not interfere with each other either physically or electrically. The polarization selectivity of each reflector is accomplished by using conductive strips parallel to the desired polarization and bonded to a paraboloid shell. The polarization orientation is normally chosen such that the vertical polarization plane is defined as the plane formed by the satellite north-south axis and the center of the earth, and the horizontal polarization is orthogonal to the vertical. The isolation between the copolarized and cross-polarized signals incident on the antenna is normally about 33 dB. To provide high gain for the coverage areas, multiple beams may be required by the communications antenna and can be designed using multihorn feed arrays. Typical antenna coverage of the continental *United States* (CONUS), Alaska, and Hawaii is shown in Fig. 2.19. These antenna beams must be accurately pointed to avoid deviation of the beam boresight from its nominal direction. Typical antenna pointing accuracies are ±0.05° for east-west (*pitch*) and north-south (*roll*) beam pointing errors and ±0.2° for beam rotation (*yaw*) errors.

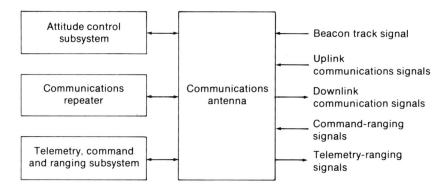

Figure 2.17 Fundamentals of a communications satellite subsystem.

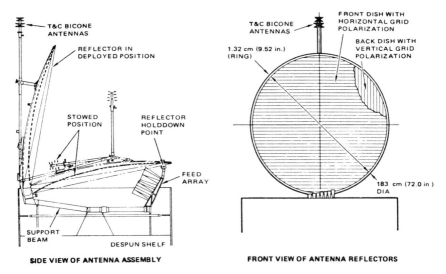

SIDE VIEW OF ANTENNA ASSEMBLY   **FRONT VIEW OF ANTENNA REFLECTORS**

Figure 2.18 Dual-polarized communications satellite antenna. (*Courtesy of Hughes Aircraft Co.*).

The second part of the communications subsystem is the communications repeater which is an interconnection of many channelized transponders. The communications repeater generally consists of the following modules, as shown schematically in Fig. 2.20:

1. A wideband communications *receiver/downconverter*
2. An *input multiplexer*
3. Channelized *traveling wave tube amplifiers* (TWTAs)
4. An *output multiplexer*

The wideband communications receiver/downconverter is designed to operate within the 500-MHz bandwidth allocated for C-band (5.9 to 6.4 GHz) and Ku-band (14 to 14.5 GHz) uplink signals and is shown schematically in Fig. 2.21 for a Ku-band uplink. The uplink signals are first filtered by a waveguide bandpass filter with about a 600-MHz bandwidth and then amplified by a parametric or a solid-state gallium arsenite field effect transistor (GaAs FET) low-noise amplifier with a typical noise figure of 2 to 4 dB. The amplified signals are then downconverted to the 11.7 to 12.2-GHz downlink Ku band (3.7 to 4.2 GHz for a downlink C band) by a microwave integrated circuit downconverter. After downconversion, the signals are again amplified by a 11.7 to 12.2-GHz GaAs FET amplifier and passed through a ferrite isolator to the input multiplexer.

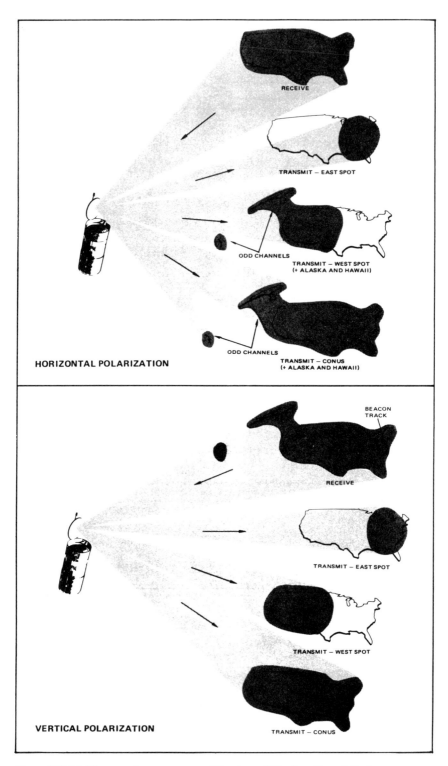

Figure 2.19 Multibeam antenna coverage. (*Courtesy of Hughes Aircraft Co.*).

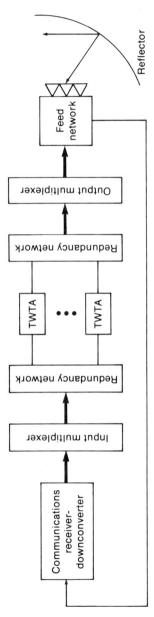

Figure 2.20 Communications repeater.

57

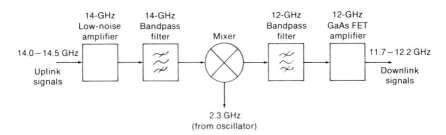

Figure 2.21 Wideband communications receiver-downconverter.

The input multiplexer is employed to separate the 500-MHz bandwidth into individual transponder channels whose bandwidth depends on the satellite's mission. For example, a 500-MHz bandwidth can be divided into 8 transponder channels with a center-to-center frequency separation of 61 MHz. With frequency reuse, there are altogether 16 transponder channels in the satellite. The input multiplexer normally consists of input circulators, input filters, group delay equalizers, amplitude equalizers, and output circulators, as shown in Fig. 2.22.

The channelized TWTAs amplify the low-level downlink signals to a high level for transmission back to earth. Driver amplifiers are normally employed in front of the high-power TWTAs to allow the communications receiver to be operated in the linear mode. The size of the TWTA depends on the mission and is about 15 to 30 W for a 61-MHz Ku-band transponder. The TWTA establishes the transponder output power and normally operates near saturation to achieve the desired output power. Thus it is the dominant nonlinear device in a transponder and can affect the link signal performance considerably.

The output downlink signals from the channelized TWTAs are combined by the output multiplexer for retransmission to earth. The output multiplexer provides the required output out-of-band attenuation, as well as the attenuation necessary to suppress signal harmonics and spurious noise generated by the TWTAs. Variable power dividers may be used at the input of the output multiplexer to provide the necessary power split to select the desired transmit antenna coverage which can be selected by ground command.

A complete communications subsystems employing frequency reuse and consisting of 16 transponders (8 transponders use the horizontal polarization and 8 transponders use the vertical polarization) is shown in Fig. 2.23 for the type of antenna coverage shown in Fig. 2.19. The number of odd and even transponders in the east and west beams can be selected by using the variable power dividers.

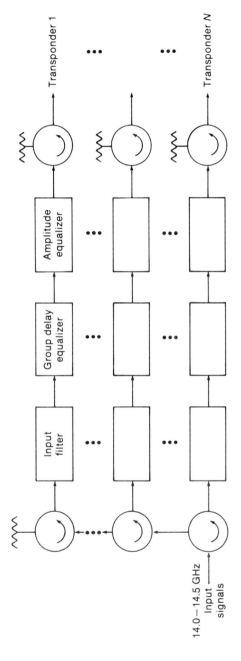

Figure 2.22 Input multiplexer.

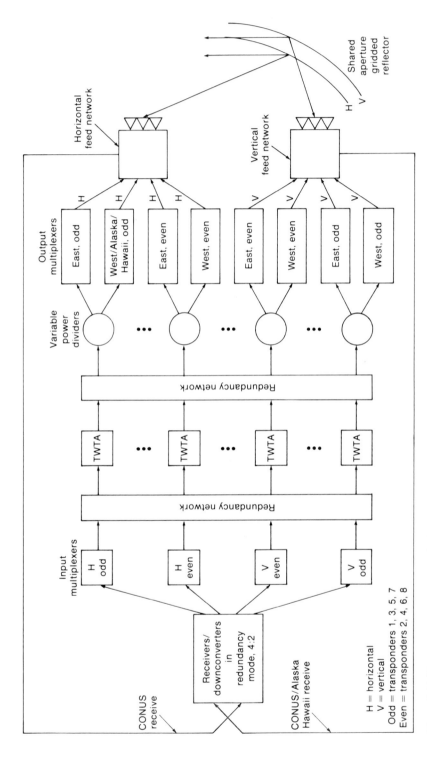

Figure 2.23 Frequency reuse communications satellite subsystem.

60

2.7.2 Telemetry, Command, and Ranging Subsystem

The telemetry subsystem monitors all satellite subsystems and continuously transmits to the earth sufficient information for determination of the satellite attitude, status, and performance as required for satellite and subsystem control. The telemetry transmitter also serves as the downlink transmitter for the ranging tones. The primary telemetry data mode is normally pulse code modulation. In normal on-station operation, telemetry data is transmitted via the communications antenna. In the transfer orbit, the telemetry transmitter is connected to a TWTA in the communications repeater selected to provide adequate power for telemetry coverage via the omni antenna.

The command subsystem controls the satellite operation through all phases of the mission by receiving and decoding commands from the ground station. It also generates a verification signal and upon receipt of an execute signal carries out the commands. The command subsystem also serves as an uplink receiver for the ranging signals. Again, the omni antenna is used in the transfer orbit for command and ranging and as an on-station backup, while the communications antenna is used on-station for command and ranging.

The ranging subsystem determines the slant range from the ground control station to the satellite for precise transfer and geostationary orbit determination. The slant range is determined by transmitting to the satellite multiple tones modulated onto the command carrier which is received by the command receiver, demodulated, and retransmitted by the telemetry transmitter to the ground control station where the phase difference is accurately measured. During on-station operation ranging is performed via the communications antenna; antenna coverage for ranging during the transfer orbit is provided by the omni antenna. A block diagram of the telemetry, command, and ranging subsystem is shown in Fig. 2.24. For reliability the command receiver, the telemetry transmitter, the decoder, and the encoder are all in a redundancy mode. (Redundancy configurations will be discussed in detail in Chap. 3.)

2.7.3 Attitude Control Subsystem

The major functions of the attitude control subsystem are to maintain accurate satellite position and communications antenna pointing for the life of the satellite and to control satellite maneuvers in the transfer orbit and during stationkeeping. The requirements of the attitude control system depend on whether the satellite is spin-stabilized or three-axis body-stabilized in a geostationary orbit. Most satellites are spin-stabilized in the transfer orbit. When rotated at a certain speed around the axis of maximum inertia, a satellite attains a high angular momentum and the spin

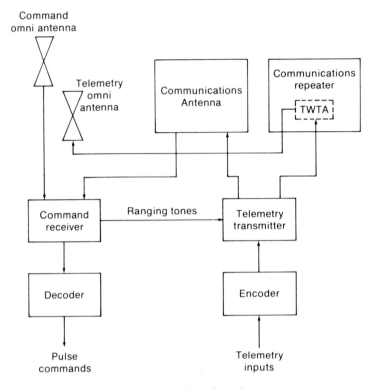

Figure 2.24 Telemetry, command, and ranging subsystem.

axis is fixed in a certain direction in space by preservation of the law of angular momentum. While in the transfer orbit, the attitude control subsystem must perform the following tasks:

1. Ensure nutational stability
2. Determine the spin axis orientation and reorientation control via the telemetry, command, and ranging subsystem and ground control

Residual nutation stemming from the launcher separation maneuver has to be damped out to maintain spin axis stabilization. For a basically stable satellite, a simple, efficient damping system such as a pendulum or a ball in a curved tube can be used. The resonance frequency and damping coefficient are tuned to the nutational frequency as determined by the satellite moments of inertia. For a basically unstable satellite, the damping has to be performed by an active nutation control loop. Nutation is sensed by a linear accelerometer with its sensitive axis parallel to the spin axis. Nutation produces a sinusoidal acceleration at the nutation frequency with a peak value proportional to the nutation amplitude. The accelerometer signal is bandpass-filtered to remove both dc and high-frequency components and then fed to a threshold circuit. When the amplitude exceeds the threshold, a thruster pulse is fired to damp out the nutation.

The spin axis orientation or attitude is determined via telemetered data from earth and sun sensors in the satellite. The earth sensor is a passive infrared device with a pencil beam field of view which operates in the 14- to 16-μm wavelength carbon dioxide absorption band. It senses the infrared rays radiating from the earth's surface by scanning the boundary of the horizon between the earth and space. The location of the horizon is indicated by the sharp increase in the sensor output caused by the temperature difference between space and the earth's horizon. Two earth sensors are canted about 5° north and 5° south of the spin plane, respectively, as shown in Fig. 2.25. The sensors provide output radiance pulses as they scan the earth. When the attitude of the satellite is correctly maintained, the output pulses of the north earth sensor and the south earth sensor are in-phase. But when the spin axis is inclined at an angle, the output pulses of the sensors are out-of-phase. The phase difference is utilized to control the attitude of the satellite. For spin-stabilized satellites in a geosynchronous orbit with a spin axis nominally parallel to the earth's polar axis, telemetry data from the earth sensors alone is adequate for spin axis attitude determination. This is because of the slow drift rate of the spin axis and the systematic variation in inertial direction to the earth due to

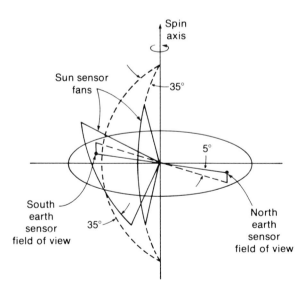

Figure 2.25 Earth and sun sensors.

orbital motion. Over a 24-h observation period, the spin axis attitude can be determined to an accuracy of ±0.02°. Additional data, used primarily in the transfer orbit, is provided by the sun sensors. The sun sensor has a fan-shaped field of view and operates in the visual spectral range. It employs a silicon photovoltaic cell and a slit aperture for detecting solar radiance as the slit is swept past the sun. There are two fan-shaped fields of view. One fan is parallel to the spin axis, and the other is canted 35° and rotated 35° with respect to the other. Each sensor outputs a short pulse as the sun passes through the field of view of the sun sensor. These pulses are telemetered on the ground to obtain a solar aspect angle measurement. The combination of the solar aspect angle, the earth aspect angle (from the earth sensors), and the spin angle between the sun and the earth determines unambiguously the spin axis orientation.

For a spin-stabilized satellite the reaction control subsystem thrusters are used in several control modes: spin axis attitude control, orbit control, spin rate control, and active nutation control. A typical arrangement of the thrusters in a satellite is shown in Fig. 2.26. Two axial thrusters have moment arms about the satellite center of mass, and two radial thrusters lie approximately in the lateral center-of-mass plane. Attitude control maneuvers are performed periodically by ground command. The update period between these maneuvers depends on the spin axis drift rate due to external disturbance torques. At a geostationary altitude, the only significant disturbance torque is that due to solar radiation pressure. The satellite configuration is designed to minimize torque by placing the center of mass close to the average center of pressure. The drift varies according to the changing solar aspect over the year. Attitude control, which holds the spin axis within a ±0.2° range, may be performed at intervals of approximately 6 days. Orbit control consists of east-west stationkeeping

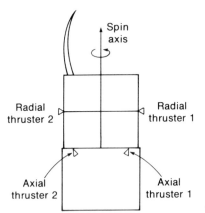

Figure 2.26 Arrangement of thrusters in a spin-stabilized satellite.

(longitude control) and north-south stationkeeping (inclination control). The frequency of stationkeeping maneuvers depends on the permissible error and the satellite orbital position. For a stationkeeping accuracy of ±0.05°, east-west maneuvers are required at intervals of about 21 days using a single radial thruster operating in a pulsed mode, and north-south maneuvers are required at longer interval of about 28 days using two axial thrusters fired together.

For a three-axis body-stabilized satellite, stabilization can be achieved with a bias-momentum system. The momentum of a high-speed wheel provides gyroscopic stability for the satellite. Magnetic torquing, wheel speed control, and wheel axis trim automatically correct attitude errors after initial acquisition. The wheel axis is the satellite pitch axis; thus pitch is controlled by causing changes in wheel speed to exchange angular momentum with the satellite body. Magnetic torquing corrects roll and yaw errors and is used for nutation damping. Both roll and pitch can be modulated about a bias setting by ground observation and programmed into the on-board control logic. When activated, this provides a bias to pitch and roll error signals of the earth sensor and commands the momentum wheel to pivot about the roll axis. Torquing currents and wheel speed signals respond to the earth sensor error signals. Control results from the interaction between the magnetic fields of the torquer and the earth. The earth sensor detects earth-space transition at about +45° latitude. Roll error is the difference in the length of the two chords; pitch error is indicated by off-centering of the longer chord.

2.7.4 Electrical Power Subsystem

The satellite generates power by using a solar array of silicon cells. In a spin-stabilized satellite such as the one shown in Fig. 2.14, the solar array consists of two concentric cylindrical panels of silicon cells. The forward panel is attached to the main structure and is divided into two arrays separated by a thermal radiator band. The aft panel is retracted over the forward panel during a transfer orbit and extended into its operating position in a geostationary orbit. In a transfer orbit, solar power is provided by the aft panel only. The disadvantage of a spin-stabilized satellite is that only one-third of the solar array is exposed to the sun at any time, resulting in power limitations. For a higher power level a larger satellite is required to provide space for body-mounted solar cells. The three-axis body-stabilized configuration can provide much more power by using deployed solar panels of wings such as those shown in Fig. 2.16. The array consists of many panels hinged together in two sets. In a transfer orbit, the panels are folded and stowed by restraint bands against the north- and south-facing sides of the satellite. The outermost panel is partially illuminated by the sun and furnishes a small amount of solar power. When the satellite reaches the geostationary orbit, the array is deployed and full power becomes available. In this operating position, the booms that fasten each set of panels to the satellite are normal to the plane of orbit, and it is only necessary to rotate the boom one revolution per day to keep the face of the array pointing toward the sun. Each array is driven by an independent but redundant motor. Figure 2.27 is a weight-versus-power comparison of three-axis body-stabilized and spin-stabilized satellites. These curves show that at a higher power level a larger power-to-weight ratio can be obtained by using deployed solar arrays. However, the weight and complexity of a deployed solar array drive motor make the spin-stabilized type more attractive for low-power satellites.

Because geostationary satellites experience 88 eclipses during 1 year on station with a maximum duration of 70 min/day, batteries must be used to deliver power during an eclipse. Nickel-cadmium batteries are commonly used, but nickel-hydrogen batteries are replacing them because of the higher energy-to-weight ratio. The batteries are charged regularly by the solar array.

Besides the communications subsystems, the telemetry, command, and ranging subsystem, the attitude control subsystem, and the electrical

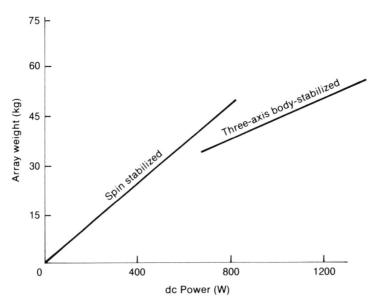

Figure 2.27 Weight-versus-power comparison of three-axis body-stabilized and spin-stabilized satellites.

power subsystem described in the above sections, the satellite also has a reaction control subsystem consisting of thrusters that provide the impulse required to perform satellite velocity and attitude control maneuvers such as the one shown in Fig. 2.26 for a spin-stabilized satellite. The thrust can be created either from gas under pressure in tanks or by a small rocket motor. The gas hydrazine, which is decomposed with a catalyst in the thrust nozzle, is commonly used. The apogee kick motor is a solid-propellant rocket motor employed to provide the velocity increment required to place the satellite in a geostationary orbit. The apogee kick motor is fired by a ground command, and after firing the spent motor remains as part of the in-orbit mass of the satellite.

REFERENCES

1. E. M. Gaposchkin and K. Lambeck, "1969 Smithsonian Standard Earth (II)," Spec. Rep. 315, Smithsonian Astrophysical Observatory, May 18, 1970.
2. M. E. Ash, "Determination of Earth Satellite Orbits," Tech. Note 1972-5, Lincoln Laboratory, MIT, April 19, 1972.
3. M. E. Ash, "Communications Terminal Orbit Calculations," Tech. Note 1973-74, Lincoln Laboratory, MIT, May 24, 1973.
4. J. V. Charyk, "Communications Satellites," Von Karman Lectureship in Astronautics, AIAA 13th Annual Meeting, Washington, D.C., 1977, Paper No. 77-323.
5. R. J. Rush et al., "INTELSAT V Spacecraft Design Summary," AIAA 7th Communications Satellite Systems Conference, San Diego, Calif., 1978, pp. 8–20.
6. R. C. Davis et al., "Future Trends in Communications Satellite Systems," *Acta Astronautica,* Vol. 5, No. 3–4, March–April 1978, pp. 275–298.
7. G. Abutaleb et al., "The COMSTAR Satellite System," *COMSAT Tech. Rev.,* Vol. 7, No. 1, Spring 1977, pp. 35–83.
8. D. C. Bakeman et al., "The Relative Merits of Three-Axis and Dual-Spin Stabilization Systems for Future Synchronous Communications Satellites," *AIAA Progress in Astronautics and Aeronautics: Communications Satellites for the 70's–Systems,* Vol. 26, 1971, pp. 605–653.
9. H. J. Dougherty et al., "Attitude Stabilization of Synchronous Communications Satellites Employing Narrow-Beam Antennas," *J. Spacecraft Rockets,* Vol. 8, No. 8, Aug. 1971, pp. 834–841.
10. J. E. Keigler et al., "Momentum Wheel Three-Axis Attitude Control for Synchronous Communications Satellites," in P. L. Bargellini (ed.), *Communications Satellite Technology, AIAA,* Vol. 33. Cambridge, Mass.: MIT Press, 1974, pp. 141–161.
11. M. H. Kaplan, *Modern Spacecraft Dynamics and Control.* New York: Wiley, 1976.
12. G. E. Schmidt, "Magnetic Attitude Control for Geosynchronous Spacecraft," AIAA 7th Communications Satellite Systems Conference, San Diego, Calif., 1978, pp. 278–284.
13. F. W. Weber, "Telemetry and Command Subsystems—The INTELSAT IV Spacecraft," *COMSAT TECH. Rev.* Vol. 2, No. 2, Fall 1972, pp. 341–358.
14. H. L. Van Trees, ed., *Satellite Communications.* New York: IEEE, 1979.
15. W. L. Pritchard et al., *Communications Satellite Systems Worldwide,* 1975–1985. Dedham, Mass.: Horizon-House Microwave, 1975.

16. K. Miya, *Satellite Communications Engineering*. Tokyo, Lattice, 1975.
17. J. Spilker, *Digital Communications by Satellite*. Englewood Cliffs, N.J.: Prentice-Hall, 1977.
18. J. Martin, *Communications Satellite Systems*. Englewood Cliffs, N.J.: Prentice-Hall, 1979.
19. G. K. O'Neill, *The Technology Edge*. New York: Simon & Schuster, 1983.

PROBLEMS

2.1 Find the velocity of a satellite at the perigee and the apogee of its elliptical orbit in terms of the semimajor axis a and the eccentricity e.

2.2 What is the longest geostationary satellite roundtrip delay?

2.3 Consider two geostationary satellites at longitudes 75°E and 75°W. Can these two satellites see each other?

2.4 Calculate the station latitude in terms of the elevation angle, the station longitude, and the geostationary satellite longitude asuming $E = 10°$ and $|\theta_S - \theta_L| = 50°$.

2.5 Consider an earth station located at longitude $\theta_L = 80°$W and latitude $\theta_l = 40°$N and a geostationary satellite at longitude $\theta_S = 120°$W. Assume that the satellite experiences an orbital inclination $i = 1°$.

(a) Find the maximum and minimum elevation angles for the earth station.

(b) Suppose the earth station has no tracking capability. What is the maximum pointing error of the antenna?

2.6 Calculate the maximum time a satellite is visible to a user on the earth's surface assuming a circular orbit.

2.7 Find the coverage area of a satellite from which it is visible at a minimum elevation angle $E_{min} = 10°$ for the following circular orbits:

(a) geostationary
(b) $H = 10,000$ km
(c) $H = 20,000$ km.

2.8 Consider a satellite at an altitude of $H = 15,000$ km. What is the time period during which it will pass over a point on the earth with an elevation angle exceeding $E = 20°$ assuming the satellite rotates in the same direction as the earth?

2.9 A satellite has an orbital period of T hours in the same direction as the earth's orbit. The time period during which an earth station can communicate with this satellite with a minimum elevation angle of $E = 10°$ is 4 h. Find T.

2.10 Consider n satellites in the same circular orbit of radius 10,000 km and an earth station working with a minimum elevation angle of 10°. How many satellites would be needed for the earth station to be able to communicate all the time?

2.11 Two geostationary satellites at longitudes 60°W and 120°W communicate via an inter-satellite link. Find the roundtrip propagation delay between them.

2.12 Find the maximum line-of-sight distance between two satellites at the same altitude H.

2.13 Consider a geostationary satellite at longitude 105°W and two earth stations at longitudes 90°W and 120°W, respectively, and at latitudes 35°N and 45°N, respectively. What is the total delay in sending 100 kbits over a 5-Mbps satellite link?

2.14 Let (x,y,z) be the geocentric coordinates of an earth station, where the z axis is the polar axis oriented from south to north, the x axis is the intersection of the equatorial plane and the 0° longitudinal plane, and the y axis completes the right-handed coordinate system.

(a) Find the coordinates (x,y,z) in terms of the station longitude θ_L and latitude θ_l.

(b) Let θ_s be the longitude of a geostationary satellite. Find the coordinates (x',y',z') of the satellite.

(c) Find the distance R between the station and the satellite.

2.15 Consider two stations A and B with the distances R_A and R_B to a geostationary satellite. Let E be the elevation angle of station A relative to the satellite and let α be its azimuth angle measured relative to a line connecting the two stations. (Assume that they are close enough to be considered in the same local horizontal plane.)

(a) Show that $R_B^2 = R_A^2 + d^2 - 2R_A d \cos \alpha \cos E$, where d is the distance between stations A and B.

(b) Assume that the angles α and E are measured with errors having an equal variance σ^2. What is the total error ϵ in the estimated value of R_B?

(c) Assuming that max $(R_A/R_B) = 1.2$, the angle measurement accuracy is $\pm 0.02°$, and $d = 100$ km, what is the maximum error for R_B?

2.16 An *Ariane* rocket (built in Western Europe) launches a satellite into a geostationary orbit at Kourou in French Guiana. Kourou's latitude is 5.3°N. Find the incremental velocity required to correct the orbit inclination and to achieve orbit circularization assuming the perigee of the elliptical transfer orbit is 450 km.

2.17 Consider an earth station located at longitude 60°W and latitude 30°S and a geostationary satellite at longitude $\theta_S = 80°$W. Find the earth station azimuth and elevation angles.

2.18 An earth station located at longitude 58°W and latitude 5°S points its antenna toward a geostationary satellite with an elevation angle of $E = 78.907°$. Find the satellite longitude assuming it is east of the earth station.

2.19 A satellite is in a circular equatorial orbit at an altitude of $H = 10,000$ km. Find the maximum eclipse time.

chapter *3*

Earth Station

In Chap. 1 we gave a functional description of the digital earth station shown in Fig. 1.4. In practice, an earth station is basically divided into two parts:

1. A RF terminal, which consists of an upconverter and a downconverter, a high-power amplifier, a low-noise amplifier, and an antenna
2. A baseband terminal, which consists of baseband equipment, an encoder and decoder, and a modulator and demodulator.

The RF terminal and the baseband terminal may be located a distance apart and connected by appropriate IF lines. In this chapter we will focus attention on a discussion of the RF terminal and will consider the baseband terminal from a system point of view only because its subsystems will be discussed separately and in detail in subsequent chapters. The baseband equipment will be discussed in Chaps. 6 and 7; the modulator and demodulator will be considered in Chaps. 9 and 10; and the encoder and decoder will be covered in Chap. 9.

3.1 EARTH STATION ANTENNA

The earth station antenna is one of the important subsystems of the RF terminal because it provides a means of transmitting the modulated RF carrier to the satellite within the uplink frequency spectrum and receiving the RF carrier from the satellite within the downlink frequency spectrum. The earth station antenna must meet three basic requirements:

1. The antenna must have a *highly directive gain;* that is, it must focus its radiated energy into a narrow beam to illuminate the satellite antenna in both the transmit and receive modes to provide the required uplink and downlink carrier power. Also, the antenna radiation pattern must have a low sidelobe level to reduce interference from unwanted signals and to minimize interference into other satellites and terrestrial systems.

2. The antenna must have a *low noise temperature* so that the effective noise temperature of the receive side of the earth station, which is proportional to the antenna temperature, can be kept low to reduce the noise power within the downlink carrier bandwidth. To achieve a low noise characteristic, the antenna radiation pattern must be controlled in such a way as to minimize the energy radiated into sources other than the satellite. Also, the ohmic losses of the antenna that contribute directly to its noise temperature must be minimized. This includes the ohmic loss of the waveguide that connects the low-noise amplifier to the antenna feed.

3. The antenna must be *easily steered* so that a tracking system (if required) can be employed to point the antenna beam accurately toward the satellite taking into account the satellite's drift in position. This is essential for minimizing antenna pointing loss.

3.1.1 Antenna Types

The two most popular earth station antennas that meet the above requirements are the paraboloid antenna with a focal point feed and the Cassegrain antenna.

A *paraboloid antenna* with a focal point feed is shown in Fig. 3.1. This type of antenna consists of a reflector which is a section of a surface formed by rotating a parabola about its axis, and a feed whose phase center is located at the focal point of the paraboloid reflector. The size of the antenna is represented by the diameter D of the reflector. The feed is connected to a high-power amplifier and a low-noise amplifier through an *orthogonal mode transducer* (OMT) which is a three-port network. The inherent isolation of the OMT is normally better than 40 dB. On the transmit side the signal energy from the output of the high-power amplifier is radiated at the focal point by the feed and illuminates the reflector which reflects and focuses the signal energy into a narrow beam. On the receive side the signal energy captured by the reflector converges on the focal point and is received by the feed which is then routed to the input of the

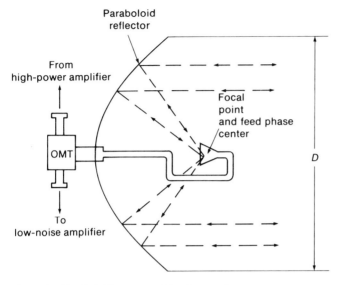

Figure 3.1 Paraboloid antenna with a focal point feed.

low-noise amplifier. This type of antenna is easily steered and offers reasonable gain efficiency in the range of 50 to 60%. The disadvantage occurs when the antenna points to the satellite at a high elevation angle. In this case, the feed radiation which spills over the edge of the reflector (spillover energy) illuminates the ground whose noise temperature can be as high as 290° K and results in a high antenna noise contribution. Paraboloid antennas with a focal point feed are most often employed in the United States for receive-only applications.

A *Cassegrain antenna* is a dual-reflector antenna which consists of a paraboloid main reflector, whose focal point is coincident with the virtual focal point of a hyperboloid subreflector, and a feed, whose phase center is at the real focal point of the subreflector, as shown in Fig. 3.2. On the transmit side, the signal energy from the output of the high-power amplifier is radiated at the real focal point by the feed and illuminates the convex surface of the subreflector which reflects the signal energy back as if it were incident from a feed whose phase center is located at the common focal point of the main reflector and subreflector. The reflected energy is reflected again by the main reflector to form the antenna beam. On the receive side, the signal energy captured by the main reflector is directed toward its focal point. However, the subreflector reflects the signal energy back to its real focal point where the phase center of the feed is located. The feed therefore receives the incoming energy and routes it to the input of the low-noise amplifier through the OMT. A Cassegrain antenna is more expensive than a paraboloid antenna because of the addition of the subreflector and the integration of the three antenna elements—the main reflector, subreflector, and feed—to produce an optimum antenna system. However, the Cassegrain antenna offers many advantages over the paraboloid antenna: low noise temperature, pointing accuracy, and flexibility in feed design. Since the spillover energy from the feed is directed toward the sky whose noise temperature is typically less than 30° K, its contribution to the antenna noise temperature is small compared to that of the paraboloid antenna. Also, with the feed located near the vertex of the main reflector, greater mechanical stability can be achieved than with the focal point feed in the paraboloid antenna. This increased stability permits very accurate pointing of high-gain narrow-beam antennas.

To minimize the losses in the transmission lines connecting the high-power amplifier and the low-noise amplifier to the feed, a beam waveguide feed system may be employed. A Cassegrain antenna with a beam waveguide feed system is shown in Fig. 3.3. The beam waveguide assembly consist of four mirrors supported by a shroud and precisely located relative to the subreflector, the feed, the elevation axis, and the azimuth axis. This mirror configuration acts as a RF energy funnel between the feed and the subreflector and, as such, must be designed to achieve minimum loss while allowing the feed to be mounted in the concrete foundation at ground level. The shroud assembly acts as a shield against

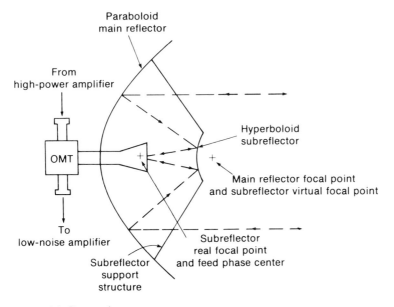

Figure 3.2 Cassegrain antenna.

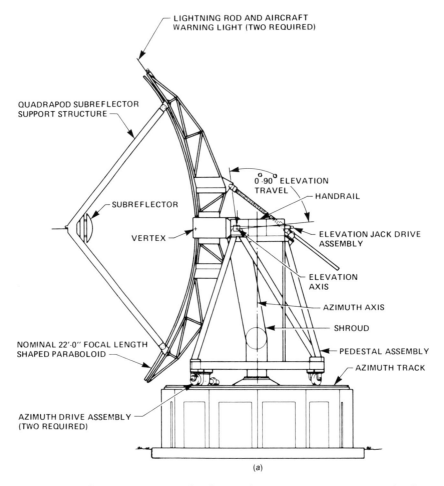

(a)

Figure 3.3 (*a*) Complete structure of a Cassegrain antenna. (*b*) Beam waveguide feed. (*Courtesy of GTE.*)

ground noise and provides a rigid structure which maintains the mounting integrity of the mirrors when the antenna is subjected to wind, thermal, or other external loading conditions. The lower section of the shroud assembly is supported by the pedestal and rotates about the azimuth axis. The upper section of the shroud assembly is supported by the main reflector support structure and rotates about the elevation axis. As seen in Fig. 3.3*b*, the beam waveguide mirror system directs the energy to and from the feed and the reflectors. The configuration utilized is based on optics, though a correction is made for diffraction effects by using slightly elliptical curved mirrors. For proper shaping and positioning of the beam

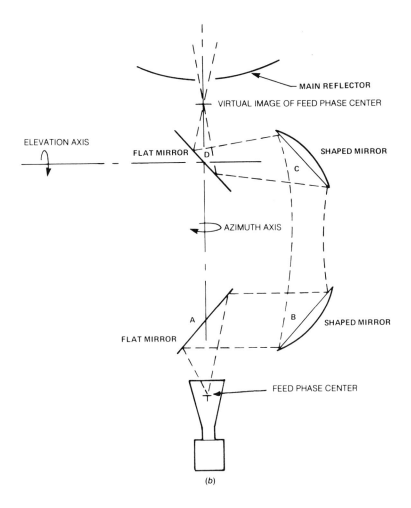

MAIN REFLECTOR

VIRTUAL IMAGE OF FEED PHASE CENTER

ELEVATION AXIS

FLAT MIRROR

SHAPED MIRROR

AZIMUTH AXIS

SHAPED MIRROR

FLAT MIRROR

FEED PHASE CENTER

(b)

waveguide mirrors, the energy from the feed located in the equipment room is refocused so that the feed phase center appears to be at the subreflector's real focal point. In operation, mirrors A, B, C, and D move as a unit when the azimuth platform rotates. Mirror D is on the elevation axis and rotates also when the main reflector is steered during elevation. In this way, the energy to and from the beam waveguide system is always directed through the opening in the main reflector vertex.

As mentioned in Chap. 1, modern communications satellites often employ dual polarizations to allow two independent carriers to be sent in the same frequency band, thus permitting frequency reuse and doubling

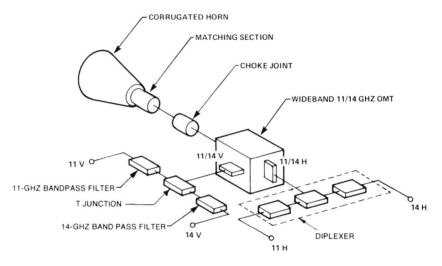

Figure 3.4 Wideband orthomode transducer-diplexer feed. (*Courtesy of GTE.*)

the satellite capacity. Figure 3.4 shows a wideband OMT diplexer type of frequency reuse feed for a Cassegrain antenna that provides horizontal and vertical polarization for a Ku-band operation.

3.1.2 Antenna Gain

Gain is perhaps the key performance parameter of an earth station antenna because it directly affects the uplink and downlink carrier power. For an antenna, the gain is given by

$$G = \eta \, \frac{4 \, \pi \, A}{\lambda^2} = \eta \, \frac{4 \, \pi \, A f^2}{c^2} \tag{3.1}$$

where A = antenna aperture area (m²)
λ = radiation wavelength (m)
f = radiation frequency (Hz)
c = speed of light = 2.997925×10^8 m/s
η = antenna aperture efficiency ($\eta < 1$)

For a circular aperture, it follows that $A = \pi D^2/4$; therefore

$$G = \eta \left(\frac{\pi D}{\lambda} \right)^2 = \eta \left(\frac{\pi f D}{c} \right)^2 \tag{3.2}$$

where D = antenna diameter (m).

The antenna efficiency η represents the percentage of the aperture area A that is used effectively in transmission or reception and is a product of various efficiency factors that reduce the antenna gain. Typical efficiency factors for a Cassegrain antenna such as the one shown in Fig. 3.3a are

$$\eta = \eta_1\eta_2\eta_3\eta_4\eta_5\eta_6 \qquad (3.3)$$

where η_1 = main reflector illumination efficiency
η_2 = spillover efficiency
η_3 = phase efficiency
η_4 = subreflector and support structure blocking efficiency
η_5 = feed system dissipative efficiency
η_6 = reflector surface tolerance efficiency

The illumination efficiency η_1 is determined by the characteristic of the field distribution across the main reflector aperture. If it is uniform over the entire aperture area, then $\eta_1 = 1$. The spillover efficiency η_2 represents not only the energy spilled over the edge of the main reflector but also the energy spilled over the edge of the subreflector. To minimize the spillover loss, a feed with low sidelobes in its radiation pattern is desired. To achieve this pattern, multiple modes are used in the design of the feed radiation section which is a horn. Furthermore, the feed angle subtended by the subreflector is chosen so that the main beam of the feed radiation pattern intersects the subreflector at a low level, minimizing the feed main beam spillover past the subreflector. However, the low edge illumination of the subreflector normally results in sharply tapered illumination across the main reflector aperture, resulting in a low illuminating efficiency η_1. With a Cassegrain antenna this condition can be improved substantially by deliberately altering the shape of the subreflector to distribute the energy essentially uniformly nearly to the edge of the main reflector but then falling off very sharply. An illumination efficiency η_1 of 0.94 to 0.96 can be achieved in practice with a main reflector spillover efficiency of as high as 0.99. With good feed design, a subreflector spillover efficiency on the order of 0.98 can be realized. Thus a spillover efficiency of $\eta_2 = 0.97$ can be achieved in a shaped system.

Distorting the shape of the subreflector to achieve uniform illumination across the main reflector results in a phase error being introduced into the main reflector. This phase error results in energy being radiated in undesired directions, thus decreasing the gain and increasing the sidelobe level of the antenna. The phase efficiency η_3 identifies this gain loss. Most of this loss can be eliminated, however, by reshaping the main reflector to correct the phase error. In a well-designed Cassegrain antenna the phase efficiency η_3 can be on the order of 0.98 and 0.99 at the design frequency and remain on the order of 0.95 over 70% of the operating 500-MHz band. Blocking of the main reflector aperture by the subreflector and support structure results in an effectively smaller aperture, hence a loss in the antenna gain. The subreflector blocking efficiency is about 0.97, and that of the support structure is about 0.95 in a well-designed antenna. The dissipative loss of the feed system also reduces the antenna gain. Depending on the feed system structure the efficiency η_5 can be as high as 0.94.

All the above-cited efficiency factors are primarily dependent on the main reflector and subreflector geometries and the feed system structure and not on the operating frequency. In practice the main reflector and subreflector cannot be built to the ideal shapes without some surface tolerance. This results in a scattering of energy in unwanted directions in a manner similar to that associated with the phase error. The surface tolerance may, in effect, be considered a special type of phase error which limits the maximum achievable gain G_M in the sense that, for a given surface tolerance and antenna diameter, increasing the operating frequency increases the antenna gain until it equals G_M. Further increasing the operating frequency will decrease the antenna gain. The surface tolerance efficiency η_6 imposes an upper limit on the maximum operating frequency, hence on the maximum antenna gain, and is fixed by current manufacturing technology. The reflector surface tolerance efficiency η_6 may be expressed [1] as

$$
\eta_6 = \exp\left[-\left(\frac{4\pi\epsilon}{\lambda}\right)^2\right]
$$
$$
= \exp\left[-\left(\frac{\epsilon}{D}\right)^2\left(\frac{4\pi fD}{c}\right)^2\right]
$$

(3.4)

where ϵ = rms reflector surface error (m) and ϵ/D = antenna surface tolerance. The factor $(4\pi\epsilon/\lambda)^2$ is in effect the mean-square phase error introduced by the surface error ϵ. The antenna surface tolerance ϵ/D within current commercial technology is as follows:

$$10^{-3} \leq \frac{\epsilon}{D} \leq 10^{-4} \qquad\qquad D \leq 1.2 \text{ m}$$

$$2 \times 10^{-4} \leq \frac{\epsilon}{D} \leq 5 \times 10^{-5} \qquad 2.5 \text{ m} \leq D \leq 6 \text{ m} \qquad (3.5)$$

$$10^{-4} \leq \frac{\epsilon}{D} \leq 2 \times 10^{-5} \qquad 9 \text{ m} \leq D \leq 24 \text{ m}$$

The performance of a well-designed Ku-band 20-m Cassegrain antenna with a beam waveguide frequency reuse feed system is shown in Table 3.1.

3.1.3 Antenna Pointing Loss

In the previous section we have discussed the gain of the antenna as given in (3.2). This is the peak or on-axis gain of the antenna which can be achieved if the antenna beam is pointed accurately toward the satellite. A loss in gain can occur if the antenna pointing vector is not in line with the

Table 3.1 Performance of a Ku-band 20-m Cassegrain antenna

Parameters	11.95 GHz	14.25 GHz
Illumination efficiency	0.96	0.94
Spillover efficiency		
Main reflector	0.99	0.99
Subreflector	0.96	0.98
Phase efficiency	0.98	0.98
Blocking efficiency		
Subreflector	0.97	0.97
Support structure	0.95	0.95
Feed system dissipative efficiency		
Basic feed	0.94	0.93
Diplexer	0.96	0.98
Beam waveguide	0.91	0.96
Surface tolerance efficiency		
Main reflector	0.87	0.83
Subreflector	0.97	0.97
Net antenna efficiency	0.57	0.54
Antenna gain (dB)	65.53	66.82

satellite position vector as shown in Fig. 3.5. The antenna pointing loss can be evaluated from the antenna gain pattern which is a function of the off-axis angle θ. For a paraboloid reflector with aperture diameter D, the normalized gain pattern for a *uniform aperture distribution* [2] is given by

$$G_{n,u}(\theta) = 4 \left| \frac{J_1(u)}{u} \right|^2 \tag{3.6a}$$

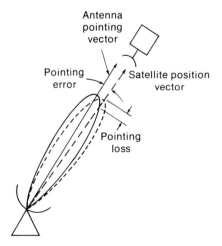

Figure 3.5 Antenna pointing error.

where

$$u = \frac{\pi D}{\lambda} \sin \theta \qquad (3.6b)$$

and θ = off-axis angle and $J_1(\bullet)$ = first-order Bessel function of the first kind (see Appendix 5B). If the aperture distribution is *parabolic* [the aperture field distribution is of the form $1 - (2r/D)^2$, where r is the radial distance from the center of the circular aperture], then the normalized gain pattern is given by

$$G_{n,p}(\theta) = 64 \left| \frac{J_2(u)}{u^2} \right|^2 \qquad (3.7)$$

where $J_2(\bullet)$ = second-order Bessel function of the first kind (see Appendix 5B).

The *half-power beamwidth* for a uniform aperture distribution is twice the value of θ at $G_{n,u}(\theta) = \frac{1}{2}$, which is $1.02\lambda/D$ radians or $58.5\lambda/D$ degrees. For a parabolic aperture distribution, the half-power beamwidth is $1.27\lambda/D$ radians or $72.7\lambda/D$ degrees. In practice the antenna pointing accuracy is normally kept within one-third of the half-power beamwidth. For example, the half-power beamwidth of a 20-m Cassegrain antenna with uniform aperture distribution is $0.062°$ at 14.25 GHz. If the pointing error is kept within $0.02°$, the normalized gain will be greater than $G_{n,u}$ $(0.02°) = 0.757$ or, equivalently, the loss in gain will be less than $-10 \log 0.757 = 1.2$ dB. (Note that the normalized peak gain at $\theta = 0$ is simply 1.)

Because the earth station antenna is subjected to a wind loading effect and the satellite drifts in orbit, an antenna tracking system is necessary for a large-diameter antenna to minimize the pointing error. The antenna tracking system is a closed-loop pointing system; that is, the antenna pointing vector which is a function of the azimuth and elevation angles is derived from the received signal. One of the commonly used antenna tracking systems for earth stations is a step track which derives the antenna pointing vector from the signal strength of a satellite beacon signal. In this type of tracking the antenna is caused to move around the edge of a square pattern in the plane normal to the axis of the antenna beam as shown in Fig. 3.6. The center of the square $ABCD$ corresponds to the assumed satellite position. Thus, if the satellite position vector is aligned with the antenna pointing vector, the amplitude of the beacon signal received at each of the four pointing positions A, B, C, and D will be equal. An error in elevation results in the off-axis angle θ for positions A and B differing from the off-axis angle θ for positions C and D. Since the received signal amplitude is proportional to the square of the off-axis angle θ as indicated in (3.6a) or (3.7), by comparing the average amplitude at A and B with that of C and D, the difference in the offset angle or elevation angle error can be determined. Similarly, by comparing the average amplitude at A and D with that at B and C, the azimuth angle error can be

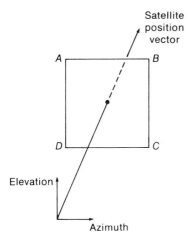

Satellite
position
vector

Elevation

Azimuth

Figure 3.6 Principle of a step track.

determined. The step track system can provide a tracking accuracy of from $\frac{1}{5}$ to $\frac{1}{3}$ of the half-power beamwidth, which results in a pointing loss of between 0.5 and 1.5 dB. For better tracking accuracy, a monopulse tracking system must be employed [2]. This type of tracking can provide a tracking accuracy of up to $\frac{1}{20}$ of the half-power beamwidth. For a fixed earth station with a small antenna (a half-power beamwidth on the order of 0.2° or more), a program tracking system can be used. This is an open-loop pointing system where the antenna pointing vector is derived from the earth station position and the satellite ephemeris data. The satellite ephemeris data is normally obtained from a *telemetry, tracking,* and *control* station (TT&C) and is stored in the memory of the tracking system. The satellite ephemeris data is updated periodically. The updated interval depends on the satellite stationkeeping characteristics.

3.1.4 Effective Isotropic Radiated Power

To express the transmitted power of an earth station or a satellite, the *effective isotropic radiated power* (EIRP) is normally employed. The earth station EIRP is simply the power generated by the high-power amplifier times the gain of the earth station antenna, taking into account the loss in the transmission line (waveguide) that connects the output of the high-power amplifier to the feed of the earth station antenna. If we let P_T denote the input power at the feed of the antenna and G_T the transmit antenna gain, the earth station EIRP is simply

$$\text{EIRP} = P_T G_T \qquad (3.8)$$

For example, consider a 2-kW high-power amplifier and a 20-m Cassegrain antenna whose transmitted gain is 66.82 dB at 14.25 GHz as given in Table 3.1. If it is assumed that the loss of the waveguide that connects the high-power amplifier to the feed is 1 dB, then the earth station EIRP in decibel-watts is (noting that P_T decibel-watts is equivalent to 10 log P_T watts).

$$\text{EIRP} = 33 + 66.82 - 1 = 98.82 \text{ dBW}$$

The uplink carrier power, that is, the power of the carrier received at the satellite, is directly proportional to the earth station EIRP. This will be discussed in Chap. 4.

3.1.5 Antenna Gain-to-Noise Temperature Ratio

The antenna gain-to-noise temperature ratio G/T is a figure of merit commonly used to indicate the performance of the earth station antenna and the low-noise amplifier in relation to sensitivity in receiving the downlink carrier from the satellite. The parameter G is the receive antenna gain referred to the input of the low-noise amplifier. For example, if the input of the low-noise amplifier is connected directly to the output port of the feed system of the 20-m Cassegrain antenna whose performance is illustrated in Table 3.1, then the receive antenna gain at 11.95 GHz is 65.53 dB. If a piece of waveguide with a 0.53-dB loss is used to connect the input of the low-noise amplifier to the output port of the feed system, the receive antenna gain referred to the input of the low-noise amplifier is simply 65 dB. The parameter T is defined as the earth station system noise temperature referred also to the input of the low-noise amplifier. We have discussed the antenna gain previously, therefore in this section we will concentrate on determination of the earth station system noise temperature.

The treatment of noise in communications systems is based on a form of noise called *white noise* whose power spectral density is $N_0/2$, as shown in Fig. 3.7, and is flat over a large range of frequencies. White noise is characterizied as a Gaussian random process with a zero mean, and it includes thermal noise produced by the random motion of electrons in conducting media, solar noise, and cosmic noise. White noise corrupts the received signal in an additive fashion and is normally referred to as *additive white Gaussian noise* (AWGN) in the analysis of communications systems. The treatment of noise is outside the scope of this book, and it is recommended that the reader consult [3] for a detailed study. (Appendix 3A gives a brief review of thermal noise.)

In electrical communications systems, the power spectral density of white noise delivered to a matched load from a noise source is customarily expressed in watts per hertz (W/Hz) as

$$\frac{N_0}{2} = \frac{kT_s}{2} \tag{3.9}$$

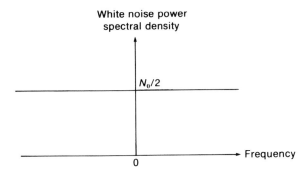

Figure 3.7 White noise power spectral density.

where k is Boltzmann's constant (1.38×10^{-23} J/K) and T_s is the noise temperature of the noise source measured in kelvins. This means that, if this noise source is connected to the input of an ideal bandpass filter with bandwidth B in hertz (Fig. 3.8) whose input resistance is matched to the source resistance, the output noise power in watts is simply

$$N = N_0 B = kT_s B \qquad (3.10)$$

Since any passive or active two-port system, such as the waveguide that connects the input of the low-noise amplifier to the antenna feed, and the low-noise amplifier itself have equivalent noise that contributes to the noise from the antenna, we must take into account their effects. Consider a two-port system with gain G and a noise source of temperature T_s connected to its input. The output noise power in a bandwidth B (Hz) is then

$$N = GkT_s B + N_n \qquad (3.11)$$

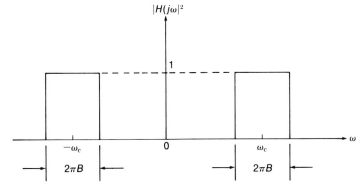

Figure 3.8 Ideal bandpass filter. $H(j\omega)$, transfer function of bandpass filter; ω_c, center frequency of bandpass filter; B, bandwidth of bandpass filter in hertz.

where N_n is the output noise power produced by the internal noise sources in the system. Equation (3.11) can be written as

$$N = GkB\left(T_s + \frac{N_n}{GkB}\right)$$

$$= GkB(T_s + T_e) \tag{3.12}$$

where

$$T_e = \frac{N_n}{GkB} \tag{3.13}$$

From (3.12) it is seen that N_n can be considered to be produced by a fictitious noise source of *equivalent noise temperature* T_e connected to the input of the system. Therefore we conclude that a noisy two-port system can be characterized by its equivalent noise temperature T_e. The parameter $T_s + T_e$ in (3.12) is defined as the system noise temperature referred to the input of the two-port system:

$$T = T_s + T_e \tag{3.14}$$

In other words, we can model a noisy two-port system as a noiseless two-port system and account for the increased noise by assigning to the input noise source a new temperature T higher than T_s by T_e. Note that $GT = G(T_s + T_e)$ is simply the noise temperature measured at the output of the two-port system. Another measure of the internal noise generated by a two-port system is the noise figure F, defined as the output noise power of the system divided by the output noise power if the system is noiseless (i.e., all the internal noise sources are absent), assuming that the noise sources at the input is at the ambient temperature T_0 (T_0 is normally taken to be 290 K). By this definition the *noise figure* F is also the ratio of the signal-to-noise ratio at the system input to the signal-to-noise ratio at the system output. Thus F is simply the ratio of N in (3.11), with $T_s = T_0$ and GkT_0B which is N when $N_n = 0$:

$$F = \frac{GkT_0B + N_n}{GkT_0B}$$

$$= 1 + \frac{T_e}{T_0} \tag{3.15}$$

From (3.15) it is seen that

$$T_e = (F - 1)T_0 \tag{3.16}$$

Now consider two 2-port systems M_1 and M_2 in cascade as shown in Fig. 3.9. Each system M_i is characterized by its gain G_i and its equivalent noise temperature T_{ei}, $i = 1,2$. The noise source at the input of the cas-

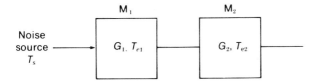

Figure 3.9 Cascaded two-port system for equivalent noise temperature analysis.

caded system is assumed to have a temperature T_s. The noise power N_1 at the output of system M_1 is given by (3.12):

$$N_1 = G_1 kB(T_s + T_{e1}) \tag{3.17}$$

This is amplified by M_2 and appears at its output as

$$N_{12} = G_1 G_2 kB(T_s + T_{e1}) \tag{3.18}$$

Thus N_{12} is the noise power produced by the input noise source and the internal noise sources at M_1. From (3.13) the noise power produced by the internal noise sources at M_2 is given by

$$N_2 = G_2 kT_{e2} B \tag{3.19}$$

The total output noise power is simply the sum of N_{12} and N_2:

$$N = N_{12} + N_2 = G_1 G_2 kB(T_s + T_{e1}) + G_2 kT_{e2} B$$

$$= G_1 G_2 kB\left(T_s + T_{e1} + \frac{T_{e2}}{G_1}\right) \tag{3.20}$$

By comparing (3.20) and (3.12) it is seen that the cascaded system can be characterized by its gain $G = G_1 G_2$, which is obvious, and its equivalent noise temperature

$$T_e = T_{e1} + \frac{T_{e2}}{G_1} \tag{3.21}$$

Equation (3.21) demonstrates clearly the contribution of the second system M_2 to the overall noise temperature. It is seen that, if the gain of the first system M_1 is large enough ($G_1 >> T_{e2}/T_{e1}$), then the second system contributes negligibly to the overall noise temperature. The results of (3.21) can be easily generalized to a cascade of n systems:

$$T_e = T_{e1} + \frac{T_{e2}}{G_1} + \frac{T_{e3}}{G_1 G_2} + \cdots + \frac{T_{en}}{G_1 G_2 \ldots G_{n-1}} \tag{3.22}$$

By using (3.15), the noise figure of n systems in cascade can be expressed as

$$F = F_1 + \frac{F_2 - 1}{G_1} + \frac{F_3 - 1}{G_1 G_2} + \cdots + \frac{F_n - 1}{G_1 G_2 \ldots G_{n-1}} \tag{3.23}$$

Before proceeding to an evaluation of the earth station system noise temperature referred to the input of the low-noise amplifier, we pause here for a moment to make the following observation concerning the ohmic loss in a transmission line such as a waveguide, coaxial cable, or other device characterized by power loss rather than power gain. For such a lossy two-port system, let $L > 1$ be the power loss (i.e., its gain $G = 1/L < 1$) and T_0 its ambient temperature; then the output noise is simply kT_0B. By using (3.12) with $T_s = T_0$ and $G = 1/L$ we have

$$kT_0B = \frac{1}{L} kB(T_0 + T_e)$$

which yields the equivalent noise temperature of a lossy two-port system:

$$T_e = (L - 1)T_0 \qquad (3.24)$$

By comparing (3.24) and (3.16) it is seen that the noise figure of a lossy two-port system is

$$F = L \qquad (3.25)$$

Now consider the receive side of the earth station, which consists of the antenna, the waveguide that connects the antenna feed to the low-noise amplifier, the low-noise amplifier, and the downconverter in cascade, as shown in Fig. 3.10. The antenna noise temperature is measured at the feed output and is denoted by T_A. The waveguide is characterized by its power loss $L_1 > 1$ (or power gain $G_1 = 1/L_1 < 1$) and its equivalent noise temperature $T_{e1} = (L_1 - 1)T_0$. The low-noise amplifier is characterized by its gain G_2 and its equivalent noise temperature T_{e2}. The equivalent noise temperature of the downconverter is T_{e3}. The equivalent noise temperature T_e of the cascaded low-noise amplifier and downconverter is given by (3.21) as

$$T_e = T_{e2} + \frac{T_{e3}}{G_2} \qquad (3.26)$$

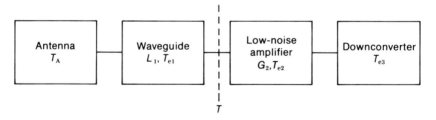

Figure 3.10 Receive side of the earth station for system noise temperature evaluation.

From (3.12) we note that the noise power at the output of the waveguide is given by

$$N = G_1 kB(T_A + T_{e1})$$

$$= \frac{1}{L_1} kB \left[T_A + (L_1 - 1)T_0 \right]$$

$$= kB \left(\frac{T_A}{L_1} + \frac{L_1 - 1}{L_1} T_0 \right) \tag{3.27}$$

Thus the noise temperature T_s measured at the output of the waveguide is

$$T_s = \frac{T_A}{L_1} + \frac{L_1 - 1}{L_1} T_0 \tag{3.28}$$

From (3.14), (3.26), and (3.28), it is seen that the earth station system noise temperature referred to the input of the low-noise amplifier is simply

$$T = T_s + T_e$$

$$T = \frac{T_A}{L_1} + \frac{L_1 - 1}{L_1} T_0 + T_{e2} + \frac{T_{e3}}{G_2} \tag{3.29}$$

Example 3.1 To illustrate evaluation of the antenna gain-to-noise temperature ratio, consider a 20-m Cassegrain antenna whose receive antenna gain is 65.53 dB at 11.95 GHz as given by Table 3.1. The receive side of the earth station is characterized by the following parameters.

Antenna noise temperature: $T_A = 60$ K
Waveguide loss: $L_1 = 1.072$ (0.3 dB)
Low-noise amplifier
   Equivalent noise temperature: $T_{e2} = 150$ K
   Gain: $G_2 = 10^6$ (60 dB)
Downconverter equivalent noise temperature: $T_{e3} = 11 \times 10^3$ K

Substituting all the above values into (3.29) yields the earth station system noise temperature referred to the input of the low-noise amplifier as ($T_0 = 290$ K):

$$T = \frac{60}{1.072} + \frac{0.072}{1.072} (290) + 150 + \frac{11 \times 10^3}{10^6}$$

$$= 225.5 \text{ K}$$

The antenna gain referred to the input of the low-noise amplifier is

$$G = 65.53 - L_1$$

$$= 65.53 - 0.3 = 65.23 \text{ dB}$$

Thus the antenna gain-to-noise temperature ratio in decibels per kelvin is

$$\frac{G}{T} = G \text{ (dB)} - 10 \log T$$

$$= 65.23 - 23.53$$

$$= 41.7 \text{ dB/K}$$

Based on the above analysis we can make the following remarks:

1. The higher the antenna gain, the higher the value of G/T, hence the higher the downlink carrier-to-noise ratio, as we will see in Chap. 4.
2. The lower the loss of the waveguide that connects the antenna feed to the input of the low-noise amplifier, the higher the value of G/T.
3. The lower the equivalent noise temperature of the low-noise amplifier, the higher the value of G/T. Also, its gain must be sufficiently large to reduce the noise contribution of the downconverter.
4. The value of G/T is invariant regardless of the reference point. We choose the input of the low-noise amplifier because this is the point where the contribution of the low-noise amplifier is clearly shown.

3.1.6 G/T Measurement

One of the precision methods for measuring the antenna gain-to-noise temperature ratio G/T of the earth station is the radio star method. This measurement employs the Y factor defined as the ratio of the noise power received with the star accurately centered on the receive beam axis to the background noise power (when the star has drifted out of the beam):

$$Y = \frac{T + T_{star}}{T} \qquad (3.30)$$

where T_{star} = effective noise temperature of star and T = system noise temperature. The receive antenna gain G is related to T_{star} [4] as follows:

$$kT_{star} = \frac{SG\lambda^2}{8\pi} = \frac{SGc^2}{8\pi f^2} \qquad (3.31)$$

where S = flux density per 1-Hz bandwidth of radio star (W/m²/Hz)
f = measurement frequency (Hz)
c = speed of light = 2.997925×10^8 m/s
k = Boltzmann's constant = 1.38×10^{-23} J/K

Substituting (3.30) into (3.31) yields

$$\frac{G}{T} = (Y - 1)\frac{8\pi k}{S\lambda^2} \qquad (3.32)$$

Equation (3.32) represents the ideal condition where there is no atmospheric loss and the radio star is a point source. The equation for a nonideal condition is given by

$$\frac{G}{T} = L_a L_s (Y - 1)\frac{8\pi k}{S\lambda^2} \qquad (3.33)$$

where L_a = atmospheric loss and L_s = correction factor for angular extent of radio star.

Table 3.2 Typical one-way clear-air total zenith attenuation values, L_a' (dB)^a

Frequency (GHz)	Altitude (km)				
	0	0.5	1.0	2.0	4.0
10	0.053	0.047	0.042	0.033	0.02
15	0.084	0.071	0.061	0.044	0.023
20	0.28	0.23	0.18	0.12	0.05
30	0.24	0.19	0.16	0.10	0.045
40	0.37	0.33	0.29	0.22	0.135
80	1.30	1.08	0.90	0.62	0.30
90	1.25	1.01	0.81	0.52	0.22
100	1.41	1.14	0.92	0.59	0.25

^{a}Mean surface conditions: 21°C, 7.5 g/m³ H_2O, U.S. std. atmos., 45°N, July.

The atmospheric loss or attenuation L_a can be calculated from existing data and is given by

$$L_a \text{ (dB)} = \frac{L_a' \text{ (dB)} + b_\rho \left(\rho_0 - 7.5 \text{ g/m}^3\right) + c_T \left(21°C - T_0\right)}{\sin E} \quad (3.34)$$

where L_a' = zenith one-way attenuation for a moderately humid atmosphere (7.5 g/m³ surface water vapor) and a surface temperature of 21°C; L_a' is plotted in Fig. 3.11 and given in Table 3.2

b_ρ = water vapor density correction coefficient; b_ρ is given in Table 3.3

c_T = temperature correction coefficient; c_T is given in Table 3.3

ρ_0 = surface water vapor density (g/m³). The factor $b_\rho(\rho_0 - 7.5$ g/m³) accounts for the difference between the local surface water vapor density and 7.5 g/m³

Table 3.3 Water vapor density and temperature correction coefficients

Frequency (GHz)	Water vapor density Correction b_ρ	Temperature correction c_T
10	2.10×10^{-3}	2.60×10^{-4}
15	6.34×10^{-3}	4.55×10^{-4}
20	3.46×10^{-2}	1.55×10^{-3}
30	2.37×10^{-2}	1.33×10^{-3}
40	2.75×10^{-2}	1.97×10^{-3}
80	9.59×10^{-2}	5.86×10^{-3}
90	1.22×10^{-1}	5.74×10^{-3}
100	1.50×10^{-1}	6.30×10^{-3}

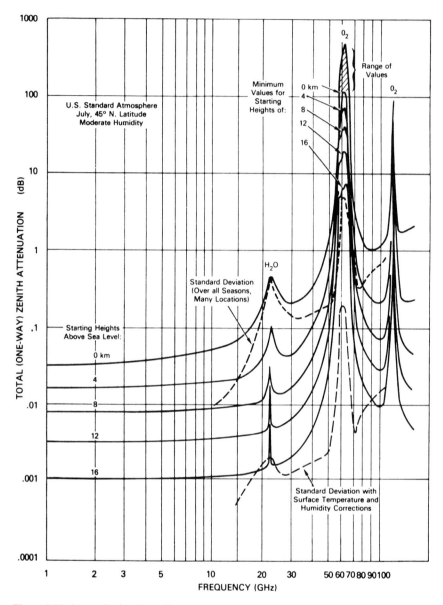

Figure 3.11 Atmospheric attenuation.

T_0 = surface temperature (°C). The factor $c_T(21°C - T_0)$ accounts for the difference between the local surface temperature and 21°C

The surface water vapor density ρ_0 may be found from the ideal gas law:

$$\rho_0 = \frac{he_s}{0.461(T_0 + 273)}$$

where h = relative humidity (e.g., $h = 60\% = 0.6$) and e_s = saturated partial pressure of water vapor (N/m^2) corresponding to local surface temperature T_0(°C); $e_s \approx 206.43 \exp[0.0354\ T_0(°F)]$.

The correction factor L_s for the angular extent of the radio star is a function of its brightness distribution and the antenna gain pattern and is plotted in Fig. 3.12 in decibels for Taurus, Cassiopeia A, and Cygnus A.

The nominal flux density for Taurus A at 3.95 GHz is 716.9×10^{-26} $W/m^2/Hz$. This star does not exhibit any flux variation with time and has a spectral index of 0.25.

The nominal flux density for Cassiopeia A at 4.08 GHz was 1086×10^{-26} $W/m^2/Hz$ in the year 1964.8 (about mid-September 1964). Its spectral index is 0.74. The flux density of Cassiopeia A decreases at a rate of 1.1% per year. The flux for the date of measurement may be computed from the expression

$$S = S_0(1 - \beta)^y \tag{3.35}$$

where S = flux density for date of measurement
   S_0 = reference flux density taken to be 1086×10^{-26} $W/m^2/Hz$ for the epoch 1964.8
   $\beta = 0.011$
   y = number of years from 1964.8 to date of measurement

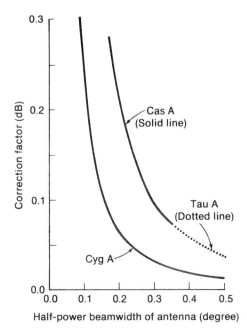

Figure 3.12 Correction factor for the angular extent of radio stars.

The nominal flux density for Cygnus A at 3.95 GHz is 494.8×10^{-26} W/m²/Hz. This star does not exhibit any flux density variation with time and has a spectral index of 1.19.

When a measurement is made at a frequenty for which the flux density is not available, the flux density at that frequency is given by

$$\frac{S}{S_0} = \left(\frac{f_0}{f}\right)^x \tag{3.36}$$

where x = spectral index of radio star and f_0 = reference frequency for reference flux density S_0.

The test setup for G/T measurement is shown schematically in Fig. 3.13, and the measurement procedure is outlined as follows:

1. Tune the oscillator to obtain the desired measurement frequency.
2. Set the attenuator to a value slightly higher than the anticipated Y factor.
3. Use the star ephemeris to move the antenna beam onto the star. Peak the meter indication by adjusting the antenna servo controls.
4. When the peak meter reading is obtained, record this value and that of the attenuator.
5. As the star drifts outside the antenna beam gradually decrease the attenuation to keep the meter reading approximately constant. When the drift is completed, adjust the attenuator until the peak reading as obtained in step 4 is reestablished and then record the new value of the attenuator. The difference in attenuator readings between steps 4 and 5 is the Y factor in decibels.
6. Repeat the measurement for a number of times and take the average.

Example 3.2 Calculate the atmospheric attenuation based on the following parameters.
Frequency: 20 GHz
Altitude: 880 m
Relative humidity: $h = 60\% = 0.6$
Surface temperature: $T_0 = 26.7°C = 80°F$
Elevation angle: $E = 47°$

1. Determine L_a' for 7.5 g/m³ surface water vapor and a 21°C surface temperature from Table 3.2:

$$L_a' = 0.20 \text{ dB}$$

2. Find the water vapor density ρ_0. The saturated partial pressure of water vapor at $T_0 = 80°F$ is $e_s = 3500$ N/m². Thus $\rho_0 = 0.6 \times 3500/0.461(26.7 + 273) = 15.2$ g/m³.
3. Determine b_ρ and c_T from Table 3.2 and 3.3:

$$b_\rho = 0.035$$

$$c_T = 1.55 \times 10^{-3}$$

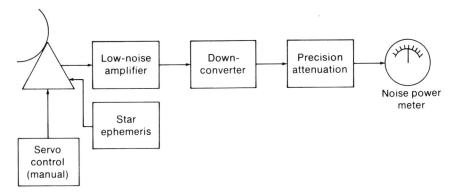

Figure 3.13 Test setup for G/T measurement.

4. Find L_a from (3.34):

$$L_a = \frac{0.2 + 0.27 - 0.009}{\sin 47°} = 0.63 \text{ db}$$

3.2 HIGH-POWER AMPLIFIER

One of the most widely used high-power amplifiers in earth stations is the *traveling wave tube amplifier* (TWTA) shown schematically in Fig. 3.14. The traveling wave tube employs the principle of velocity modulation in the form of traveling waves. The RF signal to be amplified travels down a periodic structure called a *helix*. Electrons emitted from the cathode of the tube are focused into a beam along the axis of the helix by cylindrical magnets and removed at the end by the collector after delivering their energy to the RF field. The helix slows down the propagation velocity of the RF signal (the velocity of light) to that of the electron beam, which is controlled by the dc voltage at the cathode. This results in an interaction between the electric field induced by the RF signal and the electrons,

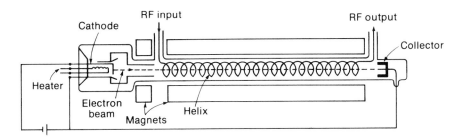

Figure 3.14 Traveling wave tube amplifier.

which results in the transfer of energy from the electron beam to the RF signal causing it to be amplified. The amplification grows as the RF signal travels down the tube. The traveling wave tube can achieve a bandwidth on the order of 10%, hence can cover the entire 500 MHz allocated for satellite communications.

Another type of high-power amplifier used in earth stations is the *klystron amplifier* which can provide higher gain and better efficiency than the traveling wave tube amplifier but at a much smaller bandwidth on the order of 2%. For low-power applications, *Impatt diode amplifiers* or *GaAs FET amplifiers* can be employed. These are solid-state amplifiers and offer much better efficiency than the above two types of amplifiers.

3.2.1 Redundancy Configurations

Reliability is of utmost importance in satellite communications. When a single high-power amplifier is used, transmission will stop upon its failure. Therefore the high-power amplifier in earth stations always employs some sort of redundancy configuration. The most basic redundancy configuration is the 1:1 redundancy shown in Fig. 3.15. The signal from the output of the upconverter is split equally by the power divider and fed to the input of HPA 1 and HPA 2. The waveguide switch allows the output signal of HPA 1 to be transmitted by the antenna while the output signal from HPA 2 is dumped to a load which dissipates the power in the form of heat. When HPA 1 fails, the switch automatically connects the output of HPA 1 to the matched load and that of HPA 2 to the antenna feed (the dotted lines in the switch in Fig. 3.15 are the connections when HPA 1 fails).

When dual polarization is employed in a frequency reuse earth station, four high-power amplifiers are needed to provide a 1:1 redundancy as shown schematically in Fig. 3.16. Since the cost of a high-power amplifier is in the tens of thousands of dollars, the 1:2 redundancy configuration shown in Fig. 3.17 is often employed for cost-effective reasons at the ex-

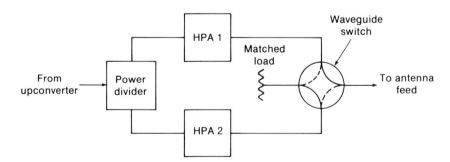

Figure 3.15 HPA in a 1:1 redundancy configuration.

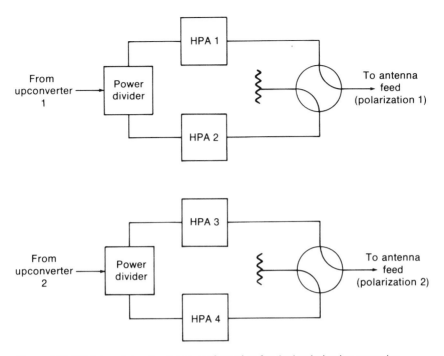

Figure 3.16 HPA in a 1:1 redundancy configuration for dual-polarization operation.

pense of less reliability. A common load is used for both switches. The 1:2 redundancy configuration shown in Fig. 3.17 is used when the earth station transmits only one carrier in each polarization at any instant in

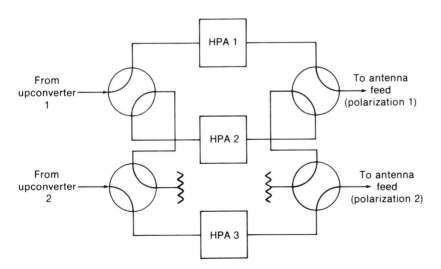

Figure 3.17 HPA in a 1:2 redundancy configuration for dual-polarization operation.

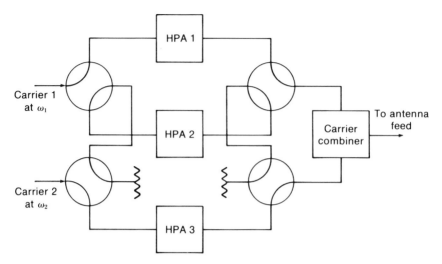

Figure 3.18 HPA in a 1:2 redundancy configuration with carrier combining.

time. When two carriers are transmitted simultaneously in the same polarization, the 1:2 redundancy configuration in Fig. 3.18 can be employed. A carrier combiner is employed to combine the two carriers. When more than two carriers are transmitted simultaneously, a more elaborate scheme must be used to combine the carriers, as will be discussed in the next section.

3.2.2 Carrier Combining

The simplest device for combining two carriers is a *directional coupler* whose coupling coefficient determines the power loss of each carrier. For example, a 3-dB coupler introduces a power loss of 3 dB for each carrier, thus reducing the power of each carrier by a factor of 2. A 4.77-dB coupler introduces a 4.77-dB power loss on one carrier and a 1.76-dB power loss on the other. [(A 4.77-dB coupler has a coupling coefficient of $\alpha = \sqrt{\log^{-1}(-4.77/10)} = 0.577$ for one input port and a coupling coefficient of $\beta = \sqrt{1 - \alpha^2} = 0.816$ for the other input port with a power loss equal to $-10 \log \beta^2 = -10 \log (1 - \alpha^2) = 1.76$ dB).] To combine N carriers, $N - 1$ directional couplers are required. For example, to combine three carriers a 3-dB directional coupler and a 4.77-dB coupler can be used as shown in Fig. 3.19. The two couplers introduce a total power loss of 4.77 dB in each carrier. For N carriers, the power loss is simply $10 \log N$ decibels. This is the disadvantage of combining carriers using directional couplers.

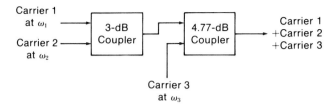

Figure 3.19 Three-carrier combiner using two directional couplers.

To reduce the power loss, a *dual-hybrid filter combiner* can be employed as shown in Fig. 3.20 for two carriers. Two identical filters tuned to the frequency of carrier 1 are connected between two 90° hybrids. The filter bandwidth is equal to the carrier bandwidth. When carrier 1 is applied to port 1, it is split equally by the first hybrid, passes through the filters, is then recombined by the second hybrid, and exits through port 3. In this mode, the power loss of carrier 1 is simply the sum of the insertion losses of the hybrids (approximately 0.1 dB for each hybrid) and of the filter (approximately 0.6 dB), which is about 0.8 dB. When carrier 2 at a frequency outside the passband of the filters is applied at port 2, it is split equally by the second hybrid, reflected by the filters, recombined by the same hybrid, and exits through port 3. The power loss of carrier 2 is simply twice the insertion loss of the hybrid, which is about 0.2 dB. To combine more than two carriers, several combiners can be joined together as shown in Fig. 3.21 for three carriers.

3.2.3 Power Combining

Power combining is of particular importance in earth stations when many power amplifiers in phase are combined to obtain higher power because a large amount of power output might not be available from a single amplifier. Also, for economic reasons, it might be desirable to obtain a specific

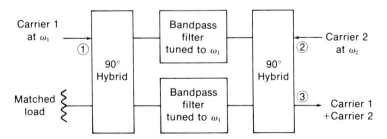

Figure 3.20 Dual-hybrid filter combiner for two carriers.

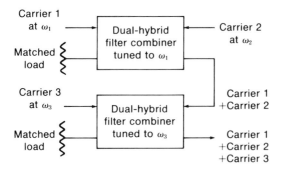

Figure 3.21 Three-carrier combiner using two dual-hybrid filter combiners.

amount of power by combing several lower-power amplifiers rather than using a higher-power amplifier. For example, low-cost helix-type traveling wave tubes are available at 14 GHz, whereas higher-power coupled cavity tubes in this band are expensive and have poorer characteristics than helix-type tubes. Besides the generation of higher power, power combining can provide graceful degradation in the case of failure of one or more amplifiers in the combined system. "Graceful degradation" means that the output power is reduced but not lost completely.

One of the commonly used configurations for obtaining higher power is the *balanced configuration* shown in Fig. 3.22. In this configuration, two identical HPAs with maximum output power P_0 and gain G are connected between two 90° hybrids. Since the HPA cannot produce more power than P_0, the input power to the HPA must be limited to $P_i = P_0/G$. When a carrier is applied to input port 1, its power P is split equally and fed to the two HPAs. Because of the above power constraint the carrier power must be limited to $P = 2P_i = 2P_0/G$. Each HPA amplifies the input signal and produces an output signal with power equal to its rated power

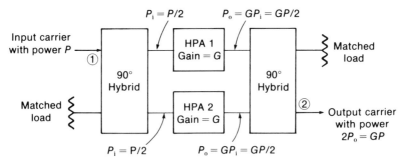

Figure 3.22 Power combining using a balanced configuration.

Table 3.4 HPA combining loss due to phase error in a balanced configuration

Phase error (deg)	Combining loss (dB)
0	0
5.63	0.009
11.25	0.040
22.50	0.170
45	0.690
90	3

P_0. The second hybrid recombines the two output signals from the HPAs and exits through port 2 the carrier whose power is equal to $2P_0$. Thus the balanced configuration is theoretically capable of producing an output power twice that of a single HPA. In practice it is impossible to have two identical HPAs (i.e., HPAs with the same gain and phase transfer characteristics), hence some of the power will be lost in the matched loads. The power loss is affected more by the phase error than the gain ripple. The combining loss due to phase errors in the balanced configuration is shown in Table 3.4. To correct the phase differentials to a certain extent, a phase shifter of low insertion loss can be placed at the input of one HPA to adjust the phase. When one HPA fails in the balanced configuration, a 6-dB reduction in output power occurs.This is due to a 3-dB power loss occurring when one HPA fails and a 3-dB hybrid combining loss occurring when the two HPA output powers no longer add together. To allow only a 3-dB power loss when one of the two HPAs fails, a modified balanced configuration using waveguide switches, as shown in Fig. 3.23, where the second hybrid can be bypassed by the two switches SW 1 and SW 2. When one HPA fails, switches SW 1, SW 2, and SW 4 are activated simultaneously and terminate all four ports of the second hybrid in matched loads. Switch SW 3 is activated depending on which HPA fails. The disadvantage of this configuration is the additional loss introduced by the switches, which is about 0.1 dB at 6 GHz and 0.2 dB at 14 GHz for each switch. Note that the modified balanced configuration also provides a 1:1 redundancy. This may be a desirable feature when compared to the 1:1 redundancy configuration in Fig. 3.15. Assume that the required output power is only P_0 instead of $2P_0$, where P_0 is the maximum output power of the HPA. Then we can operate each HPA in the modified balanced configuration with a 3-dB output backoff by placing a two-step variable 3-dB attenuator at the input of each HPA. When one of the HPAs fails, the variable attenuator is commanded to reduce the attenuation to 0 dB simultaneously with activation of the switches. By operating each HPA at a

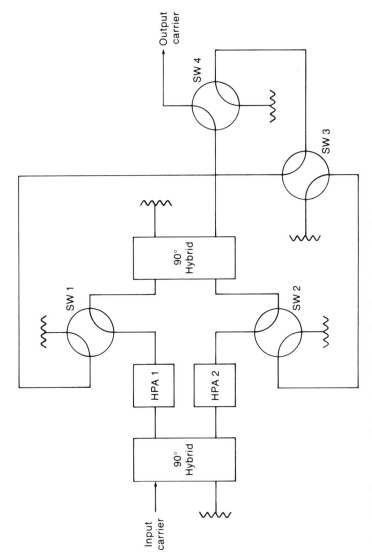

Figure 3.23 Power combining using a balanced configuration and switches.

3-dB output backoff one can reduce the nonlinearity of the HPA, which results in a lower sidelobe level of the PSK carrier in TDMA operation and less adjacent channel interference. Also, the life of the HPA will be longer at 3-dB output back-off operation.

From Fig. 3.23 it is seen that combination of the second hybrid and the switches acts as a three-port variable power combiner in the sense that it can restore a full 6-dB power loss to the theoretical 3-dB power loss when one HPA fails. In practice a variable power combiner which does not use switches has been built and used in earth stations to combine two or even three HPAs. When one of the two HPAs fails, the output level is reduced by 6 dB for 10 to 20 ms. Then the logic control circuit commands the variable power combiner to vary its internal phase shifter to change the output phase of the on-line HPA in such a way as to raise the output power level to 3 dB below normal within 80 to 90 ms. When the failed HPA is restored, its power is coupled to the dummy load until it can be switched back on-line. The variable power combiner is less costly than the hybrid switches combination, and the insertion loss is lower when one HPA fails (about 0.2 dB).

A *variable power combiner* that can combine three HPAs is shown in Fig. 3.24. In-phase signals applied to ports 1 and 2 combine into a single linearly polarized vector perpendicular to the center grating and thus pass through unattenuated. At the output of the center grating this vector $E_1 + E_2$ combines with a third in-phase signal applied via port 3 which is orthogonal to the center grating. The combined vector $E_1 + E_2 + E_3$ is then rotated by polarizer 2 (polarization rotator 2) until it lines up with the antenna output port, the latter being orthogonal to the matched load 2. If desired, the combined output vector $E_1 + E_2 + E_3$ can be diverted to the matched load 2 by rotating the polarization 90° from its normal position.

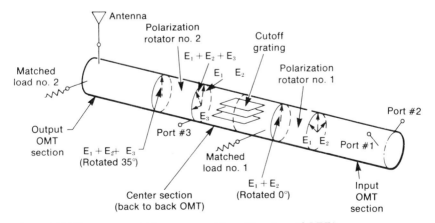

Figure 3.24 Three-port variable power combiner. (*Courtesy of GTE.*)

Should any one of the three HPAs fail, the instantaneous loss in power will be 3.5 dB. For example, if HPA 3 at port 3 fails, the matched polarization incident on the output port will become mismatched by 35.30° and the combined vector will drop in amplitude from 3 to 2. Thus the output power drops by 20 log $\frac{2}{3}$ = −3.5 dB. The same loss occurs no matter which HPA fails, the only difference being that power winds up in both matched loads rather than in just one. The power loss by the remaining two HPAs due to the failure of one HPA can be redirected to the antenna port as fast as one can reset the two polarizers, and this can be done in a matter of milliseconds. After restoration the loss is reduced from 3.5 to 1.76 dB, the theoretical loss. It should be noted that, from the instant of failure through the complete rotation of the polarizers, there is no interruption of output power such as would occur in the conventional switching 1:2 redundancy configuration in Fig. 3.17. The recoverable power in the variable combiner is transferred continuously from the matched loads to the antenna output port as the polarizers rotate.

3.3 LOW-NOISE AMPLIFIER

The two most commonly used low-noise amplifiers in earth stations are the *parametric amplifier* and the *GaAs FET amplifier*. The parametric amplifier has been employed since the inception of satellite communication and is capable of providing very low noise temperatures. The recent development of GaAs FETs with a very short gate length (0.5 μm), which yield very low noise temperatures, enables many receive-only stations to employ GaAs FET low-noise amplifiers and to take advantage of their stability, reliability, and low cost. The thermoelectrically cooled GaAs FET low-noise amplifier is now available at 11.7 to 12.2 GHz with a noise temperature of 150 K when cooling is done at 223 K.

An equivalent circuit of a parametric amplifier is shown in Fig. 3.25, which consists of two resonant circuits, one at the signal frequency ω_1 and the other at the idling frequency ω_2. The two resonant circuits are coupled by a voltage-variable capacitor provided by a variable-capacitance diode called a *varactor*. The capacitance is a sinusoidal function of the pump frequency $\omega_p = \omega_1 + \omega_2$. A circulator is employed to route the input signal to be amplified at port 1 to the resonant circuits at port 2 and to transfer the reflected amplified signal from the resonant circuits from port 2 to the load at port 3. Amplification is achieved because the parametric amplifier operates as a negative resistance amplifier. The equivalent noise temperature of a parametric amplifier can be approximated by

$$T_e \approx \frac{\omega_1}{\omega_2} T_v$$

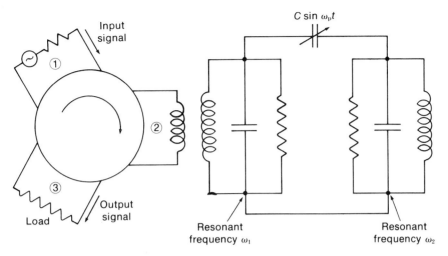

Figure 3.25 Equivalent circuit of a parametric amplifier.

where T_v is the operating temperature of the varactor. The pump frequency $\omega_p = \omega_1 + \omega_2$ is fixed by the current limit of technology and is about 350×10^9 rad/s. Thus the effective noise temperature of the parametric amplifier is practically determined by its operating temperature T_v. A parametric amplifier can be categorized by its operating temperature as follows:

1. Uncooled: $T_v \approx$ ambient temperature $+ 10° \approx 27°C = 300$ K
2. Thermoelectrically cooled: $T_v \approx -50°C = 223$ K
3. Cryogenically cooled: $T_v \approx -250°C = 23$ K.

For a GaAs FET amplifier, a typical low-noise front-end stage is shown in Fig. 3.26. The noise figure of a GaAs FET amplifier can be

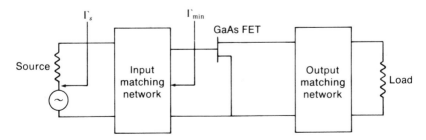

Figure 3.26 GaAs FET amplifier.

expressed by

$$F = F_{\min} + 4r_n \frac{|\Gamma_s - \Gamma_{\min}|^2}{(1 - |\Gamma_s|^2)\,|1 + \Gamma_{\min}|^2}$$

where $F_{\min}$ = minimum noise figure of GaAs FET
Γ_s = source reflection coefficient
$\Gamma_{\min}$ = source reflection coefficient that produces minimum noise figure
r_n = normalized equivalent noise resistance of GaAs FET.

An input matching network is employed to match the input of the GaAs FET to the source to provide a noise figure F as close to $F_{\min}$ as possible over the amplifier bandwidth, and an output matching network is employed to match the output of the GaAs FET to the load to provide the highest possible gain. GaAs FET amplifiers with an effective noise temperature of 100 K over the band 3.7 to 4.2 GHz at the ambient operating temperature (290 K) are now commercially available.

3.3.1 Redundancy Configurations

The basic redundancy configuration for the low-noise amplifier in earth stations is the 1:1 redundancy shown in Fig. 3.27 where the two low-noise amplifiers are connected in parallel by two waveguide switches. When the on-line low-noise amplifier fails, the switches automatically activate the standby low-noise amplifier.

For dual-polarization operation, the 1:2 redundancy configuration in Fig. 3.28 may be attractive from a cost standpoint. The standby low-noise amplifier can be switched to any polarization.

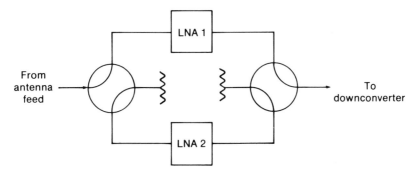

Figure 3.27 LNA in a 1:1 redundancy configuration.

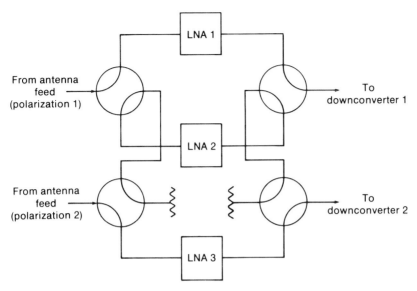

Figure 3.28 LNA in a 1:2 redundancy configuration for dual-polarization operation.

3.3.2 Nonlinearity

One of the concerns in signal detection is the nonlinearity of the receiver, because it determines the level of spurious signals generated within the received bandwidth. In earth stations the dominant nonlinear effect in the receiver is normally the amplitude nonlinearity of the low-noise amplifier, which is characterized by the power gain transfer characteristic shown in Fig. 3.29. Two parameters commonly used in practice to describe the degree of nonlinearity of the gain transfer characteristic are the *1-dB gain compression point* and the *third-order intercept point*. The 1-dB gain compression point is referred to the point on the gain transfer character- istic where the power level of the input signal causes a 1-dB gain compres- sion, that is, where the gain of the amplifier is reduced by 1 dB relative to the linear gain. The third-order intercept point is a hypothetical power level where the output power of the third-order intermodulation product (generated by the amplifier itself on application of two equal-level sinusoidal input signals at frequencies ω_1 and ω_2 and occurring at frequency $2\omega_1 - \omega_2$) would equal the power of the output signal at ω_1 or ω_2 if the low-power level where the gain is linear were extrapolated. Intui- tively it is seen that the higher the input signal power level at the 1-dB gain compression point or the higher the third-order intercept point, the less nonlinear the amplifier.

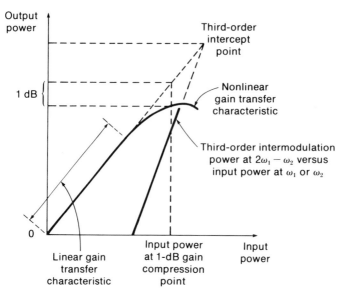

Figure 3.29 Nonlinear characteristic of a low-noise amplifier.

3.4 UPCONVERTER

The upconverter (UC) accepts the modulated IF carrier from the carrier modulator and translates its IF frequency ω_0 to the uplink RF frequency ω_u in the uplink frequency spectrum of the satellite by mixing ω_0 with a local oscillator (LO) frequency ω_1. The upconversion may be accomplished with a *single-conversion* process or with a *dual-conversion* process which is a cascade of two single conversions as shown conceptually in Fig. 3.30a–c.

3.4.1 Conversion Process

To illustrate the single-conversion process in Fig. 3.30a, consider the IF carrier $\cos(\omega_0 t + \phi)$ and the LO carrier $\cos \omega_1 t$. The resulting mixing process yields the following product (assuming $\omega_1 > \omega_0$):

$$\cos (\omega_0 t + \phi) \cos \omega_1 t = \tfrac{1}{2} \cos [(\omega_1 - \omega_0)t - \phi]$$
$$+ \tfrac{1}{2} \cos [(\omega_1 + \omega_0)t + \phi] \qquad (3.37)$$

Thus one can select the LO frequency ω_1 such that $\omega_u = \omega_1 + \omega_0$, that is, $\omega_1 = \omega_u - \omega_0$, and use the bandpass filter at the output to select the upper sideband. Now consider the dual-conversion process in Fig. 3.30b. The first conversion of the IF carrier $\cos(\omega_0 t + \phi)$ by the first LO carrier cos

$\omega_{11}t$ yields (assuming $\omega_{11} > \omega_0$)

$$\cos(\omega_0 t + \phi) \cos \omega_{11}t = \tfrac{1}{2} \cos [(\omega_{11} - \omega_0)t - \phi]$$

$$+ \tfrac{1}{2} \cos [(\omega_{11} + \omega_0)t + \phi] \qquad (3.38)$$

The first bandpass filter selects the upper sideband $\cos[(\omega_{11} + \omega_0)t + \phi]$, and the second conversion by the second LO carrier $\cos \omega_{12}t$ yields the following (assuming $\omega_{12} > \omega_{11} + \omega_0$):

$$\cos [(\omega_{11} + \omega_0)t + \phi] \cos \omega_{12}t = \tfrac{1}{2} \cos [(\omega_{12} - \omega_{11} - \omega_0)t - \phi]$$

$$+ \tfrac{1}{2} \cos [(\omega_{12} + \omega_{11} + \omega_0)t + \phi] \quad (3.39)$$

The second bandpass filter selects the upper sideband $\cos[(\omega_{12} + \omega_{11} + \omega_0)t + \phi]$, and thus $\omega_u = \omega_{12} + \omega_{11} + \omega_0$. This means that ω_{11} and ω_{12} can be arbitrarily chosen provided they satisfy the constraints $\omega_{11} + \omega_{12} = \omega_u - \omega_0$, $\omega_{12} > \omega_{11} + \omega_0$, and $\omega_{11} > \omega_0$.

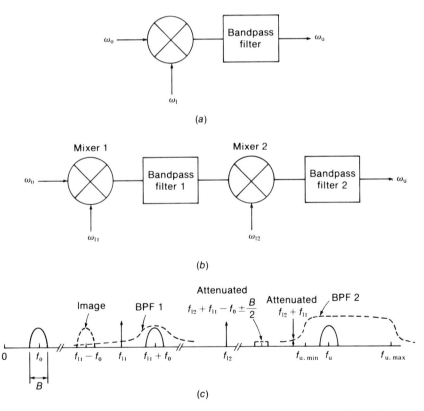

Figure 3.30 Schematic diagram of an upconverter. (*a*) Single conversion. (*b*) Dual conversion. (*c*) Frequency spectrum of dual-upconversion.

In the single-conversion process, it is seen that a change in the uplink carrier frequency ω_u (e.g., transmitting to a different transponder) requires a change in the LO frequency ω_l and a change in the bandpass filter (retuning), the latter being inflexible and unacceptable in many earth stations. On the contrary, in the dual-conversion process, the second IF frequency $\omega_{l1} + \omega_0$ is fixed and is selected such that it is sufficiently larger than the uplink frequency spectrum $\omega_{u,max}$ to $\omega_{u,min}$ which is 500 MHz at the C band (5.9 to 6.4 GHz) and the Ku band (14 to 14.5 GHz). Therefore the unwanted lower sideband $\omega_{l2} - \omega_{l1} - \omega_0$ is always outside the uplink frequency spectrum, and thus the second bandpass filter can be designed to cover the whole uplink frequency spectrum and any change in the uplink carrier frequency ω_u will require only a change in the second LO frequency ω_{l2}. (This is normally called *frequency agile upconversion.*)

In order to avoid changing the sign of the phase of the modulated carrier (which for phase modulation results in an error), the same sideband must be selected after each conversion in the dual-conversion process. The first bandpass filter in Fig. 3.30b must be designed with a bandwidth equal to the bandwidth of the modulated carrier. Furthermore, it must suppress the first LO signal at ω_{l1} and the image sideband at $\omega_{l1} - \omega_0 \pm B/2$, where B (rad/s) is the carrier bandwidth to an acceptable level (normally in the range of 100 dB below the transmit signal). Also, this filter must contribute negligible group delay at the edge of the transmit bandwidth (normally less than 4 ns peak-to-peak variation). In practice an amplifier can be placed between the first bandpass filter and the second mixer to provide the proper gain for the upconverter.

Example 3.3 In this example we discuss the design of a dual upconverter. Modern technology allows the design of local oscillators with very low harmonic outputs at frequency nf_1, $n = 2,3,\ldots$. Furthermore, the output signal of the modulator that feeds the upconverter is filtered to eliminate the harmonics at mf_0, $m = 2,3,\ldots$. Therefore we assume the spurious products at $mf_0 + nf_1$ are small and can be neglected. We use the following parameters.

Uplink frequency: $f_u = \omega_u/2\pi = 6$ GHz
IF frequency: $f_0 = \omega_0/2\pi = 70$ MHz
Uplink frequency spectrum: 5.9 to 6.4 GHz (500-MHz bandwidth)
Carrier bandwidth: $B = 36$ MHz
The first LO frequency $f_{l1} = \omega_{l1}/2\pi$ must be selected such that $f_{l1} + f_0 > 500$ MHz. Therefore we choose

$$f_{l1} = 930 \text{ MHz}$$

Thus the center frequency of the first bandpass filter is

$$f_{l1} + f_0 = 1 \text{ GHz} > 500 \text{ MHz}$$

The first bandpass filter has a bandwidth of $B = 36$ MHz and must be designed to attenuate the LO signal at $f_{l1} = 930$ MHz and the image sideband at $f_{l1} - f_0 + B/2 = 878$ MHz to acceptable levels. The second LO frequency is selected as

$$f_{l2} = f_u - f_{l1} - f_0 = 5 \text{ GHz}$$

Note that the unwanted lower sideband occupies a spectrum from $f_{12,\min} - f_{11} - f_0 = f_{u,\min} - 2(f_{11} + f_0) = 3.9$ GHz to $f_{12,\max} - f_{11} - f_0 = f_{u,\max} - 2(f_{11} + f_0) = 5.4$ GHz, hence is outside the frequency spectrum 5.9 to 6.4 GHz. The upconversion process is illustrated in Fig. 3.30c.

3.4.2 Transponder Hopping, Polarization Hopping, and Redundancy Configuration for Upconverter

Modern digital earth stations employing time division multiple access often have to transmit more than one traffic burst during one TDMA frame length to a number of designated transponders according to the transmit traffic assignment, or transmit burst time plan as it is often called. The ability of the earth station to hop its carrier from transponder to transponder during each frame is referred to as *transponder hopping*. Also, if dual polarization is employed, the earth station might have to hop its carrier from one transponder in one polarization to another transponder in the other polarization. This is known as *polarization hopping*. Transponder hopping and polarization hopping must be achieved within a fraction of the guard time between bursts (Fig. 1.13). In practice this can be on the order of 100 to 200 ns. Also, for reliability reasons, the upconverter must employ some sort of redundancy configuration. In the following discussion we will consider various solutions for an upconversion that can meet the above three requirements, namely, transponder hopping, polarization hopping, and redundancy.

The first solution is to use frequency agile upconversion where hopping of the carrier is achieved by switching the second LO frequency ω_{12} (Fig. 3.30b). This can be done by using as many local oscillators as there are hopping frequencies and a fast switch (a *pin* diode switch). This type of upconverter is called a hopping LO upconverter. A fast polarization switch (a *pin* diode switch) is employed at the output of the upconverter to direct the RF uplink carrier to the proper polarization. The LO hopping upconverter shown in Fig. 3.31 also provides a 1:1 redundancy. Since the amplitude and group delay variations in the transmit side of the earth station and the satellite contribute to the distortion of the modulated carrier, and the path length differences between upconverters result in various signal delays, equalizers (EQL) must be employed. The number of equalizers is equal to the number of hopping transponders. A fast IF switch (a *pin* diode switch) is used at the input of the upconverter to select a designated equalizer for a hopping carrier. The IF switch, the LO switch, and the polarization switch are controlled by signals from the TDMA baseband equipment. Because of the path length differences between the TDMA baseband equipment and the three above-mentioned switches, an appropriate time delay must be employed for each control signal.

The second solution for achieving transponder hopping is to use as

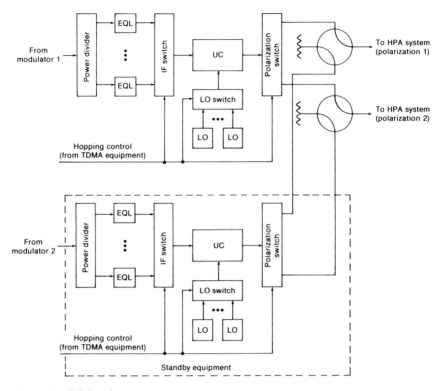

Figure 3.31 LO hopping upconverter.

many upconverters as there are frequencies. Two RF switches are employed, one for each polarization as shown in Fig. 3.32, which also provides a 1:1 redundancy. The switching can also be done at the IF frequency as shown in Fig. 3.33. The advantage of IF switching is that only one fast switch is required for each chain. Also, the IF switches can be put in the baseband TDMA terminal, which is an advantage when there is a long connection between the baseband terminal and the RF terminal, since then only the time delay due to the electrical path length of the modulator needs to be taken into account. But combining at the output of the upconverters may increase the noise output and the spurious signals which, although not correlated at the transmit output, could be correlated at the satellite if every station in the network uses the same configuration and the same second IF frequency in the upconverter.

A multiple upconverter provides easier path length equalization than a LO hopping upconverter and is less costly when the number of hopping transponders is less than three, but it is expensive when the number of hopping transponders is large.

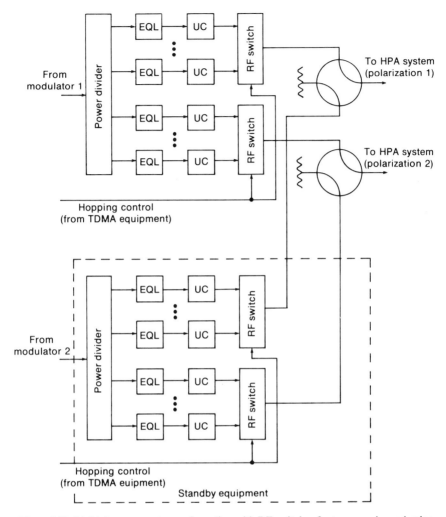

Figure 3.32 Multiple upconverter configuration with RF switches for transponder and polarization hopping.

3.5 DOWNCONVERTER

The downconverter (DC) receives the modulated RF carrier from the low-noise amplifier and translates its radio frequency ω_d in the downlink frequency spectrum of the satellite to the intermediate frequency ω_0. Like upconversion, downconversion may be achieved with a *single-conversion* process or with a *dual-conversion* process, as shown in Fig. 3.34 *a–c*.

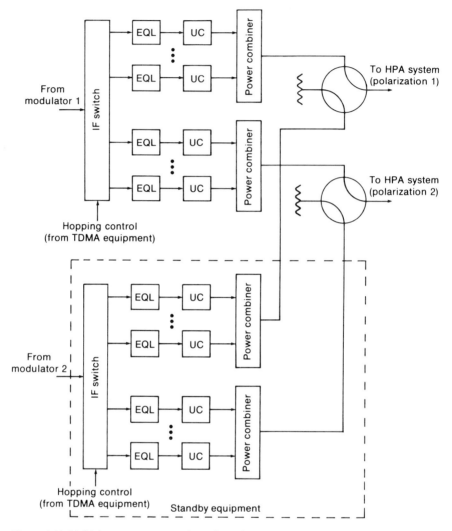

Figure 3.33 Multiple upconverter configuration with IF switches for transponder hopping and polarization hopping.

3.5.1 Conversion Process

A single downconversion is seldom used in earth stations except sometimes in low-cost receive-only stations. For the dual downconversion in Fig. 3.34*b*, consider the first conversion of the RF carrier $\cos(\omega_d t + \phi)$ by the second LO carrier $\cos \omega_{12} t$ (assuming $\omega_{12} < \omega_d$):

$$\cos(\omega_d t + \phi) \cos \omega_{12} t = \tfrac{1}{2} \cos \left[(\omega_d - \omega_{12}) t + \phi \right]$$
$$+ \tfrac{1}{2} \cos \left[(\omega_d + \omega_{12}) t + \phi \right] \qquad (3.40)$$

The first bandpass filter selects the lower sideband $\cos[(\omega_d - \omega_{12})t + \phi]$, and the second conversion by the first LO carrier $\cos \omega_{11}t$ yields (assuming $\omega_{11} < \omega_d - \omega_{12}$)

$$\cos[(\omega_d - \omega_{12})t + \phi]\cos \omega_{11} \ t = \tfrac{1}{2} \cos[(\omega_d - \omega_{12} - \omega_{11})t + \phi]$$
$$+ \tfrac{1}{2} \cos[(\omega_d - \omega_{12} + \omega_{11})t + \phi]$$

$$(3.41)$$

The output IF carrier is of course the lower sideband $\cos[(\omega_d - \omega_{12} - \omega_{11})t + \phi]$ and thus yields $\omega_0 = \omega_d - \omega_{12} - \omega_{11}$. The LO frequencies ω_{11} and ω_{12} are arbitrarily chosen provided they satisfy the constraints $\omega_{11} + \omega_{12} = \omega_d - \omega_0$, $\omega_{12} < \omega_d$, and $\omega_{11} < \omega_d - \omega_{12}$.

As in upconversion, the second intermediate frequency $\omega_d - \omega_{12}$ is fixed and is selected to be sufficiently larger than the downlink frequency spectrum. Thus the second bandpass filter can be designed to cover the whole downlink frequency spectrum, and any change in the downlink carrier frequency ω_d will require only a change in the second LO frequency ω_{12}. The design of the first bandpass filter in a dual downconverter is important because it must suppress the image sideband within

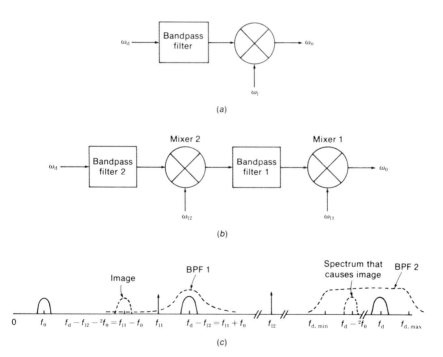

(a)

(b)

(c)

Figure 3.34 Schematic diagram of a downconverter. (a) Single conversion. (b) Dual conversion. (c) Frequency spectrum of dual-downconversion.

$\omega_{11} - \omega_o \pm B/2$ to an acceptable value, where B is the carrier bandwidth in radians per second. This is the case because there are many other carriers in the downlink frequency spectrum besides the desired carrier at ω_d, and any spectrum within the radio frequencies $\omega_{12} + \omega_{11} - \omega_0 \pm B/2$ will be downconverted to within the bandwidth B of the desired signal. The filter must also introduce negligible group delay at the edge of the receive bandwidth. Typical image suppression is 70 dB, and typical peak-to-peak group delay variation is 4 ns. In practice an amplifier is placed between the second mixer and the first bandpass filter to provide the gain for the downconverter.

Example 3.4 In this example we discuss the design of a dual downconverter. We use the following parameters.

Downlink frequency: $f_d = \omega_d/2\pi = 4$ GHz
Intermediate frequency: $f_0 = \omega_0/2\pi = 70$ MHz
Downlink frequency spectrum: 3.7 to 4.2 GHz (500-MHz bandwidth)
Carrier bandwidth: $B = 36$ MHz

The sum of the first and second LO frequencies must be

$$f_{11} + f_{12} = f_d - f_0 = 3.93 \text{ GHz}$$

The second LO frequency must be selected such that $f_d - f_{12} > 500$ MHz. Therefore we choose

$$f_{12} = 3 \text{ GHz}$$

Thus the center frequency of the first bandpass filter is

$$f_d - f_{12} = 1 \text{ GHz}$$

and the first LO frequency is given by $f_{11} = f_d - f_0 - f_{12}$:

$$f_{11} = 930 \text{ MHz}$$

The first bandpass filter with a center frequency at 1 GHz and a bandwidth of $B = 36$ MHz must be designed to attenuate the image sideband within $f_d - f_{12} - 2f_0 \pm B/2 = f_{11} - f_0 \pm B/2 = 860 \pm 18$ MHz to acceptable levels. Note that this image is caused by signals in the transponder bandwidth $f_d - 2f_0 \pm B/2 = 3860 \pm 18$ MHz. To avoid leakage of the first LO signal, a circulator or buffer amplifier may be placed at the output of the first bandpass filter. The downconversion process is illustrated in Fig. 3.34c.

3.5.2 Transponder Hopping, Polarization Hopping, and Redundancy Configuration for Downconverter

In TDMA operations, the earth station may be required to receive nonoverlapping traffic bursts from different transponders in different polarizations during a frame. As with upconverters, transponder hopping and polarization hopping can be achieved by a LO hopping downconverter, as shown in Fig. 3.35, or by multiple downconverters with IF switching, as shown in Fig. 3.36. The latter is more cost-effective when the number of hopping transponders is less than three, while the former is more attractive when the number of hopping transponders is large.

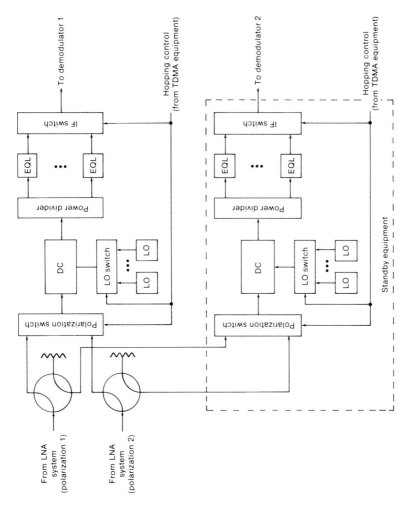

Figure 3.35 LO hopping downconverter.

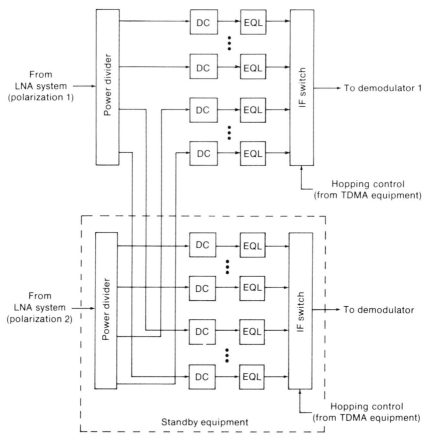

Figure 3.36 Multiple downconverter configuration with IF switches for transponder and polarization hopping.

3.6 MONITORING AND CONTROL

As we have seen in previous sections, except for the antenna all earth station systems namely, the high-power amplifier, the low-noise amplifier, the upconverter, and the downconverter, must employ some sort of redundancy to maintain high reliability which is of utmost importance. When the on-line equipment in the redundancy configuration fails, the standby (redundant) equipment is automatically switched over and becomes the on-line equipment. The process of detecting critical failure modes and resolving all these failure modes by automatic switchover from the failed to the redundant system is called *monitoring and control* (M&C).

In a satellite communications network, which consists of many earth stations and a network control center (NCC), monitoring and control is extremely important and involves many levels such as

1. RF terminal monitoring and control
2. Baseband terminal monitoring and control
3. Remote monitoring and control of the entire earth station from the network control center via the satellite link with a backup terrestrial telephone line

The monitoring and control system must have the capability to

Collect status data for classification
Convey status data to network operators
Interpret status data
Initiate fault isolations
Switch over redundant equipment on command
Convey control data to the baseband equipment for traffic assignment, antenna pointing (satellite ephemeris for program tracking), and so forth
Maintain surveillance of equipment shelter facilities.

In this section we concentrate on RF terminal monitoring and control (RFT M&C) and postpone a discussion of the baseband equipment and remote monitoring and control to Chap. 6 where the TDMA system is considered. The overall monitoring and control of a typical TDMA earth station is shown in Fig. 3.37.

RFT M&C typically has the following three main components:

1. Redundancy switching logic for the high-power amplifier, the low-noise amplifier, the upconverter, and the downconverter
2. A data acquisition unit (DAU)
3. An antenna control unit (ACU)

The redundancy switching logic of each subsystem is itself part of the subsystem. Inherent in this concept is the understanding that the subsystems being monitored and controlled are redundant and self-sustaining. That is, an automatic switchover from the failed to the redundant system is executed by the subsystem logic.

Failure alarms, status indications, and possibly analog values are then collected by the data acquisition unit which inputs them to a M&C minicomputer that may be the central processor of the TDMA equipment (which also handles other TDMA functions besides monitoring and control) or an entirely separate minicomputer that handles only earth station monitoring and control. (In the latter case the M&C computer and the

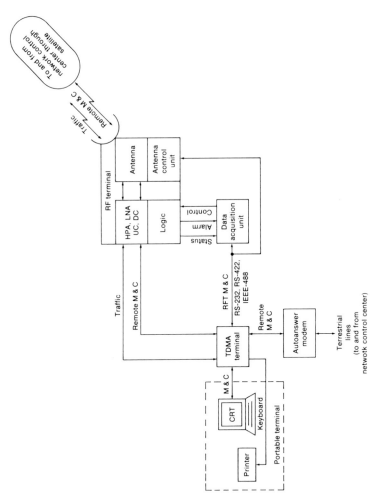

Figure 3.37 Monitoring and control of the earth station.

TDMA central processor must be appropriately interfaced for the purpose of remote monitoring and control by the NCC through a satellite link.) The DAU is a microprocessor-based device which continuously scans the input status and alarm points, stores them internally after each scan, and reports them to the M&C minicomputer which scans the DAU at a predetermined rate (normally once every few seconds). The M&C minicomputer then adds the points to the data base, checks for alarms, status changes, and priority and outputs the appropriate message to a local portable cathode ray tube (CRT) terminal and printer at the earth station or transmits it via a M&C channel through the satellite to the NCC. The DAU also accepts switching control information from the local portable terminal or from the NCC via the M&C minicomputer and inputs control information to the appropriate subsystem logic for a redundancy switchover. The DAU is interfaced with the M&C minicomputer via standard RS-232 or RS-422 data links or an IEEE-488 bus.

The antenna control unit is also interfaced with the M&C minicomputer via standard RS-232 or RS-422 data links or an IEEE-488 bus. Alarm, status, and tracking information is transmitted to the M&C minicomputer following a data scan request received by the ACU. When program tracking is employed, the ACU must be able to accept the tracking table downloaded to it from the M&C minicomputer responsible for generating the tracking information based on satellite ephemeris data received from the NCC. The satellite ephemeris is updated periodically by the NCC based on the latest TT&C data.

As mentioned above, the status and alarm data in the M&C minicomputer is continuously compared to the data received from the previous scan. Any changes in state detected are noted, placed in a message queue, and outputted at the local portable terminal or transmitted to the NCC for appropriate action. The alarm event normally assumes two priorities: 1 and 2, or major and minor, which represent the state of alarm from the subsystem; a change in state is immediately brought to the operator's attention. The status event generally assumes priority 3 and reflects a state of the subsystem not considered an alarm and which may be changed to the opposite state via a control signal. When the M&C minicomputer detects a status change, the following actions will be considered:

If the status change is due to the result of a control signal sent to it recently, the change is interpreted as being authorized and the new state is simply stored in the current data base.

If the status change is not due to a recent control signal, the change is considered unauthorized; the event priority is then changed from 3 to 1, and it is treated as a major alarm.

Figure 3.38 shows the logic control of a low-noise amplifier system in 1:1 redundancy configuration. The switching logic can accept commands

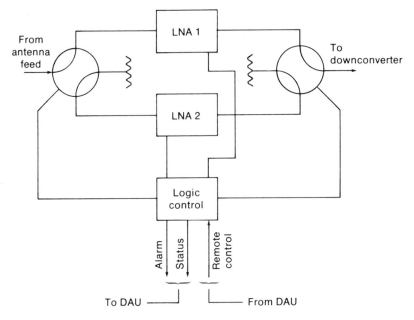

Figure 3.38 Logic control of a low-noise amplifier.

from individual LNAs or remote commands from the NCC downline-loaded to the M&C minicomputer. Typical alarms, status, and controls for the 1:1 redundant subsystems of the RF terminal are given in Table 3.5. It is worth noting that in the worst case of a logic control failure the situation at the RF terminal is unknown but that the RF terminal is not out of service.

3.7 RELIABILITY

As we know by now, every subsystem of the earth station except the antenna employs some sort of redundancy configuration to increase its reliability. In general the reliability of hardware is the probability that it will perform its task without failure under designed conditions and within defined periods. If we assume that failures occur randomly, the probability of the hardware being operated longer than the time interval t is given by the exponential distribution

$$R(t) = \exp(-\rho t) \tag{3.42}$$

where ρ is the average failure rate. The average failure time, commonly

Table 3.5 Alarms, status, and controls for the 1:1 redundant subsystems of the RF terminal

Subsystem	Alarm	Status	Control
Antenna			
Deicer on		X	X
Deicer off		X	X
Rain blower on		X	X
Rain blower off		X	X
Antenna control	X	X	X
High-power amplifier			
Helix overcurrent	X		
Helix overvoltage	X		
High-termperature tube	X		
High-temperature power supply	X		
HPA 1 on-line		X	X
HPA 1 off-line		X	X
HPA 2 on-line		X	X
HPA 2 off-line		X	X
HPA 1 prime power		X	X
HPA 2 prime power		X	X
HPA 1 fault reset			X
HPA 2 fault reset			X
Input switch position 1			X
Input switch position 2			X
Output switch position 1			X
Output switch position 2			X
Low-noise amplifier			
Power supply 1	X		
Power supply 2	X		
LNA 1 on-line		X	
LNA 1 off-line		X	
LNA 2 on-line		X	
LNA 2 off-line		X	
Input switch position 1			X
Input switch position 2			X
Output switch position 1			X
Output switch position 2			X
Upconverter-downconverter			
Power supply 1	X		
Power supply 2	X		
Local oscillators	X		
UC 1/DC 1 on-line		X	
UC 1/DC 1 off-line		X	
UC 2/DC 2 on-line		X	
UC 2/DC 2 off-line		X	
Input switch position 1			X
Input switch position 2			X
Output switch position 1			X
Output switch position 2			X

called the *mean time to failure* (MTTF), is thus given by

$$\text{MTTF} = \int_0^\infty R(t)\, dt = \frac{1}{\rho} \tag{3.43}$$

If many independent subsystems are connected in *cascade* and any one of them fails, the cascaded system will also fail. Thus the reliability of a cascaded system is simply the product of the reliabilities of the individual subsystem, $R_i(t) = \exp(-\rho_i t)$:

$$R_c(t) = R_1(t)R_2(t) \cdots R_n(t)$$

$$= \exp\left(-\sum_{i=1}^n \rho_i t\right) \tag{3.44}$$

Therefore the mean time to failure of a cascaded system is

$$\text{MTTF}_c = \int_0^\infty R_c(t)\, dt = \frac{1}{\sum\limits_{i=1}^n \rho_i} = \frac{1}{\sum\limits_{i=1}^n (\text{MTTF}_i)^{-1}} \tag{3.45}$$

where $\text{MTTF}_i = 1/\rho_i$, and the corresponding average failure rate is

$$\rho_c = \sum_{i=1}^n \rho_i \tag{3.46}$$

Examples of cascaded systems are the low-noise amplifier–downconverter chain and the high-power amplifier–upconverter chain.

When identical subsystems are connected in *parallel*, the system will operate al long as any one system operates. Therefore the reliability of the parallel system is given by

$$R_p(t) = 1 - \prod_{i=1}^n [1 - R_i(t)] = 1 - [1 - R(t)]^n$$

$$= 1 - [1 - \exp(-\rho t)]^n$$

$$= 1 - \sum_{i=0}^n (-1)^i \binom{n}{i} \exp(-i\rho t)$$

$$= \sum_{i=1}^n (-1)^{i+1} \binom{n}{i} \exp(-i\rho t)$$

$$\tag{3.47}$$

where

$$\binom{n}{i} = \frac{n!}{(n-i)!\, i!}$$

Thus the mean time to failure is simply

$$\text{MTTF}_p = \int_0^\infty R_p(t)\, dt$$

$$= \sum_{i=1}^n (-1)^{i+1} \binom{n}{i} \int_0^\infty e^{-i\rho t}\, dt$$

$$= \sum_{i=1}^n (-1)^{i+1} \binom{n}{i} \frac{1}{i\rho} \qquad (3.48)$$

and the corresponding average failure rate is

$$\rho_p = \frac{1}{\displaystyle\sum_{i=1}^n (-1)^{i+1} \binom{n}{i} \frac{1}{i\rho}} \qquad (3.49)$$

Examples of parallel systems are the 1:1 and 1:2 redundancy configurations discussed in previous sections. It is obvious that a parallel system has a higher reliability then a cascaded system.

Once the average failure rate or the mean time to failure of the subsystems in an earth station is known, the mean time to failure of the earth station can be evaluated by appropriate applications of the results in (3.45) and (3.48) or (3.46) and (3.49). The average probability of availability of the earth station depends not only on its mean time to failure but also on how fast the failed subsystem (or subsystems) can be repaired or replaced. Let MTTF_1 and MTTR_1 denote the mean time to failure and mean time to repair, respectively, of the transmit side, and let MTTF_2 and MTTR_2 denote the mean time to failure and *mean time to repair* of the receive side, respectively. Then the earth station availability is given by

$$P_A = P_{A1} P_{A2} \qquad (3.50a)$$

where

$$P_{A1} = \frac{\text{MTTF}_1}{\text{MTTF}_1 + \text{MTTR}_1} \qquad (3.50b)$$

$$P_{A2} = \frac{\text{MTTF}_2}{\text{MTTF}_2 + \text{MTTR}_2} \qquad (3.50c)$$

Large earth stations are normally designed to have a very high degree of reliability, that is, with an average probability of availability on the order of 0.9990 to 0.9998.

Example 3.5 In this example the MTTF of the redundant LNA shown in Fig. 3.27 is calculated. Typical parameters are
   MTTF of LNA $= 3.2 \times 10^4$ h
   MTTF of switch $= 5 \times 10^5$ h
The 3-port switch can be modeled by two 2-port switches each with the same MTTF. The equivalent redundant scheme is shown in Fig. 3.39. By using (3.45) the MTTF_c of

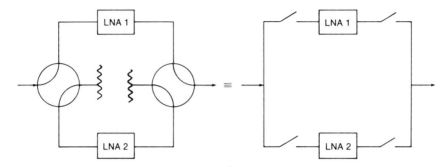

Figure 3.39 Redundant LNA and its equivalence.

one leg of the parallel system is

$$MTTF_c = \frac{1}{(5 \times 10^5)^{-1} + (3.2 \times 10^4)^{-1} + (5 \times 10^5)^{-1}}$$

$$= 28,369 \text{ h}$$

The MTTF of the parallel system can be obtained from (3.48):

$$MTTF_p = \binom{2}{1} MTTF_c - \binom{2}{2} \frac{MTTF_c}{2}$$

$$= \frac{3}{2} MTTF_c = 42,553 \text{ h}$$

Now assuming that the MTTR is 2 h, the probability of availability of the parallel system is

$$P_A = \frac{MTTF_p}{MTTF_p + MTTR} = 0.999953$$

REFERENCES

1. J. Ruze, "Antenna Tolerance Theory—A Review," *Proc. IEEE*, Vol. 54, No. 4, Apr. 1966, pp. 633–640.
2. M. I. Skolnik, *Introduction to Radar Systems*, 2d ed., New York: McGraw-Hill, 1980.
3. M. Schwartz, *Information Transmission, Modulation, and Noise*, 2d ed., New York: McGraw-Hill, 1970.
4. R. W. Kreutel, Jr., et al., "Satellite System Measurements," *Proc. IEEE*, Vol. 66, No. 4, Apr. 1978, pp. 472–482.
5. R. C. Hansen, "Low Noise Antennas," *Microwave J.*, June 1959, pp. 19–24.
6. J. Dijk et al., "Antenna Noise Temperature," *Proc. IEE*, Vol. 115, No. 10, Oct. 1968, pp. 1403–1410.
7. W. L. Stutzman and G. Thiele, *Antenna Theory*. New York: Wiley, 1981.
8. R. Stegens, "Design and Performance of Low-Cost Integrated MIC Up- and Downconverters for Earth Station Applications," *COMSAT Tech. Rev.*, Vol. 9, No. 1, Spring 1979, pp. 121–155.

APPENDIX 3A THERMAL NOISE SOURCE

Random motion of free electrons in a resistor whose temperature is above
absolute zero causes a noise voltage $v(t)$ to be generated at the terminals.
The power spectrum of the noise voltage [3] is

$$S_v(f) = 2R \left[\frac{h|f|}{2} + \frac{h|f|}{\exp(h|f|/kT_s) - 1} \right] \qquad (3.A1)$$

where R = resistor value (ohms)
   h = Planck's constant = 6.2×10^{-34} J-s
   k = Boltzmann's constant = 1.38×10^{-23} J/K
   K = kelvins
   T_s = temperature of resistor

At room temperature $T_s = 293$ K and, for frequencies below 1000 GHz,
$h|f|/kT_s < 0.15$, hence $\exp(h|f|/kT_s) \approx 1 + h|f|/kT_s$ and (3.A1) can be approximated by

$$S_v(f) \approx 2RkT_s \qquad (3.A2)$$

Therefore the thermal noise power generated in a resistor R over a
bandpass bandwidth B (Fig. 3.8) is

$$P_v = 4RkT_sB \qquad (3.A3)$$

The maximum thermal noise power that this source delivers to a matched
load of resistance R (noise-free) is one-fourth of P_v divided by R:

$$N = kT_sB$$

N is then the available noise power from the noise source $v(t)$ and is independent of the source resistance R. The available power spectral density of a thermal noise source is therefore

$$S_a(f) = \frac{N}{2B} = \frac{kT_s}{2}$$

PROBLEMS

3.1 Find the antenna gain loss due to surface tolerance at the maximum achievable gain. Is
this gain loss acceptable?

3.2 Show that the normalized gain pattern of the main lobe of an antenna can be approximated by the Gaussian function

$$G_n(\theta) \approx \exp\left(\frac{-2.78\theta^2}{\theta_B^2} \right)$$

where θ_B is the half-power beamwidth of the antenna. What is the error at the 6-dB level down from the peak gain? Compare the pointing loss at $\theta = 0.02°$ with that from (3.7) for a 20-m antenna at 14.25 GHz.

3.3 Using the result in Prob. 3.2, find the maximum achievable antenna gain for a given pointing error. What is the corresponding pointing loss? Compare this result with that in Prob. 3.1.

3.4 Consider a 20-m Cassegrain antenna with a parabolic aperture distribution operating at 14.25 GHz with an efficiency of $\eta = 0.54$, a surface tolerance $\epsilon/D = 2 \times 10^{-5}$ and a pointing error of 0.02°. Assume that the loss of the waveguide that connects the output of the high-power amplifier to the feed of the antenna is 0.7 dB. What is the rated power of the high-power amplifier needed to achieve an EIRP of 98 dBW?

3.5 Consider the receive side of the earth station shown in Fig. 3.10. Find the system noise temperatures referred to the feed of the antenna (i.e., the input of the waveguide) and referred to the downconverter input, assuming that the equipment ambient temperature is $T_0 = 290$ K, the antenna noise temperature is $T_A = 60$ K, the waveguide loss is $L_1 = 0.3$ dB, the equivalent noise temperature and gain of the low-noise amplifier are $T_{e2} = 150$ K and $G_2 = 60$ dB, respectively, and the equivalent noise temperature of the downconverter is $T_{e3} = 11 \times 10^3$ K. Let the antenna gain be 65.53 dB. Find the antenna gain-to-noise temperature ratio for each case.

3.6 Consider the receive side of an earth station (Fig. 3.10). Find the system noise temperature referred to the input of the low-noise amplifier using the following parameters.

  Antenna gain: 52.6 dB
  Waveguide loss: 0.1 dB
  Low-noise amplifier
   Gain: 55 dB
   Equivalent noise temperature: 40 K
  Downconverter equivalent noise temperature: 213,600 K
  Ambient temperature: 300 K
  Antenna noise temperature contributors
   Main beam: 25.6 K
   Subreflector spillover: 3.3 K
   Main reflector spillover: 0.7 K
   Blockage: 9.8 K
   Surface tolerance: 1.5 K
   Feed loss: 13.9 K

3.7 Let T_g be the cosmic noise temperature; T_a and L_a are the temperature and absorbing loss of the atmosphere, respectively. Assuming that the noise enters the antenna via the main beam only, what is the antenna noise temperature?

3.8 The noise temperature of an antenna is given by

$$T_a = \frac{1}{4\pi} \int_{\theta=0}^{\pi} \int_{\phi=0}^{2\pi} G(\theta,\phi)T(\theta,\phi) \sin \theta \, d\phi \, d\theta$$

where $T(\theta,\phi) =$ distribution of temperature over all angles about antenna

  $G(\theta,\phi) =$ gain of antenna in direction (θ,ϕ)

  $(\theta,\phi) =$ spherical coordinates

Find the noise temperature of an isotropic antenna, assuming the average atmospheric noise temperature in kelvins can be approximated by

$$T(\theta,\phi) = T(\theta) = \frac{2.2}{\cos\theta} + 0.8 \qquad 0° \leq \theta \leq 87.5°$$

$$T(\theta,\phi) = T(\theta) = 290 - 95.5(90 - \theta) \qquad 87.5 < \theta < 90°$$

$$T(\theta,\phi) = T(\theta) = 290 \qquad 90° < \theta < 180°$$

3.9 Consider the receive side of an earth station shown schematically in Fig. 3.10 and assume that the noise contribution by the downconverter is negligible. Let the antenna gain be 64.3 dB, the waveguide loss be 0.65 dB, and the switch loss of the redundant low-noise amplifier subsystem (Fig. 3.27) be 0.2 dB. It is desired to have G/T margin of 0.3 dB/K. Find the required equivalent noise temperature of the low-noise amplifier in the redundancy configuration given that $G/T = 39.7$ dB/K.

3.10 Consider the transmit side of an earth station with the following parameters: EIRP = 87.5 dBW; antenna gain = 66.2 dB; the high-power amplifier subsystem employs the 1:1 redundancy configuration shown in Fig. 3.15, where the switch loss is 0.2 dB; the individual HPA internal loss is 0.8 dB; and the HPA is required to operate with an output backoff of 3 dB from its maximum power. Find the required maximum output power of the individual HPA, assuming an EIRP margin of 0.5 dBW is necessary.

3.11 How many 3-dB couplers are needed to combine four carriers? Show the combining scheme. If the loss of a 3-dB coupler is 0.1 dB, what is the combining loss for each carrier?

3.12 Consider the dual-hybrid filter combiner shown in Fig. 3.21 and assume that the insertion loss of each hybrid is 0.1 dB and the insertion loss of each filter is 0.6 dB. What is the power loss of carrier 1, carrier 2, and carrier 3?

3.13 Consider the power combining scheme in Fig. 3.23 and assume that the switch loss is 0.2 dB, the combining loss due to phase error is 0.1 dB, and the hybrid loss is 0.1 dB. Find the maximum output power in watts of the individual HPA needed to produce an output power of 30 dBW. If the individual HPA has to operate with an output power back-off of 2 dB, what is its required maximum output power in watts?

3.14 Find the theoretical equivalent noise temperature of a cryogenically cooled parametric amplifier operating at 12.2 GHz, assuming the pump frequency is 60 GHz.

3.15 Consider the upconverter shown in Fig. 3.30*b*. Find the constraints on ω_{l1} and ω_{l2} if the lower sidebands are selected.

3.16 Consider a dual upconverter with the following parameters.
  Uplink frequency spectrum: 14 to 14.5 GHz
  First intermediate frequency: 140 MHz
  Carrier bandwidth: 72 MHz
  BPF 1 center frequency (Fig. 3.30*b*) = 1.19 GHz
  (a) Find the first local oscillator frequency.
  (b) Find the range of the second local oscillator frequency.
  (c) Find the frequency spectrum of the unwanted sideband.
  (d) Find the image spectrum at the output of the upconverter.
  (e) Find the order of the Chebyshev bandpass filter
BPF 1 needed to provide an attenuation of 100 dB for the first local oscillator.

3.17 Consider a dual upconverter with the following parameters.
  Uplink frequency spectrum: 14 to 14.5 GHz
  First intermediate frequency: 70 MHz
  Carrier bandwidth: 49 MHz
  Number of channels: 10
  Lower edge of image spectrum of first channel: 13.9305 GHz

Lower sideband center frequency of first channel: 11.679 GHz

(a) Find the second LO frequency.

(b) Find the first LO frequency

3.18 Modify the multiple upconverter configuration in Fig. 3.32 to increase its reliability using the same number of upconverters, equalizers, and fast RF switches. (Hint: Put the individual cascaded equalizer-upconverter chain in a 1:1 redundancy configuration.) Give a simple explanation why the modified configuration has a higher degree of reliability. If one decides to use an upconverter equipped with a frequency synthesizer that can tune the transmit RF frequency to any carrier frequency in the system as a standby upconverter, show the 1:n redundancy configuration, where n is the number of on-line upconverters.

3.19 Modify the multiple upconverter configuration in Fig. 3.33 to increase its reliability using the same number of upconverters, equalizers, and fast IF switches. Give a simple explanation why the modified configuration has a higher degree of reliability. (Hint: Put the individual cascaded equalizer-upconverter chain in a 1:1 redundancy configuration.)

3.20 Consider the downconverter shown in Fig. 3.34b. Find the constraints on ω_{l1} and ω_{l2} if $\omega_{l2} > \omega_d$.

3.21 Consider a dual downconverter for 10 channels, each with a 49-MHz bandwidth, with the following parameters.

Downlink frequency spectrum: $11,725 + (n - 1)49$ MHz, $n = 1, 2, \ldots, 10$

First intermediate frequency: 70 MHz

Second intermediate frequency: 1.173 GHz

(a) Find the first local oscillator frequency.

(b) Find the range of the second local oscillator frequency (Fig. 3.34b).

(c) Find the image frequency spectrum that must be rejected by the bandpass filter (BPF 1) at the second intermediate frequency.

3.22 Consider a dual downconverter with the following parameters.

Downlink frequency spectrum: 10.7 to 11.7 GHz

First intermediate frequency: 140 MHz

First local oscillator frequency: 2.1 GHz

Carrier bandwidth: 72 MHz

(a) Assume that the second local oscillator frequency is generated from stable voltage-controlled oscillators in the range 1582.5 to 1707.5 MHz using a frequency multiplier. What is the multiplicative factor?

(b) Find the image frequency spectrum.

3.23 Using a downconverter equipped with a frequency synthesizer that can tune the receive RF frequency to any carrier frequency in the system as standby equipment, modify the 1:1 redundancy configuration in Fig. 3.36 to a 1:n redundancy configuration, where n is the number of on-line downconverters. (Hint: Use two additional polarization switches.)

3.24 What happens if the data acquisition unit in the RF terminal M&C system (Fig. 3.37) fails? What happens if the logic control of a subsystem fails? Suggest a RF terminal M&C system that avoids these problems. Justify your answer by comparing the advantages and disadvantages to those of the system discussed in the text.

3.25 Find the MTTF of the HPA systems in Fig. 3.15 and Fig. 3.17, assuming that the average failure rates of the individual devices are 0 for the power divider, 10^{-4} for the HPA, and 2×10^{-6} for the waveguide switch.

3.26 Find the MTTF for the hopping LO upconverter configuration shown in Fig. 3.31 given that the average failure rates of the individual devices are 0 for the power divider, 2×10^{-5} for the equalizer, 1.2×10^{-5} for the IF or polarization switch, 4×10^{-5} for the hopping LO upconverter, and 2×10^{-6} for the waveguide switch.

chapter 4

Satellite Link

As in any other communications system, the ultimate goal of a satellite system is to provide a satisfactory transmission quality for signals relayed between earth stations. We know by now that a satellite channel is a bandpass channel, and modulation must be employed to transmit baseband information. In an analog satellite system using frequency modulation the signal-to-noise ratio of a voice channel at the FM demodulator output is a measure of the signal's fidelity. This output signal-to-noise ratio is a function of the carrier-to-noise ratio C/N of the satellite link (Chap. 5). In a digital satellite system the performance of the satellite signal received at an earth station is measured in terms of the average probability of bit error, which is a function of the link carrier-to-noise ratio C/N, the information bit duration T_b (or, equivalently, the information bit rate $R = 1/T_b$), and the noise bandwidth B of the satellite channel. For a given information bit rate, the signal quality results in a trade-off between the type of digital modulation, which we will discuss in Chap. 9, and the carrier-to-noise ratio over the satellite link, which is the main subject of this chapter.

We know that a satellite link consists of an uplink and a downlink. Signal quality over the uplink depends on how strong the signal is when it leaves the originating earth station and how the satellite receives it. On the downlink, the signal quality depends on how strongly the satellite can retransmit the signal and how the destination earth station receives it.

Because of the great distances between a geostationary satellite and the earth stations (compared to terrestrial communications terminals) and because the power of the radiated signal diminishes as the square of the distance it travels, the uplink signal received by the satellite and the downlink signal received by the earth station are very weak and can be easily disturbed by the ever-present AWGN. In addition, the uplink signal may be contaminated by signals transmitted by other earth stations to adjacent satellites, and the downlink signal may be contaminated by signals coming from adjacent satellites. Furthermore, rain can severely attenuate satellite signals above 10 GHz and reduce the isolation between orthogonally polarized signals in a frequency reuse system. In this chapter these effects on the performance of a satellite link will be discussed in detail.

4.1 BASIC LINK ANALYSIS

Consider a basic satellite link shown in Fig. 4.1. The transmit earth station transmits the carrier $s(t)$ whose power is simply the EIRP of the carrier given by (3.8):

$$EIRP = P_T G_T \qquad (4.1)$$

where P_T = carrier power at antenna feed and G_T = transmit antenna gain. For the time being we assume that the transmission occurs under clear-sky conditions and that the only attenuation the carrier $s(t)$ suffers is the uplink free space loss, the atmospheric attenuation discussed in Sec. 3.1.6, and the antenna tracking loss. The uplink *free space loss* is given by

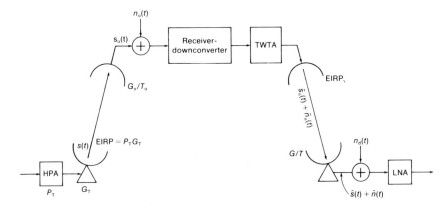

Figure 4.1 Basic satellite link.

$$L_u = \left(\frac{4\pi d_u}{\lambda_u}\right)^2 = \left(\frac{4\pi f_u d_u}{c}\right)^2 \tag{4.2}$$

where d_u = uplink slant range (m) and can be calculated from (2.16)
 λ_u = uplink wavelength (m)
 f_u = uplink carrier frequency (Hz)
 c = speed of light = 2.997925×10^8 m/s

Let $s_u(t)$ be the carrier received at the satellite and $n_u(t)$ the AWGN with zero mean that contaminates the uplink carrier $s_u(t)$. Then the received carrier plus noise at the satellite is $s_u(t) + n_u(t)$. If the satellite antenna gain is G_u, the uplink carrier power, that is, the power of $s_u(t)$, is

$$
\begin{aligned}
C_u &= E[s_u^2(t)] \\
&= \frac{(\text{EIRP}) \, G_u}{L_u L} \\
&= \frac{(\text{EIRP})}{L} \left(\frac{c}{4\pi f_u d_u}\right)^2 G_u
\end{aligned}
\tag{4.3}
$$

where $E[\cdot]$ denotes the expected value and L is the sum of the *antenna tracking loss* and *atmospheric attenuation*. The uplink noise power is given in (3.10) as

$$
\begin{aligned}
N_u &= E[n_u^2(t)] \\
&= kT_u B
\end{aligned}
\tag{4.4}
$$

where T_u = satellite system noise temperature (K) and can be calculated
 from the noise temperature of the satellite antenna (which is
 about 290 K since the antenna always sees a hot earth) and
 the equivalent noise temperature of the satellite communications repeater shown in Fig. 2.13
 B = noise bandwidth of satellite channel (Hz)
 k = Boltzmann's constant = 1.38×10^{-23} J/K

The parameter G_u/T_u is the satellite antenna gain-to-noise temperature ratio. From (4.3) and (4.4) the uplink carrier-to-noise ratio is

$$
\begin{aligned}
\left(\frac{C}{N}\right)_u &= \frac{C_u}{N_u} = \left(\frac{\text{EIRP}}{L_u L}\right) \left(\frac{G_u}{T_u}\right) \left(\frac{1}{kB}\right) \\
&= \frac{(\text{EIRP})}{L} \left(\frac{c}{4\pi f_u d_u}\right)^2 \left(\frac{G_u}{T_u}\right) \left(\frac{1}{kB}\right)
\end{aligned}
\tag{4.5}
$$

Note that $\text{EIRP}/4\pi d_u^2 L$ is simply the carrier *power flux density* (W/m²) at the satellite:

$$\Omega = \frac{\text{EIRP}}{4\pi d_u^2 L} \tag{4.6}$$

Also, the term $4\pi f_u^2/c^2$ is simply the gain of an ideal antenna whose aperture area is 1 m^2 as seen from (3.1) when $\eta = 1$ and $A = 1$ m^2:

$$G_{1m^2} = \frac{4\pi f_u^2}{c^2} \tag{4.7}$$

Substituting (4.6) into (4.5) yields

$$\left(\frac{C}{N}\right)_u = \Omega \left(\frac{c^2}{4\pi f_u^2}\right) \left(\frac{G_u}{T_u}\right) \left(\frac{1}{kB}\right) \tag{4.8}$$

In summary, the uplink carrier-to-noise ratio can be calculated by (4.5) if the carrier EIRP is given, and by (4.8) if the power flux density at the satellite is given.

As we have indicated previously, the received carrier plus noise at the satellite is $s_u(t) + n_u(t)$. This carrier plus noise is amplified and downconverted by the satellite communications receiver-downconverter and then amplified again by the satellite TWTA and retransmitted back to earth by the satellite antenna. Denote the retransmitted carrier plus noise by $\hat{s}_u(t) + \hat{n}_u(t)$ which possesses the same carrier-to-noise ratio $(C/N)_u$ given in (4.5) as the received carrier plus noise $s_u(t) + n_u(t)$.† Let EIRP$_s$ be the satellite EIRP (or power) of the retransmitted carrier $\hat{s}_u(t)$; that is, $\hat{C}_u = E[\hat{s}_u^2(t)] = $ EIRP$_s$. Then the power of the accompanied uplink noise $\hat{n}_u(t)$ is

$$\frac{\hat{C}_u}{\hat{N}_u} = \left(\frac{C}{N}\right)_u \tag{4.9a}$$

or

$$\hat{N}_u = \frac{\text{EIRP}_s}{(C/N)_u} \tag{4.9b}$$

The received carrier plus noise at the receive earth station is $\hat{s}(t) + \hat{n}(t) + n_d(t)$, where $\hat{s}(t)$ and $\hat{n}(t)$ are the attenuated versions of $\hat{s}_u(t)$ and $\hat{n}_u(t)$, respectively, and $n_d(t)$ is the additional independent downlink AWGN with zero mean that further contaminates $\hat{s}(t)$. After taking into account the free space loss, the antenna tracking loss, the atmospheric attenuation on the downlink, and the receive antenna gain G of the earth station, the power of the carrier $\hat{s}(t)$ at the receive earth station is

$$C = E[\hat{s}^2(t)]$$

$$= \frac{(\text{EIRP}_s)G}{L_d L'}$$

$$= \frac{(\text{EIRP}_s)}{L'} \left(\frac{c}{4\pi f_d d_d}\right)^2 G \tag{4.10}$$

---

†In considering the frequency-translating transponder, we neglect the frequency translation between uplink and downlink and simply convert the former to the latter through an ideal amplifier of gain g. Therefore $\hat{C}_u/\hat{N}_u = gC_u/gN_u = C_u/N_u = (C/N)_u$.

where d_{d} = downlink slant range (m)

f_{d} = downlink carrier frequency (Hz)

L' = antenna tracking loss and atmospheric attenuation

The power of the accompanied uplink noise $\hat{n}(t)$ is

$$\hat{N} = E[\hat{n}^2(t)]$$

$$= \frac{\hat{N}_{\text{u}}G}{L_{\text{d}}L'}$$

$$= \frac{(\text{EIRP}_{\text{s}})G}{L'L_{\text{d}}(C/N)_{\text{u}}}$$

$$= \frac{(\text{EIRP}_{\text{s}})}{L'}\left(\frac{c}{4\pi f_{\text{d}}d_{\text{d}}}\right)^2\left(\frac{C}{N}\right)_{\text{u}}^{-1}G \qquad (4.11)$$

The downlink noise power is again given by (3.10) as

$$N_{\text{d}} = E[n_{\text{d}}^2(t)]$$

$$= kTB \qquad (4.12)$$

where T = earth station system noise temperature (K) and can be evaluated from (3.29).

Thus the composite noise power at the receive earth station is†

$$N = E\{[\hat{n}(t) + n_{\text{d}}(t)]^2\} = E[\hat{n}^2(t)] + E[n_{\text{d}}^2(t)]$$

$$= \hat{N} + N_{\text{d}}$$

$$= \frac{(\text{EIRP}_{\text{s}})}{L'}\left(\frac{c}{4\pi f_{\text{d}}d_{\text{d}}}\right)^2\left(\frac{C}{N}\right)_{\text{u}}^{-1}G + kTB \qquad (4.13)$$

The carrier-to-noise ratio of the overall satellite link (uplink and downlink) is therefore given by (4.10) and (4.13) as follows:

$$\frac{C}{N} = \frac{(\text{EIRP}_{\text{s}})(c/4\pi f_{\text{d}}d_{\text{d}})^2 G/L'}{(\text{EIRP}_{\text{s}})(c/4\pi f_{\text{d}}d_{\text{d}})^2(C/N)_{\text{u}}^{-1}G/L' + kTB}$$

$$= \left\{\left(\frac{C}{N}\right)_{\text{u}}^{-1} + \left[\frac{(\text{EIRP}_{\text{s}})}{L'}\left(\frac{c}{4\pi f_{\text{d}}d_{\text{d}}}\right)^2\left(\frac{G}{T}\right)\left(\frac{1}{kB}\right)\right]^{-1}\right\}^{-1} \qquad (4.14)$$

Let

$$\left(\frac{C}{N}\right)_{\text{d}} = \frac{(\text{EIRP}_{\text{s}})}{L'}\left(\frac{c}{4\pi f_{\text{d}}d_{\text{d}}}\right)^2\left(\frac{G}{T}\right)\left(\frac{1}{kB}\right) \qquad (4.15)$$

By comparing (4.15) to (4.5) it is seen that $(C/N)_{\text{d}}$ is simply the downlink

---

†Since the two wide-sense stationary AWGN processes $\hat{n}(t)$ and $n_{\text{d}}(t)$ are independent and have zero means, $E[\hat{n}(t)\,n_{\text{d}}(t)] = E[n_{\text{d}}(t)\,\hat{n}(t)] = E[\hat{n}(t)]\,E[n_{\text{d}}(t)] = 0$. Therefore $E\{[\hat{n}(t) + n_{\text{d}}(t)]^2\} = E[\hat{n}^2(t)] + E[n_{\text{d}}^2(t)] + E[\hat{n}(t)\,n_{\text{d}}(t)] + E[n_{\text{d}}(t)\,\hat{n}(t)] = E[\hat{n}^2(t)] + E[n_{\text{d}}^2(t)]$.

carrier-to-noise ratio, and the familiar parameter G/T is the antenna gain-to-noise temperature ratio of the receive earth station. Substituting (4.15) into (4.14) yields the link carrier-to-noise ratio:

$$\frac{C}{N} = \left[\left(\frac{C}{N}\right)_u^{-1} + \left(\frac{C}{N}\right)_d^{-1} \right]^{-1} \tag{4.16}$$

Equation (4.16) provides the fundamental analysis of a satellite link where the satellite transponder is a classical frequency-translating repeater. In this type of satellite transponder the uplink noise adds directly to the downlink noise, and the uplink is said to be *coupled* to the downlink. From (4.16) we note that, if $(C/N)_u \gg (C/N)_d$, then $C/N \approx (C/N)_d$. In this case the satellite link is said to be *downlink-limited*. This is the common case in satellite communications. When the reverse situation occurs, that is, $(C/N)_u \ll (C/N)_d$, then $C/N \approx (C/N)_u$ and the satellite link is said to be *uplink-limited*.

In link analysis, the carrier EIRP, or its power flux density Ω at the satellite, and the satellite $\mathrm{EIRP_s}$ for the retransmitted carrier are normally given with respect to the operating point of the transponder TWTA whose typical normalized power gain characteristic for single carrier amplification is shown in Fig. 4.2. The operating point where the TWTA

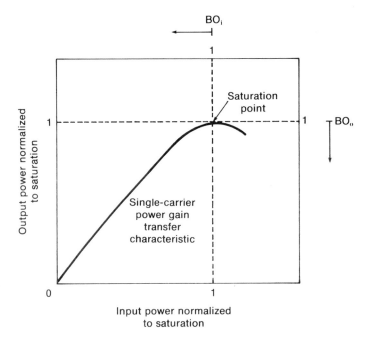

Figure 4.2 Nonlinear characteristic of a satellite TWTA.

output power is maximum is called the *saturation point*. A further increase in the TWTA input power will result in a decrease in the output power. In many cases the TWTA is operated below the saturation point to avoid nonlinear distortions, especially when there are many carriers per transponder (more than one), to reduce the power of the generated intermodulation products that act as interference signals (we will discuss intermodulation interference in Chap. 5). Let BO_i and BO_o be the input and output *back-offs* of the TWTA, respectively, which are defined as

$$BO_i = \frac{EIRP_{sat}}{EIRP} > 1 \qquad (4.17)$$

$$BO_i = \frac{\Omega_{sat}}{\Omega} > 1 \qquad (4.18)$$

$$BO_o = \frac{EIRP_{s,sat}}{EIRP_s} > 1 \qquad (4.19)$$

where $EIRP_{sat}$ = carrier EIRP required to saturate satellite TWTA
Ω_{sat} = saturation power flux density
$EIRP_{s,sat}$ = satellite saturation EIRP
$BO_o = f(BO_i)$ and is a nonlinear function of BO_i

Substituting (4.17) into (4.5), (4.18) into (4.8), and (4.19) into (4.15) yields

$$\left(\frac{C}{N}\right)_u = (EIRP_{sat}) \left(\frac{c}{4\pi f_u d_u}\right)^2 \left(\frac{G_u}{T_u}\right) \left(\frac{1}{kB}\right) BO_i^{-1} L^{-1} \qquad (4.20)$$

$$\left(\frac{C}{N}\right)_u = \Omega_{sat} \left(\frac{c^2}{4\pi f_u^2}\right) \left(\frac{G_u}{T_u}\right) \left(\frac{1}{kB}\right) BO_i^{-1} \qquad (4.21)$$

$$\left(\frac{C}{N}\right)_d = (EIRP_{s,sat}) \left(\frac{c}{4\pi f_u d_u}\right)^2 \left(\frac{G}{T}\right) \left(\frac{1}{kB}\right) BO_o^{-1} L'^{-1} \qquad (4.22)$$

The uplink and downlink carrier-to-noise ratios in decibels are 10 times the logarithm (base 10) of the corresponding quantities:

$$\left(\frac{C}{N}\right)_u = EIRP_{sat} \text{ (dBW))} - 20 \log \left(\frac{4\pi f_u d_u}{c}\right) + \frac{G_u}{T_u} \text{ (dB/K)}$$
$$-10 \log k - 10 \log B - BO_i \text{ (dB)} - L \text{ (dB)} \qquad (4.23)$$

$$\left(\frac{C}{N}\right)_u = \Omega_{sat} \text{ (dBW/m}^2) - 10 \log \left(\frac{4\pi f_u^2}{c^2}\right) + \frac{G_u}{T_u} \text{ (dB/K)}$$
$$- 10 \log k - 10 \log B - BO_i \text{ (dB)} \qquad (4.24)$$

$$\left(\frac{C}{N}\right)_d = EIRP_{s,sat} \text{(dBW)} - 20 \log \left(\frac{4\pi f_d d_d}{c}\right) + \frac{G}{T} \text{ (dB/K)}$$
$$- 10 \log k - 10 \log B - BO_o \text{ (dB)} - L' \text{ (dB)} \qquad (4.25)$$

Example 4.1 To illustrate the evaluation of the total carrier-to-noise ratio of a satellite link, consider a Ku-band (14/12-GHz) satellite system operating in the single-carrier-per-transponder TDMA mode (Fig. 1.13) and using QPSK carrier modulation (this type of modulation will be discussed in Chap. 9). The system parameters are as follows.

Carrier modulation parameters
    Bit rate: 60 Mbps
    Bit duration-bandwidth product: 0.6
    Noise bandwidth: 36 MHz
Satellite parameters
    Antenna gain-to-noise temperature ratio: 1.6 dB/K
    Satellite saturation EIRP: 44 dBW
    TWTA input back-off: 0 dB
    TWTA output back-off: 0 dB
Earth station parameters
    Antenna diameter: 7 m
    Transmit antenna gain at 14 GHz: 57.6 dB
    Receive antenna gain at 12 GHz: 56.3 dB
    Carrier power into antenna: 174 W
    Maximum uplink and downlink slant range: 37,506 km
    Tracking loss: 1.2 dB (uplink) and 0.9 dB (downlink)
    System noise temperature: 160 K.

Based on these parameters, the link calculation using (4.23) and (4.25) is given in Table 4.1.

Example 4.2 The above example illustrates a single-carrier-per-transponder operation where neither the uplink or downlink is limited. In this example, we will consider the case of a multiple-carriers-per-transponder operation for a C-band (6/4-GHz) satellite system operating in the FDMA mode (Fig. 1.12) and also using QPSK modulation. The system parameters are as follows.

Carrier modulation parameters
    Bit rate: 64 kbps
    Bit duration-bandwidth product: 0.625
    Noise bandwidth: 40 kHz
Satellite parameters
    Antenna gain-to-noise temperature ratio: -7 dB/K
    Satellite saturation EIRP: 36 dBW
    TWTA input back-off: 11 dB
    TWTA output back-off: 6 dB
    Number of carriers per transponder: 200
    Power flux density at satellite for transponder saturation: -80 dBW/m^2
Earth station parameters
    Saturation power flux density per carrier: $-80 - 10 \log 200 = -103$ dBW/m^2
    Transmit antenna gain: 47 dB
    Receive antenna gain: 44.5 dB
    Antenna gain-to-noise temperature ratio: 22 dB/K
    Maximum downlink slant range: 37,506 km

The link calculation using (4.24) and (4.25) based on these parameters is given in Table 4.2. It is seen that the satellite link is downlink-limited.

4.2 INTERFERENCE ANALYSIS

As mentioned previously, a carrier may be impaired by other unintended interference signals besides the ever-present AWGN. By using the fun-

Table 4.1 Link calculation of a single-carrier-per-transponder system

Uplink (14.25 GHz)	
Carrier EIRP	80 dBW
Free space loss	206.9 dB
Antenna tracking loss	1.2 dB
Satellite G/T	1.6 dB/K
Boltzmann's constant	−228.6 dBW/K-Hz
Noise bandwidth	75.6 dB-Hz
$(C/N)_u$	26.5 dB
Downlink (11.95 GHz)	
Satellite EIRP	44 dBW
Free space loss	205.5 dB
Antenna tracking loss	0.9 dB
Earth station G/T	34.3 dB/K
Boltzmann's constant	−228.6 dBW/K-Hz
Noise bandwidth	75.6 dB-Hz
$(C/N)_d$	24.9 dB
Total carrier-to-noise ratio	22.6 dB

damental link equation (4.16) we can generalize the result to include their effect on both the uplink and the downlink. To do so we have to make the assumption that all interference signals including the AWGN are statistically independent wide-sense stationary random processes of zero means.

Table 4.2 Link calculation of a multiple-carriers-per-transponder system

Uplink (6 GHz)	
Saturation power flux density per carrier	−103 dBW/m²
Gain of an ideal 1-m² antenna	37 dB
Satellite G/T	−7 dB/K
Boltzmann's constant	−228.6 dBW/K-Hz
Noise bandwidth	46 dB-Hz
TWTA input back-off	11 dB
$(C/N)_u$	24.6 dB
Downlink (4 GHz)	
Saturation EIRP per carrier	13 dBW
	$(36 - 10 \log 200)$
Free space loss	196 dB
Earth station G/T	22 dB/K
Boltzmann's constant	−228.6 dBW/K-Hz
Noise bandwidth	46 dB-Hz
TWTA output back-off	6 dB
$(C/N)_d$	15.6
Total carrier-to-noise ratio	15 dB

4.2.1 Carrier-to-Noise plus Interference Ratio

Let $i_{1,u}(t)$, $i_{2,u}(t)$, ..., $i_{p,u}(t)$ be the additive interference signals on the uplink with respective powers $I_{1,u}$, $I_{2,u}$, ..., $I_{p,u}$ within the bandwidth of the desired carrier. Then the total uplink interference plus noise power is

$$\mathcal{N}_u = E\left\{\left[n_u(t) + \sum_{k=1}^{p} i_{k,u}(t)\right]^2\right\}$$

$$= N_u + \sum_{k=1}^{p} I_{k,u} \tag{4.26}$$

and thus the uplink carrier-to-noise plus interference ratio is given as

$$\left(\frac{C}{\mathcal{N}}\right)_u = \frac{C_u}{\mathcal{N}_u} = \left[\left(\frac{C_u}{N_u}\right)^{-1} + \sum_{k=1}^{p}\left(\frac{C_u}{I_{k,u}}\right)^{-1}\right]^{-1}$$

$$= \left[\left(\frac{C}{N}\right)_u^{-1} + \sum_{k=1}^{p}\left(\frac{C}{I}\right)_{k,u}^{-1}\right]^{-1}$$

$$= \left[\left(\frac{C}{N}\right)_u^{-1} + \left(\frac{C}{I}\right)_u^{-1}\right]^{-1} \tag{4.27a}$$

where

$$\left(\frac{C}{I}\right)_u = \left[\sum_{k=1}^{p}\left(\frac{C}{I}\right)_{k,u}^{-1}\right]^{-1} \tag{4.27b}$$

and where $(C/I)_{k,u} = C_u/I_{k,u}$ = uplink carrier-to-kth interference ratio and $(C/I)_u$ = uplink carrier-to-interference ratio.

Similarly, if we let $i_{1,d}(t)$, $i_{2,d}(t)$, ..., $i_{q,d}(t)$ be the additive interference signals on the downlink with power $I_{1,d}$, $I_{2,d}$, ..., $I_{q,d}$, respectively, then the downlink carrier-to-interference plus noise ratio can be expressed as

$$\left(\frac{C}{\mathcal{N}}\right)_d = \left[\left(\frac{C}{N}\right)_d^{-1} + \sum_{k=1}^{q}\left(\frac{C}{I}\right)_{k,d}^{-1}\right]^{-1}$$

$$= \left[\left(\frac{C}{N}\right)_d^{-1} + \left(\frac{C}{I}\right)_d^{-1}\right]^{-1} \tag{4.28a}$$

where

$$\left(\frac{C}{I}\right)_d = \left[\sum_{k=1}^{q}\left(\frac{C}{I}\right)_{k,d}^{-1}\right]^{-1} \tag{4.28b}$$

and where $(C/I)_{k,d}$ = downlink carrier-to-kth interference ratio and $(C/I)_d$ = downlink carrier-to-interference ratio. By replacing $(C/N)_u$ and $(C/N)_d$ in (4.16) with $(C/\mathcal{N})_u$ in (4.27a) and $(C/\mathcal{N})_d$ in (4.28a), respectively, the carrier-to-noise plus interference ratio of the overall satellite link becomes

$$\frac{C}{\mathcal{N}} = \left[\left(\frac{C}{N} \right)_u^{-1} + \left(\frac{C}{N} \right)_d^{-1} + \left(\frac{C}{I} \right)_u^{-1} + \left(\frac{C}{I} \right)_d^{-1} \right]^{-1}$$

$$= \left[\left(\frac{C}{N} \right)^{-1} + \left(\frac{C}{I} \right)^{-1} \right]^{-1} = \left[\left(\frac{C}{\mathcal{N}} \right)_u^{-1} + \left(\frac{C}{\mathcal{N}} \right)_d^{-1} \right]^{-1} \qquad (4.29a)$$

where

$$\left(\frac{C}{I} \right)^{-1} = \left(\frac{C}{I} \right)_u^{-1} + \left(\frac{C}{I} \right)_d^{-1} \qquad (4.29b)$$

and where C/N = carrier-to-noise ratio of the overall link and C/I = carrier-to-interference ratio of the overall link.

Equation (4.29a) is the most widely used equation in satellite system engineering. Once the type of modulation is selected, the total carrier-to-noise plus interference ratio can be employed to predict the performance of the link. In digital satellite systems, the performance is measured in terms of the average probability of bit error, which is a function of the link carrier-to-noise ratio, as we will discuss in Chap. 9 where the signal is assumed to be contaminated by AWGN only. When *non-Gaussian* interference signals are taken into account, the result might not be correct. Therefore the application of the carrier-to-noise plus interference ratio in (4.29a) must be carried out under certain conditions. When the interferences are non-Gaussian and none of them has a dominant effect, their joint probability density function approaches the Gaussian density function with zero mean and variance equal to the sum of the individual variances as stated by the *central limit theorem*. Their effect can be approximated like that of a single AWGN process which produces the same carrier-to-interference ratio. The treatment of non-Gaussian interferences as equivalent AWGN in a digital satellite link using (4.29a) is valid for a large carrier-to-interference ratio C/I and $C/I > C/N$, where C/N is the carrier-to-noise ratio. (By a "large" C/I we mean that it is on the order of 20 dB or more and is at least 3 dB larger than C/N. Such a satellite link is said to be *noise-dominant*. When $C/I < C/N$, the link is said to be *interference-dominant*.) Otherwise the replacement of non-Gaussian interferences by Gaussian noise of the same power will result in a higher probability of bit error; that is, the result will be pessimistic. Figures 4.3 and 4.4 show the average probability of bit error P_b for PSK and QPSK carriers, respectively, in the presence of AWGN and one non-Gaussian interference signal [1,2]. Consider the case of a QPSK carrier in Fig. 4.4; it is seen that, when there is no interference ($C/I = \infty$), a carrier-to-noise ratio of $C/N = 13.5$ dB is required for $P_b = 10^{-6}$. When $C/I = 20$ dB, it is required that $C/N = 14.3$ dB to achieve $P_b = 10^{-6}$. If the interference is treated as AWGN, then according to (4.29a) the total carrier-to-noise ratio will be 13.26 dB which according to (9.51), or the curve labeled

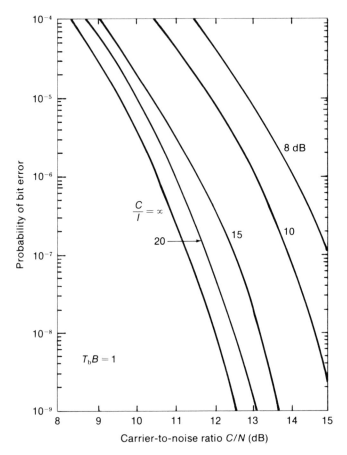

Figure 4.3 Average probability of bit error for PSK with one interference signal.

$C/I = \infty$ in Fig. 4.4, yields $P_b = 2 \times 10^{-6}$, a slightly pessimistic result as compared to the actual $P_b = 10^{-6}$. Now consider the case where $C/I = 15$ dB; then it is required that $C/N = 15.3$ dB to achieve $P_b = 10^{-6}$. If the interference is treated as AWGN, again (4.29a) states that the total carrier-to-noise should be 12.14 dB which, according to (9.51), or the curve labeled $C/I = \infty$ in Fig. 4.4., yields $P_b = 4 \times 10^{-5}$ which clearly overestimates the effect of the interference. The result is even more pessimistic at a high P_b.

The above discussion considers the effect of only one interference. Fig. 4.5 shows the average probability of bit error P_b for a QPSK carrier in the presence of AWGN and four equal-amplitude interferences [3], where C/I is the total carrier-to-interference ratio for all four interferences. Also plotted in Fig. 4.5 is the case of a single interference with an equal C/I. It is seen that, for the same C/I, four equal-amplitude interfer-

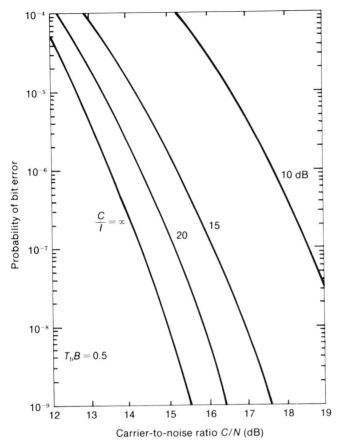

Figure 4.4 Average probability of bit error for QPSK with one interference signal.

ences degrade the system's performance more than a single interference. The degradation is more severe at low C/I, enough to wipe out the overestimation which results in the treatment of interferences as AWGN. Such a situation occurs frequently where severe rain attenuation may reduce the carrier power (rain attenuation will be discussed in Sec. 4.3) considerably, and it seems in these systems that the treatment of interferences as AWGN in link design is not conservative at all.

The consideration of interference in satellite communications systems is of utmost importance. In the United States, it is required by the *Federal Communications Commission* that a proposed transmit and receive satellite system will not unduly interfere with any type of existing satellite system (interference from earth stations into adjacent satellites). Each system must meet the allowable interference requirements (interference from adjacent satellites into the earth station, terrestrial interference

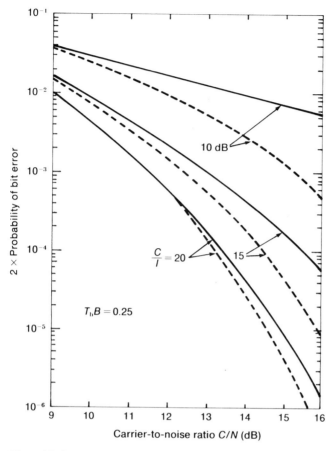

Figure 4.5 Average probability of bit error for QPSK with one (broken lines) and with four (solid lines) interference signals.

caused by line-of-sight microwave systems). In the following sections we will investigate these interference sources and others that are necessary for link design.

4.2.2 Interference into or from Adjacent Satellite Systems

The interference generated by an earth station into an adjacent satellite comes from antenna sidelobes such as the one shown in Fig. 4.6. [The reader may verify that the normalized gain patterns in (3.6a) and (3.7) yield the first sidelobe levels of 17.5 and 24.6 dB, respectively, below the peak gain.] The FCC regulation specifies a sidelobe envelope level relative to the normalized peak gain (1 or 0 dB) as follows:

$$32 - 25 \log \theta \text{ (dB)} \qquad 1° \leq \theta \leq 48°$$
$$-10 \text{ dB} \qquad\qquad 48° \leq \theta \leq 180° \qquad (4.30)$$

where θ is the antenna off-axis angle (deg). For example, at $\theta = 4°$, the sidelobe envelope level must not exceed 17 dB above the 0-dB level. Thus with the peak antenna gain of 50 dB, the sidelobe level at $\theta = 4°$ must be at least 33 dB down from the on-axis gain.

In order to find the interference power generated by or received from the sidelobe of the earth station antenna into or from an adjacent satellite, it is necessary to know the angular separation between two geostationary satellites as seen by an earth station. This is depicted in Fig. 4.7 where the parameters are defined as follows:

θ = angular separation between two geostationary satellites as seen by earth station antenna

β = angular separation between two geostationary satellites ($\beta = |\theta_{S,A} - \theta_{S,B}|$, where $\theta_{S,A}$ and $\theta_{S,B}$ are the longitudes of satellites A and B, respectively).

d_i = slant range between earth station and satellite i $(i = A,B)$ and can be calculated from (2.12) and (2.16)

r = geostationary orbit = 42,164.2 km

d = separation distance between two satellites

We note that θ and β can be related to d as follows:

$$d^2 = d_A^2 + d_B^2 - 2d_A d_B \cos \theta \qquad (4.31)$$

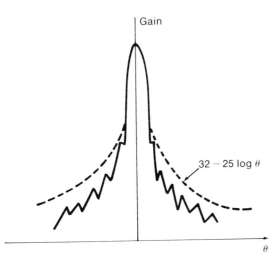

Figure 4.6 Antenna radiation pattern with a FCC-specified sidelobe level.

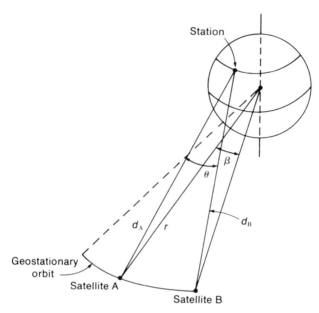

Figure 4.7 Separation of two satellites as seen by an earth station.

$$d^2 = 2r^2 - 2r^2 \cos \beta = 2r^2(1 - \cos \beta) \qquad (4.32)$$

By comparing (4.31) and (4.32) we get

$$\theta = \cos^{-1} \left[\frac{d_A^2 + d_B^2 - 2r^2(1 - \cos \beta)}{2d_A d_B} \right] \qquad (4.33)$$

If the minimum separation between satellites operating at the C band (6/4 GHz) is 4°, then based on an east-west stationkeeping accuracy of ±0.05°, the worst-case orbital separation is 3.9°. For earth stations in the continental United States, the worst-case viewing angle θ is about 4°. Thus for C-band satellite systems, the antenna off-axis angle can be taken to be $\theta = 4°$.

To analyze the interference into or from an adjacent satellite system consider the satellite link and interference paths (dotted lines) between two satellite systems A and B in Fig. 4.8. Let A be the existing satellite system and B be the proposed satellite system. Then the satellite link between the transmit earth station A_2 and the receive earth station A_1 is affected by two interference sources: the uplink interference signals from earth stations in the proposed system B, and the downlink interference signals coming from satellite B. The total carrier-to-interference ratio due to these two interference sources represents the interference generated by the proposed satellite system B into the adjacent satellite system A. If we interchange the roles of A and B, then the situation represents the interference generated by the adjacent satellite system B into the proposed satellite system A.

Based on the analysis given in Sec. 4.1, the uplink interference power is given by

$$I_u = (\text{EIRP}') \left(\frac{c}{4\pi f'_u d'_u}\right)^2 G'_u \tag{4.34}$$

where $\text{EIRP}' =$ EIRP of interference signal in direction of interfered satellite A

$f'_u =$ uplink interference frequency

$d'_u =$ uplink slant range between interfered satellite A and interfering earth station B_1

$G'_u =$ antenna gain of interfered satellite A in direction of interfering earth station B_1

By using (4.3) the uplink carrier-to-interference ratio is given by (assuming $f'_u \approx f_u$, and $d'_u \approx d_u$ which is the case in practice)

$$\left(\frac{C}{I}\right)_u = \frac{C_u}{I_u} = \left(\frac{\text{EIRP}}{\text{EIRP}'}\right) \left(\frac{f'_u d'_u}{f_u d_u}\right)^2 \left(\frac{G_u}{G'_u}\right)$$

$$\approx \left(\frac{\text{EIRP}}{\text{EIRP}'}\right) \left(\frac{G_u}{G'_u}\right) \tag{4.35}$$

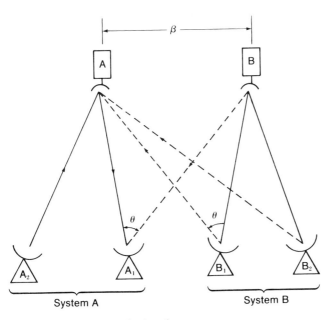

Figure 4.8 Adjacent satellite interference.

or, equivalently,

$$\left(\frac{C}{I}\right)_u \approx \text{EIRP (dBW)} - \text{EIRP}' \text{ (dBW)} + G_u \text{ (dB)} - G_u' \text{ (dB)} \quad (4.36)$$

Also, by using (4.6) we have

$$\left(\frac{C}{I}\right)_u \approx \left(\frac{\Omega}{\Omega'}\right)\left(\frac{G_u}{G_u'}\right) \quad (4.37)$$

where $\Omega' = \text{EIRP}'/4\pi d_u'^2$ and is the power flux density of the interfering signal at the interfered satellite A or, equivalently,

$$\left(\frac{C}{I}\right)_u \approx \Omega \text{ (dBW/m}^2) - \Omega' \text{ (dBW/m}^2) + G_u \text{ (dB)} - G_u' \text{ (dB)}$$

$$(4.38)$$

Both parameters EIRP′ and Ω' can be expressed in terms of their operating EIRP* and Ω^*, respectively, assuming the sidelobe envelope level relative to 0 dB in (4.30):

$$\text{EIRP}' \text{ (dBW)} = \text{EIRP}^* \text{ (dBW)} - G_i \text{ (dB)} + 32 - 25 \log \theta$$

$$(4.39)$$

$$\Omega' \text{ (dBW/m}^2) = \Omega^* \text{ (dBW/m}^2) - G_i \text{ (dB)} + 32 - 25 \log \theta$$

$$(4.40)$$

where EIRP* = EIRP of interference signal in direction of interfering satellite B

Ω^* = power flux density of interference signal at interfering satellite B

G_i = on-axis transmit antenna gain of interfering earth station B_1

Substituting (4.39) into (4.36) and (4.40) into (4.38) yields

$$\left(\frac{C}{I}\right)_u \approx \text{EIRP (dBW)} - \text{EIRP}^* \text{ (dBW)} + G_i \text{ (dB)} - (32 - 25 \log \theta)$$
$$+ G_u \text{ (dB)} - G_u' \text{ (dB)} \quad (4.41)$$

or

$$\left(\frac{C}{I}\right)_u \approx \Omega \text{ (dBW/m}^2) - \Omega^* \text{ (dBW/m}^2) + G_i \text{ (dB)} - (32 - 25 \log \theta)$$
$$+ G_u \text{ (dB)} - G_u' \text{ (dB)} \quad (4.42)$$

Similarly, the downlink carrier-to-interference ratio is given by (assuming the downlink interference frequency f_d' is approximately equal to

f_d and the downlink slant range d_d' between the interfering satellite B and the interfered earth station A_1 is approximately equal to d_d)

$$\left(\frac{C}{I}\right)_d \approx \text{EIRP}_s \text{ (dBW)} - \text{EIRP}_s' \text{ (dBW)} + G \text{ (dB)} - (32 - 25 \log \theta)$$

(4.43)

where $\text{EIRP}_s = $ EIRP of interfered satellite A in the direction of interfered earth station A_1

$\text{EIRP}_s' = $ EIRP of interfering satellite B in the direction of interfered earth station A_1

$G = $ on-axis receive antenna gain of interfered earth station A_1

The total carrier-to-interference ratio caused by the adjacent satellite system is

$$\frac{C}{I} = \left[\left(\frac{C}{I}\right)_u^{-1} + \left(\frac{C}{I}\right)_d^{-1}\right]^{-1}$$

(4.44)

Example 4.3 To illustrate the interference calculation, consider the proposed multiple-carriers-per-transponder system B shown in Table 4.2. The existing system A is a 6/4-GHz single-carrier-per-transponder system with the following parameters

Carrier modulation parameters

Data rate: 60 Mbps

Bit duration-bandwidth product: 0.6

Noise bandwidth: 36 MHz

Satellite A parameters

Satellite separation:4°

Transponder saturation power flux density: −80 dBW/m²

Transponder saturation EIRP: 35 dBW

Transponder input back-off: 4 dB

Transponder output back-off: 1 dB

Differential antenna gain: $G_u - G_u' = -3$ dB

Earth station A parameters

Transmit antenna gain: 53 dB

Receive antenna gain: 51 dB

Using (4.42) and (4.43) the interference generated by system B into system A is calculated and is given in Table 4.3.

Example 4.4 To calculate the interference received by the multiple-carriers-per-transponder system B from the single-carrier-per-transponder system A, it is necessary to know the power spectral density of the interference. Figure 4.9 shows the power spectral density of three types of single carriers, namely, a 60-Mbps QPSK carrier, a 1200-channel FDM-FM carrier, and a TV-FM carrier. Note that the peak level occurs at the carrier frequency. For the 60-Mbps QPSK carrier shown, the maximum interfering power in the 1-Hz bandwidth is 73 dB below the carrier power in the 36-MHz bandwidth or, equivalently, the maximum interfering power in the 40-kHz bandwidth is 27 dB below the carrier power in the same bandwidth. This 27 dB must be subtracted from Ω^* in (4.42) and from EIRP_s in (4.43) to obtain the interference for the

Table 4.3 Calculation of interference into adjacent satellite system

Uplink	
Power flux density Ω at the interfered satellite A	-84 dBW/m² ($-80 - 4$)
Power flux density Ω^* at the interfering satellite B (for 200 carriers)	-91 dBW/m² ($-80 - 11$)
Transmit antenna gain of interfering earth station B_1	47 dB
$32 - 25 \log \theta$	17 dB
Differential antenna gain $G_u - G'_u$	-3 dB
$(C/I)_u$	34 dB
Downlink	
EIRP of the interfered satellite A	34 dBW ($35 - 1$)
EIRP of the interfering satellite B (200 carriers)	30 dBW
Receive antenna gain of the interfered earth station A_1	51 dB
$32 - 25 \log \theta$	17 dB
$(C/I)_d$	38 dB
Total carrier-to-interference ratio	32.5 dB

Table 4.4 Calculation of interference from adjacent satellite system

Uplink	
Power flux density Ω at the interfered satellite B per carrier	-114 dBW/m² ($-103 - 11$)
Power flux density Ω^* at the interfering satellite A	-84 dBW/m²
Maximum interference level in 40-kHz bandwidth relative to interfering carrier power	27 dB
Transmit antenna gain of the interfering earth station A_1	53 dB
$32 - 25 \log \theta$	17 dB
Differential antenna gain $G_u - G'_u$	-3 dB
$(C/I)_u$	30 dB
Downlink	
EIRP of the interfered satellite B per carrier	7 dBW
EIRP of the interfering satellite A	34 dBW
Maximum interference level in 40-kHz bandwidth relative to interfering carrier power	27 dB
Receive antenna gain of the interfered earth station B_1	44.5 dB
$32 - 25 \log \theta$	17 dB
$(C/I)_d$	27.5 dB
Total carrier-to-interference ratio	25.6 dB

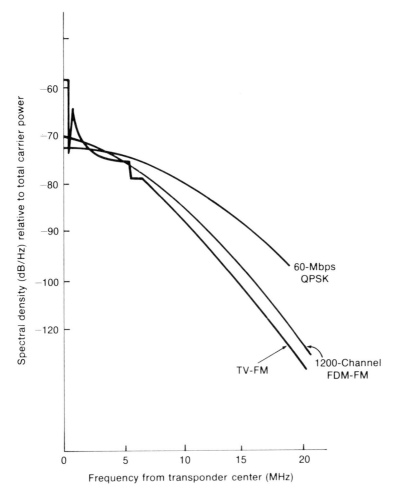

Figure 4.9 Power spectral densities of QPSK, FDM-FM, and TV-FM carriers.

multiple-carriers system B as shown in Table 4.4. As seen from Fig. 4.9 the power spectral density peaks at the carrier frequency; therefore it is necessary to place the multiple carriers outside the ±1-MHz bandwidth centered around the single-carrier frequency to avoid excessive interference, especially in the case of the FDM-FM carrier.

Finding the power spectral density of a modulated carrier is by no means an easy task. Fortunately, the power spectral density of a QPSK carrier can be represented (Appendix 9B) by a simple expression:

$$S(f) = CT_b \left\{ \left[\frac{\sin 2\pi (f - f_c) T_b}{2\pi (f - f_c) T_b} \right]^2 \right.$$

$$\left. + \left[\frac{\sin 2\pi (f + f_c) T_b}{2\pi (f + f_c) T_b} \right]^2 \right\} \tag{4.45}$$

where C = carrier power in an infinite bandwidth

T_b = bit duration = $1/R_b$, where R_b is the bit rate

f_c = carrier frequency (Hz)

It is seen that the one-sided power spectral density level at the carrier frequency is $2CT_b$ (W/Hz) or $10 \log 2CT_b$ (dBW/Hz) which is obtained from (4.45) by setting $f = f_c$ and $f = -f_c$. When the bit rate is 60 Mbps, this level is $10 \log 2T_b = -74.77$ dB/Hz relative to the carrier power C. In satellite communications the QPSK carrier is normally filtered such that its bandwidth satisfies the relation $T_b B = 0.6$, thus yielding $B = 0.6/T_b = 0.6R_b = 36$ MHz for $R_b = 60$ Mbps. The total carrier power within this 36-MHz bandwidth is about 67% of the carrier power C in an infinite bandwidth. Therefore the spectral density level at the carrier frequency relative to $0.67C$ is $10 \log(2T_b/0.67) = -73$ dB/Hz.

The spectral density of the FDM-FM carrier can be approximated by a Gaussian function as

$$S(f) = \frac{C}{2\sqrt{2\pi\sigma^2}} \left\{ \exp\left[\frac{-(f - f_c)^2}{2\sigma^2} \right] \right.$$

$$\left. + \exp\left[\frac{-(f + f_c)^2}{2\sigma^2} \right] \right\} \tag{4.46}$$

where C = carrier power in an infinite bandwidth

f_c = carrier frequency (Hz)

σ = rms multichannel deviation (Hz) (which is the product of the rms test-tone deviation and the multichannel loading factor or the ratio of the maximum frequency deviation of the baseband signal to the multichannel peak factor. Section 5.1 will address this problem)

For the example considered in Fig. 4.9, σ is taken to be 4 MHz, yielding a power spectral density level of -70 dB/Hz relative to the carrier power at the carrier frequency which is obtained from (4.46) by setting $f = f_c$ and $f = -f_c$. Unlike the QPSK carrier, most of the FDM-FM carrier power is captured within the 36-MHz bandwidth.

The power spectral density of the TV-FM carrier can also be approximated by a Gaussian function from $f_c + 1$ MHz to $f_c + 18$ MHz (and from $f_c - 1$ MHz to $f_c - 18$ MHz). The region ± 1 MHz around the carrier frequency can be approximated by a constant value equal to -59 dB/Hz relative to the carrier power. Also, like the FDM-FM carrier, most of the TV-FM carrier power is captured within the 36-MHz bandwidth.

4.2.3 Terrestrial Interference

The 6/4-GHz frequency bands allocated to satellite communications are also allocated to terrestrial microwave links. The terrestrial microwave networks in these bands have developed over the years into vast, complex route networks spanning the whole United States. In areas of heavy population density the terrestrial link congestion can be so great that it may be impossible to locate an earth station. Since the earth station receives in the 4-GHz band, it is susceptible to interference from terrestrial microwave transmission at 4 GHz. Also, the earth station transmits into the 6-GHz band and thus generates interference into terrestrial microwave reception at 6 GHz.

The mutual interference between an earth station and a terrestrial microwave system is a function of the carrier power, the carrier spectral density, and the frequency offset between the two carriers. The interfering power within the bandwidth of the satellite signal received by the earth station depends on the spectral density of the terrestrial interfering carrier. For a broadband satellite carrier whose bandwidth includes the interfering carrier frequency, the interfering carrier power is used. For a narrowband satellite carrier, the interfering carrier power is reduced by an interference reduction factor which is the ratio between the total carrier power and the power in the narrow bandwidth. Figure 4.10 shows the interference reduction factor for a narrowband carrier with a 40-kHz bandwidth as a function of the frequency offset from the center frequency of a FDM-FM terrestrial microwave carrier in the 4-GHz band.

Similarly frequency offset advantage is available for interference into terrestrial microwave systems from narrowband transmit earth stations. The amount of interference is determined by the frequency separation between the interfering carrier frequency and the terrestrial carrier frequency. The interference reduction factor can be obtained by convolving the spectral densities of the interfering carrier and the terrestrial carrier. Figure 4.11 shows the interference reduction factor into an FDM-FM terrestrial microwave carrier from a narrowband satellite carrier.

The interference power objective from terrestrial microwave carriers has been established at 25 dB below the satellite carrier power. For the multiple-carriers satellite system considered in Table 4.2, the received carrier power is −144.5 dBW (saturation EIRP per carrier minus transponder output back-off minus free space loss plus receive antenna gain). Therefore, in order to obtain a 25-dB carrier-to-terrestrial interference

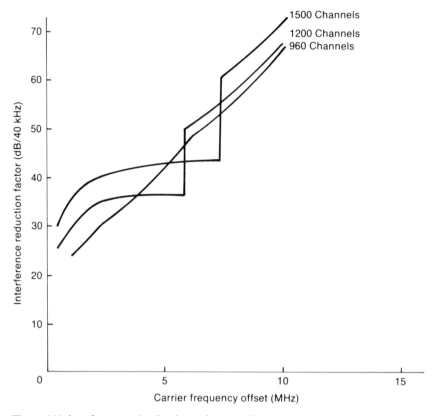

Figure 4.10 Interference reduction factor for a satellite carrier.

ratio, the maximum allowable interference power must be −169.5 dBW in the 40 kHz bandwidth of the satellite carrier.

The interference power objective into terrestrial microwave systems is −154 dBW/4 kHz and should not be exceeded for more than 20% of the time, and −131 dBW/4 kHz and should not be exceeded for more than 0.01% of the time.

4.2.4 Cross-polarization Interference

Frequency reuse satellite communications systems which employ orthogonal linear polarizations (vertical and horizontal linear polarizations) or orthogonal circular polarizations (left-hand and right-hand circular polarizations) encounter another major source of interference, namely, the coupling of energy from one polarization state to the other orthogonal polarization state. This results from the finite cross-polarization discrimination of the earth station and satellite antennas and by the depolarization caused by rain, which we will discuss in a later section. At the 6/4-GHz bands, the effect of rain is negligible, hence the cross-polarization interference is pretty much determined by the discrimination provided by the earth station and satellite antennas. The *cross-polarization discrimination* is defined as the ratio of the power received in the principal polarization to the power received in the orthogonal polarization from the same incident signal and thus represents the carrier-to-cross-polarization interference ratio when the two polarized signals have the same power. High-quality antennas can achieve 30 to 40 dB cross-polarization discrimination along the antenna axis. The net cross-polarization discrimination of a satellite link is the combined effect of the earth station and satellite antennas for both the uplink and the downlink. Let X_e and X_s be the cross-polarization discrimination of the earth station and satellite antennas, respectively. Then the minimum net link cross-polarization discrimination is given by

$$X_{min} = \tfrac{1}{2} (X_e^{-1} + X_s^{-1})^{-1} \tag{4.47}$$

For example, when $X_e = 33$ dB and $X_s = 33$ dB, $X_{min} = 27$ dB. The minimum net link cross-polarization discrimination represents the worst-case carrier-to-cross-polarization interference ratio $(C/I)_X = X_{min}$ and can be used in (4.29) as an additional interference source to obtain the total carrier-to-noise plus interference ratio.

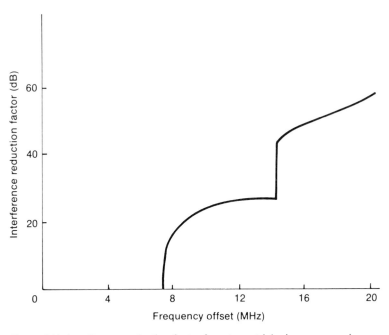

Figure 4.11 Interference reduction factor for a terrestrial microwave carrier.

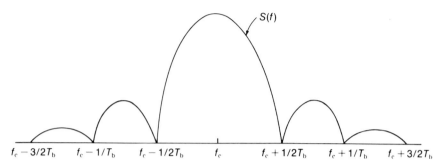

Figure 4.12 Power spectral density of a QPSK carrier.

4.2.5 Adjacent Channel Interference

Another source of interference in a satellite link is the adjacent channel interference (or adjacent transponder interference) which arises in band-limited satellite channels. For example, consider the power spectral density of a QPSK signal given in (4.45) and plotted in Fig. 4.12 (for positive frequencies only). It is seen that the main lobe, where most of the energy is concentrated, occupies a bandwidth of $B = 1/T_b$. However, in practice, the QPSK signal is normally filtered to a bandwidth $B = 0.6/T_b$, resulting in a band-limited channel as shown in Fig. 4.13. The interference arises when some of the energy of a band-limited signal falls into adjacent channels because of the overlapping amplitude characteristics of the channel filters, as depicted in Fig. 4.14. The situation is more severe in a single-carrier-per-transponder TDMA system using QPSK modulation where the earth station HPA and/or the satellite TWTA operates near or at saturation. The nonlinear characteristic of these power amplifiers (Fig. 4.2) regenerates the sidelobes of the filtered QPSK spectrum and thus causes interference into adjacent channels. The phenomenon is termed *spectrum spreading*, as shown in Fig. 4.15. The degree of spectrum spread-

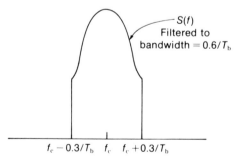

Figure 4.13 Band-limited power spectral density of a QPSK carrier.

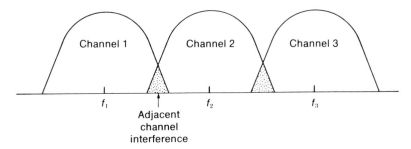

Figure 4.14 Concept of adjacent channel interference.

ing varies according to the operating point (back-off) of the power amplifier. The spectrum spreading of the earth station HPA determines the amount of uplink adjacent channel interference that can be reduced by the HPA back-off. The spectrum spreading of the satellite TWTA determines the amount of downlink adjacent channel interference that can be controlled by out-of-band rejection of the satellite output multiplexer (Fig. 2.13).

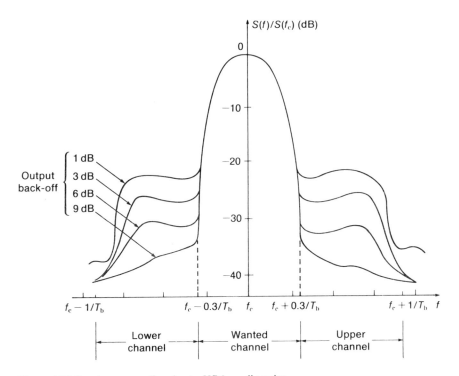

Figure 4.15 Spectrum spreading due to HPA nonlinearity.

The uplink carrier-to-adjacent channel interference ratio (for interference from two adjacent channels) is given by

$$
\left(\frac{C}{I}\right)_u = \frac{\displaystyle\int_{fc-B/2}^{fc+B/2} S_u(f)\, df}{\displaystyle 2\int_{fc-\Delta}^{fc+B/2-\Delta} S_u(f)\, df}
$$

where $S_u(f)$ = power spectral density of QPSK carrier at output of earth station HPA (W/Hz) and Δ = separation of adjacent carrier frequencies (Hz).

The downlink carrier-to-adjacent channel interference ratio (for interference from two adjacent channels) is given by

$$
\left(\frac{C}{I}\right)_d = \frac{\displaystyle\int_{fc-B/2}^{fc+B/2} S_d(f)\,|H(f)|^2\, df}{\displaystyle 2\int_{fc-\Delta}^{fc+B/2-\Delta} S_d(f)\,|H(f)|^2\, df}
$$

where $S_d(f)$ = power spectral density of QPSK carrier at output of satellite TWTA (W/Hz) and $H(f)$ = amplitude response of satellite output multiplexer.

These equations can be used in (4.29a) as additional interference sources to obtain the total carrier-to-noise plus interference ratio. As shown in Fig. 4.15, a HPA output back-off of 5 to 6 dB is necessary to achieve a carrier-to-adjacent channel interference (from two adjacent channels) of better than 25 dB.

4.2.6 Intermodulation Interference

This type of interference is caused by the intermodulation products generated within a satellite transponder as a result of the amplification of multiple carriers by the TWTA that exhibits both amplitude nonlinearity and phase nonlinearity. To avoid excessive intermodulation interference the TWTA must be operated with a large output back-off, and this could result in a downlink-limited system and reduce the transponder's capacity. This is why single-carrier-per-transponder TDMA systems have become increasingly popular because the satellite TWTA can be operated at or close to saturation to provide maximum EIRP for the downlink. Intermodulation interference once calculated can be used in (4.29a) as an additional interference source to obtain the total carrier-to-noise plus interference ratio. Intermodulation interference is an inherent problem of FDMA systems and will be studied in Chap. 5.

4.2.7 Intersymbol Interference

This type of interference does not come from outside sources such as those discussed previously but instead is generated within the channel itself, as a result of filtering and the nonlinear characteristic of the satellite TWTA normally operated close to saturation, especially in single-carrier-per-transponder satellite systems. In a linear channel where the bandwidth available for transmitting data at a rate of R bits per second is between $R/2$ and R hertz, intersymbol interference can be eliminated by the use of *Nyquist pulse shaping criteria*. This is not possible for nonlinear satellite channels where intersymbol interference cannot be eliminated and degradation of the carrier-to-noise ratio results. This degradation, in decibels, should be subtracted directly from the link carrier-to-noise plus interference ratio, also in decibels, in (4.29a) to obtain the receive carrier-to-noise plus interference ratio used to calculate the average probability of bit error for the satellite link. We will discuss this subject further in Chap. 9 where we study nonlinear satellite channels.

4.3 RAIN-INDUCED ATTENUATION

Besides the ever-present free space loss (Sec. 4.1) and the atmospheric absorption (Sec. 3.16), which is significant only in bands centered at frequencies of 22.2 GHz (water vapor), 60 GHz and 118.8 GHz (oxygen) as seen in Fig. 3.11, satellite communications above 10 GHz must deal with another type of attenuation caused by rain. Although rain is not a problem at the 6/4-GHz band, it is a major concern that strongly influences link design as satellite communications move into the 14/12- and 30/20-GHz bands to avoid orbital congestion. A reliable prediction of attenuation by rain is therefore necessary for a system designer to realistically determine link availability, establish the link margin, and provide means to combat rain effects.

4.3.1 Prediction of Attenuation

Although the prediction of rain attenuation is a *statistical process*, many models have been developed which yield results that agree well with experimental observations. The four main models are those of Rice-Holmberg [4], Dutton-Dougherty [5], Lin [6], and Crane [7]. The *Crane global model* will be discussed here because of its accuracy and the ease with which it can be used with a calculator. The model gives an estimate of a "*total time*," over a 1-year period, that attenuation by rain can be expected to exceed a given amount for a propagation path length or slant path. The total path attenuation that may be exceeded for P percent of the year is a function of the point rain rate distribution, the vertical extent of the rain, the raindrop size distribution, and the rain rate distribution along the slant path; and it is given by the following composite expressions:

$$L_r \text{ (dB)} = \frac{aR_p^b L}{D} \left[\frac{\exp(ubD) - 1}{ub} \right] \qquad 0 \le D \le d \tag{4.48a}$$

$$L_r \text{ (dB)} = \frac{aR_p^b L}{D} \left[\frac{\exp(ubd) - 1}{ub} - \frac{x^b \exp(vbd)}{vb} + \frac{x^b \exp(vbD)}{vb} \right]$$

$$d \le D \le 22.5 \text{ km} \tag{4.48b}$$

where

$$d = 3.8 - 0.6 \ln R_p \tag{4.48c}$$

$$x = 2.3 \, R_p^{-0.17} \tag{4.48d}$$

$$v = 0.026 - 0.03 \ln R_p \tag{4.48e}$$

$$u = \frac{\ln[x \exp(vd)]}{d} \tag{4.48f}$$

$$D = \frac{H - H_0}{\tan E} \qquad E \ge 10° \tag{4.48g}$$

$$D = (r_e + H_0) \, \psi \qquad E < 10° \tag{4.48h}$$

$$\psi = \sin^{-1} \left\{ \frac{\cos E}{r_e + H} \left[- (r_e + H_0) \sin E \right. \right.$$
$$\left. \left. + \sqrt{(r_e + H_0)^2 \sin^2 E + 2r_e(H - H_0) + H^2 - H_0^2} \right] \right\} \tag{4.48i}$$

$$L = \frac{D}{\cos E} \qquad E \ge 10° \tag{4.48j}$$

$$L = - (r_e + H_0) \sin E$$
$$+ \sqrt{(r_e + H_0)^2 \sin^2 E + 2r_e(H - H_0) + H^2 - H_0^2}$$
$$E < 10° \tag{4.48k}$$

The parameters given in (4.48) are defined as follows:

H_0 = height of earth station (km)
H = height of 0°C isotherm = vertical extent of rain (km)
E = earth station elevation angle
r_e = effective earth's radius = 8500 km
ψ = central angle (rad)
D = surface projected path length (km)
L = slant path (km)
R_p = point rain rate that may be exceeded for P percent of year (mm/h)

a,b = frequency-dependent coefficients based on raindrop characteristics

When $D > 22.5$ km, R_p is replaced by R_p', where R_p' is the point rain rate that may be exceeded for P' percent of the year, where $P'$is given by

$$P' = \frac{22.5}{D} P \qquad D > 22.5 \text{ km} \qquad (4.49)$$

The frequency-dependent coefficients a and b are given in Table 4.5 [8] and can be approximated by the following analytical expressions, where f is in gigahertz.

$$a = 4.21 \times 10^{-5}f^{2.42} \qquad 2.9 \leq f \leq 54 \text{ GHz}$$

$$a = 4.09 \times 10^{-2}f^{0.699} \qquad 54 \leq f \leq 180 \text{ GHz}$$

$$b = 1.41f^{-0.0779} \qquad 8.5 \leq f \leq 25 \text{ GHz}$$

$$b = 2.63f^{-0.272} \qquad 25 \leq f \leq 164 \text{ GHz} \qquad (4.50)$$

The point rain rate R_p depends on the rain climate regions. The Crane global model provides eight rain climate regions A through H covering the whole globe, as shown in Fig. 4.16. Figure 4.17 presents expanded regions for the United States, where region D has further been subdivided into regions D_1, D_2, and D_3. The value of R_p may be obtained from the point rain rate distribution shown in Table 4.6.

The 0°C isotherm height H, which is the vertical extent of rain, is a function of latitude and probability of occurrence given in percent of year.

Table 4.5 Values of coefficients a and b for rain attenuation

Frequency (GHz)	a		b	
	$R_p \leq 30$ mm/h	$R_p > 30$ mm/h	$R_p \leq 30$ mm/h	$R_p > 30$ mm/h
10	0.0117	0.0114	1.178	1.189
11	0.0150	0.0152	1.171	1.167
12	0.0186	0.0196	1.162	1.150
15	0.0321	0.0347	1.142	1.119
20	0.0626	0.0709	1.119	1.083
25	0.105	0.132	1.094	1.029
30	0.162	0.226	1.061	0.964
35	0.232	0.345	1.022	0.907
40	0.313	0.467	0.981	0.864
50	0.489	0.669	0.907	0.815
60	0.658	0.796	0.850	0.794
70	0.801	0.869	0.809	0.784
80	0.924	0.913	0.778	0.780
90	1.020	0.945	0.756	0.776
100	1.080	0.966	0.742	0.774

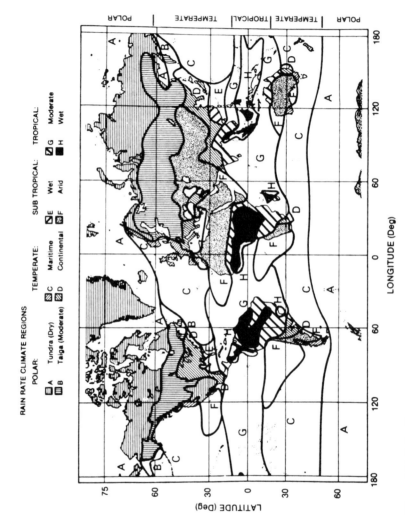

Figure 4.16 Rain climate regions in a Crane global model.

Table 4.6 Point rain rate distribution values (mm/h) versus percent of year rain rate is exceeded

Percent of year P%	Rain climate region												Minutes per year	Hours per year
	A	B₁	B	B₂	C	D₁	D = D₂	D₃	E	F	G	H		
0.001	28.5	45	57.5	70	78	90	108	126	165	66	185	253	5.26	0.09
0.002	21	34	44	54	62	72	89	106	144	51	157	220.5	10.5	0.18
0.005	13.5	22	28.5	35	41	50	64.5	80.5	118	34	120.5	178	26.3	0.44
0.01	10.0	15.5	19.5	23.5	28	35.5	49	63	98	23	94	147	52.6	0.88
0.02	7.0	11.0	13.5	16	18	24	35	48	78	15	72	119	105	1.75
0.05	4.0	6.4	8.0	9.5	11	14.5	22	32	52	8.3	47	86.5	263	4.38
0.1	2.5	4.2	5.2	6.1	7.2	9.8	14.5	22	35	5.2	32	64	526	8.77
0.2	1.5	2.8	3.4	4.0	4.8	6.4	9.5	14.5	21	3.1	21.8	43.5	1,052	17.5
0.5	0.7	1.5	1.9	2.3	2.7	3.6	5.2	7.8	10.6	1.4	12.2	22.5	2,630	43.8
1.0	0.4	1.0	1.3	1.5	1.8	2.2	3.0	4.7	6.0	0.7	8.0	12.0	5,260	87.7
2.0	0.1	0.5	0.7	0.8	1.1	1.2	1.5	1.9	2.9	0.2	5.0	5.2	10,520	175
5.0	0.0	0.2	0.3	0.3	0.5	0.0	0.0	0.0	0.0	0.0	1.8	1.2	26,298	438

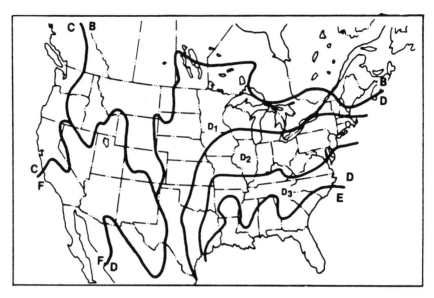

Figure 4.17 Rain climate regions in the United States.

The value of H is given in Fig. 4.18. The relationship between H, the earth station height H_0, the slant path L, the surface projected path length D, and the elevation angle is depicted in Fig. 4.19 for $E \geq 10°$ and in Fig.

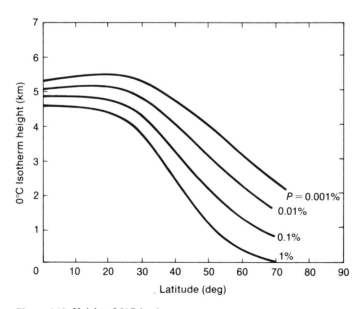

Figure 4.18 Height of 0°C isotherm.

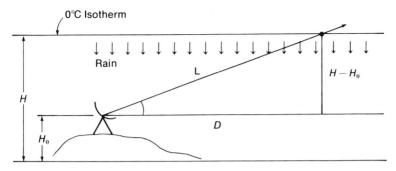

Figure 4.19 Surface projected path length D for $E \geq 10°$.

4.20 for $E < 10°$. From Fig. 4.19 it is seen that

$$L = \frac{D}{\cos E} \qquad E \geq 10°$$

For low elevation angles $E < 10°$, the curvature of the earth surface is taken into account. From Fig. 4.20 we have

$$\frac{L}{\sin \psi} = \frac{r_e + H}{\sin (90° + E)} = \frac{r_e + H}{\cos E}$$

which yields

$$\psi = \sin^{-1} \left(\frac{L \cos E}{r_e + H} \right) \qquad (4.51)$$

Also,

$$L^2 + (r_e + H_0)^2 - 2(r_e + H_0) L \cos (E + 90°) = (r_e + H)^2$$

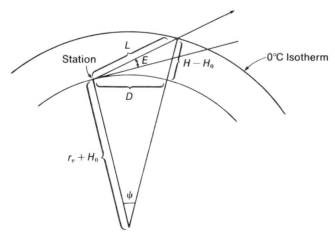

Figure 4.20 Surface projected path length D for $E < 10°$.

which yields (4.48k). Substituting (4.48k) into (4.51) yields (4.48i). Once the path attenuation L_r(dB) that may be exceeded for P percent of the year is computed for the uplink or downlink, it can be subtracted directly from the uplink or downlink carrier-to-noise ratio in (4.24) or (4.25).

Example 4.5 In this example the downlink attenuation by rain is calculated for an earth station with the following parameters.

Latitude: 35°N
Longitude: 83°W
Height above sea level: $H_0 = 0.9$ km
Antenna elevation angle: $E = 47°$
Downlink frequency: $f = 20$ GHz

1. Select the rain climate region: D_3.
2. Select the surface point rain rate.

P%	R_p (mm/h)
0.01	63
0.02	48
0.05	32
0.10	22
0.20	14.5
0.50	7.8
1.00	4.7

3. Determine the 0°C isotherm height; the height varies correspondingly with the probability of occurrence P. To interpolate, plot H versus log P and use a straight line to relate H to P.

P%	H (km)
0.01	4.40
0.02	4.20
0.05	3.95
0.10	3.75
0.20	3.55
0.50	3.30
1.00	3.20

4. Compute the surface projected path length from (4.48g).

P%	D (km)
0.01	3.26
0.02	3.08
0.05	2.84
0.10	2.66
0.20	2.47
0.50	2.24
1.00	2.14

5. Select the coefficients a and b from (4.50) or Table 4.5. For $0.01 \leq P\% \leq 0.05$, $R_p > 30$ mm/h, so $a = 0.0709$ and $b = 1.083$. For $0.10 \leq P\% \leq 1.00$, $R_p < 30$ mm/h, so $a = 0.0626$ and $b = 1.119$.

6. Compute the constants d, x, v, and u in (4.48c) to (4.48f).

$P\%$	d	x	v	u
0.01	1.314	1.137	−0.098	−0.00029
0.02	1.477	1.191	−0.090	0.028
0.05	1.721	1.276	−0.078	0.064
0.10	1.945	1.360	−0.067	0.091
0.20	2.196	1.460	−0.054	0.118
0.50	2.568	1.622	−0.036	0.152
1.00	2.871	1.768	−0.020	0.178

7. Compute the attenuation from (4.48). For $0.01 \leq P\% \leq 0.2$, $D > d$, so (4.48b) is used. For $0.5 \leq P\% \leq 1.00$, $D < d$ and (4.48a) is used. Note that L_r in (4.48) is in decibels.

$P\%$	L_r (dB)
0.01	28.4
0.02	21.1
0.05	13.4
0.10	8.8
0.20	5.3
0.50	2.5
1.00	1.4

4.3.2 Effect of Rain Attenuation on System Noise Temperature

Besides the attenuation that directly reduces the signal power, rain also increases the sky noise temperature significantly. Since the antenna noise temperature is a function of the sky noise temperature, rain in effect increases the system noise temperature of the earth station. Let T_r denote the rain temperature and L_r the rain attenuation. The noise power available over a bandwidth B is simply kT_rB. The noise power after passing through rain with an attenuation factor of L_r is kT_rB/L_r. Thus the amount of power absorbed by the rain is $kT_rB(1 - 1/L_r)$. The increase in noise temperature caused by the attenuation factor L_r is therefore

$$\Delta T = T_r \left(1 - \frac{1}{L_r}\right) \tag{4.52}$$

When the attenuation L_r is high, ΔT is nearly equal to the rain temperature T_r. In practice T_r is usually taken to be 273 K. The increase in noise temperature due to rain is added directly to the earth station system noise temperature and further reduces the downlink carrier-to-noise ratio. Note that the increase in noise temperature due to rain does not affect the system noise temperature of the satellite because its antenna always looks at a hot earth at 290 K. Once ΔT in (4.52) is determined, it can be added to the system noise temperature T of the earth station in (4.25).

Example 4.6 For the earth station considered in Example 4.5 the increase in sky noise temperature due to rain is given in the following table, where $L_r = \log^{-1} [L_r$ (dB)/10].

$P\%$	L_r (dB)	L_r	ΔT (K)
0.01	28.4	691.8	273
0.02	21.1	128.8	271
0.05	13.4	21.9	261
0.10	8.8	7.6	237
0.20	5.3	3.4	193
0.50	2.5	1.8	121
1.00	1.4	1.4	78

4.3.3 Carrier-to-Noise plus Interference Ratio Including Rain-Induced Attenuation

The effect of rain-induced attenuation on the uplink and downlink carrier-to-noise plus interference ratios is summarized in this section. Once the total path attenuation factor L_r that may be exceeded for P percent of the year is determined, the uplink and downlink carrier-to-noise ratios that may not be achieved for P percent of the year is given by

$$\left(\frac{C}{N}\right)_{u,r} = \text{EIRP}_{\text{sat}} \text{ (dBW)} - 20 \log \left(\frac{4\pi f_u d_u}{c}\right)$$

$$+ \frac{G_u}{T_u} \text{ (dB/K)} - 10 \log k - 10 \log B - BO_i \text{ (dB)}$$

$$- L \text{ (dB)} - L_{r,u} \text{ (dB)} \tag{4.53}$$

$$\left(\frac{C}{N}\right)_{d,r} = \text{EIRP}_{\text{s,sat}} \text{ (dBW)} - 20 \log \left(\frac{4\pi f_d d_d}{c}\right)$$

$$+ \frac{G}{T + 273 \{1 - 1/\log^{-1} [L_{r,d} \text{ (dB)}/10]\}} \text{ (dB/K)}$$

$$- 10 \log k - 10 \log B - BO_o^* \text{ (dB)} - L' \text{ (dB)} - L_{r,d} \text{ (dB)}$$

$$\tag{4.54}$$

where $BO^* =$ output back-off associated with uplink attenuation $L_{r,u}$;
$BO_o^* = f(BO_i + L_{r,u})$ and is a nonlinear function of $BO_i + L_{r,u}$

$BO_i =$ clear-sky input back-off (dB)

$L_{r,u} =$ uplink rain-induced attenuation (dB)

$L_{r,d} =$ downlink rain-induced attenuation (dB)

The uplink carrier-to-interference ratio (dB) is also reduced by the amount equal to the uplink attenuation $L_{r,u}$ (dB) if we assume that rain does not occur at the interfering sources (worst case). Therefore

$$\left(\frac{C}{I}\right)_{u,r} = \left(\frac{C}{I}\right)_u - L_{r,u} \tag{4.55}$$

where $(C/I)_u =$ clear-sky carrier-to-interference ratio. Assume that rain does not occur simultaneously on both the uplink and downlink of a satellite link between earth stations A and B. Then the downlink carrier-to-interference ratio is reduced by $BO_o^* - BO_o$, where BO_o (dB) is the clear-sky output back-off:

$$\left(\frac{C}{I}\right)_d^* = \left(\frac{C}{I}\right)_d - (BO_o^* - BO_o) \tag{4.56}$$

When rain occurs on the downlink only, the carrier and interference are attenuated equally; therefore the downlink carrier-to-interference ratio remains at the clear-sky value:

$$\left(\frac{C}{I}\right)_{d,r} = \left(\frac{C}{I}\right)_d \tag{4.57}$$

where $(C/I)_d =$ clear-sky carrier-to-interference ratio.

4.3.4 Path Diversity

As seen from Example 4.5, rain-induced attenuation at 20 GHz can degrade the satellite link significantly. Fades in excess of 10 dB can be encountered for many hours per year. In overcoming this communication outage problem, one solution is to use path diversity (or site diversity) involving two earth stations located 5 to 30 km apart. This is based on experiments indicating that intense rain that causes severe attenuation is limited spatially and thus severe rain-induced attenuation becomes uncorrelated at separated earth stations and the probability of outage at a single site is reduced to the joint probability of outage. The path diversity gain may be defined as the difference between the attenuation, in decibels, exceeded on a single path and that exceeded jointly on two separated paths for a given percent of the time. For single-site attenuation below 15 dB, an empirical expression for path diversity gain is given in [9] as

$$G_D \text{ (dB)} = a[1 - \exp(-bs)] \tag{4.58a}$$

$$a = L_r - 3.6[1 - \exp(-0.24L_r)] \tag{4.58b}$$

$$b = 0.46[1 - \exp(-0.26L_r)] \tag{4.58c}$$

where s = separation distance between two stations (km) and L_r = single-site rain attenuation (dB).

For example, let $L_r = 14$ dB and $s = 10$ km; then $a = 10.53$, $b = 0.448$, and $G_D = 10.4$ dB. The path diversity gain can be subtracted directly from the attenuation by rain in the link calculation given in (4.53) or (4.54), depending on whether path diversity is applied on the uplink or the downlink. Here we assume that the connecting link between two stations is not affected by rain; for example, the two stations can be connected via microwave terrestrial links operated in the 6/4-GHz band where rain is not a problem.

4.3.5 Uplink Power Control

Although path diversity is effective against rain-induced attenuation, it is not cost-effective because investment in an extra earth station is required. For TDMA earth stations operating with one carrier in a saturated or near-saturated satellite transponder, uplink power control seems to be attractive. Uplink power control is achieved by varying the earth station transmit power to keep the satellite flux density from falling below a certain level as a result of rain-induced attenuation on the uplink to maintain a specified performance. To do so, the earth station needs to know the uplink attenuation fairly accurately so that the power flux density at the satellite will not be exceeded. Indeed, it is normally desired to maintain the power flux density at the satellite within -2 to -4 dB of the clear-sky value to the extent that it is possible with the total power control range available. The uplink attenuation can be calculated from the downlink attenuation obtained from the signal strength of the satellite beacon signal, for example, and from the estimated uplink-to-downlink attenuation ratio which can be found from the Crane global model. Many measurements at a number of frequencies have shown that there is a good long-term correlation between the attenuations measured at two frequencies along the same path. As the averaging time for comparing the attenuations is reduced, however, the variation in the correlation ratio increases. Averaging times of 10 s or less show large fluctuations in the correlation ratio, particularly at low attenuation levels. The theoretical cumulative maximum error in estimating the uplink attenuation is given in Table 4.7 and has also been confirmed by experiments [10]. The cumulative maximum errors shown are conservative. Many errors which are too small to es-

Table 4.7 Errors in uplink attenuation when it is estimated from downlink attenuation and attenuation ratio

Rain attenuation at 14 GHz (dB)	Errors (dB)		
	Elevation angle		
	5°	15°	25°
$L_r < 1$	±1.1	±1	±0.9
$1 \leq L_r \leq 6$	±1.5	±1.3	±1.1
$L_r > 6$	±1.8	±1.6	±1.3

timate, such as antenna tracking errors, spacecraft nutation and pointing errors, antenna gain degradations, and refractive effects below a 10° elevation angle, have not been included in the above-mentioned errors. Also, differential attenuation across the 500-MHz satellite bandwidth can vary over 1 dB in a 10-dB downlink attenuation at 12 GHz. Open-loop uplink power control at the lower band edge will therefore suffer an additional potential ±0.5-dB error if a satellite beacon signal in the upper band edge is used as the level indicator in changing the uplink power setting. This error is over and above the error cited previously. The overall error in estimating uplink attenuation from downlink attenuation and the attenuation ratio can be on the order of ±2 dB.

Another major problem in using uplink power control for single-carrier-per-transponder TDMA systems is the adjacent channel interference due to spectrum spreading of the carrier as it passes through the nonlinear high-power amplifier of the earth station, as shown in Fig. 4.15. In a TDMA system the earth station HPA normally operates at a 5- to 6-dB output back-off to control the adjacent channel interference. Roughly speaking, a 1-dB decrease in the HPA output back-off induces about a 2-dB increase in the carrier-to-adjacent channel interference. Therefore, if the HPA output back-off is decreased to compensate for uplink attenuation, the decrease should be half the attenuation to the extent of the output back-off range, taking into account the error cited above. For example, let BO_0 (dB) be the HPA output back-off under clear-sky conditions. Then, for an uplink attenuation $L_{r,u}$ (dB), the HPA output back-off setting should be

$$BO_{o,r} \text{ (dB)} = BO_0 \text{ (dB)} - \tfrac{1}{2}[L_{r,u} \text{ (dB)} - 2] \qquad BO_{o,r} \geq 0 \quad \text{dB}$$

$$(4.59)$$

where the error in uplink attenuation estimation is taken to be ±2 dB.

If we assume that the the satellite beacon signal is available for

Table 4.8 Typical input-output back-off relationship of earth station HPA

Input back-off (dB)	Output back-off (dB)
20	14.5
15	9.5
12	7
10	5
8	3.5
7	2.8
6	2
5	1.3
4	0.8
3	0.5
2	0.2
1	0.05
0	0

measuring the downlink attenuation, the following uplink power control procedure can be used:

1. Measure the satellite beacon signal every 200 ms and compare each measurement with the reference level.
2. If the value changes by 1 dB or more, calculate the new uplink attenuation $L_{r,u}$ (dB) from the measured downlink attenuation and the attenuation ratio (from a rain model or from actual measured data if available).
3. Find the HPA output back-off according to (4.59).
4. Set the HPA input back-off to achieve the calculated output back-off in step 3 according to the typical HPA input-output back-off relationship shown in Table 4.8 (the normalized power transfer characteristic). Figure 4.21 is a block diagram of an uplink power control system.

4.4 RAIN-INDUCED CROSS-POLARIZATION INTERFERENCE

Frequency reuse with orthogonal polarizations is often employed in modern satellites to double the capacity. However, the isolation between orthogonal polarizations is always degraded as a result of the static depolarization effect of satellite and earth station antennas and the time-varying depolarization effect of the propagation medium such as rain.

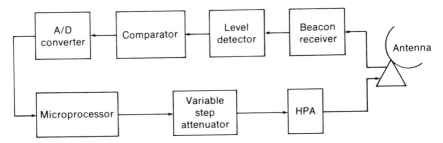

Figure 4.21 Schematic diagram of an uplink power control system.

These effects cause a portion of the signal energy transmitted in one polarization to be transferred to the signal in the orthogonal polarization, causing cross-polarization interference between two satellite channels. Cross-polarization interference caused by the depolarization effect of satellite and earth station antennas was considered in Sec. 4.2.4. In this section we study the cross-polarization interference induced by rain.

Raindrops falling through the atmosphere take on an oblate shape as a result of the effects of air resistance. A linearly polarized wave polarized parallel to an axis of symmetry of a raindrop has its amplitude attenuated and its phase changed but retains its polarization state. When the same linearly polarized wave is incident at an angle, each symmetry axis of the raindrop produces a different amplitude attenuation and a different phase change. The differential attenuation and phase shift alter the polarization state of the wave. Let the tilt angle of the oblate raindrop with respect to the local horizontal axis be ϕ (referred to as the *canting angle*) and let two orthogonal linearly polarized waves E_H and E_V be aligned in the horizontal and vertical axis as shown in Fig. 4.22. If $\phi = 0°$, that is, the symmetry axes of the raindrops are parallel to the horizontal and vertical axes, the

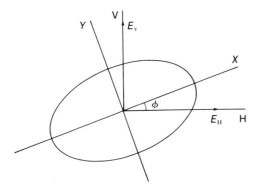

Figure 4.22 Oblate raindrop.

polarization states of the two waves E_H and E_V are unchanged after passing through the raindrops and can be represented by

$$\begin{bmatrix} E'_H \\ E'_V \end{bmatrix} = \begin{bmatrix} T_X & 0 \\ 0 & T_Y \end{bmatrix} \begin{bmatrix} E_H \\ E_V \end{bmatrix} \tag{4.60a}$$

where T_X, T_Y = transmission coefficients over an effective rainfall path L_e (km) [see (4.65d)] and E'_H, E'_V = attenuated waves after passing through raindrops. These transmission coefficients are

$$T_X = \exp\left[-(A_X - j\psi_X)L_e\right] \tag{4.60b}$$

$$T_Y = \exp\left[-(A_Y - j\psi_Y)L_e\right] \tag{4.60c}$$

where A_X, A_Y = attenuation coefficients (nepers/km) and ψ_X, ψ_Y = phase shift coefficients (rad/km) in the major and minor axes (of the raindrop) directions, respectively. For $\phi \neq 0°$, the waves E'_H and E'_V can be obtained from E_H and E_V by a coordinate transformation:

$$\begin{bmatrix} E'_H \\ E'_V \end{bmatrix} = \begin{bmatrix} \cos\phi & -\sin\phi \\ \sin\phi & \cos\phi \end{bmatrix} \begin{bmatrix} T_X & 0 \\ 0 & T_Y \end{bmatrix} \begin{bmatrix} \cos\phi & \sin\phi \\ -\sin\phi & \cos\phi \end{bmatrix} \begin{bmatrix} E_H \\ E_V \end{bmatrix} \tag{4.61}$$

or, equivalently,

$$E'_H = (T_X \cos^2\phi + T_Y \sin^2\phi)E_H + [(T_X - T_Y)\sin\phi\cos\phi]E_V$$

$$E'_V = [(T_X - T_Y)\sin\phi\cos\phi]E_H + (T_X \sin^2\phi + T_Y \cos^2\phi)E_V \tag{4.62}$$

The cross-polarization discrimination is defined as the ratio of the power received in the principal polarization to the power received in the orthogonal polarization from the same signal and thus is the carrier-to-cross-polarization interference ratio when the two waves E_H and E_V have the same power. Let X_H and X_V denote the horizontal and vertical cross-polarization discrimination in decibels, respectively; then

$$X_H = 10 \log \frac{|(T_X \cos^2\phi + T_Y \sin^2\phi)E_H|^2}{|[(T_X - T_Y)\sin\phi\cos\phi]E_H|^2}$$

$$= 20 \log \frac{|(T_X/T_Y) + \tan^2\phi|}{|(T_X/T_Y - 1)\tan\phi|} \tag{4.63}$$

$$X_V = 10 \log \frac{|(T_X \sin^2\phi + T_Y \cos^2\phi)E_V|^2}{|[(T_X - T_Y)\sin\phi\cos\phi]E_V|^2}$$

$$= 20 \log \frac{|(T_X/T_Y)\tan^2\phi + 1|}{|(T_X/T_Y - 1)\tan\phi|} \tag{4.64}$$

where

$$\frac{T_X}{T_Y} = \exp\{-[(A_X - A_Y) - j(\psi_X - \psi_Y)]L_e\}$$

$$= \exp[-(\Delta A - j\Delta\psi)L_e] \tag{4.65a}$$

$$\Delta A = A_X - A_Y \tag{4.65b}$$

$$\Delta\psi = \psi_X - \psi_Y \tag{4.65c}$$

$$L_e = \frac{L_r\,(\mathrm{dB})}{aR_p^b}$$

$$= \begin{cases} \dfrac{L}{D}\left[\dfrac{\exp(ubD) - 1}{ub}\right] & 0 \le D \le d \\[3mm] \dfrac{L}{D}\left[\dfrac{\exp(ubd) - 1}{ub} - \dfrac{x^b\exp(vbd)}{vb} + \dfrac{x^b\exp(vbD)}{vb}\right] & \\[2mm] d \le D \le 22.5\ \mathrm{km} \end{cases} \tag{4.65d}$$

ΔA = differential attenuation (nepers/km) and is commonly given in dB/km ($=8.686$ nepers/km)

$\Delta\psi$ = differential phase shift (rad/km) and is commonly given in deg/km $[=(180/\pi)$ rad/km$]$

L_e = effective path length (km), which is defined as the ratio of the total path attenuation L_r (dB) to the rain attenuation per unit length aR_p^b (dB/km); for the Crane global model, L_e in (4.65d) is calculated from (4.48a) and (4.48b)

When linearly polarized waves are not aligned in the local horizontal and vertical directions (only earth stations at the subsatellite point have horizontal and vertical polarizations like those of the satellite), the rain canting angle ϕ is replaced by the rain canting angle τ with respect to the polarization direction, where $\tau = \phi - \theta$ and θ is the polarization tilt angle with respect to the local horizontal direction. For example, if the polarization tilt angle $\theta = 15°$ and the rain canting angle $\phi = 25°$, then $\tau = 10°$. The horizontal and vertical cross-polarization discriminations are now called quasi-horizontal and quasi-vertical cross-polarization discriminations X_{QH} and X_{QV} and can be obtained from (4.63) and (4.64) with ϕ replaced by τ.

In the case of orthogonal circular polarizations, note that a circularly polarized wave is two linearly polarized waves superimposed and 90° out-of-phase. Let E_R and E_L denote the right-hand and left-hand circularly polarized waves, respectively. Then

$$E_R = E_H - jE_V \tag{4.66}$$

$$E_L = E_H + jE_V \tag{4.67}$$

Substituting (4.66) and (4.67) into (4.61) yields the waves E'_R and E'_L which are the versions of E_R and E_L, respectively, obtained after propagation through the raindrops:

$$E'_R = (T_X + T_Y)E_R + [(T_X - T_Y)\exp(-j2\phi)]E_L \qquad (4.68)$$

$$E'_L = [(T_X - T_Y)\exp(j2\phi)]E_R + (T_X + T_Y)E_L \qquad (4.69)$$

The right- and left-hand cross-polarization discriminations are all the same and are given by

$$X_C = 20\log\frac{|T_X + T_Y|}{|T_X - T_Y|}$$

$$= 20\log\frac{|T_X/T_Y + 1|}{|T_X/T_Y - 1|} \qquad (4.70)$$

When the waves E_R and E_L are of the same power, the circular cross-polarization X_C represents the carrier-to-cross-polarization interference ratio due to rain.

In practice, raindrops are not equally oriented and the canting angle ϕ tends to distribute around a mean value. Wave components scattered by oppositely canted raindrops tend to cancel out, thus reducing the depolarization effect. Plots of *differential attenuation* ΔA (dB/km) and *differential phase shift* $\Delta\psi$ (deg/km) are presented in Figs. 4.23 and 4.24, respectively [11] for the case where the angle α between the direction of propagation and the raindrop symmetry Y axis is 90°. For an earth-satellite path, the following approximation is quite accurate:

$$\Delta A = (\sin^2\alpha)\ \Delta A|_{\alpha=90°}$$

$$= (\cos^2 E)\ \Delta A|_{\alpha=90°}\ \text{(dB/km)} \qquad (4.71)$$

$$\Delta\psi = (\sin^2\alpha)\ \Delta\psi|_{\alpha=90°}$$

$$= (\cos^2 E)\ \Delta\psi|_{\alpha=90°}\ \text{(deg/km)} \qquad (4.72)$$

where E = elevation angle of earth station antenna.

In the calculation of cross-polarization discriminations for orthogonal linear polarizations, the canting angle can be assumed to have a mean value of 25° [11].

Example 4.7 Consider the earth station in Example 4.5. We wish to find the cross-polarization discriminations for both orthogonal linear and circular polarizations that may not be exceeded for $P = 0.1\%$ of the year.

For $P = 0.1\%$ the surface point rain rate is $R_p = 22$ mm/hr. At $f = 20$ GHz, this surface point rain rate yields a differential attenuation ΔA and a differential phase $\Delta\psi$ at $\alpha = 90°$ (or elevation angle $E = 0°$) as follows (Figs. 4.23 and 4.24):

$$\Delta A|_{\alpha=90°} = 0.45\ \text{dB/km}$$

$$\Delta\psi|_{\alpha=90°} = 4°/\text{km}$$

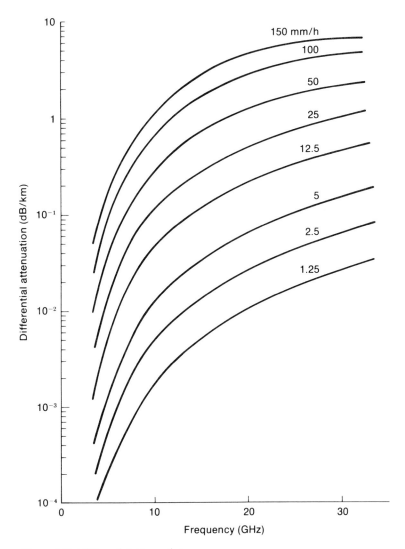

Figure 4.23 Differential attenuation.

The corresponding differential attenuation and differential phase at elevation angle $E = 47°$ are

$$\Delta A = (\cos^2 47°) \, \Delta A|_{\alpha = 90°} = 0.21 \text{ dB/km} = 0.024 \text{ neper/km}$$

$$\Delta \psi = (\cos^2 47°) \, \Delta \psi|_{\alpha = 90°} = 1.86°/\text{km} = 0.032 \text{ rad/km}$$

The effective path length L_e can be evaluated from (4.65d) using the data in Example 4.5, with $L_r = 8.8$ dB:

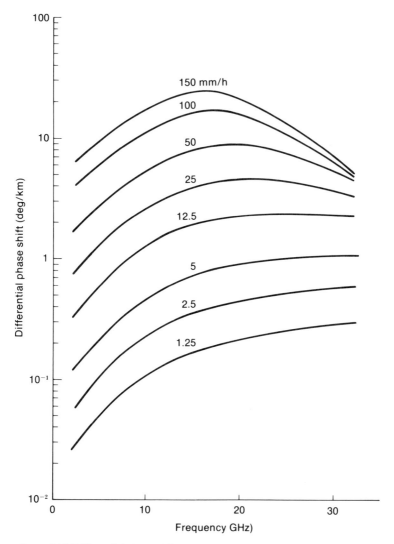

Figure 4.24 Differential phase shift.

$$L_e = \frac{L_r \text{ (dB)}}{aR_p^b} = \frac{8.8}{(0.0626)22^{1.119}} = 4.42 \text{ km}$$

From (4.65a), the ratio T_X/T_Y is given as

$$\frac{T_X}{T_Y} = \exp\left[-(0.024 - j0.032)\,4.42\right] = \exp\left(-0.106 + j0.141\right)$$

$$= 0.89 + j0.126$$

(a) For orthogonal linear polarizations with a rain canting angle of $\phi = 25°$:

$$X_H = 20 \log \frac{|0.89 + j0.126 + \tan^2 25°|}{|(0.89 + j0.126 - 1) \tan 25°|} = 23.1 \text{ dB}$$

$$X_V = 20 \log \frac{|(0.89 + j0.126) \tan^2 25° + 1|}{|(0.89 + j0.126 - 1) \tan 25°|} = 23.7 \text{ dB}$$

(b) For orthogonal circular polarizations:

$$X_C = 20 \log \frac{|0.89 + j0.126 + 1|}{|0.89 + j0.126 - 1|} = 21.1 \text{ dB}$$

It is seen that linear polarizations perform better than circular polarizations at the same attenuation level (8.8 dB in this case).

4.5 SYSTEM AVAILABILITY

From the results in Sec. 4.3 and 4.4 it is seen that rain-induced attenuation and cross-polarization interference can significantly reduce the link carrier-to-noise ratio and the link carrier-to-interference ratio. For a digital satellite link, this means an increase in the average probability of bit error. The rain effect is expressed in terms of the attenuation that is exceeded for P percent of the year and/or the cross-polarization discrimination that is not exceeded for P percent of the year. In link design, the system engineer often faces the problem of determining this P percent of the year (called *outage*) where the average probability of bit error exceeds a specified value P_b (called the P_b *threshold*). Since a satellite link involves the uplink (transmit earth station to satellite) and the downlink (satellite to receive earth station), the P_b threshold outage is the sum of the uplink P_b threshold outage (rain on uplink only), the downlink P_b threshold outage (rain on downlink only), and the joint P_b threshold outages (rain on both uplink and downlink). In practice the joint outages are normally not available, especially for mobile terminals, but when the sum of the uplink outage and the downlink outage is small, say less than 1%, then the joint outage is considered to be very small and can be neglected. This case pertains to large and fixed earth stations in the 14/12-GHz band. For small and mobile earth stations in the 30/20-GHz band this may not be the case, and the sum of the uplink outage and the downlink outage represents the lower bound for the link outage. It is up to the system designer to allocate an extra margin in the link design for such a situation. There is no rule for assigning the joint outages, but in general joint outages are more when the transmit and receive earth stations are in the same rainy climate regions and less when they are not. In any case it is believed that the joint outages are several order of magnitude less than the sum of the uplink and downlink outages.

If P in percent of the year is the link outage with respect to the P_b

threshold, the average probability of availability for the link, or *link availability,* is

$$P_{AL} = 1 - \frac{P}{100} \tag{4.73}$$

The *system availability* is usually defined as the availability of the link from the transmit earth station up through the satellite and down to the receive earth station. With the assumption that an earth station failure, a satellite failure, and a link outage occur mutually independently, the system availability is

$$P_A = P_{A1} \, P_{A2} \, P_{AL} \, P_{AS} \tag{4.74}$$

where P_{A1} = availability of the transmit earth station's transmit side, as expressed by (3.50b)

P_{A2} = availability of the receive earth station's receive side as expressed by (3.50c)

P_{AL} = link availability

P_{AS} = satellite availability

Since P_{AL} is normally expressed in percent of the year it is also customary to express P_A in percent of the year. This means that P_A is percent of the year the end-to-end performance is expected to be above or equal to the P_b threshold.

4.6 SATELLITE LINK DESIGN

From the discussions in Secs. 4.1 to 4.4 it is seen that the design of a satellite link, meaning the establishment of a carrier-to-noise plus interference ratio to meet specified performance criteria, is a complex task. The performance criteria may vary with systems and applications. For example, it is required that the average probability of bit error exceeds 10^{-7} for 0.1% of the year and 10^{-3} for 0.01% of the year at most. The system engineer hence must determine the earth station EIRP and G/T (i.e., the antenna gain and noise temperature, the output power of the HPA, and the noise temperature of the LNA) within specified cost constraints to achieve the desired performance. If it is assumed that the type of multiple access and carrier modulation have been selected, the remaining factors that influence the link design are as follows:

1. Adjacent satellite interference that can be minimized only by using a low sidelobe antenna, as seen in (4.42) and (4.43)
2. Terrestrial interference that can be minimized by site selection and earth station shielding or by using a higher frequency band, say 14/12 GHz.

3. Cross-polarization interference caused by the satellite and earth station dual-polarization antennas that can be minimized by good antenna design

4. Adjacent channel interference that can be minimized by operating the HPA with appropriate output back-off and with appropriate filtering by the satellite output multiplexer and the demodulator filter

5. Intermodulation interference that can be minimized by appropriate transponder TWTA output back-off or by using a single carrier per transponder such as TDMA

6. Intersymbol interference that can be minimized by appropriate selection of modulator and demodulator filters

7. Rain-induced attenuation that can be minimized by operating the satellite TWTA at or close to saturation and by using uplink power control

8. Rain-induced cross-polarization interference that can be minimized by using orthogonal linear polarizations instead of orthogonal circular polarizations

9. Antenna pointing loss that can be minimized by using an appropriate tracking system.

4.6.1 Link without Frequency Reuse

In the following discussion we will present a systematic approach to the design of a digital satellite link without frequency reuse to meet the performance criterion that the average probability of bit error exceeds P_b for P percent of the year at most. Consider a 14/12-GHz satellite link between station A–satellite (A uplink) and satellite–station B (B downlink). Such a link is called a *simplex link*. Two simplex links between A–satellite–B and B–satellite–A form a *duplex link* (a voice circuit, for example). The total outage time for an A–satellite–B simplex link is approximately equal to the sum of the A-uplink outage and B-downlink outage (assuming again that the joint outage is negligibly small compared to this sum). The total outage time for the A–satellite–B duplex link is approximately equal to the sum of the A-uplink and downlink outages and the B-uplink and downlink outages. Therefore we explore the simplex link design only. The satellite link parameters are as follows:

Carrier modulation parameters
 Type of modulation: QPSK
 Bit rate: 60 Mbps
 Bit duration-bandwidth product: 0.6
 Noise bandwidth: 36 MHz
Satellite parameters
 Power flux density at satellite for transponder saturation: -81.5
  dBW/m²

Antenna gain-to-noise temperature ratio: 3.1 dB/K
Satellite saturation EIRP: 46.2 dBW
Clear-sky TWTA input back-off: 3 dB
Clear-sky TWTA output back-off: 0.3 dB
TWTA power gain characteristic as given in Table 4.9
Uplink slant range: 37,506 km
Downlink slant range: 37,000 km
Antenna tracking and atmospheric losses: 1.5 dB (uplink) and 1.2 dB (downlink)
Interference parameters
Carrier-to-adjacent satellite interference ratio: 32 dB (uplink) and 32 dB (downlink)
Carrier-to-adjacent channel interference ratio: 29 dB (uplink) and 29 dB (downlink)
Margin for intersymbol interference: 3 dB
Rain attenuation characteristics

$P\%$	Uplink attenuation (dB)	Downlink attenuation (dB)
0.01	13.5	12
0.02	11.0	9.3
0.03	9.5	8.0
0.04	8.5	7.0
0.05	7.8	6.2
0.06	7.3	5.8
0.07	7.0	5.5
0.08	6.8	5.2
0.09	6.5	4.8
0.10	6.2	4.5
0.14	5.5	3.5

Link requirement: $P_b \geq 10^{-4}$ for $P\% = 0.15\%$ of the year at most

1. *Allocation of outages:* The first task should be the allocation of outages to the uplink and the downlink. Since most satellites are downlink-limited, one should allocate more outage on the downlink than on the uplink. A 2:1 ratio would be a good start. Therefore let $P_u\% = 0.05\%$ and $P_d\% = 0.1\%$, so that $P_u\% + P_d\% = P\% = 0.15\%$. From the rain attenuation characteristic, the corresponding uplink attenuation is $L_{r,u} = 7.8$ dB, and the corresponding downlink attenuation is $L_{r,d} = 4.5$ dB.

2. *Sky noise temperature:* The downlink rain attenuation increases the sky noise temperature by an amount equal to $\Delta T = 273 (1 - 1/\log^{-1}[L_{r,d} \text{ (dB)}/10]) = 176$ K. This adds to the system noise temperature of earth station B.

Table 4.9 Typical input-output back-off relationship of satellite TWTA

Input back-off (dB)	Output back-off (dB)
20	10
18	9
16	7.5
14	6
12	4.7
9	2.5
6	1
3	0.3
0 saturation	0
-2 } overdrive	0.2
-4	0.7

3. *Uplink carrier-to-noise ratio with uplink rain only.* This can be evaluated from (4.53), noting that EIRP_{sat} (dBW) $= \Omega_{\text{sat}}$ (dBW) $+ 10 \log(4\pi d_u^2) + L$ (dB) $= 82.5$ dBW. Thus

$$\left(\frac{C}{N}\right)_{\text{u,r}} = 19.5 \text{ dB}$$

4. *Uplink carrier-to-noise plus interference ratio with uplink rain only.* We assume, in the worst case, that rain does not occur at the interfering sources. Therefore the uplink carrier-to-interference ratio with rain attenuation of $L_{\text{r,u}} = 7.8$ dB at station A is

$$\left(\frac{C}{I}\right)_{\text{u,r}} = \left(\frac{C}{I}\right)_{\text{u}} - L_{\text{r,u}}$$

$$= 27.24 - 7.8 = 19.44 \text{ dB}$$

Hence the uplink carrier-to-noise plus interference ratio with rain attenuation of $L_{\text{r,u}} = 7.8$ dB at station A is given by

$$\left(\frac{C}{\mathcal{N}}\right)_{\text{u,r}} = \left[\left(\frac{C}{N}\right)_{\text{u,r}}^{-1} + \left(\frac{C}{I}\right)_{\text{u,r}}^{-1}\right]^{-1} = 16.46 \text{ dB}$$

5. *Downlink carrier-to-noise plus interference ratio with uplink rain only.* The 10^{-4} threshold requires a link carrier-to-noise plus interference ratio equal to 10.6 dB [see (9.51)]. A margin of 3 dB is required for intersymbol interference; thus the total link $C/\mathcal{N}$ must be

$$\left(\frac{C}{\mathcal{N}}\right) = 13.6 \text{ dB}$$

This yields a downlink carrier-to-noise plus interference ratio:

$$\left(\frac{C}{\mathcal{N}}\right)_d = \left[\left(\frac{C}{\mathcal{N}}\right)^{-1} - \left(\frac{C}{\mathcal{N}}\right)_{u,r}^{-1}\right]^{-1} = 16.77 \text{ dB}$$

With an uplink attenuation of 7.8 dB, the total TWTA input back-off is 10.8 dB, which corresponds to an output back-off of 3.7 dB. Thus the additional output back-off due to an uplink attenuation of 7.8 dB is 3.4 dB. Hence the downlink carrier-to-interference ratio is $(C/I)_d = 27.24 - 3.4 = 23.84$ dB. Therefore the required downlink carrier-to-noise ratio with uplink rain only is

$$\left(\frac{C}{N}\right)_d = \left[\left(\frac{C}{\mathcal{N}}\right)_d^{-1} - \left(\frac{C}{I}\right)_d^{-1}\right]^{-1} = 17.72 \text{ dB}$$

6. *G/T with uplink rain only.* From (4.25) the system gain-to-noise temperature G/T is given by

$$\frac{G}{T} = 28.77 \text{ dB/K}$$

7. *EIRP.* The uplink EIRP of earth station A is

$$\text{EIRP} = \text{EIRP}_{\text{sat}} - \text{BO}_i = 82.5 - 3 = 79.5 \text{ dBW}$$

8. *Uplink carrier-to-noise plus interference ratio with downlink rain only.*

$$\left(\frac{C}{\mathcal{N}}\right)_u = \left(\frac{C}{\mathcal{N}}\right)_{u,r} + 7.8 = 16.46 + 7.8 = 24.26 \text{ dB}$$

9. *Downlink carrier-to-noise plus interference ratio with downlink rain only.* Since the required total link $C/\mathcal{N} = 13.6$ dB, the corresponding $(C/\mathcal{N})_{d,r}$ with $L_{r,d} = 4.5$ dB is

$$\left(\frac{C}{\mathcal{N}}\right)_{d,r} = \left[\left(\frac{C}{\mathcal{N}}\right)^{-1} - \left(\frac{C}{\mathcal{N}}\right)_u^{-1}\right]^{-1} = 14 \text{ dB}$$

When rain occurs on the downlink only, the carrier-to-interference ratio remains at the clear-sky value; thus $(C/I)_{d,r} = (C/I)_d = 27.24$ dB. Therefore the downlink carrier-to-noise ratio with a downlink rain attenuation of 4.5 dB is

$$\left(\frac{C}{N}\right)_{d,r} = \left[\left(\frac{C}{\mathcal{N}}\right)_{d,r}^{-1} - \left(\frac{C}{I}\right)_{d,r}^{-1}\right]^{-1} = 14.2 \text{ dB}$$

10. $(G/T)_r$ *with downlink rain only.* From (4.54) the system gain-to-noise temperature $(G/T)_r$ with downlink rain only is $G/(T + \Delta T)$

$$\left(\frac{G}{T}\right)_r = 26.35 \text{ dB/K}$$

By comparing $(G/T)_r$ to G/T with uplink rain only in item 6 and using

$\Delta T = 176$ K in item 2 we have

$$10 \log \frac{T + \Delta T}{T} = 2.42 \text{ dB}$$

This yields the system's clear-sky noise temperature:

$$T = 236 \text{ K}$$

11. *Receive antenna gain.* Since $G/T = 28.77$ dB/K and $T = 236$ K, the receive antenna gain is

$$G_r = 52.5 \text{ dB}$$

12. *Transmit antenna gain.* Assume the same aperture efficiency for both the transmit and receive modes; then the transmit antenna gain is

$$G_t = G_r \left(\frac{f_u}{f_d}\right)^2 = 53.8 \text{ dB}$$

13. *Antenna diameter.* Assume an aperture efficiency of 0.56; then the antenna diameter is 4.5 m.
14. *HPA output power.* The EIRP for the uplink is 79.5 dBW as in item 7. With a transmit antenna gain of 53.8 dB, the required power to the antenna is 25.7 dBW or 371.5 W, including the waveguide loss. If a waveguide loss of 0.3 dB is assumed, the HPA output power must be 26 dBW or 400 W.
15. *Link availability.* Assume that the satellite availability is 0.9999 and the earth stations availability is 0.999; then the simplex link availability is $(1 - 0.15/100)(0.9999)(0.999) = 0.9974$ or 99.74% of the year for a 10^{-4} threshold. For the duplex link the availability is $(1 - 0.15/100^2(0.9999)(0.999) = 0.9959$ or 99.59% of the year for a 10^{-4} threshold, assuming stations A and B have the same rain attenuation characteristics.
16. *Remarks.* To optimize the antenna diameter, system noise temperature, the HPA size to meet cost requirements or equipment availability on the market, more iterations may be carried out for a range of outage allocations. Also, note that any combination of antenna diameter, system noise temperature, and HPA size such that EIRP = 79.5 dBW, clear-sky $G/T = 28.77$ dB/K, and $(G/T)_r \geq 26.35$ dB/K meets the required link performance. For example, if a 5-m antenna is selected instead of a 4.5-m antenna, T can be increased to 293 K and the HPA output power will then be 317 W.

4.6.2 Link with Frequency Reuse

In this section we take a look at the same digital simplex link considered in Sec. 4.6.1 but assume that it operates in a frequency reuse system with the following additional interference parameters: (a) The satellite antenna cross-polarization discrimination is 35 dB at 14 and 12 GHz. (b) The earth station antenna cross-polarization discrimination is 35 dB at 14 and 12 GHz. (c) There is an uplink interfering earth station C operating in the copolarized transponder with antenna cross-polarization discrimination of 32 dB. (d) The rain depolarization characteristic at earth station B for 0.1% of the year is 20 dB.

1. *Allocation of outages.* For this link we also let $P_u\% = 0.05\%$ and $P_d\% = 0.1\%$, so that $P_u\% + P_d\% = 0.15\%$. From the rain attenuation and depolarization characteristics, the corresponding uplink attenuation $L_{r,u} = 7.8$ dB, the downlink attenuation $L_{r,d} = 4.5$ dB, and the downlink carrier-to-cross-polarization interference ratio due to rain $X_B = 20$ dB.

2. *Sky noise temperature.* The increase in sky noise temperature by $L_{r,d} = 4.5$ dB is $\Delta T = 176\ K$. This will add to the system noise temperature of station B.

3. *Uplink carrier-to-noise ratio with uplink rain only.* From (4.53) we have

$$\left(\frac{C}{N}\right)_{u,r} = 19.5\ \text{dB}$$

4. *Uplink carrier-to-noise plus interference ratio with uplink rain only.* We assume in the worst case that rain does not occur at the interfering source. The carrier-to-cross-polarization interference consists of that of the station C antenna and that of the satellite antenna, which amounts to 30.24 dB. The total uplink $(C/I)_{u,r}$ is then equal to

$$\left(\frac{C}{I}\right)_{u,r} = 25.47 - 7.8 = 17.67\ \text{dB}$$

Hence the uplink carrier-to-noise plus interference ratio with an uplink rain attenuation of 7.8 dB is

$$\left(\frac{C}{\mathcal{N}}\right)_{u,r} = \left[\left(\frac{C}{N}\right)_{u,r}^{-1} + \left(\frac{C}{I}\right)_{u,r}^{-1}\right]^{-1} = 15.5\ \text{dB}$$

5. *Downlink carrier-to-noise plus interference ratio with uplink rain only.* The required total link $C/\mathcal{N} = 13.6$ dB at $P_b = 10^{-4}$. This yields

$$\left(\frac{C}{\mathcal{N}}\right)_d = \left[\left(\frac{C}{\mathcal{N}}\right)^{-1} - \left(\frac{C}{\mathcal{N}}\right)_{u,r}^{-1}\right]^{-1} = 18.1\ \text{dB}$$

From item 5 in Sec. 4.6.1 and taking into account the carrier-to-cross-polarization interference of the satellite antenna and the station B antenna we have $(C/I)_d = 23.22$ dB. Hence the required downlink carrier-to-noise ratio with uplink rain only is

$$\left(\frac{C}{N}\right)_d = \left[\left(\frac{C}{\mathcal{N}}\right)_d^{-1} - \left(\frac{C}{I}\right)_d^{-1}\right]^{-1} = 19.7\ \text{dB}$$

6. *G/T with uplink rain only.* From (4.25) the gain-to-system noise temperature G/T is

$$\frac{G}{T} = 30.75 \text{ dB/K}$$

7. *EIRP.* EIRP is the same as in item 7 in Sec. 4.6.1 and is

$$\text{EIRP} = 79.5 \text{ dBW}$$

8. *Uplink carrier-to-noise plus interference ratio with downlink rain only.*

$$\left(\frac{C}{\mathcal{N}}\right)_{u} = \left(\frac{C}{\mathcal{N}}\right)_{u,r} + 7.8 = 15.5 + 7.8 = 23.3 \text{ dB}$$

9. *Downlink carrier-to-noise plus interference ratio with downlink rain only.*

$$\left(\frac{C}{\mathcal{N}}\right)_{d,r} = \left[\left(\frac{C}{\mathcal{N}}\right)^{-1} - \left(\frac{C}{\mathcal{N}}\right)_{u}^{-1}\right]^{-1} = 14.1 \text{ dB}$$

The carrier-to-cross-polarization interference ratio consists of that of the satellite and the station B antennas, which amounts to 32 dB, and the rain depolarization effect at station B, which is 20 dB. The total $(C/I)_{d,r}$ is 19 dB. Therefore

$$\left(\frac{C}{N}\right)_{d,r} = \left[\left(\frac{C}{\mathcal{N}}\right)_{d,r}^{-1} - \left(\frac{C}{I}\right)_{d,r}^{-1}\right]^{-1} = 15.8 \text{ dB}$$

10. *$(G/T)_r$ with downlink rain only.* From (4.54) the system gain-to-noise temperature ratio with $L_{r,d} = 4.5$ dB is

$$\left(\frac{G}{T}\right)_{r} = 27.95 \text{ dB/K}$$

By comparing $(G/T)_r$ to G/T with uplink rain only in item 6 and using $\Delta T = 176$ K in item 2 we have

$$10 \log \frac{T + \Delta T}{T} = 2.8 \text{ dB}$$

This yields the system's clear-sky noise temperature:

$$T = 194.4 \text{ K}$$

11. *Receive antenna gain.* Since $G/T = 30.75$ dB/K and T = 194.4K, the receive antenna gain is

$$G_r = 53.64 \text{ dB}$$

12. *Transmit antenna gain.* Assuming identical aperture efficiency for both the transmit and receive modes we have

$$G_t = 54.98 \text{ dB}$$

13. *Antenna diameter.* Assume an aperture efficiency of 0.56; then the antenna diameter is 5.1 m.
14. *HPA output power.* The HPA output power is 303.4 W, assuming a waveguide loss of 0.3 dB.
15. *Remarks.* So far we have not included the effect of uplink cross-polarization interference by rain at station A. In practice one should avoid this situation by not allowing simultaneous traffic in two copolarized transponders.

REFERENCES

1. V. K. Prabhu, "Error Rate Considerations for Coherent Phase-Shift-Keyed Systems with Co-channel Interference," *Bell Syst. Tech. J.,* Vol. 48, Mar. 1969, pp. 743–767.
2. A. S. Rosenbaum, "PSK Error Performance with Gaussian Noise and Interference, *Bell Syst. Tech. J.,* Vol. 47, Feb. 1969, pp. 413–442.
3. R. Fang and O. Shimbo, "Unified Analysis of a Class of Digital Systems in Additive Noise and Interference," *IEEE Trans. Commun.,* Vol. COM-21, No. 10, Oct. 1973, pp. 1075–1091.
4. P. L. Rice and N. R. Holmberg, "Cumulative Time Statistics of Surface Point-Rainfall Rates," *IEEE Trans. Commun.,* Vol. COM-21, No. 10, Oct. 1973, pp. 1131–1136.
5. E. J. Dutton and H. T. Dougherty, "Modeling the Effects of Cloud and Rain upon Satellite-to-Ground System Performance," Office of Telecommunications, Boulder, Colo., OT Report 73-5, Mar. 1973.
6. S. H. Lin, "Empirical Rain Attenuation Model for Earth-Satellite Paths," *IEEE Trans. Commun.,* Vol. COM-27, No. 5, May 1979, pp. 812–817.
7. R. K. Crane, "Prediction of Attenuation by Rain," *IEEE Trans. Commun.,* Vol. COM-28, No. 9, Sept. 1980, pp. 1717–1735.
8. R. L. Olsen et al., "The aR^b Relation in the Calculation of Rain Attenuation," *IEEE Trans. Antennas Propagation,* Vol. AP-26, Mar. 1978, pp. 318–329.
9. D. B. Hodge, "An Empirical Relationship for Path Diversity Gain," *IEEE Trans. Antennas Propagation,* Vol. AP-24, Mar. 1976, p. 250.
10. INTELSAT BG/T-36-11E, W/2/81, Jan. 30, 1981.
11. T. S. Chu, "Rain-Induced Cross-polarization at Centimeter and Millimeter Wavelengths," *Bell Syst. Tech. J.,* Vol. 53, No. 8, Oct. 1974, pp. 1557–1579.
12. T. S. Chu, "A Semi-empirical Formula for Microwave Depolarization versus Rain Attenuation on Earth-Space Paths," *IEEE Trans. Commun.,* Vol. COM-30, No. 12, Dec. 1982, pp. 2550–2554.

PROBLEMS

4.1 Give a simple explanation why the carrier-to-noise ratio of the retransmitted carrier is equal to the carrier-to-noise ratio of the received carrier for a frequency-translating satellite transponder in the model in Fig. 4.1.

4.2 Calculate the power flux density at a geostationary satellite located at 100°W longitude for an earth station with the following parameters.

　　Antenna diameter: 5 m
　　Antenna efficiency: 0.55

Carrier frequency: 14 GHz
HPA saturation power: 600 W
HPA output back-off: 3 dB
Loss of waveguide between HPA output and antenna input: 0.5 dB
Earth station location: 90°W longitude, 40°N latitude

4.3 Consider a satellite system with the following parameters.

Saturation transponder EIRP: 36 dBW
Transponder output back-off: 6 dB
Number of carriers per transponder: 200
Noise bandwidth: 40 kHz
Earth station antenna
   Diameter: 5 m
   Efficiency: 0.55
   Downlink slant range: 37,506 km
   Receive frequency: 4 GHz
Terrestrial interference power: −156 dBW/MHz

Calculate the carrier-to-terrestrial interference ratio assuming the system is downlink-limited.

4.4 Consider three adjacent satellites A, B, and C located at longitudes 96°W, 100°W, and 104°W, respectively, with the following parameters.

	A	B	C
Saturation power flux density, min-max (dBW/m²)	−75 to −72	−83 to −80	−80 to −77
Saturation transponder EIRP, min-max (dBW)	33 to 37	33 to 36	32 to 36
Earth station antenna			
Diameter (m)	10	10	10
Efficiency	0.55	0.55	0.55
Transmit frequency (GHz)	6	6	6
Receive frequency (GHz)	4	4	4

Assuming a differential antenna gain of −3 dB, calculate the worst-case carrier-to-adjacent satellite interference ratio of satellite system B.

4.5 Calculate the carrier-to-noise plus interference ratio for the following 14/12-GHz geostationary satellite system.

Carrier modulation parameters
  Bit rate: 64 kbps
  Bit duration-bandwidth product: 0.625
Satellite parameters
  Antenna gain-to-noise temperature ratio: 1.6 dB/K
  Transponder saturation EIRP: 44 dBW
  Transponder input back-off: 10 dB
  Transponder output back-off: 5 dB
  Number of carriers per transponder: 400
  Satellite stationkeeping: 0.05° east-west and north-south
  Satellite location: 100°W longitude
Earth station parameters
  Power flux density at satellite for transponder saturation: −73.5 dBW/m²
  Antenna diameter: 7.5 m

Antenna efficiency: 0.55
System noise temperature: 150 K
Earth station location: 90°W longitude, 35°N latitude
Interference parameters
Carrier-to-intermodulation ratio: 21 dB
Total carrier-to-adjacent satellite interference ratio: 25 dB
Earth station cross-polarization discrimination: 33 dB
Satellite cross-polarization discrimination: 32 dB

4.6 Find the rain-induced attenuations at 12 GHz that may be exceeded for $P = 0.001, 0.01,$ and 0.1% of the year for earth stations A and B with the following parameters.
Earth station A
Climate region: E
Latitude: 32°N
Earth station height: 0.5 km,
Elevation angle: 5°
Earth station B
Climate region: C
Latitude: 45°N
Earth station height: 1 km
Elevation angle: 20°

4.7 Find the rain-induced attenuation at 20 GHz that may be exceeded for $P = 0.01\%$ of the year for an earth station with the following parameters.
Climate region: F,
Latitude: 30°N
Earth station height: 0.2 km,
Elevation angle: 15°

4.8 Find the rain-induced attenuation at 44 GHz that may be exceeded for $P = 0.1\%$ of year for an earth station with the following parameters.
Climate region: D_1
Latitude: 35°N
Earth station height $= 0.4$ km
Elevation angle: 30°

4.9 Find the rain-induced cross-polarization discrimination at 12 GHz that may not be exceeded for 0.05% of the year for an earth station with the following parameters.
Latitude: 35°N
Longitude: 83°W
Earth station height: 0.9 km
Elevation angle: 47°
Polarization state: circular polarization

4.10 Find the rain-induced cross-polarization discrimination at 30 GHz that may not be exceeded for 0.01% of the year for an earth station with the following parameters.
Climate region: D_1
Latitude: 35°N
Earth station height: 0.5 km
Elevation angle: 30°
Polarization state: horizontal linear polarization

4.11 Find the rain-induced cross-polarization discrimination at 20 GHz that may not be exceeded for 0.01% of the year for an earth station with the following parameters:
Climate region: E
Earth station height: 0 km
Elevation angle: 30°

Latitude: 32°N

Polarization state: quasi-vertical with a 10° tilt angle

4.12 Consider a 14/12-GHz satellite system with the following parameters.

Carrier modulation parameters

Modulation: QPSK

Bit rate: 90 Mbps

Bit duration-bandwidth product: 0.6

Satellite parameters

Satellite location: 103°W longitude

Satellite stationkeeping: 0.05° east-west and north-south

Antenna gain-to-noise temperature ratio: 1.6 dB/K

Satellite saturation EIRP: 44 dBW

Clear-sky TWTA input back-off: 3dB

Clear-sky TWTA output back-off: 0.4 dB

TWTA power gain characteristic as follows:

TWTA input back-off (dB)	0	2	3	4	5	6	8	10	12	14	16	18
TWTA output back-off (dB)	0	0.1	0.4	0.8	1	1.7	3	4.5	6.2	8	10	12

Interference parameters

Carrier-to-adjacent channel interference ratio: 26 dB (uplink) and 27 dB (downlink)

Carrier-to-cross-polarization interference ratio by satellite and earth station antennas: 30 dB (uplink) and 30 dB (downlink)

Degradation of received C/N due to intersymbol interference: 2 dB

Earth station parameters

Earth station A location

Climate region: E

Height: 0 km

Longitude: 87°W

Latitude: 33°N

Earth station B location

Climate region: D_1

Height: 0.5 km

Longitude: 90°W

Latitude: 45°N

Antenna

Diameter: 7.7 m

Efficiency: 0.55

Aperture distribution: parabolic

Tracking accuracy: one-third of half-power beamwidth

HPA

Size: 1.2 kW

Combining loss: 0.5 dB

Waveguide loss: 0.5 dB

Output back-off: 3 dB

System noise temperature: 160 K

(a) Find the A-to-B link availability for $P_b = 10^{-6}$, 10^{-3}

(b) Find the B-to-A link availability for $P_b = 10^{-6}$, 10^{-3}

(c) Find the Duplex link availability between A and B for $P_b = 10^{-6}$, 10^{-3}.

4.13 Consider a 14/12-GHz satellite link with the following parameters.

Carrier-to-adjacent channel interference ratio: 26 dB (uplink) and 27 dB (downlink)

Earth station cross-polarization discrimination: 30 dB

Satellite cross-polarization discrimination: 32 dB

Rain-induced cross-polarization: 20 dB (uplink) and 18 dB (downlink)

Find the carrier-to-noise ratio required to achieve an average probability of bit error $P_b = 10^{-6}$ given that the carrier modulation is QPSK with a bit duration-bandwidth product of $T_b B = 0.5$. Is it possible to achieve $P_b = 10^{-6}$? (The theoretical P_b versus C/N for a QPSK carrier with $T_b B = 0.5$ is given by the curve labeled $C/I = \infty$ in Fig. 4.4.)

4.14 Consider a 14/12-GHz satellite link using QPSK modulation with $T_b B = 0.5$ between two earth stations A and B operated in quasi-horizontal polarization with the following parameters:

Carrier-to-interference ratio: 26 dB (uplink) and 28 (downlink)

Carrier-to-noise ratio: 22 dB (uplink) and 20 dB (downlink)

An earth station C operating in the cross-polarized transponder using quasi-vertical polarization with a tilt angle of $\tau = 20°$ interferes with the uplink between stations A and B. Assume that the elevation angle of earth station C is $E = 20°$ and find the average probability of bit error for the satellite link between stations A and B that may be exceeded for 0.01% of the year due to rain on earth station C only (assuming that the rain-induced attenuation that may exceed 0.10% at earth station C is 10 dB) using the result in [12].

4.15 A duplex satellite link between earth stations A and B has the following outages.

Uplink of A: 0.05%

Downlink of B: 0.06%

Uplink of B: 0.08%

Downlink of A: 0.04%

Find the outage of the duplex link.

4.16 A nonlinear satellite link employing QPSK modulation with $T_b B = 0.6$ has been designed to work at a bit rate of $R = 90$ Mbps and $C/N = 12.81$ dB. The nonlinear channel degradation is 1.6 dB, and the intersymbol interference is 1.4 dB. Later the bit rate is reduced to 70 Mbps. Assume that each 10^6-bps reduction in the bit rate reduces the nonlinear channel degradation by 1% and the intersymbol interference by 1.5%. Find the bit error rate of the link.

4.17 Consider a satellite link with clear-sky $(C/\mathcal{N})_u = 20.9$ dB and $(C/\mathcal{N})_d = 17.5$ dB. The link is designed to work with PSK modulation $(T_b B = 1)$. The PSK demodulator design takes into account 3-dB degradation in the carrier-to-noise ratio due to channel nonlinearities and intersymbol interference. The minimum bit error rate required for the link is 10^{-6}, and the earth station G/T is 25 dB/K. The TWTA operating point and characteristic are given in Sec. 4.6.1.

(a) Find the maximum uplink rain attenuation that can be tolerated assuming no rain on the downlink.

(b) Find the maximum downlink rain attenuation that can be tolerated assuming no rain on the uplink.

4.18 Repeat Prob. 4.17 assuming the following.

(a) The earth station EIRP can vary from 74 dBW under clear-sky conditions to 86.4 dBW during rain.

(b) The satellite EIRP can vary from 54.3 to 60 dBW.

4.19 Consider the satellite link discussed in Probs. 4.17 and 4.18. Find the minimum satellite EIRP required to maintain a minimum bit error rate of 10^{-6} for a downlink fade of 2 dB (assuming no uplink fade).

4.20 Consider the link analysis in Sec. 4.6.1 assuming that the link requirement is $P_b \geq 10^{-3}$

for $P_u\% = 0.05\%$ and $P_d\% = 0.05\%$. Find the earth station antenna size, system noise temperature, and HPA output power assuming a 0.3-dB waveguide loss.

4.21 Consider the link analysis in Sec. 4.6.1, where the following new parameters are used.
Power flux density at satellite for transponder saturation: -78 dBW/m²
Satellite saturation EIRP: 49 dBW
Design the earth station so that $P_b \geq 10^{-5}$ for $P_u\% = 0.03\%$ and $P_d\% = 0.07\%$ assuming a 0.3-dB waveguide loss.

4.22 Repeat Prob. 4.20 for $P_u\% = 0.03\%$ and $P_d\% = 0.07\%$.

4.23 Repeat Prob. 4.21 for $P_u\% = 0.05\%$ and $P_d\% = 0.05\%$.

4.24 Consider the link analysis in Sect. 4.6.2 assuming that the link requirement is $P_b \geq 10^{-3}$ for $P_u\% = 0.05\%$ and $P_d\% = 0.05\%$ and the rain depolarization at earth station B for 0.05% of the year is 22 dB. Design the earth station assuming a waveguide loss of 0.3 dB.

4.25 Repeat Prob. 4.24 for $P_u\% = 0.03\%$ and $P_d\% = 0.07\%$ assuming the rain depolarization at earth station B for 0.07% of the year is 24 dB.

4.26 Repeat Prob. 4.24 for the new link requirement $P_b \geq 10^{-5}$ for $P_u\% = 0.03\%$ and $P_d\% = 0.07\%$ and the new satellite parameters given in Prob. 4.21.

chapter 5

Frequency Division Multiple Access

The simplest and most widely used multiple access technique of satellite communications is *frequency division multiple access,* where each earth station in a satellite network transmits one or more carriers at different center frequencies to the satellite transponder. Each carrier is assigned a frequency band with a small guard band to avoid overlapping between adjacent carriers. The satellite transponder receives all the carriers in its bandwidth, amplifies them, and retransmits them back to earth. The earth station in the satellite beam served by the transponder can select the carrier that contains messages intended for it. A frequency division multiple access system is shown schematically in Fig. 5.1. In this type of system each carrier can employ either analog modulation, such as *frequency modulation*, or digital modulation, such as *phase-shift keying.* A major problem in the operation of FDMA satellite systems is the presence of intermodulation products in the carrier bandwidth generated by the amplification of multiple carriers by a common TWTA in the satellite transponder that exhibits both amplitude nonlinearity and phase nonlinearity, as shown in Fig. 5.2.† As the number of carriers

---

†For constant-amplitude carriers, the carrier power is simply one-half the square of the amplitude. Therefore the corresponding amplitude transfer characteristic is the square root of the TWTA power transfer characteristic. (A multiplicative factor of $1/\sqrt{2}$ is employed to normalize the input and output voltages at saturation to 1 V when the corresponding normalized power at saturation is 1 W.)

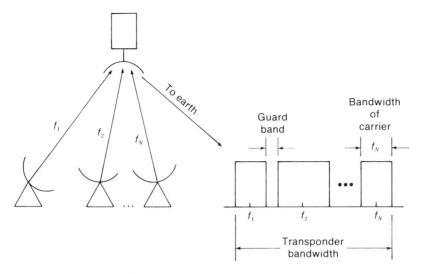

Figure 5.1 Concept of a FDMA system.

increases, it becomes necessary to operate the TWTA close to saturation in order to supply the required power per carrier to reduce the effect of downlink thermal noise. But near saturation the input/output amplitude transfer characteristic of the TWTA is highly nonlinear, and consequently the level of intermodulation products is increased and affects the overall performance. In terms of amplifier design, the output power rating of the

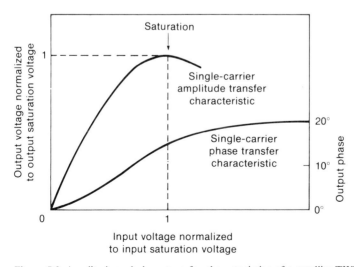

Figure 5.2 Amplitude and phase transfer characteristics of a satellite TWTA.

TWTA must be increased so that it can be backed off from saturation sufficiently to reduce intermodulation effects. At the same time the TWTA must supply adequate transmitting power for each carrier. Determination of the carrier-to-intermodulation ratio is crucial in any FDMA system because it directly sets a limit on the overall carrier-to-noise ratio of the satellite link. This is the main subject of this chapter.

5.1 FDM-FM-FDMA

Since the inception of satellites analog modulation, such as frequency modulation, has been used for carrier modulation in satellite communications using FDMA; it will probably be employed in existing equipment for years to come despite advances in the development of digital satellite systems. There are two main FDMA techniques in operation today:

1. *Multichannel-per-carrier* transmission, where the transmitting earth station frequency division-multiplexes several single-sideband suppressed carrier telephone channels into one carrier baseband assembly which frequency-modulates a RF carrier and is transmitted to a FDMA satellite transponder. This type of operation is referred to as FDM-FM-FDMA.
2. *Single-channel-per-carrier* (SCPC) transmission, where each telephone channel independently modulates a separate RF carrier and is transmitted to a FDMA satellite transponder. The modulation can be analog, such as FM, or digital, such as PSK. This type of operation will be studied in Sec. 5.2.

Because frequency modulation is discussed in any text on communications theory we do not intend to repeat the analysis here but present instead a review of frequency modulation and describe the performance of a FDM-FM-FDMA system in terms of the signal-to-noise ratio S/N of a telephone channel at the FM demodulator output.

Frequency modulation is a form of modulation in which the modulated signal is represented by

$$s(t) = A \cos \left[\omega_c t + 2\pi k_f \int_0^t m(\tau) \, d\tau \right] \tag{5.1}$$

where A = carrier amplitude
ω_c = carrier frequency
k_f = a constant
$m(t)$ = modulating baseband signal

From (5.1) it is seen that frequency modulation is a nonlinear modulation process, and it is difficult to relate the bandwidth of the FM signal to that

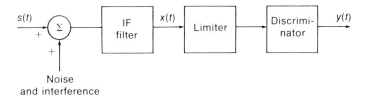

Figure 5.3 FM demodulator.

of the message. Therefore a sinusoidal tone is commonly employed in the analysis. When $m(t) = A_m \cos \omega_m t$, where ω_m is the modulating frequency, (5.1) becomes

$$s(t) = A \cos\left(\omega_c t + \frac{2\pi k_f A_m}{\omega_m} \sin \omega_m t\right)$$

$$= A \cos\left(\omega_c t + \frac{\Delta f}{f_m} \sin \omega_m t\right) \tag{5.2a}$$

where

$$\Delta f = k_f A_m \tag{5.2b}$$

is the *peak frequency deviation* (Hz) and $f_m = \omega_m / 2\pi$. The ratio $M = \Delta f / f_m$ is the *modulation index*.

To find the output *signal-to-noise ratio* of a telephone channel in a FDM-FM system, consider the FM demodulator shown in Fig. 5.3 where the signal $s(t)$ is given in (5.1) and the noise† is assumed to have zero mean and a power spectral density of $N_0/2$. The IF filter is assumed to be ideal with center frequency f_c and bandwidth B and passes the signal $s(t)$ without distortion. The limiter eliminates amplitude fluctuations due to noise. The signal plus noise at the output of the IF filter is

$$\begin{aligned} x(t) &= s(t) + n(t) \\ &= A \cos[\omega_c t + \phi(t)] + n(t) \end{aligned} \tag{5.3a}$$

where

$$\phi(t) = 2\pi k_f \int_0^t m(\tau) \, d\tau \tag{5.3b}$$

The narrowband noise $n(t)$ is represented in terms of its in-phase and quadrature components $n_c(t)$ and $n_s(t)$ by [1]

$$n(t) = n_c(t) \cos \omega_c t - n_s(t) \sin \omega_c t \tag{5.4}$$

---

†The noise considered here may include AWGN and other types of statistically independent interferences considered in Chap. 4 and the intermodulation interference considered in Sec. 5.3 and 5.4. These interferences are treated as equivalent AWGN.

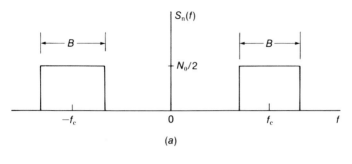

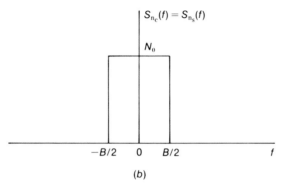

Figure 5.4 Power spectral density of noise. (*a*) Bandpass noise. (*b*) Equivalent lowpass noise.

The power spectral densities of $n(t)$, $n_c(t)$, and $n_s(t)$ are shown in Fig. 5.4. Equivalently, we may express $n(t)$ in terms of its envelope and phase as

$$n(t) = r(t) \cos[\omega_c t + \psi(t)] \tag{5.5a}$$

where

$$r(t) = \sqrt{n_c^2(t) + n_s^2(t)} \tag{5.5b}$$

$$\psi(t) = \tan^{-1}\left[\frac{n_s(t)}{n_c(t)}\right] \tag{5.5c}$$

$$n_c(t) = r(t) \cos \psi(t) \tag{5.5d}$$

$$n_s(t) = r(t) \sin \psi(t) \tag{5.5e}$$

By using the signal term in (5.3) as a reference, we can represent $x(t)$ by the phasor diagram in Fig. 5.5. The phase of the phasor representing $x(t)$ is

$$\theta(t) = \phi(t) + \tan^{-1}\left\{\frac{r(t) \sin[\psi(t) - \phi(t)]}{A + r(t) \cos[\psi(t) - \phi(t)]}\right\} \tag{5.6}$$

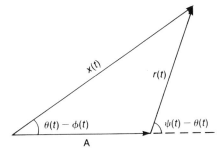

Figure 5.5 Phasor representation of a narrowband input signal plus noise.

For the low-noise case where $A >> r(t)$, we find that, for almost all t, $\theta(t)$ in (5.6) can be approximated by

$$\theta(t) \approx \phi(t) + \Delta\phi(t) \qquad (5.7a)$$

where

$$\Delta\phi(t) = \frac{r(t)}{A} \sin[\psi(t) - \phi(t)] \qquad (5.7b)$$

Because the baseband signal $\phi(t)$ varies much more slowly than the noise $\psi(t)$, we can approximate $\phi(t)$ in (5.7b) by a constant ϕ. Hence

$$\Delta\phi(t) \approx \frac{r(t)}{A} \sin[\psi(t) - \phi]$$

$$\approx \frac{r(t)}{A} \sin \psi(t) \cos \phi - \frac{r(t)}{A} \cos \psi(t) \sin \phi$$

$$\approx \frac{n_s(t)}{A} \cos \phi - \frac{n_c(t)}{A} \sin \phi \qquad (5.8)$$

The discriminator differentiates the angle $\theta(t)$ to extract the information. Therefore

$$y(t) = \frac{1}{2\pi} \left[\frac{d\theta(t)}{dt} \right]$$

$$= \frac{1}{2\pi} \left[\frac{d\phi(t)}{dt} + \frac{d\,\Delta\phi(t)}{dt} \right]$$

$$= k_f m(t) + \frac{1}{2\pi} \left[\frac{d\Delta\phi(t)}{dt} \right] \qquad (5.9)$$

Thus the baseband signal $m(t)$ is recovered (the constant $1/2\pi$ being just a scaling term). The second term on the right side of (5.9) represents the output noise.

Since $n_c(t)$ and $n_s(t)$ are uncorrelated for AWGN, the power spectral density of the noise term $\Delta\phi(t)$ in (5.8) at the input of the discriminator is

$$S_{\Delta\phi} (f) = \frac{\cos^2\phi}{A^2} S_{n_s}(f) + \frac{\sin^2 \phi}{A^2} S_{n_c}(f)$$

$$= \frac{S_{n_c} (f)}{A^2} = \frac{S_{ns} (f)}{A^2} = \frac{N_0}{A^2} \tag{5.10}$$

The power spectral density of the output noise $n_0(t) = (1/2\pi) \, d \, \Delta\phi(t)/dt$ is simply equal to

$$S_{n_0} (f) = \left| \frac{j2\pi f}{2\pi} \right|^2 S_{\Delta\phi} (f)$$

$$= \frac{N_0 f^2}{A^2} \tag{5.11}$$

Note that the transfer function of the time differentiator is $j2\pi f/2\pi = jf$. We see from (5.11) that the channel noise power is proportional to the square of the frequency of the FDM baseband signal, and thus it becomes maximum at the top channel of the baseband FDM signal. Because the baseband signal is limited to the maximum frequency f_m, we need only to find the output noise power N for a telephone channel of bandwidth b located at f_m ($b << f_m$) where the noise is strongest. [Since FDM uses single-sideband transmission (see Fig. 1.8), the corresponding single-sideband noise bandwidth is b.] Consequently,

$$N = 2bS_{n_0} (f_m) = \frac{2N_0 f_m^2 \, b}{A^2} \tag{5.12}$$

The power of the output signal $k_f m(t)$ in (5.9) is

$$S = k_f^2 E \{m^2(t)\} = \frac{k_f^2 A_m^2}{2}$$

$$= \frac{\Delta f^2}{2}$$

where the peak frequency deviation Δf is given by (5.2b). If we let f_r be the *rms frequency deviation*, then

$$f_r = \frac{\Delta f}{\sqrt{2}}$$

and, consequently,

$$S = f_r^2 \tag{5.13}$$

Because the power level in a telephone channel varies with time and from speaker to speaker, a reference point is defined for each telephone channel so that the average power is 1 mW or 0 dBm. In a FDM-FM system each channel is alloted a 0-dBm test tone with a rms frequency deviation of f_r as a reference point. Therefore the *output signal-to-noise ratio* of a telephone channel at the top FDM baseband frequency f_m can be calculated from (5.12) and (5.13) as

$$\frac{S}{N} = \frac{A^2 f_r^2}{2N_0 f_m^2 \, b} \tag{5.14}$$

From (5.3a) we note that the FM carrier power is

$$C = \frac{A^2}{2}$$

and the noise power in the IF bandwidth is

$$\mathcal{N} = N_0 B$$

as seen from Fig. 5.4. Hence the *carrier-to-noise ratio* of the received FDM-FM-FDMA signal at the input of the FM demodulator is

$$\frac{C}{\mathcal{N}} = \frac{A^2}{2N_0 B}$$

Therefore we can express the signal-to-noise ratio of a telephone channel at the top FDM baseband frequency f_m in terms of the received carrier-to-noise ratio as

$$\frac{S}{N} = \frac{C}{\mathcal{N}} \left(\frac{B}{b} \right) \left(\frac{f_r}{f_m} \right)^2 \tag{5.15}$$

and, consequently†,

$$\frac{C}{\mathcal{N}} = \frac{S}{N} \left(\frac{b}{B} \right) \left(\frac{f_m}{f_r} \right)^2 \tag{5.16}$$

From (5.16) it is seen that, once the number of channels per carrier n (hence the bandwidth B) and the desired top channel signal-to-noise ratio at the FM demodulator output are selected, the total carrier-to-noise ratio $C/\mathcal{N}$ can be traded for RF bandwidth B to maximize the transponder capacity. The only constraint in this trade-off is that the carrier-to-noise ratio $C/\mathcal{N}$ cannot fall below a minimum of 10 dB (typical) to avoid operation in the threshold region of the FM demodulator. Note that the IF bandwidth B is the bandwidth of the FM carrier transmitted by the earth station and is given by Carson's rule as

$$B = 2(f_\Delta + f_m)$$

where f_m = maximum FDM baseband frequency (discussed above) and f_Δ = FDM peak frequency deviation given by

---

† The carrier-to-noise ratio $C/\mathcal{N}$ considered here is a combination of the carrier-to-noise (AWGN) plus interference ratio in (4.29) and the carrier-to-intermodulation ratio to be discussed in Secs. 5.3 and 5.4.

$$f_\Delta = g l f_r$$

The parameter l is the FDM *loading factor* and according to the CCITT is

$$l = \begin{cases} \log^{-1}[(-1 + 4 \log n)/20] & n < 240 \\ \log^{-1}[(-15 + 10 \log n)/20] & n \geq 240 \end{cases}$$

where n is the number of telephone channels in the FDM-FM carrier. Because the FDM-FM waveform approximates noise, the peak-to-rms ratio or FDM peak factor g is taken to be

$$g = 3.16 = 10 \text{ dB}$$

It is seen that the FDM *rms frequency deviation* is simply $f_\Delta/g = l f_r$.

The multichannel loading factor $l = 20 \log l$ (dB) is defined as the ratio of the FDM multichannel equivalent power to a test-tone level of 0 dBm. The FDM multichannel peak factor g is defined as the ratio of the overload voltage (peak voltage) to the rms voltage for a given number of active channels at constant volume. (The average active channels are about 25% of the number of channels per carrier n.) In FDM-FM-FDMA operations, the peak factor is taken to be 10 dB independent of the number of active channels.

The allocated satellite bandwidth for the same carrier with bandwidth B in a FDMA operation must take into account the guard bands of 10 or 20% of B. The maximum baseband frequency f_m corresponding to the number of channels per FDM-FM carrier is

n	24	36	60	72	96	132	192
f_m (kHz)	108	156	252	300	408	552	804

n	252	312	432	612	792	972	1092
f_m (kHz)	1052	1300	1796	2540	3284	4028	4812

The interfering effect of noise on speech telephony is a function of the response of the human ear to specific frequencies in the voice channel as well as the type of telephone subset. When noise measurement units were first defined, it was decided that it would be convenient to measure the relative interfering effect of noise on the listener as a positive number. The 1000-Hz test-tone signal at the −90-dBm level was chosen because a tone whose level is less than 1 pW is not ordinarily audible. The telephone subset was the Western Electric 144 handset. With the 144-type handset as a test receiver and with a wide distribution of average listeners, it was found that a 3000-Hz signal required an 18-dB increase to have the same interfering effect on an average listener over the 1000-Hz reference. A curve showing the relative interfering effects of sinusoidal tones compared to a reference frequency is called a *weighting curve*. In FDM-FM-FDMA operations a CCITT psophometric weighting curve is used. The noise measurement unit associated with this curve is picowatts psophometrically weighted (pWp). The reference frequency in this case is 800 Hz rather than 1000 Hz. For example, when measured by a noise measurement set using CCITT weighting networks (filters whose responses resemble the weighting curve), a 3100-Hz band of white noise (flat, i.e., not weighted) is attenuated by 2.5 dB. In other words, if the white noise power level is measured in a 3100-Hz bandwidth with a flat attenuation frequency characteristic, the noise power level must be reduced by 2.5 dB to obtain the psophometric power level. (Note that the bandwidth *b* of the telephone channel is 3100 Hz.)

The noise power spectral density within the bandwidth determined by the maximum baseband frequency f_m is proportional to the square of the frequency (5.11). In other words, high-frequency channels are noisier than low-frequency channels in FDM-FM-FDMA operations. To reduce the difference in performance between individual channels emphasis is employed to equalize the noise difference between high- and low-frequency channels and improves the signal-to-noise ratio at the FM demodulator output [1]. The emphasis weighting factor *P* for FDM-FM-FDMA operations gives an improvement of about 4 dB. The combined effect of emphasis and psophometric weighting is an improvement of about 5.5 to 6.5 dB in the highest frequency channel.

In the above discussion we considered the relationship between the signal-to-noise ratio *S/N* at the FM demodulator and the total carrier-to-noise ratio $C/\mathcal{N}$ available in the satellite downlink. This total $C/\mathcal{N}$ takes into account the thermal noise in the uplink and downlink and the transponder intermodulation noise (Sec. 5.3). Besides this, one has to consider the effects of adjacent transponder interference and cochannel interference when employing frequency reuse as discussed in Chap. 4. In addition, the earth station equipment noise must also be taken into account. This type of noise includes earth station out-of-band emission, earth station intermodulation noise due to the high-power amplifier, and noise due to total system group delay distortion. Furthermore, one has to consider the interference from terrestrial systems and adjacent satellites sharing the same frequency bands.

Once the *S/N* ratio has been selected, the noise per channel (pWp) can be determined:

$$\text{Noise power} = \log^{-1}\left(\frac{90 - S/N}{10}\right) \text{ (pWp)}$$

The noise budget for a FDM-FM-FDMA link using *INTELSAT IV* and *IVA* is

Thermal noise (uplink plus downlink) plus satellite intermodulation noise	7,500 pWp
Earth station equipment noise	1,500 pWp
Interference noise	1,000 pWp
Total	10,000 pWp

Example 5.1 Consider a 60-channel FDM system with a maximum baseband frequency of $f_m = 252$ kHz and a specified top channel signal-to-noise ratio of $S/N = 52$ dB. Assume that a FDM multichannel rms frequency deviation of $lf_r = 546$ kHz is used.

(a) The bandwidth of the FDM-FM-FDMA carrier is given by Carson's rule:

$$B = 2(f_\Delta + f_m) = 2(glf_r + f_m)$$

$$= 2(3.16 \times 546 \text{ kHz} + 252 \text{ kHz})$$

$$= 3.955 \text{ MHz}$$

(b) The FDM multichannel loading factor for 60 channels is

$$l = \log^{-1}\left(\frac{-1 + 4 \log 60}{20}\right) = 2$$

(c) The 0-dBm test-tone rms frequency deviation is

$$f_r = \frac{lf_r}{l} = \frac{546 \text{ kHz}}{2} = 273 \text{ kHz}$$

(d) Assume a 6.5-dB improvement in emphasis and psophometric weighting; then the FDM-FM-FDMA carrier-to-noise ratio in decibels is

$$\frac{C}{N} = 10 \log \left(\frac{S}{N}\right) + 10 \log \left(\frac{b}{B}\right) + 20 \log \left(\frac{f_m}{f_r}\right) - 10 \log PW$$

where P = emphasis weighting factor
$\quad\ W$ = psophometric weighting factor
$\quad\ b$ = telephone channel bandwidth = 3.1 kHz

Then

$$\frac{C}{N} = 52 + 10 \log \left(\frac{3.1 \text{ kHz}}{3955 \text{ kHz}}\right) + 20 \log \left(\frac{252 \text{ kHz}}{273 \text{ kHz}}\right) - 6.5$$

$$= 13.75 \text{ dB}$$

5.2 SINGLE CHANNEL PER CARRIER

Unlike FDM-FM-FDMA systems which serve large-capacity links, *single-channel-per-carrier* systems are more suitable for applications that require only a few channels per link. In these systems each telephone channel independently modulates a separate RF carrier and is transmitted to the satellite transponder on a FDMA basis. A 36-MHz transponder can carry as many as 800 voice channels or more. If the carrier modulation is digital, the performance is measured in terms of the average probability of bit error (Chap. 9). For analog carrier modulation, FM is employed. FM-SCPC systems are the most commonly used systems because of their attractiveness in terms of cost and simplicity. The design of a FM-SCPC link can be expressed in terms of the signal-to-noise ratio at the FM demodulator output, as in FDM-FM-FDMA. Because there is only one channel per carrier, the output noise power is obtained by integrating (5.11) from $-f_m$ to f_m, where f_m is the maximum frequency of the channel:

$$N = \frac{N_0}{A^2} \int_{-f_m}^{f_m} f^2 \, df$$

$$= \frac{2N_0 f_m^3}{3A^2} \tag{5.17}$$

Let Δf be the peak test-tone deviation; then

$$S = \frac{\Delta f^2}{2} \tag{5.18}$$

and

$$\frac{S}{N} = \frac{3A^2 \, \Delta f^2}{4N_0 f_m^3}$$

$$= \frac{3}{2} \left(\frac{C}{N}\right) \left(\frac{B}{f_m}\right) \left(\frac{\Delta f}{f_m}\right)^2 \tag{5.19}$$

where $C/N = A^2/2N_0 B$ = carrier-to-noise ratio and $B = 2(\Delta f + f_m)$ = carrier bandwidth. For a speech signal, the lowest frequency is 300 Hz and the highest frequency is $f_m = 3.4$ kHz, giving a channel bandwidth of $b = 3.1$ kHz.

Example 5.2 Consider a SCPC-FM-FDMA system with a specific channel signal-to-noise ratio $S/N = 33$ dB. Let the test-tone peak frequency deviation be $\Delta f = 9.1$ kHz. (a) The bandwidth of the SCPC-FM-FDMA is given by Carson's rule:

$$B = 2(\Delta f + f_m)$$

$$= 2(9.1 \text{ kHz} + 3.4 \text{ kHz})$$

$$= 25 \text{ kHz}$$

(b) The SCPC-FM-FDMA carrier-to-noise ratio is

$$\frac{C}{N} = 10 \log \left(\frac{2}{3}\right) + 10 \log \left(\frac{f_m}{B}\right) + 20 \log \left(\frac{f_m}{\Delta f}\right) + 10 \log \left(\frac{S}{N}\right)$$

$$= 14 \text{ dB}$$

5.3 FM-FDMA TELEVISION

Television broadcasting via satellite in the United States is among the most highly developed in the world. TV programming is distributed on the *fixed satellite service* portion of the C and Ku bands. In 1983 the *Federal Communications Commission* approved a frequency band for domestic *direct broadcast satellite* services (DBS) to provide direct-to-home television: an uplink frequency of 17.3 to 17.8 GHz and a downlink frequency of 12.2 to 12.7 GHz. The DBS downlink portion of the Ku band is adjacent to the 11.7- to 12.2-GHz downlink frequency of the FSS portion of the Ku band. High-power direct broadcast satellites have many characteristics similar to those of communications satellites, except that the DBS downlink radiated power is about 10 dB more per transponder. The powerful television signal lets individual users receive programs with antennas as small as 0.7 m in diameter, which can be mounted on the roof of an average house. The nominal carrier-to-noise ratio is about 14 to 15 dB when used with an earth station G/T of 10 dB/K. The television receive-only terminal employs an offset-fed antenna with an efficiency in the range of 70 to 75% and a receiver using a GaAs FET low-noise amplifier with noise figures ranging from 2.5 to 3.2 dB.

The performance of a FM-FDMA television channel is expressed in terms of the peak-to-peak luminance signal-to-noise ratio. For a sinusoidal wave, the peak-to-peak power is $(2\sqrt{2})^2$ times the rms power. Also, the peak-to-peak value of the luminance signal is $1/\sqrt{2}$ times the peak-to-peak value of the composite television signal. Therefore the *peak-to-peak luminance signal-to-noise ratio* for a FM-FDMA television signal is

$$\left(\frac{S}{N}\right)_{p-p} = (2\sqrt{2})^2 \left(\frac{1}{\sqrt{2}}\right)^2 \frac{3}{2} \left(\frac{C}{N}\right) \left(\frac{B}{f_m}\right) \left(\frac{f_\Delta}{f_m}\right)^2 PW$$

$$= 6 \left(\frac{C}{N}\right) \left(\frac{B}{f_m}\right) \left(\frac{f_\Delta}{f_m}\right)^2 PW$$

where C/N = carrier-to-noise ratio
P = preemphasis-deemphasis factor
W = noise weighting factor
f_Δ = peak frequency deviation
f_m = maximum frequency
$B = 2(f_\Delta + f_m)$ = Carson's rule bandwidth

Typical FM-FDMA television parameters in fixed satellite service are $B = 36$ MHz, $f_m = 4.2$ MHz, $f_\Delta = B/2 - f_m = 13.8$ MHz, and $PW = 12.8$ dB. For DBS television channels, typical parameters are $B = 24$ MHz, $f_m = 4.2$ MHz, $f_\Delta = B/2 - f_m = 7.8$ MHz, and $PW = 12.8$ dB. Using these parameters we can express $(S/N)_{p-p}$ in terms of C/N as

$$\left(\frac{S}{N}\right)_{p-p} = 40.24 + \frac{C}{N} \text{ (dB)} \qquad \text{FSS}$$

$$\left(\frac{S}{N}\right)_{p-p} = 33.53 + \frac{C}{N} \text{ (dB)} \qquad \text{DBS}$$

With a clear-sky carrier-to-noise ratio of 14 dB, the receiver signal-to-noise ratio for a DBS television channel is 47.5 dB. A more detailed calculation which allows for an audio subcarrier above the video baseband yields a 2-dB reduction in the video S/N. A 45.5-dB S/N indicates a subjective impairment grade of approximately 4 (perceptible impairment but not annoying). A description of DBS systems can be found in [2].

5.4 COMPANDED FDM-FM-FDMA AND SSB-AM-FDMA

The transponder capacity in FDM-FM-FDMA operations can be improved by the use of *syllabic compandors*. The traditional use of syllabic compandors has been to improve the quality of signal transmission over poor channels. A compandor consists of a compressor at the transmit side of the satellite channel and an expandor at the receive side. The compressor is a variable-gain amplifier that gives more gain to weak signals than to strong signals. This results in an improved overall signal-to-noise ratio because the low-level speech signals are increased in power above the channel noise. On the receive side, the expandor restores the signal's level by attenuating the low-level speech signals. During pauses in the speech signal, channel noise is reduced by the expandor, giving further improvement in the overall subjective signal-to-noise ratio. A 36-MHz transpondor can accommodate a single FDM-FM-FDMA carrier of 1100 uncompanded channels. On companding the channels, the capacity is increased to about 2100 channels. With overdeviation beyond its allocated bandwidth (with no loss in the quality of the channels), such a transponder can carry about 2900 channels.

Recent use of solid-state power amplifiers with sufficiently linear characteristics to replace nonlinear TWTAs allows the use of companded single-sideband–amplitude modulation–frequency division multiple access (SSB-AM-FDMA) to achieve 6000 channels per transponder of 36-MHz bandwidth for a single carrier. Besides the high capacity, SSB-AM-FDMA offers another major advantage over FDM-FM-FDMA from a multiple access point of view. The capacity of a satellite transponder using SSB-AM-FDMA is not decreased by multiple access, unlike FDM-FM-FDMA. Also, the capacity of small FDM-FM-FDMA carriers cannot be increased by overdeviation, because of the crosstalk among the carriers. A transponder carrying 6000 SSB-AM-FDMA channels can be accessed, say, by four earth stations with 1500 channels, each with no loss in capacity. On the other hand, a four-carrier companded FDM-FM-FDMA transponder can carry a total of about 1500 channels. The design principles of FDM-FM-FDMA and SSB-AM-FDMA can be found in [3] and [4].

5.5 INTERMODULATION PRODUCTS RESULTING FROM AMPLITUDE NONLINEARITY

In this section the analysis of intermodulation products is carried out based on the amplitude nonlinearity of the TWTA only, with the assumption that the level of intermodulation products caused by phase nonlinearity (conversion of multiple-carrier amplitude fluctuations to phase modulation of all carriers, hence generating intermodulation products) is small.† Also, it is assumed that the input consists of n carriers having equal frequency separation and equal amplitude. (This is the case in long-haul transmissions, such as voice transmissions by satellite, to avoid the dominant effect of intermodulation products on small-amplitude carriers.) The multicarrier amplitude transfer characteristic of a TWTA such as the one in Fig. 5.2 can be represented by an odd polynomial:

$$v_0 = a_1 v_i + a_3 v_i^3 + a_5 v_i^5 + \cdots \qquad (5.20)$$

where v_0 = output voltage of TWTA

v_i = input voltage

a_i = constants having positive and negative values alternatively

The equal-amplitude n-carrier input signal is given by

$$v_i = \sum_{i=1}^{n} A \cos \left[\omega_i t + \theta_i(t) \right] \qquad (5.21)$$

hence the total input power P_{ti} is simply

$$P_{ti} = \frac{nA^2}{2} \qquad (5.22)$$

or

$$A = \sqrt{\frac{2P_{ti}}{n}} \qquad (5.23)$$

Substituting v_i into the expression of v_0 yields

---

†This may be the case where solid-state GaAs FET power amplifiers are used in lieu of TWTAs in satellite transponders. The phase nonlinearity of these power amplifiers in the vicinity of a 1- to 6-dB output back-off is small and may contribute an additional 1 to 2 dB to intermodulation.

$$v_0 = \sum_{i=1}^{n} B_n \cos\left[\omega_i t + \phi_i(t)\right] + IM + H \tag{5.24}$$

where H consists of harmonics $k\omega_i$ and can be filtered and IM is the intermodulation products that fall into the bandwidth of the carriers. These intermodulation products are located at frequencies $2\omega_i - \omega_{i+1}$, $\omega_i + \omega_{i+1} - \omega_{i+2}$, and so on. They fall within the transponder bandwidth and interfere with other carriers. The amplitude B_n of the individual output carrier [5] is given by

$$B_n = a_1 \sqrt{\frac{2P_{ti}}{n}} \left\{ 1 + 3\frac{a_3}{a_1}\left(\frac{P_{ti}}{n}\right)\left(n - \frac{1}{2}\right) \right.$$

$$+ 15\frac{a_5}{a_1}\left(\frac{P_{ti}}{n}\right)^2\left[\frac{1}{6} + (n-1)(n-2) + \frac{3}{2}(n-1)\right]$$

$$+ 105\frac{a_7}{a_1}\left(\frac{P_{ti}}{n}\right)^3\left[(n-1)(n-2)(n-3) + 3(n-1)(n-2)\right.$$

$$\left.+ \frac{34}{24}(n-1) + \frac{1}{24}\right] + \cdots \right\}$$

$$\tag{5.25}$$

where the term $a_1\sqrt{2P_{ti}/n}$ represents the output amplitude due to the first term of the amplitude transfer characteristic and the term inside the braces is the amplitude compression factor due to amplitude nonlinearity. Among the intermodulation products, the third-order types at frequency $2\omega_i - \omega_{i+1}$ and $\omega_i + \omega_{i+1} - \omega_{i+2}$ are the most dominant, and to a lesser extent the fifth-order types at frequency $3\omega_i - 2\omega_{i+1}$. The amplitudes $V_{2.1}$, $V_{1.1.1}$, and $V_{3.2}$ of the intermodulation products $2\omega_i - \omega_{i+1}$, $\omega_i + \omega_{i+1} - \omega_{i+2}$, and $3\omega_i - 2\omega_{i+1}$ for n-carrier input [5] are given by

$$V_{2.1} = \frac{3}{4}a_3\left(\frac{2P_{ti}}{n}\right)^{3/2}\left\{ 1 + \frac{2a_5}{3a_3}\left(\frac{P_{ti}}{n}\right)[12.5 + 15(n-2)]\right.$$

$$\left.+ 105\frac{a_7}{a_3}\left(\frac{P_{ti}}{n}\right)^2\left[(n-2)(n-3) + \frac{13}{6}(n-2) + \frac{7}{12}\right] + \cdots \right\}$$

$$\tag{5.26}$$

$$V_{1.1.1} = \frac{3}{2}a_3\left(\frac{2P_{ti}}{n}\right)^{3/2}\left\{ 1 + 10\frac{a_5}{a_3}\left(\frac{P_{ti}}{n}\right)\left[\frac{3}{2} + (n-3)\right]\right.$$

$$\left.+ 210\frac{a_7}{a_3}\left(\frac{P_{ti}}{n}\right)^2\left[1 + \frac{7}{4}(n-3) + \frac{1}{2}(n-3)(n-4)\right] + \cdots \right\}$$

$$\tag{5.27}$$

$$V_{3.2} = \frac{5}{8}a_5\left(\frac{2P_{ti}}{n}\right)^{5/2}\left\{ 1 + \frac{49a_7}{4a_5}\left(\frac{P_{ti}}{n}\right)\left[1 + \frac{12}{7}(n-2)\right] + \cdots \right\}$$

$$\tag{5.28}$$

where each expression is valid only when $n - x$ is positive or zero. The term outside the braces in the above three equations represents the contribution of the intermodulation products resulting from the cubic term in the transfer characteristic. The term inside the braces—the intermodulation compression factor—represents the intermodulation resulting from higher-order nonlinearities. It is observed that the amplitude of the third-order intermodulation product $\omega_i + \omega_{i+1} - \omega_{i+2}$ is 3 dB higher than that of the type $2\omega_i - \omega_{i+1}$. Also, since $|a_5| \ll |a_3|$ in practice and the compression factor for the fifth-order intermodulation product decreases rapidly with increasing n, it can be ignored in comparison with third-order types and so are all other higher-order products in comparison with third-order types.

To evaluate the amplitude B_n of the output carrier and associated third-order intermodulation amplitudes $V_{1,1,1}$ and $V_{2,1}$ it is necessary to know the coefficients a_i of the n-carrier amplitude transfer characteristic. In practice the single-carrier ($n = 1$) transfer characteristic is normally given. To calculate a_i we note that, when the input consists of only one carrier of amplitude

$$A_0 = \sqrt{2P_{ti}}$$

the output carrier has an amplitude B_1 given by

$$B_1 = a_1 \sqrt{2P_{ti}} \left(1 + \frac{3a_3}{2a_1} P_{ti} + \frac{5a_5}{2a_1} P_{ti}^2 + \frac{35a_7}{8a_1} P_{ti}^3 + \cdots \right)$$

$$= a_1 A_0 + \tfrac{3}{4} a_3 A_0^3 + \tfrac{5}{8} a_5 A_0^5 + \tfrac{35}{64} a_7 A_0^7 + \cdots$$

$$= a_1' A_0 + a_3' A_0^3 + a_5' A_0^5 + a_7' A_0^7 + \cdots \tag{5.29}$$

where $a_1' = a_1$, $a_3' = 3a_3/4$, $a_5' = 5a_5/8$, $a_7' = 35a_7/64$, ..., can be found by least squares fitting (see Appendix 5A) the measured data, such as the data presented graphically in Fig. 5.2 where the normalized single-carrier amplitude transfer characteristic is given, that is, $B_1/B_{1,sat}$ versus $A_0/A_{0,sat}$. (Note that the input amplitude A_0 is normalized to the saturation input amplitude $A_{0,sat}$ where the saturation output amplitude $B_{1,sat}$ occurs; beyond this point a further increase in the input amplitude will produce a lower output amplitude; also, the output amplitude B_1 is normalized to $B_{1,sat}$.) Once $a_1', a_3', a_5', a_7', \ldots$, are found, $a_1, a_3, a_5, a_7, \ldots$, can be computed, hence B_n and $V_{1,1,1}$ and $V_{2,1}$ can be evaluated.

As analyzed above, the intermodulation products produced by the amplitude nonlinearity of the TWTA operating near saturation are predominantly third-order types whose amplitudes can be evaluated from the single-carrier amplitude transfer characteristic. Therefore, if the number of third-order intermodulation products which fall in the bandwidth of the respective carriers is known, the carrier-to-(third-order) intermodulation ratio C/I for n-carrier input can be calculated. This C/I can be used in conjunction with the carrier-to-noise ratio on the uplink and the downlink to determine the overall carrier-to-intermodulation

Table 5.1 Distribution of the number of the third-order intermodulation products $2\omega_i - \omega_{i+1}$

				r				
n	1	2	3	4	5	6	7	8
1	0							
2	0	0						
3	1	0	1					
4	1	1	1	1				
5	2	1	2	1	2			
6	2	2	2	2	2	2		
7	3	2	3	2	3	2	3	
8	3	3	3	3	3	3	3	3

plus noise ratio and will be discussed in a later section. For the case of equally spaced equal-amplitude n-carrier input, the number of third-order intermodulation products that fall right on the rth carrier [5] can be shown to be

$$D_{2,1}(r,n) = \tfrac{1}{2}\{n - 2 - \tfrac{1}{2}[1 - (-1)^n]\,(-1)^r\} \tag{5.30}$$

for the $2\omega_i - \omega_{i+1}$ third-order type, and

$$D_{1,1,1}(r,n) = \frac{r}{2}\,(n - r + 1) + \frac{1}{4}\,[(n - 3)^2 - 5]$$

$$- \frac{1}{2}\,[1 - (-1)^n]\,(-1)^{n+r} \tag{5.31}$$

for the $\omega_i + \omega_{i+1} - \omega_{i+2}$ third-order type. Tables 5.1 and 5.2 show these distributions for n from 1 to 8. It is observed that, as n increases,

Table 5.2 Distribution of the number of the third-order intermodulation products $\omega_i + \omega_{i+1} - \omega_{i+2}$

				r				
n	1	2	3	4	5	6	7	8
1	0							
2	0	0						
3	0	1	0					
4	1	2	2	1				
5	2	4	4	4	2			
6	4	6	7	7	6	4		
7	6	9	10	11	10	9	6	
8	9	12	14	15	15	14	12	9

the $\omega_i + \omega_{i+1} - \omega_{i+2}$ third-order intermodulation products become dominant in number. Also, the $2\omega_i - \omega_{i+1}$ third-order intermodulation products are distributed equally among the carriers, while the $\omega_i + \omega_{i+1} - \omega_{i+2}$ types are concentrated more at the central carriers. For example, when $n = 8$, the difference between the center carriers $r = 4,5$ and the edge carriers $r = 1,8$ is $10 \log (\frac{15}{9}) = 2.2$ dB in carrier-to-intermodulation ratio. The carrier-to-(third-order) intermodulation ratio $(C/I)_r$ of any output carrier r is therefore given by (assuming that the modulated phases θ_i are independent and that the various intermodulation products are also independent and can be added on the power basis)

$$\left(\frac{C}{I}\right)_r = \frac{B_n^2}{D_{2,1}\ (r,n)\ V_{2,1}^2 + D_{1,1,1}\ (r,n)\ V_{1,1,1}^2} \tag{5.32}$$

The most meaningful indicator of the power drive level of the TWTA is the *output back-off* BO_o which is defined as the ratio of the peak single-carrier output power $(n = 1)$ to the total desired n-carrier output power as a function of the total input power P_{ti}. Thus

$$BO_0 = \frac{B_{1,\text{sat}}^2}{nB_n^2} \tag{5.33a}$$

If we wish to use the normalized amplitude transfer characteristic (Fig. 5.2), then $B_{1,\text{sat}} = 1$ and

$$BO_0 = \frac{1}{nB_n^2} \tag{5.33b}$$

The concept of output back-off is shown in Fig. 5.6. It is seen that the total n-carrier output power $nB_n^2/2$ is always smaller than the total single-carrier output $B_1^2/2$ because power is lost in intermodulation products. Moreover, the input back-off is the ratio of the single-carrier input power which yields the saturated output power $B_{1,\text{sat}}^2/2$ to the total input power of the n-carrier input at its operating point. In practice $B_{1,\text{sat}}^2/2$ is given (e.g., a 15-W TWTA means $B_{1,\text{sat}}^2/2 = 15$ W), and it is a straightforward matter to relate all the normalized quantities to their real values. Equations (5.32) and (5.33) show that $(C/I)_r$ and BO_o are both functions of the total input power P_{ti}; $(C/I)_r$ versus BO_0 can be easily plotted by using the above two equations.

Example 5.3 In this example we will calculate the carrier-to-(third-order)-intermodulation product for a FDMA system having 200 equally spaced equal-amplitude carriers. The normalized single-carrier amplitude transfer characteristic is

Input voltage	0	0.10	0.20	0.30	0.40	0.50	0.60
Output voltage	0	0.20	0.40	0.60	0.73	0.84	0.92

Input voltage	0.70	0.80	1	1.2	1.4	1.6	2
Output voltage	0.967	0.995	1	0.98	0.93	0.88	0.77

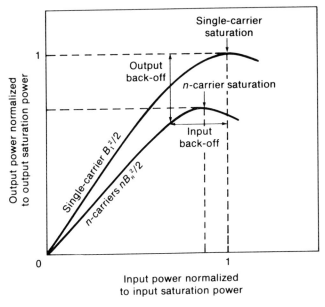

Figure 5.6 Concept of input and output back-offs of a TWTA.

Using least-squares fitting in Appendix 5A, we have $a_1' = 1.9817$, $a_3' = -1.3638$, $a_5' = 0.4507$, and $a_7' = -0.0524$.

Input voltage	Output voltage by fitting
0.0	0
0.1	0.1968
0.2	0.3856
0.3	0.5588
0.4	0.7099
0.5	0.8340
0.6	0.9280
0.7	0.9908
0.8	1.0237
1.0	1.0161
1.2	0.9550
1.4	0.9035
1.6	0.9036
2.0	0.7677

The carrier-to-intermodulation ratio versus total input power, input back-off, and output back-off are given in Table 5.3. Note that input back-off is given by

$$\mathrm{BO_i} = 10 \log \frac{B_{i,\mathrm{sat}}^2/2}{P_{ti}} = 10 \log \frac{0.5}{P_{ti}}$$

When the effect of phase nonlinearity is taken into account, it can be shown to a first-order approximation [5] that the amplitude of the dominant third-order intermodulation products relative to the amplitude of the desired output signal is given by

Table 5.3 Carrier-to-intermodulation ratios in Example 5.3

P_{ti}	BO_i	BO_o	$(C/I)_{1,200}$	$(C/I)_{100}$
0.020	13.98	8.50989	31.501	29.729
0.040	10.97	5.95807	25.729	23.957
0.060	9.21	4.64188	22.484	20.712
0.080	7.97	3.82250	20.289	18.517
0.100	6.99	3.26790	18.681	16.909
0.120	6.21	2.87449	17.451	15.679
0.140	5.53	2.58701	16.485	14.713
0.160	4.97	2.37263	15.712	13.940
0.180	4.44	2.21050	15.082	13.310
0.200	3.99	2.08682	14.557	12.785
0.220	3.57	1.99233	14.104	12.332
0.240	3.21	1.92085	13.693	11.921
0.260	2.84	1.86844	13.296	11.525
0.280	2.53	1.83287	12.887	11.115
0.300	2.22	1.81327	12.437	10.665
0.320	1.97	1.80989	11.924	10.152
0.340	1.67	1.82399	11.326	9.554
0.360	1.44	1.85771	10.626	8.854
0.380	1.19	1.91407	9.813	8.041
0.400	0.99	1.99696	8.881	7.109
0.420	0.76	2.11126	7.830	6.058
0.440	0.57	2.26295	6.662	4.890
0.460	0.36	2.45936	5.383	3.612
0.480	0.21	2.70958	3.998	2.227
0.500	0.00	3.02501	2.510	0.738

$0.1516K_0/2n$ for the $2\omega_1 - \omega_2$ products and $0.1516K_0/n$ for the $\omega_1 + \omega_2 - \omega_3$ products, where K_0 is the amplifier amplitude modulation–phase modulation (AM/PM) conversion coefficient (deg/dB) and can be calculated from the phase nonlinearity. Typical values of K_0 range from 1.5 to 3°/dB at an input back-off of 11 to 7 dB. For example, let $K_0 = 2°/dB$ at $BO_i = 11$ dB for the TWTAs in Table 5.3. Then, $0.1516K_0/2n = -31.2$ dB and $0.1516K_0/n = -28.2$ dB. Therefore the additional carrier-to-intermodulation ratio resulting from phase nonlinearity is 26.4 dB. Since the carrier-to-intermodulation ratio resulting from amplitude nonlinearity is 24 dB, the total carrier-to-intermodulation ratio is 22 dB.

Example 5.4 Consider the FDMA system discussed in Example 4.2 where 200 QPSK carriers are involved. Table 4.2 shows the carrier-to-noise ratio is

$$\frac{C}{N} = 15 \text{ dB}$$

for an input back-off of $BO_i = 11$ dB and an output back-off of $BO_o = 6$ dB. Suppose the TWTA considered in Example 4.2 is also the one discussed in Example 5.3. Then the carrier-to-intermodulation ratio for the center carrier 100 is

$$\left(\frac{C}{I}\right)_{100} = 24 \text{ dB}$$

Assume that the carrier-to-terrestrial interference ratio $(C/I)_t = 26$ dB and the carrier-to-adjacent satellite interference ratio is $(C/I)_a = 22$ dB. Then, the total carrier-to-noise plus interference ratio of the FDMA system is

$$\frac{C}{\mathcal{N}} = \left[\left(\frac{C}{N}\right)^{-1} + \left(\frac{C}{I}\right)^{-1}_{100} + \left(\frac{C}{I}\right)^{-1}_t + \left(\frac{C}{I}\right)^{-1}_a\right]^{-1}$$

$$= 13.5 \text{ dB}$$

The corresponding E_b/N_0 for $T_b B = 0.625$ is (see 9.51)

$$\frac{E_b}{N_0} = T_b B\left(\frac{C}{\mathcal{N}}\right) = 11.5 \text{ dB}$$

The average probability of bit error for the center channel can be computed from (9.51):

$$P_b \approx 10^{-7}$$

5.6 INTERMODULATION PRODUCTS RESULTING FROM BOTH AMPLITUDE AND PHASE NONLINEARITIES

In the above analysis, the carrier-to-(third order)-intermodulation ratio is calculated based on the assumption that the amplitude of the intermodulation products caused by the phase nonlinearity is small. In this section this effect is included in the analysis and, furthermore, other intermodulation products besides the above third-order types are also considered. Consider the multiple-carrier input signal.

$$v = \sum_{i=1}^{n} A_i \cos\left[\omega_i t + \theta_i(t)\right] \tag{5.34}$$

where $\theta_i(t)$ is the modulated phase. This can be expressed as a carrier having a mean angular frequency of $\omega_0 = (\omega_1 + \omega_n)/2$ with envelope $R(t)$ and phase $\theta(t)$ as

$$v = R(t) \cos\left[\omega_0 t + \theta(t)\right] \tag{5.35a}$$

where

$$R^2(t) = \left\{\sum_{i=1}^{n} A_i \cos\left[(\omega_i - \omega_0)t + \theta_i(t)\right]\right\}^2$$

$$+ \left\{\sum_{i=1}^{n} A_i \sin\left[(\omega_i - \omega_0)t + \theta_i(t)\right]\right\}^2 \tag{5.35b}$$

$$\theta(t) = \tan^{-1}\left\{\frac{\sum\limits_{i=1}^{n} A_i \sin\left[(\omega_i - \omega_0)t + \theta_i(t)\right]}{\sum\limits_{i=1}^{n} A_i \cos\left[(\omega_i - \omega_0)t + \theta_i(t)\right]}\right\} \tag{5.35c}$$

In the amplitude-phase model, the corresponding output signal $V(t)$ of the

TWTA is expressed as

$$V(t) = G[R(t)] \cos \{\omega_0 t + \theta(t) + F[R(t)]\} \qquad (5.36)$$

where $G[R]$ and $F[R]$ represent the amplitude and phase transfer characteristics of the TWTA, respectively, as shown in Fig. 5.2. Also, $V(t)$ can be written as

$$V(t) = G[R(t)] \cos F[R(t)] \cos [\omega_0 t + \theta(t)]$$
$$-G[R(t)] \sin F[R(t)] \sin [\omega_0 t + \theta(t)]$$
$$= P[R(t)] \cos [\omega_0 t + \theta(t)] - Q[R(t)] \sin [\omega_0 t + \theta(t)]$$

where

$$P[R] = G[R] \cos F[R] \qquad (5.37)$$

$$Q[R] = G[R] \sin F[R] \qquad (5.38)$$

represent the in-phase and quadrature transfer characteristics of the TWTA, respectively (quadrature model).

The amplitude of the intermodulation products contained in $V(t)$ whose angular frequency is $k_1\omega_1 + k_2\omega_2 + \cdots + k_n\omega_n$ has been derived in [6]:

$$V_m = |M(k_1, k_2, \ldots, k_n)| \qquad (5.39)$$

where k_i are integers with the constraint

$$k_1 + k_2 + \cdots + k_n = 1 \qquad (5.40)$$

to guarantee that all the intermodulation products considered fall in the respective carrier bandwidth, and m is the order of the intermodulation product given by

$$m = |k_1| + |k_2| + \cdots + |k_n| \qquad (5.41)$$

which by the above constraint in (5.40) assumes only odd values and $M(k_1, k_2, \ldots, k_n)$ is the complex quantity given by

$$M(k_1, k_2, \ldots, k_n) = \int_0^\infty r \prod_{i=1}^n J_{k_i} (A_i r) \, dr \int_0^\infty \rho G[\rho] \exp\{jF[\rho]\} \, J_1 (r\rho) \, d\rho$$

$$= \int_0^\infty r \prod_{i=1}^n J_{k_i} (A_i r) \, dr \int_0^\infty \rho\{P[\rho] + jQ[\rho]\} J_1(r\rho) \, d\rho$$

$$(5.42)$$

where $J_k =$ Bessel function of the first kind of order k (Appendix 5B).

Note that the amplitude of the carrier with angular frequency ω_l can be found from (5.39) by letting $k_l = 1$ and $k_i = 0$ for all $i = 1, 2, \ldots, n$ and $i \neq l$.

To evaluate the amplitude V_m of the mth-order intermodulation products, it is necessary to have a model for $G[R(t)]$ and $F[R(t)]$ using the measured data of the single-carrier input such as the curves in Fig. 5.2 and letting $R(t) = A_0$, where A_0 is the single-carrier amplitude. One can represent the amplitude model $G[A_0]$ by an odd polynomial of A_0, and the phase model $F[A_0]$ by an even polynomial of A_0. However, simple two-parameter models have been proposed for $G[A_0]$ and $F[A_0]$ which fit the measured data of the single-carrier amplitude and phase transfer characteristics very well [7], and these are

$$G[A_0] = \frac{\alpha_G A_0}{1 + \beta_G A_0^2} \tag{5.43}$$

$$F[A_0] = \frac{\alpha_F A_0^2}{1 + \beta_F A_0^2} \tag{5.44}$$

Again the coefficients α_G, β_G and α_F, β_F can be found from the measured single-carrier amplitude and phase transfer functions by least-squares fitting (see Appendix 5A). In a similar way $P[A_0]$ and $Q[A_0]$ in (5.37) and (5.38) can also be represented by two-parameter models:

$$P[A_0] = \frac{\alpha_P A_0}{1 + \beta_P A_0^2} \tag{5.45}$$

$$Q[A_0] = \frac{\alpha_Q A_0^3}{(1 + \beta_Q A_0^2)^2} = -\frac{\partial P[A_0]}{\partial \beta_P}\Big|_{\alpha_P \to \alpha_Q, \beta_P \to \beta_Q} \tag{5.46}$$

Substituting $P[\cdot]$ and $Q[\cdot]$ into the expression of $M(k_1, k_2, \ldots, k_n)$ in (5.42) and using the Hankel-type integral given in [8] yields [9]

$$\begin{aligned}
M(k_1, k_2, \ldots, k_n) = \int_0^\infty r \Big\{ &\alpha_P \beta_P^{-3/2} K_1(r\beta_P^{-1/2}) \\
&+ j \alpha_Q \Big[\beta_Q^{-5/2} K_1(r\beta_Q^{-1/2}) \\
&- \Big(\frac{r}{2\beta_Q^3}\Big) K_0(r\beta_Q^{-1/2}) \Big] \Big\} \prod_{i=1}^n J_{k_i}(A_i r) \, dr
\end{aligned} \tag{5.47}$$

where K_0 and K_1 are modified Bessel functions of the second kind of order 0 and 1, respectively (Appendix 5B).

One interesting special case occurs when all the carrier amplitudes in (5.34) are equal:

$$A_i = A \qquad i = 1, 2, \ldots, n \tag{5.48}$$

and when the number of carriers n is large such that for any m of interest

$$n \gg m \tag{5.49}$$

In this case the mth-order intermodulation products whose amplitudes can be related to $M(k_1, k_2, \ldots, k_n)$ in (5.42) or (5.47) as [10]

$$M_{m,n} = X_{m,n} + jY_{m,n} \qquad m = 1, 3, 5, \ldots$$
$$= M \underbrace{(1, 1, \ldots, 1,}_{\frac{m+1}{2}} \underbrace{-1, -1, \ldots, -1,}_{\frac{m-1}{2}} \underbrace{0, 0, \ldots, 0)}_{n-m} \qquad (5.50)$$

totally dominate all other intermodulation products, basically because of their larger number, but also, to a lesser degree, because of their larger magnitudes. Because all the carriers are assumed to have equal amplitude, the order of $1, -1$, and 0 in (5.50) is not important. Note that $M_{1,n}$ given by

$$M_{1,n} = M(1, 0, 0, \ldots, 0)$$

is the amplitude of the desired output signal of any carrier, since $A_i = A$, for $i = 1, 2, \ldots, n$, and that $M_{m,n}$, $m \geq 3$, is the amplitude of the interfering intermodulation products. A simple expression for (5.50) is obtained in [9] and [11]:

$$M_{m,n} = X_{m,n} + jY_{m,n} = n^{-m/2} \sum_{k=0}^{(m-1)/2} S_{k,m} R_k (\sqrt{n}A) \qquad (5.51a)$$

where $\sqrt{n}A = \sqrt{2P_{ti}}$ (P_{ti} is the total input power) and

$$R_k(\sqrt{n}A) = 2 \int_0^\infty \{P[\sqrt{n}Ax] + jQ[\sqrt{n}Ax]\} \exp(-x^2) x^{2(k+1)} dx$$
$$= 2 \int_0^\infty G[\sqrt{n}Ax] \exp\{jF[\sqrt{n}Ax]\} \exp(-x^2) x^{2(k+1)} dx$$

$$(5.51b)$$

$$S_{k,m} = (-1)^k \binom{\frac{m-1}{2}}{k} \frac{[(m+1)/2]!}{(k+1)!} \qquad (5.51c)$$

The upper limit in (5.51b) should be at least 10 dB above $\sqrt{n}A$.

An important question concerning the validity of the analysis is: What is the least number of carriers n such that (5.49) still holds? According to experiments, intermodulation products on the order of $m > 7$ are not significant (recall that only third-order intermodulation products are treated in Sec. 5.1). Therefore it is reasonable to conclude that the smallest valid number of n is about 15. It was shown in [10] that the results are accurate within 1 dB for n as low as 7.

In order to evaluate the carrier-to-intermodulation ratio it is necessary to know the distribution of the intermodulation products. For most practical cases where all carriers are equally spaced in frequency, that is,

$$\omega_{i+1} - \omega_i = \omega_{j+1} - \omega_j \qquad i, j = 1, 2, \ldots, n-1$$

the number of dominant mth-order intermodulation products that fall at the normalized output frequency ν associated with the output angular frequency ω_l

$$\nu = \frac{\omega_l - (\omega_1 + \omega_n)/2}{\omega_1 - \omega_n} = \frac{\omega_l - \omega_0}{\omega_1 - \omega_n} \tag{5.52}$$

is given by [10]:

$$D_{m,n}(\nu) = \frac{n^{m-1}}{\left(\frac{m+1}{2}\right)! \left(\frac{m-1}{2}\right)! (m-1)!} \sum_{k=0}^{\lfloor m/2 - \nu \rfloor}$$

$$(-1)^k \binom{m}{k} \left(\frac{m}{2} - k - \nu\right)^{m-1} \qquad |\nu| \leq \frac{m}{2} \tag{5.53}$$

where $\lfloor m/2 - \nu \rfloor$ above the summation sign means the integer part of $m/2 - \nu$ and $D_{m,n}(\nu) = 0$ for $|\nu| > m/2$. Note that $\nu = 0$ corresponds to the center channel, $\nu = \pm\frac{1}{2}$ to an edge channel, and $|\nu| > \frac{1}{2}$ to a frequency falling outside the output frequency band.

Assume that the modulated phases $\theta_i(t)$ are independent so that all the dominant mth-order intermodulation products are independent and can be added on the power basis; then the carrier-to-intermodulation ratio at any desired output signal at the normalized frequency ν is given by

$$\left(\frac{C}{I}\right)_\nu = \frac{|M_{1,n}|^2}{\sum_{m=3,5,7,\dots}^{\infty} D_{m,n}(\nu) |M_{m,n}|^2} \tag{5.54}$$

Similarly one can compute the output backoff of the TWTA as a function of the total input power $P_{\text{ti}} = A_0^2/2 = nA^2/2$:

$$BO_0 = \frac{\max_{A_0} \{G^2[A_0]\}}{n|M_{1,n}|^2} = \frac{\max_{A_0} \{P^2[A_0] + Q^2[A_0]\}}{n|M_{1,n}|^2} \tag{5.55a}$$

If one wishes to use the normalized amplitude and phase transfer characteristics (Fig. 5.2), then $\max_{A_0} \{G^2[A_0]\} = 1$ and

$$BO_0 = \frac{1}{n|M_{1,n}|^2} \tag{5.55b}$$

Note that $(C/I)_\nu$ and BO_0 are both functions of the carrier amplitude A or, equivalently, of the total input power P_{ti}, since $A_0 = \sqrt{2P_{\text{ti}}}$ and $\sqrt{n}A = \sqrt{2P_{\text{ti}}}$. They are also independent of n. Therefore one can easily plot $(C/I)_\nu$ versus BO_0 using the above two equations. The plot of $(C/I)_\nu$ versus BO_0 for the center carrier $\nu = 0$ using a TWTA with $\alpha_G = 2.16$, $\beta_G = 1.15$, $\alpha_F = 4$, and $\beta_F = 9.1$ is in Fig. 5.7.

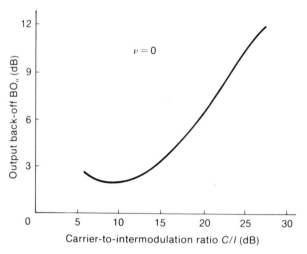

Figure 5.7 Typical C/I versus BO_o.

5.7 OPTIMIZED CARRIER-TO-INTERMODULATION PLUS NOISE RATIO

The intermodulation generated from the amplification of multiple carriers passing through a nonlinear TWTA may affect the overall satellite link performance considerably. When the number of carriers in a transponder is sufficiently large such that the intermodulation noise in each channel is essentially Gaussian, the total carrier-to-noise plus interference ratio of the satellite link (Chap. 4) can be expressed by

$$\frac{C}{\mathcal{N}} = \left[\left(\frac{C}{\mathcal{N}} \right)_u^{-1} + \left(\frac{C}{\mathcal{N}} \right)_d^{-1} + \left(\frac{C}{I} \right)^{-1} \right]^{-1} \tag{5.56}$$

where $(C/\mathcal{N})_u$, $(C/\mathcal{N})_d$, and C/I are the carrier-to-noise plus interference ratio in the uplink, carrier-to-noise plus interference ratio in the downlink, and carrier-to-intermodulation ratio, respectively. From the link equations in Chap. 4 (4.24 and 4.25) we note that $(C/\mathcal{N})_u$ is a function of the TWTA input back-off and $(C/\mathcal{N})_d$ is a function of the TWTA output back-off. Also, Fig. 5.6 shows that the TWTA input back-off BO_i relative to single-carrier saturation is a function of the corresponding output back-off BO_o; therefore the carrier-to-noise plus interference ratio in the uplink is also a function of BO_o. Hence the total carrier-to-noise plus interference ratio of the satellite link is a function of the TWTA output back-off BO_o and by the above equation possesses a maximum value with respect to an optimum BO_o. Figure 5.8 is a graphical presentation of $(C/\mathcal{N})_u$, $(C/\mathcal{N})_d$, C/I, and $C/\mathcal{N}$ as functions of BO_o. The optimum BO_o is the point where $C/\mathcal{N}$ reaches its maximum.

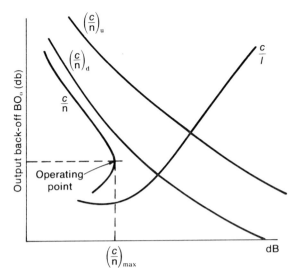

Figure 5.8 Typical $(C/N)_u$, $(C/N)_d$, C/I, and C/N versus BO_o.

REFERENCES

1. S. Haykin, *Communication Systems*, 2d ed. New York: Wiley, 1983.
2. Special issues on satellite communications. *IEEE Commun. Mag.,* Vol. 22, No. 3, Mar. 1984.
3. K. Jonnalagadda and L. Schiff, "Improvements in Capacity of Analog Voice Multiplex Systems Carried by Satellite," *Proc. IEEE,* Vol. 72, No. 11, Nov. 1984, pp. 1537–1547.
4. E. Laborde and P. J. Freedenberg, "Analytical Comparisons of CSSB and TDMA/DSI Satellite Transmission and Techniques," *Proc. IEEE,* Vol. 72, No. 11, Nov. 1984, pp. 1548–1555.
5. R. J. Westcott, "Investigation of Multiple FM/FDM Carriers through a Satellite TWT Operating near to Saturation," *Proc. IEE (London),* Vol. 114, No. 6, June 1967, pp. 726–740.
6. O. Shimbo, "Effects of Intermodulation, AM-PM Conversion, and Additive Noise in Multicarrier TWT Systems," *Proc. IEEE,* Vol. 59, No. 2, Feb. 1971, pp. 230–238.
7. A. A. M. Saleh, "Frequency-Independent and Frequency-Dependent Nonlinear Models of TWT Amplifiers," *IEEE Trans. Commun.,* Vol. COM-29, No. 11, Nov. 1981, pp. 1715–1719.
8. I. S. Gradshteyn and I. M. Ryzhik, *Tables of Integrals, Series and Products.* New York: Academic, 1965, p. 686, eq. (6565.4).
9. A. A. M. Saleh, "Intermodulation Analysis of FDMA Systems Employing Compensated and Uncompensated TWT's," *IEEE Trans. Commun.,* Vol. COM-30, No. 5, May 1982, pp. 1233–1242.
10. F. E. Bond and H. F. Meyer, "Intermodulation Effects in Limiter Amplifier Repeaters," *IEEE Trans. Commun. Technol.,* Vol. COM-18, No. 4, Apr. 1970, pp. 127–135.
11. G. R. Stette, "Calculation of Intermodulation from a Single Carrier Amplitude Characteristics," *IEEE Trans. Commun.,* Vol. COM-22, No. 3, Mar. 1974, pp. 319–323.

12. J. L. Dicks and M. P. Brown, Jr., "Frequency Division Multiple Access (FDMA) for Satellite Communications Systems," EASCON 1974, pp. 167–178.

13. J. L. Dicks, P. H. Schultze, and C. H. Schmitt, "System Planning," COMSAT Tech. Rev., Vol. 2, Fall 1972, pp. 452–469.

14. M. E. Ferguson, "Design of FM Single-Channel-per-Carrier Systems," Int. Conf. Commun. (ICC), Vol. 1, 1975, pp. 12.11–12.16.

15. C. C. Sanderson and L. G. Ludwig, "Single-Channel-per-Carrier Voice Transmission via Communications Satellites," Proceedings of the 5th AIAA Communications Satellite Systems Conference, 1974.

16. J. C. Fuenzalida et al., "Time-Domain Analysis of Intermodulation Effects Caused by Nonlinear Amplifiers," COMSAT Tech. Rev., Vol. 3, Spring 1973, pp. 89–143.

17. A. L. Berman and C. E. Mahle, "Nonlinear Phase Shift in Traveling Wave Tubes as Applied to Multiple Access Communications Satellites," IEEE Trans. Commun. Technol., Vol. COM-18, No. 2, Feb. 1970, pp. 37–48.

18. R. F. Pawula et al., "Intermodulation Distortion in Frequency Multiplexed Satellite Repeaters," Proc. IEEE, Vol. 59, Feb. 1971, pp. 213–218.

19. E. D. Sunde, "Intermodulation Distortion in Multicarrier FM Systems," 1965 IEEE Int. Conv. Rec., Vol. 13, Pt. 2, March 1965, pp. 130–146.

20. F. Rieger and S. Gover, "Predict C/I Performance with Simple Measurements," Microwaves, Vol. 20, No. 8, Aug. 1981, pp. 107–110.

APPENDIX 5A LEAST-SQUARES FITTING

The method of least-squares fitting is a process of finding the best possible values for a set of m unknowns, say $x_1, x_2, \ldots, x_m$ connected by $n > m$ linear equations:

$$a_{11}x_1 + a_{12}x_2 + \cdots + a_{1m}x_m = b_1$$

$$a_{21}x_1 + a_{22}x_2 + \cdots + a_{2m}x_m = b_2$$

$$a_{n1}x_1 + a_{n2}x_2 + \cdots + a_{nm}x_m = b_n \qquad (5A.1)$$

Because the number of equations exceeds the number of unknowns, there is no solution for $x_1, x_2, \ldots, x_m$ such that (5A.1) is satisfied. Therefore it makes sense to find the values for $x_1, x_2, \ldots, x_m$ to minimize the error $e_i = a_{i1}x_1 + a_{i2}x_2 + \cdots + a_{im}x_m - b_i = 0$ for $i = 1, 2, \ldots, n$. The criterion for least-squares fitting is to minimize the sum of the square of e_i, that is, to minimize

$$E = \sum_{i=1}^{n} e_i^2$$

$$= \sum_{i=1}^{n} (a_{i1}x_1 + a_{i2}x_2 + \cdots + a_{im}x_m - b_i)^2 \qquad (5A.2)$$

The square of e_i is necessary to prevent large positive and large negative e_i from canceling each other, resulting in a false impression of accuracy.

The error quantity E in (5A.2) is minimized when

$$\frac{\partial E}{\partial x_1} = 0 \qquad \frac{\partial E}{\partial x_2} = 0 \qquad \cdots \qquad \frac{\partial E}{\partial x_m} = 0$$

which yields

$$x_1 \sum_{i=1}^{n} a_{ij}a_{i1} + x_2 \sum_{i=1}^{n} a_{ij}a_{i2} + \cdots$$

$$+ x_m \sum_{i=1}^{n} a_{ij}a_{im} = \sum_{i=1}^{n} a_{ij}b_i \qquad j = 1, 2, \ldots, m \qquad (5A.3)$$

Equation (5A.3) can be written in terms of a single matrix equation (the superscript T denotes a matrix transpose):

$$(A^T A)\, x = A^T b \qquad (5A.4)$$

where

$$A = [a_{ij}] \qquad i = 1, 2, \ldots, n;\ j = 1, 2, \ldots, m$$

$$x = \begin{bmatrix} x_1 \\ x_2 \\ \cdot \\ \cdot \\ \cdot \\ x_m \end{bmatrix} \qquad b = \begin{bmatrix} b_1 \\ b_2 \\ \cdot \\ \cdot \\ \cdot \\ b_n \end{bmatrix}$$

Therefore the least-squares solution of (5A.1) is

$$x = (A^T A)^{-1} A^T b$$

assuming that the product $A^T A$ is nonsingular.

APPENDIX 5B BESSEL FUNCTIONS

A Bessel function of the first kind of order k and argument x is defined by

$$J_k(x) = \frac{1}{2\pi} \int_{-\pi}^{\pi} \exp(jx \sin \theta - jk\theta)\, d\theta$$

$$= \frac{1}{\pi} \int_{0}^{\pi} \cos(x \sin \theta - k\theta)\, d\theta$$

$$= \sum_{n=0}^{\infty} \frac{(-1)^n \left(\frac{1}{2}x\right)^{k+2n}}{n!\,(n+k)!}$$

The Bessel function $J_k(x)$ has the following properties:

$$J_k(x) = (-1)^k J_{-k}(x) \qquad k \text{ integer}$$

$$J_k(x) = (-1)^k J_k(-x) \qquad k \text{ integer}$$

$$J_{k-1}(x) + J_{k+1}(x) = \frac{2k}{x} J_k(x)$$

$$J_k(x) \approx \frac{x^k}{k! \, 2^k} \qquad x \to 0$$

$$J_k(x) \approx \sqrt{\frac{2}{\pi x}} \cos\left(x - \frac{\pi}{4} - \frac{k\pi}{2}\right) \qquad x \to \infty$$

$$\sum_{k=-\infty}^{\infty} J_k(x) \exp(jky) = \exp(jx \sin y)$$

$$\sum_{k=-\infty}^{\infty} J_k^2(x) = 1$$

A Bessel function of the second kind of order k and argument x is defined by

$$Y_k(x) = \frac{J_k(x) \cos(k\pi) - J_{-k}(x)}{\sin(k\pi)} \qquad k \text{ not integer}$$

When $k = m =$ an integer, then $\sin(m\pi) = 0$, $\cos(m\pi) = (-1)^m$, and $J_m(x) = (-1)^m J_{-m}(x)$. Therefore $Y_m(x)$ is indeterminate. It can also be shown that

$$Y_m(x) = \lim_{k \to m} Y_k(x) = \lim_{k \to m} \frac{J_k(x) \cos(k\pi) - J_{-k}(x)}{\sin(k\pi)}$$

A modified Bessel function of the first kind of order k and argument x is defined by

$$I_k(x) = j^{-k} J_k(jx)$$

$$= \frac{1}{2\pi} \int_{-\pi}^{\pi} \exp(x \cos \theta) \cos(k\theta) \, d\theta$$

$$= \sum_{n=0}^{\infty} \frac{\left(\frac{1}{2}x\right)^{k+2n}}{n! \, (n+k)!}$$

A modified Bessel function of the second kind of order k and argument x is defined by

$$K_k(x) = \frac{\pi}{2} \frac{I_{-k}(x) - I_k(x)}{\sin(k\pi)} \qquad k \text{ not integer}$$

When $k = m =$ an integer, it can be shown that

$$K_m(x) = \lim_{k \to m} K_k(x) = \lim_{k \to m} \frac{\pi}{2} \frac{I_{-k}(x) - I_k(x)}{\sin(k\pi)}$$

PROBLEMS

5.1 Consider a 60-channel FDM-FM-FDMA system with a baseband expanding from 12 to 252 kHz and a specified noise per channel of 10,000 pWp. Assume a 200-kHz test-tone rms frequency deviation and a 5.5-dB weighting improvement.

(a) Find the carrier-to-noise density ratio.
(b) Find the FDM-FM-FDMA bandwidth.
(c) Find the carrier-to-noise ratio.

5.2 An existing 60-channel FDM-FM-FDMA system is modified to reduce its bandwidth to half the original bandwidth. To maintain the same transmission quality, the link carrier-to-noise ratio is increased by 6 dB. The original test-tone rms frequency deviation is 273 kHz. Find the required new FDM multichannel rms frequency deviation.

5.3 Consider an FDM-FM-FDMA carrier with the following parameters.

Channel bandwidth: $b = 3100$ Hz
Multichannel peak factor: 10 dB
Psophometric weighting factor: 2.5 dB
Emphasis weighting factor: 4 dB
Signal-to-noise ratio: 51.2 dB

Find the carrier bandwidth when the carrier-to-noise ratio is $C/N = 15$ dB and the number of channels per carrier is 192.

5.4 Verify (5.25) to (5.28).

5.5 Calculate the carrier-to-intermodulation ratio (third-order products) for the center and edge channels versus the output backoff of a TWTA having the following normalized single-carrier amplitude transfer characteristic

Input voltage	0	0.10	0.20	0.30	0.40	0.50	0.60
Output voltage	0	0.20	0.40	0.60	0.73	0.84	0.92

Input voltage	0.70	0.80	1	1.2.	1.4	1.6	2
Output voltage	0.967	0.995	1	0.98	0.93	0.88	0.77

for an equally spaced equal-amplitude n-carrier input signal with $n = 7, 100, 667, 1333$.

5.6 Write a computer program for calculating the carrier-to-intermodulation ratio (for intermodulation products up to the seventh) versus the output back-off of a TWTA having the normalized single-carrier amplitude transfer characteristic shown in Prob. 5.5 and the following phase transfer characteristic

Input voltage	0	0.10	0.20	0.30	0.40	0.50	0.60
Output phase (rad)	0	0.04	0.12	0.20	0.26	0.31	0.34

Input voltage	0.70	0.80	0.90	1	1.2	1.5	2
Output phase (rad)	0.36	0.375	0.39	0.396	0.41	0.42	0.43

for an equally spaced equal-amplitude n-carrier input signal with $n = 100, 667, 1333$.

5.7 Let the phase transfer characteristic of a TWTA be $F[A(t)] = kA^2(t)$. For an input signal $v(t) = V(1 + m \cos \omega_m t) \cos \omega_c t = A(t) \cos \omega_c t$, where $A(t) = V(1 + m \cos \omega_m t)$,

(a) Find the peak output phase deviation for $m \ll 1$.

(a) Find the peak output phase error in degrees per decibel

5.8 Consider the following FDM-FM-FDMA system: $n = 24$ channels, $f_m = 108$ kHz, $S/N = 51$ dB, and $C/N = 12.5$ dB. The combined psophometric noise weighting factor and preemphasis improvement factor is 6.5 dB. Find the rms frequency deviation of the test tone.

5.9 Consider the cascade of a cuber predistortion linearizer (CPL) with the single-carrier amplitude transfer function

$$S[A] = k_1 A + k_2 A^3$$

and a TWTA with the following in-phase and quadrature transfer function

$$P[A] + jQ[A] = \frac{\alpha_P A}{1 + \beta_P A^2} + j \frac{\alpha_Q A^3}{(1 + \beta_Q A^2)^2}$$

(a) Find the transfer function $P'[A] + jQ'[A]$ of the CPL-TWTA cascade.

(b) Expand $P'[A] + jQ'[A]$ in an odd-power series in A and show that the cubic term responsible for most of the intermodulation distortion can be made to vanish if

$$k_2 = |k_1|^2 k_1 \left(\beta_P - \frac{j\alpha_Q}{\alpha_P} \right)$$

5.10 Let $G[x]$ be the amplitude nonlinearity of a TWTA such that for an input signal $x(t) = \operatorname{Re} z(t)$, where $z(t) = A(t) \exp[j(\omega_c t + \theta)]$, the output signal is

$$G[x(t)] = \operatorname{Re}\{z \exp (jF[A])\} + K_1 \operatorname{Re}\{z \exp(jF[A])\}$$

where $F[A] = K_2 A^2$ is the phase nonlinearity of the TWTA. Find the magnitude of the intermodulation products at $2\omega_1 - \omega_2$ and $2\omega_2 - \omega_1$ for $z(t) = A_1 \cos[(\omega_0 + \omega_1)t] + A_2 \cos[(\omega_0 + \omega_2)t]$, where A_1 and A_2 are constant.

5.11 Consider a satellite TWTA with the following amplitude nonlinearity:

$$V = a_1 v + a_3 v^3$$

and assume that the satellite transponder bandwidth is from $f_1 = 3.7$ GHz to $f_u = 4.2$ GHz. Find the amplitude and location of all intermodulation products within the transponder bandwidth resulting from the amplification of a three-carrier input signal

$$v(t) = A_1 \cos \omega_1 t + A_2 \cos \omega_2 t + A_3 \cos \omega_3 t$$

where $f_1 = 3.8$ GHz, $f_2 = 3.95$ GHz, $f_3 = 4.05$ GHz.

5.12 Consider a geostationary satellite at 105°W longitude and an earth station at 120°W longitude and 45°N latitude. The satellite operates in the FDMA mode with the following parameters.

Satellite saturation EIRP per carrier: 22 dBW

Frequency of downlink carrier received by earth station: 4 GHz

Earth station gain-to-noise temperature ratio at downlink frequency: 40.7 dB/K

Assume that the satellite is downlink power-limited, that is, $(C/N)_d \ll (C/N)_u$, and that the carrier-to-intermodulation ratio versus the satellite TWTA output backoff is given by

C/I (dB)	8	10	14	16	18	20	25	30
BO_o (dB)	2.3	2	2.1	2.8	3.4	4.2	6.5	9

Find the optimized carrier-to-intermodulation plus noise and the corresponding operating output back-off of the satellite TWTA, assuming the carrier bandwidth is 5 MHz.

5.13 Consider an SCPC system consisting of 200 carriers that operates at a 6-GHz uplink and a 4-GHz downlink with the following link parameters.

Saturation power flux density per carrier: -103 dBW/m^2

Satellite G/T: -7 dB/K

Noise bandwidth: 40 kHz

Satellite saturation EIRP per carrier: 13 dBW

Downlink slant range: 37,506 km

Earth station G/T: 22 dB/K

TWTA single-carrier amplitude transfer characteristic given in Prob. 5.5.

Find the optimum output back-off of the satellite TWTA.

5.14 Consider the satellite link discussed in Prob. 4.5. Find the signal-to-noise ratio of the top baseband channel under clear-sky conditions if the link between satellites A and B is used for an FDM system with the following parameters.

Number of channels: 1872

Maximum baseband frequency: 8.12 MHz

Test-tone rms frequency deviation: 413.3 kHz

Weighting improvement: 6.5 dB

chapter **6**

Time Division Multiple Access

Time division multiple access is a multiple access protocol in which many earth stations in a satellite communications network use a single carrier for transmission via each satellite transponder on a time division basis; that is, all earth stations operating on the same transponder are allowed to transmit traffic bursts in a periodic time frame—the *TDMA frame*. Over the length of the burst, each earth station has the entire transponder bandwidth available to it for transmission. The transmit timing of the bursts is carefully synchronized so that all the bursts arriving at the satellite transponder from a community of earth stations in the network are closely spaced in time but do not overlap. The satellite transponder receives one burst at a time, amplifies it, and retransmits it back to earth. Thus every earth station in the satellite beam served by the transponder can receive the entire burst stream and extract the bursts intended for it. A simplified diagram of a TDMA operation is shown in Fig. 6.1. In this chapter we analyze the TDMA protocol and the TDMA system for implementing this protocol.

6.1 TDMA FRAME STRUCTURE

In a TDMA network each earth station periodically transmits one or more bursts to the satellite. The input signal to the satellite transponder carrying TDMA traffic thus consists of a set of bursts originating from a

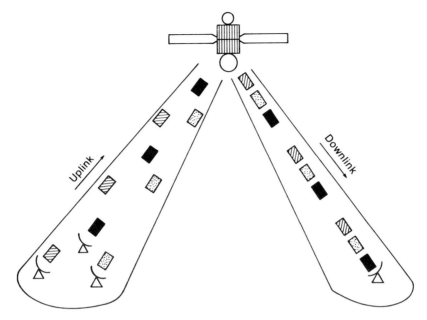

Figure 6.1 Time division multiple access.

number of earth stations. This set of bursts is referred to as a *TDMA frame* and is illustrated in Fig. 6.2. It consists of two *reference bursts* RB1 and RB2, *traffic bursts,* and the *guard time* between bursts. The TDMA frame is the period between RB1 reference bursts.

6.1.1 Reference Burst

Each TDMA frame normally consists of two *reference bursts* RB1 and RB2 for reliability. The *primary reference burst* (PRB), which can be either RB1 or RB2, is transmitted by one of the stations in the network designated as the primary reference station (PRS). A *secondary reference*

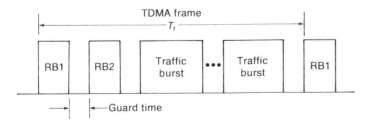

Figure 6.2 TDMA frame structure.

burst (SRB), which also can be either RB1 (if PRB = RB2) or RB2 (if PRB = RB1), is transmitted by a secondary reference station (SRS) which allows *automatic switchover* in the event of primary reference station failure to provide undisrupted service for the TDMA network. The reference bursts carry no traffic information and are used to provide *timing references* for all the stations accessing a particular satellite transponder. This allows satisfactory interleaving of bursts within a TDMA frame. The TDMA traffic stations take their timing reference from the primary reference burst or from the secondary reference burst when there is a failure of the primary reference station.

6.1.2 Traffic Burst

The *traffic bursts* (TBs) transmitted by the traffic stations carry digital information. Each station accessing a transponder may transmit one or more traffic bursts per TDMA frame and may position them anywhere in the frame according to a *burst time plan* that coordinates traffic between stations. The length of the traffic burst depends on the amount of information it carries and can be changed if required. The location of the traffic bursts in a frame is referenced to the time of occurrence of the primary reference burst. By detecting the primary reference burst, a traffic station can locate and extract the traffic bursts or portions of traffic bursts intended for it. Also, it can derive the transmit timing of its bursts precisely, so that they arrive at the satellite transponder within their allocated positions in the TDMA frame and avoid overlapping with bursts from other stations.

6.1.3 Guard Time

A short *guard time* is required between bursts originating from several stations that access a common transponder to ensure that the bursts never overlap when they arrive at the transponder. The guard time must be long enough to allow differences in transmit timing accuracy and in the range rate variation of the satellite. The guard time is normally equal to the time interval used to detect the receive timing pulse marking the start of a receive TDMA frame at a station. There is no transmission of information during the guard time.

The TDMA frame length is normally selected to be in the range $0.75 \leq T_f \leq 20$ ms for voice service. It is usually a multiple of 0.125 ms, which is the sampling period of PCM (8000-Hz sampling rate). The frame length is chosen at the outset and remains constant for a TDMA system. However, in the event that a new service requires a change in frame length, it may be altered by redefining the number of bits per frame and storing this count in the network memory.

6.2 TDMA BURST STRUCTURE

In general the structure of the reference burst and the traffic burst are as shown schematically in Fig. 6.3. In the traffic burst, information bits are preceded by a group of bits referred to as a *preamble* that is used to synchronize the burst and to carry management and control information. The

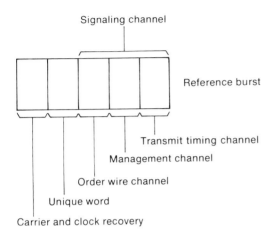

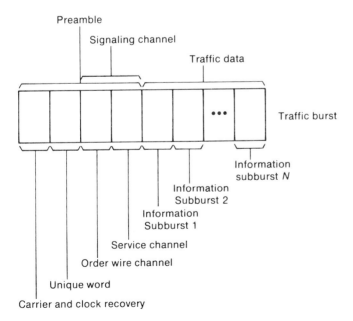

Figure 6.3 TDMA burst structure.

reference burst contains only the preamble, that is, no traffic data. Normally the preamble consists of three contiguous parts: the *carrier and clock recovery sequence* (CCR), the *unique word* (UW), and the *signaling channel*.

6.2.1 Carrier and Clock Recovery Sequence

Each burst begins with a sequence of bits or symbols (for modulation such as QPSK) which enable the earth station demodulator to recover the carrier phase and regenerate the bit or symbol timing clock for data demodulation. Normally, the length of the carrier and clock recovery sequence depends on the carrier-to-noise ratio at the input of the demodulator and the acquisition range (carrier frequency uncertainty). A high carrier-to-noise ratio and a small aquisition range require a short CCR sequence, and vice versa. Typically, a high-bit-rate TDMA system requires a long CCR sequence, for example, 300 to 400 bits (150–200 symbols) for 120-Mbps TDMA.

6.2.2 Unique Word

The *unique word* that follows the carrier and clock recovery sequence is used in the reference burst to provide the receive frame timing that allows a station to locate the position of a traffic burst in the frame. The unique word in the traffic burst marks the time of occurrence of the traffic burst and provides the receive burst timing that allows the station to extract only the wanted subbursts within the traffic burst. The unique word is a sequence of ones and zeros selected to exhibit good correlation properties to enhance detection. At the demodulator, the unique word enters a unique word detector, like the digital correlator shown in Fig. 6.4, where it is correlated with a stored pattern of itself. The correlator consists of two N-stage shift registers (where N is the length of the unique word), N modulo-2 adders, a summer, and a threshold detector. The received data is shifted in the shift register in synchronization with the data clock rate. Each stage in the shift register is applied to a modulo-2 adder whose output is a logical zero when the data bit or symbol in the stage is in agreement with the stored unique word bit or symbol in the same position. All the modulo-2 adder outputs are summed, and the sum is compared to a preset threshold by the threshold detector. The output of the summer is thus a step function representing the number of agreements or disagreements between the input data and the stored unique word pattern. The maximum number of errors allowed in the unique word detection is called the *detection threshold* ϵ. When the correlation errors are equal to or below ϵ, the detection of the unique word is declared. The unique word detection occurs at the instant of reception of the last bit or symbol of the

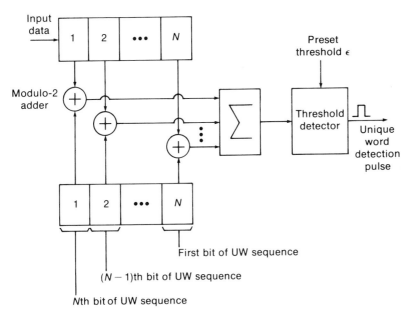

Figure 6.4 Unique word detector.

unique word and is used to mark the *receive frame timing* if the unique word belongs to the primary reference burst, or to mark the *receive traffic burst timing* if the unique word belongs to the traffic burst. The position of every burst in the frame is defined with respect to the receive frame timing, and the position of every subburst in a traffic burst is defined with respect to the burst's receive burst timing. It is seen that accurate detection of the unique word is of utmost importance in a TDMA system. For example, when the unique word of a traffic burst is missed, the entire traffic burst is lost. This causes impulses or clicks in voice transmission. In data transmission, a block is lost and consequently the bit error rate is increased. A *false detection* of the primary reference burst unique word generates the wrong receive frame timing and consequently incorrect transmit frame timing, causing the earth station to transmit out of synchronization and resulting in overlapping with other bursts at the satellite. A false detection is generated whenever data or noise agrees with the stored unique word pattern to the extent that the number of bits or symbols in disagreement fall below the detection threshold ϵ. A unique word miss occurs when channel noise causes more than ϵ errors in the receive unique word sequence, making the number of bits or symbols in disagreement exceed the detection threshold ϵ. In general, for a given unique word length, increasing the detection threshold ϵ makes the miss detection probability smaller but raises the false detection probability. On the other hand, lowering ϵ to improve the false detection probability increases the miss detection probability.

Based on the above discussion the miss detection probability for a unique word of length N is the probability of its having $\epsilon + 1$ or more errors. If p is the average probability of error for the receive data, then the probability $P(i)$ that i bits or symbols out of N will be in error is given by the binomial distribution

$$P(i) = \binom{N}{i} p^i (1 - p)^{N-i} \tag{6.1}$$

where

$$\binom{N}{i} = \frac{N!}{i! \, (N - i)!}$$

The probability of a correct detection is thus the sum of the probabilities of $0, 1, 2, \ldots, \epsilon$ errors:

$$P_C = \sum_{i=0}^{\epsilon} \binom{N}{i} p^i (1 - p)^{N-i} \tag{6.2}$$

Consequently, the miss detection probability P_M is simply $P_M = 1 - P_C$ or

$$P_M = \sum_{i=\epsilon+1}^{N} \binom{N}{i} p^i (1 - p)^{N-i} \tag{6.3}$$

The miss detection probability P_M is plotted in Fig. 6.5 for $p = 10^{-3}$. which is the typical threshold error probability for link data. It is seen that, for a given link error probability, the unique word miss detection probability P_M can be reduced by decreasing the unique word length N or by increasing the detection threshold value ϵ. In any case, N and ϵ should be selected such that $P_M \ll p$.

The false detection probability P_F is given by the probability that random data (the data bits 1 and 0 are assumed to be generated with equal probability) accidentally corresponds to the stored unique word pattern to the extent that the number of bits or symbols in disagreement does not exceed the detection threshold ϵ. For a unique word of length N, there are 2^N combinations in which random data can occur, hence the probability of the occurrence of one unique combination that corresponds to the stored unique word pattern is $1/2^N$, which is also the false detection probability when $\epsilon = 0$. For a given value of ϵ, the total number of possible combinations in which ϵ or fewer errors can occur is $\sum_{i=0}^{\epsilon} \binom{N}{i}$. Thus the probability that N random data bits or symbols will be decoded as the unique word, or the false detection probability P_F, is

$$P_F = \frac{1}{2^N} \sum_{i=0}^{\epsilon} \binom{N}{i} \tag{6.4}$$

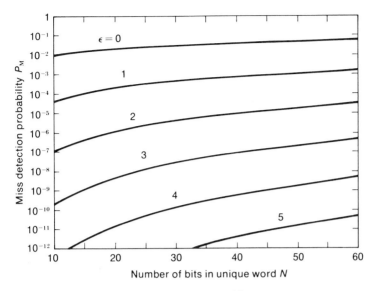

Figure 6.5 Unique word miss detection probability.

and is independent of the link error probability. The false detection probability is plotted in Fig. 6.6. It is seen that P_F can be reduced by increasing the unique word length N or decreasing the detection threshold ϵ.

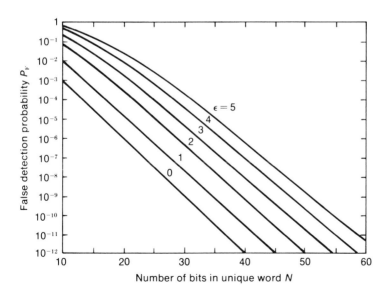

Figure 6.6 Unique word false detection probability.

As an example, consider a unique word of length $N = 40$ and a detection threshold of $\epsilon = 5$. The miss detection probability P_M for a given link error probability of $p = 10^{-3}$ is given in Fig. 6.5 as

$$P_M \approx 5 \times 10^{-12}$$

If the data rate is $R = 60$ Mbps, then a miss detection will occur once every 3333 s. The corresponding false detection probability P_F is given in Fig. 6.6 as

$$P_F \approx 10^{-6}$$

If the data rate is $R = 60$ Mbps, then a false detection can be expected to occur at least 60 times every second.

The above example shows that a unique word miss occurs very infrequently, in contrast to a false detection. To avoid this problem, the *aperture technique* is employed to suppress false detections. The aperture timing period is started by the unique word detection pulse, and one TDMA frame later an aperture window is formed at the expected occurrence of the unique word detection pulse, as shown in Fig. 6.7. All correlation pulses which do not occur within the aperture are suppressed. The aperture window allows detection of the unique word within the specified time interval. The length of the aperture window must be sufficient to compensate for drift of the unique word from its expected position as a result of timing uncertainty in the TDMA system resulting from satellite motion. Figure 6.8 shows the position of the aperture window relative to the detection pulse resulting from correlation of the carrier and clock recovery sequence that precedes the unique word and the unique word itself with the stored unique word pattern as they are shifted sequentially into the shift register. The aperture window is W bits, and the occurrence of the unique word detection relative to the end of the aperture is X bits. If $X = 0$, the aperture is in the most advanced position. Further advancing

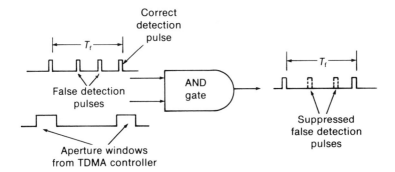

Figure 6.7 Aperture technique for reducing false detection.

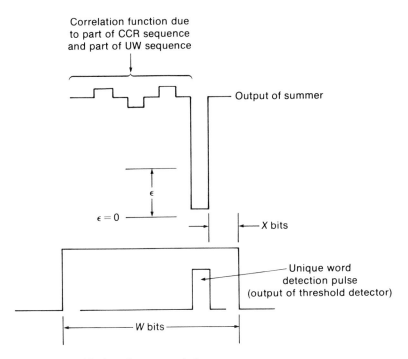

Figure 6.8 Positioning of aperture window.

the aperture to the left will cause the unique word detection to be missed. For $X = W - 1$, the aperture window is in the least advanced position. If the aperture is set back further, the unique word detection will be missed. In practice $X = W/2$ is the nominal position for the aperture relative to the occurrence of the unique word detection. Note that the occurrence of the unique word detection completes the unique word detection process, hence the bit or symbol pattern of the preamble that follows the unique word sequence does not play a role in the correlation process and only the carrier and clock recovery sequence takes part in the correlation. Note that in the worst case, when the aperture is in the most advanced position, that is, $X = 0$, the correlation function at the output of the summer in the interval $W - 1$ bits before the unique word detection can cause a false detection if the number of bits in disagreement during any 1-bit correlation interval happen to be less than the threshold level ϵ. Therefore it is important to select the last $W - 1$ bits of the carrier and clock recovery sequence in such a way that they do not exhibit a strong correlation with the unique word pattern. An example of such a selection is illustrated in Fig. 6.9, where the correlation function within the aperture window $W = 23$ bits and $X = 2$ bits is shown. The amplitude of the correlation

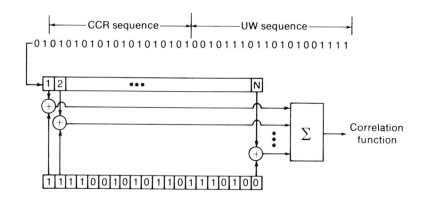

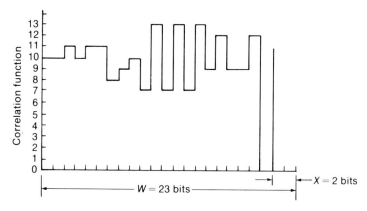

Figure 6.9 Correlation function of CCR, UW, and stored UW sequences.

functions depends on the number of bits in the carrier and clock recovery sequence that reside in the shift register. The correlation of any bit b_i in the shift register with the corresponding bit b_i' of the stored unique word is the output of the modulo-2 adder and is given by the following rule:

$b_i \oplus b_i'$	0	1
0	0	1
1	1	0

It is seen that, when the last 10, 8, and 6 bits of the carrier and clock recovery sequence and the first 10, 12 and 14 bits of the unique word reside in the shift register, the correlation functions have the next lowest amplitude of 7. When all 20 bits of the unique word reside in the shift register, the correlation function has an amplitude of 0. If the detection threshold ϵ is selected to be $\epsilon < 7$, then the unique word is detected. For this particular carrier and clock recovery sequence and this particular unique word, the mean value of the correlation amplitude due to combinations of the carrier and clock recovery sequence and the unique word is $N/2 = 10$ with δ amplitude fluctuations, where $-3 \leq \delta \leq 3$. The false detection probability P_F clearly depends on the width W of the aperture and the amplitude of the correlation function. It also depends on the position X of the end of the aperture relative to the occurrence of the unique word detection. As an illustration consider the case in Fig. 6.9 with $X = 0$. If $W = 1$, then $P_F = 0$, since the only correlation function that appears within the aperture is the unique word correlation. When $W = 2$, there is an additional correlation function involving the last bit of the carrier and clock recovery sequence plus $N - 1 = 19$ bits of the unique word in the shift register. If there are no errors, the amplitude of this pulse will be $N/2 + \delta = 12$, where $\delta = 2$, and no false detection will occur before the occurrence of the unique word correlation pulse. However, if the above sequence contains between $N/2 + \delta - \epsilon = 12 - \epsilon$ and $N/2 + \delta = 12$ errors ($\epsilon < 7$), the correlation threshold ϵ will be reached and a false detection will occur. The false detection probability is the probability that i bits, $N/2 + \delta - \epsilon \leq i \leq N/2 + \delta$, out of $N/2 + \delta$ disagreement bits will be in error:

$$P_{N/2 + \delta} = \sum_{i = N/2 + \delta - \epsilon}^{N/2 + \delta} \binom{N/2 + \delta}{i} p^i (1 - p)^{N/2 + \delta - i} \tag{6.5}$$

where p is the link error probability. In the case $W = 2$ the false probability is $P_{N/2 + 2}$.

Now consider the case $W = 3$; there are two correlation functions that exist before the occurrence of the unique word detection. This yields two opportunities for a false detection. The total probability of a false detection is the sum of the individual probabilities. The probability of a false detection of the first correlation (the last bit in the CCR sequence and the first $N - 1$ bits in the UW sequence) is just $P_{N/2 + \delta}$ as given in (6.5). The probability of a false detection of the second correlation (the last 2 bits in the CCR sequence and the first $N - 2$ bits in the UW sequence) is the joint probability that the first bit is correct, which is $1 - p$, and that the second correlation contains between $N/2 + \delta - \epsilon = 9 - \epsilon$ and $N/2 + \delta = 9$ errors ($\delta = -1$), which is $P_{N/2 + \delta}$ in (6.5) with $\delta = -1$. Thus the joint probability is $(1 - p)P_{N/2 - 1}$. The total false detection probability is

$$P_2 = P_{N/2 + 2} + (1 - p)P_{N/2 - 1} \tag{6.6}$$

Proceeding in a similar manner, the overall probability of false detection is

$$P_F = P_{N/2+2} + (1-p)P_{N/2-1} + (1-p)^2 P_{N/2-1} + (1-p)^3 P_{N/2+2}$$
$$+ (1-p)^4 P_{N/2-1} + (1-p)^5 P_{N/2+3} + (1-p)^6 P_{N/2-3}$$
$$+ (1-p)^7 P_{N/2+3} + (1-p)^8 P_{N/2-3} + (1-p)^9 P_{N/2+3}$$
$$+ (1-p)^{10} P_{N/2-1} + (1-p)^{11} P_{N/2} + (1-p)^{12} P_{N/2-1}$$
$$+ (1-p)^{13} P_{N/2-2} + (1-p)^{14} P_{N/2} + (1-p)^{15} P_{N/2+1}$$
$$+ (1-p)^{16} P_{N/2} + (1-p)^{17} P_{N/2+1} + (1-p)^{18} P_{N/2}$$
$$+ (1-p)^{19} P_{N/2}$$

$$(6.7)$$

For $p < 10^{-3}$, $P_{N/2+\delta} \approx P_{N/2}$ ($-3 \leq \delta \leq 3$) and

$$P_F \approx \frac{1-(1-p)^{W-X-1}}{p} P_{N/2} \qquad W = 23, X = 2 \qquad (6.8)$$

The above false detection probability will be reduced even more when the aperture window is positioned such that $X = W/2$.

The selection of the aperture window W directly determines the guard time between bursts. Even though the satellite is said to be stationary in a geostationary orbit, the situation is approximate because of the attraction of the moon and sun. Analysis of the departure from a geostationary orbit reveals that, for the worst case, the round trip range rate variation (Doppler shift) is approximately 40 ns/s. For a TDMA system with a burst bit rate of R bits per second, the Doppler shift is equivalent to $40 \times 10^{-9}R$ bits per second. Since the aperture window is W bits centered around the expected unique word detection pulse, the time required for it to drift out of its aperture is approximately $W/(80 \times 10^{-9}R)$ seconds. Therefore each station must correct its transmit timing at least once every $W/(80 \times 10^{-9}R)$ seconds. Also, the guard time between bursts must be at least as long as the aperture width to guarantee nonoverlapping bursts if the unique word detection pulses of two adjacent bursts drift to the opposite extremes of their respective apertures. To provide an additional degree of protection against the possibility of overlapping bursts, the guard time can be selected to be longer than the aperture window.

6.2.3 Signaling Channel

In general the *signaling channel* of the reference burst consists of the following subbursts:

1. An *order wire channel* carrying voice (telephone), and data (teletype) traffic via which instructions are passed to and from earth stations.

Order wire is a term used in manual telephone switching to describe a circuit on which operators and maintenance personnel can talk to one another. Operators use the order wire for placing calls.

2. A *management channel* which is sent by the reference stations to all traffic stations carrying frame management instructions such as burst time plan changes. The burst time plan describes the coordination of traffic between stations. It identifies the boundaries of the time slots of the frame allocated to the stations, that is, burst positions. It also identifies the position, length, and source or destination stations corresponding to subbursts in the bursts. This channel also carries monitoring and control messages to the traffic stations when the reference station wants to obtain a status report (monitoring) and/or to control the switchover of subsystems at the traffic stations remotely.

3. A *transmit timing channel* carrying acquisition and synchronization information to the traffic stations which enables them to adjust their transmit burst timing so that transmitted bursts arrive at the satellite transponder within the correct time slots in the TDMA frame. It also carries the status codes which allow the traffic stations to identify the primary reference burst and the secondary reference burst from RB1 and RB2 as shown in Fig. 6.2.

The signaling channel of the traffic burst consists of the following subbursts:

1. An *order wire channel* which is the same as the reference burst order wire channel.
2. A *service channel* carrying the traffic station's status to the reference station, or other information such as the high bit error rate and unique word loss alarms to other traffic stations.

Besides these subbursts in the preamble, both reference and traffic bursts can carry additional subbursts containing the *frame identification number* (for frame management purposes), *station identification number,* and *type of transmitting bursts* (primary reference burst, secondary reference burst, traffic burst). Different types of unique words can be employed to provide burst identification.

6.2.4 Traffic Data

Traffic information is carried by the traffic burst immediately following the preamble. The length of a traffic subburst depends primarily on the type of services and the total number of channels required for each service being supported in the burst. This portion contains information from the calling user being communicated to the called user, whether it be voice, data, video, or facsimile signals. The information for each channel is transmitted as a continuous subburst. The size of each subburst may be selected to be any number of bits to specifically accommodate the actual speed of the voice, data, video, or facsimile signal. For example, one PCM voice channel is equivalent to 64 kbps; if the frame length $T_f = 2$ ms, the resulting subburst of one PCM voice channel is 128 bits long. Each station in the TDMA network normally can transmit many traffic bursts containing different numbers of subbursts per frame and is also capable of receiving many traffic bursts or subbursts per frame.

6.3 TDMA FRAME EFFICIENCY

The TDMA *frame efficiency* depends on the percentage of the frame length T_f allocated to traffic data. The higher this percentage, the higher the system's efficiency. In order to achieve this goal, the *overhead portion* of the frame (e.g., guard times and preambles) has to be lowered, but not to the point of making the design of the system difficult. The carrier and clock recovery sequence must be long enough to provide enough time for stable acquisition of the carrier and to minimize the effect of interburst interference (the tail of the preceding burst interfering with the head of the succeeding burst, hence degrading the carrier-to-noise ratio of the latter) caused by a finite filter response in the demodulator. Furthermore, the guard time between bursts must be long enough to allow synchronization tolerance due to the uncertainty of the satellite position and the method of frame synchronization employed. Therefore, trade-offs between TDMA efficiency and system implementation must be carefully considered in any TDMA design.

The TDMA frame efficiency η is usually defined as

$$\eta = 1 - \frac{T_x}{T_f} \qquad (6.9)$$

where T_x is the *overhead* portion of the frame. If there are n bursts in a frame, then T_x can be expressed as

$$T_x = nT_g + \sum_{i=0}^{n} T_{p,i} \qquad (6.10)$$

where T_g = guard time between bursts and $T_{p,i}$ = preamble of burst i.

It is obvious that the frame efficiency can be increased without lowering the overhead simply by increasing the frame length. But this in turn increases the amount of memory needed to store the incoming terrestrial data at a continuous rate for one frame, to transmit the data at a much higher burst bit rate to the satellite, and to store the receive traffic bursts and convert them to lower continuous outgoing terrestrial data. Furthermore, the frame length has to be kept small compared to the maximum satellite roundtrip delay of about 274 ms (5° elevation angle) to avoid adding a significant delay to the transmission of voice traffic. For voice traffic the frame length is normally selected to be less than 20 ms.

As an example, consider a TDMA system with the frame and burst structures shown in Figs. 6.2 and 6.3, respectively. Calculation of the frame efficiency is based on the following parameters:

1. The TDMA frame length is 15 ms.
2. The TDMA burst bit rate is 90 Mbps.
3. Each of the 10 stations transmits 2 traffic bursts for a total of 20 traffic bursts in the frame plus 2 reference bursts.
4. The length of the carrier and clock recovery sequence is 352 bits.
5. The length of the unique word is 48 bits.
6. The order wire channel has 510 bits.
7. The management channel has 256 bits.
8. The transmit timing channel has 320 bits.
9. The service channel has 24 bits.
10. The guard time is assumed to be 64 bits.

From the above assumptions, we have

Number of bits in the reference burst preamble: 1486
Number of bits in the traffic burst preamble: 934
Total number of overhead bits: 23,060
Total number of bits in a frame (15 ms $\times$ 90 Mbps): 1.35×10^6
Frame efficiency: 98.29%

Assume that all the traffic data is PCM-encoded voice. Each voice channel data rate is 64 kbps and each channel is carried by a subburst in the traffic burst. The number of bits in a 15-ms frame for a voice subburst is 64 kbps $\times$ 15 ms = 960. The maximum number of PCM voice channels carried in a frame is $0.9829 \times 1.35 \times 10^6/960 \approx 1382$.

6.4 TDMA SUPERFRAME STRUCTURE

The two most critical functions in a TDMA network are control of the burst position in the frame and coordination of the traffic between stations in such a way that any rearrangement of the position and length of bursts does not cause service disruption or burst overlapping. Control of the position of bursts may be carried out by the reference station using the transmit timing channel, while coordination of traffic is achieved through the management channel of the reference burst.

To provide control and coordination, the reference station has to address all the traffic stations in the network. If there are N stations to be addressed in the network, there will be N messages in the transmit timing channel and N messages in the management channel of the reference burst. Furthermore, to provide almost error-free communication for these critical control and coordination messages, some form of coding is normally employed. The most commonly used coding for these channels is the 8:1 redundancy coding algorithm where an information bit is repeated eight times according to a predetermined pattern and then decoded using majority decision logic at the receive end. This effectively increases the time slot allocated to each message eight times and further reduces the frame efficiency. The same reasoning applies to the service channel of the traffic bursts.

In order to reduce the length of the preamble of the reference bursts and the traffic bursts, the reference station can send one message to one station per frame instead of N messages to N stations per frame. To address N stations in the network, the process takes N frames. For example, station 1 is addressed by the reference station in frame 1, station 2 by the one in frame 2, so on, and finally station N by the one in frame N. The procedure is repeated in the same fashion for the next N frames until completion. Similarly, if the status report sent by the traffic station to the reference station, or other information sent to other traffic stations, is sent over N frames and repeated until completion, the length of the traffic burst preamble will also be reduced, hence the frame efficiency will be increased.

In this way, N frames can be put into one group called a *superframe*, where N is the number of stations addressed by the reference station as shown in Fig. 6.10. To identify the frames in a superframe, a frame identification number may be carried in the management channel or in a separate channel in the reference burst for each frame. Normally the identification number of frame 1 serves as the superframe marker. Alternatively, different unique words can be employed by the reference bursts and the traffic bursts to distinguish the superframe marker from the frame markers.

When the number of stations N in the network is fixed, or its max-

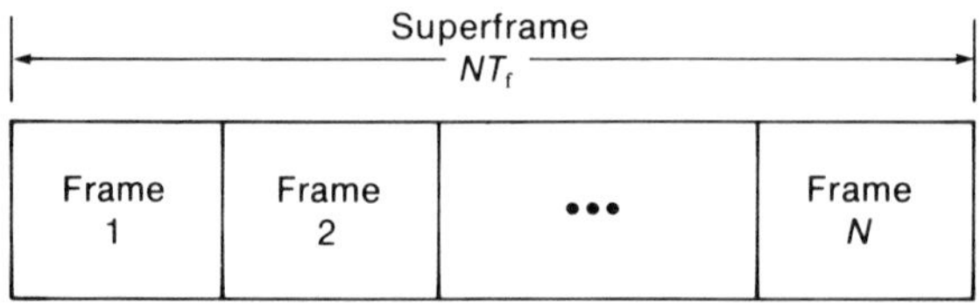

Figure 6.10 Superframe.

imum is known, it is easy to design the service channel of the traffic bursts so that its message can be transmitted over N frames. For example, any message transmitted by the service channel of the traffic bursts is limited to a maximum of 40 bits. If the 8:1 redundancy coding algorithm is used for the message, it will take 320 bits to transmit it. Suppose $N = 10$ (i.e., a superframe consists of 10 frames): then a superframe would be needed to transmit the 320-bit message with 32 bits per frame. That is, the service channel occupies a time slot of only 32 bits. Although the rate of message data transmission is now only 4 bits per frame, the frame efficiency is increased significantly as compared to transmitting 320 bits per frame (40 bits of message data per frame).

When the number of stations N in the network is variable, that is, the network can grow, and if demand assignment is employed (to be discussed in Chap. 7), it might be appropriate to transmit the messages in the service channel of the traffic bursts and demand assignment messages in a separate *superframe short burst* (SSB) at the superframe rate. That is, each of the N stations in the network transmits a superframe short burst once per superframe. In other words, each frame of a superframe contains a superframe short burst from a designated station, as illustrated in Fig. 6.11. For the above example, the superframe short burst would be allocated a time slot of 320 bits for a 40-bit message with 8:1 redundancy coding. Note that message data rate is still 40 bits per superframe, as in the case where a service channel with 4 bits of message data per frame is used in a traffic burst. The advantage of putting the service channel in the superframe short burst instead of in the traffic burst is to increase the frame efficiency when a station transmits more than one traffic burst per frame. Since the messages in the service channel of all the traffic bursts in the same frame that originate from the same station are normally identical for ease of design, the redundancy of messages reduces the frame efficiency. A typical superframe short burst is shown in Fig. 6.12.

6.5 FRAME ACQUISITION AND SYNCHRONIZATION

In a TDMA system, a traffic station must perform two functions:

On the receive side, the traffic station must be able to receive traffic bursts addressed to it from a satellite transponder (or transponders) periodically every frame.

On the transmit side, the traffic station must be able to transmit traffic bursts destined for other stations periodically every frame in such a way that the bursts arrive at a satellite transponder (or transponders) without overlapping with bursts from other traffic stations.

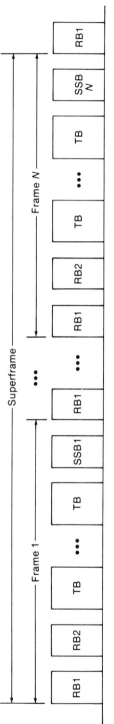

Figure 6.11 Superframe short burst position in a superframe.

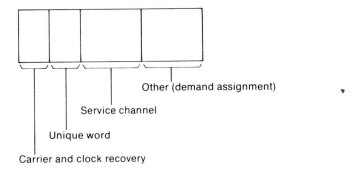

Figure 6.12 Structure of a superframe short burst.

As mentioned before, the timing reference in a TDMA system is provided by the primary reference burst. By detecting the unique word of the primary reference burst, the traffic station can establish the *receive frame timing* (RFT) which is defined as the instant of occurrence of the last bit or symbol of the primary reference burst's unique word. Also, the last bit or symbol of the traffic burst's unique word marks its *receive burst timing* (RBT). Since the receive frame timing marks the start of a received frame, the position of a traffic burst in a received frame is determined by the off-set between the receive frame timing and the receive burst timing. This offset (in bits or symbols) is contained in a receive burst time plan which is stored in the foreground memory of the traffic station. Using the receive burst time plan, the traffic station can extract any traffic burst intended for it in a received frame.

To transmit a traffic burst so that it arrives at the satellite transponder within the allocated position in the frame, the traffic station must establish a *transmit frame timing* (TFT), which marks the start of the station's transmit frame, and a *transmit burst timing* (TBT), which marks the start of transmission of the traffic burst to the satellite. The position of the traffic burst in a transmitted frame is determined by the offset between the transmit frame timing and the transmit burst timing. This offset is contained in a transmit burst time plan stored in the foreground memory of the traffic station. If the traffic station transmits a traffic burst at the transmit frame timing, it will arrive at the satellite transponder at the same time as the primary reference burst that marks the start of a frame at the transponder. Any traffic burst transmitted at its transmit burst timing will fall into its appropriate position in the TDMA frame at the transponder. In this way, traffic bursts from many stations that access a particular transponder will fall into their preassigned positions in the frame at the transponder and burst overlapping will not occur.

The processes of acquiring the receive frame timing and the transmit frame timing are called *receive frame acquisition* (RFA) and *transmit frame acquisition* (TFA), respectively. The processes of maintaining these timings are called *receive frame synchronization* (RFS) and *transmit frame synchronization* (TFS). The acquisition process is needed when the traffic station enters or reenters operation. The synchronization process is necessary because of satellite movement in orbit. A geostationary satellite is subjected to small perturbations caused by the moon and sun and moves about a region centered at its nominal position. A geostationary orbit can be specified in terms of its inclination angle relative to the equatorial plane (north-south drift), its orbital eccentricity, and its east-west drift. The orbital eccentricity can cause a peak-to-peak altitude variation of 0.2% of the geostationary orbit radius (42,164 km) or about 85 km. The east-west or north-south peak-to-peak variation at the end of a satellite as life is typically about 0.2° or 150 km (stationkeeping of ±0.1°). All these variations introduce a maximum range variation of $(85^2 + 150^2)^{1/2} = 172$ km or about 0.575 ms in the one-way propagation delay between the earth station and the satellite. The maximum roundtrip delay variation is thus about 1.15 ms. As a result, the maximum Doppler shift is 1.15 ms/8 h $\approx$ 40 ns/s, where 8 h is the time during which the satellite moves from its nominal position to the position where maximum delay variation occurs. This Doppler shift undoubtedly causes errors in the burst positions as they arrive at the satellite transponder. Therefore frame synchronization is necessary to maintain correct reception and transmission of the traffic burst. At most, synchronization can be carried out only once per roundtrip delay; otherwise it can lead to errors. In summary, all the traffic stations in a TDMA network must perform the following four procedures in order to synchronize their traffic bursts with the reference bursts:

Receive frame acquisition
Receive frame synchronization
Transmit frame acquisition
Transmit frame synchronization

Receive frame acquisition and receive frame synchronization are achieved by detecting the unique word of the reference burst. There are two unique word detection modes: the search mode and the track mode. In the search mode, a continuous or fully open aperture is employed by the traffic station to detect the unique word of the reference burst. In this mode the unique word detection threshold ϵ is set to 0; that is, the unique word is considered detected only when its received UW sequence exactly matches its stored pattern. When the unique word of the reference burst is detected, the search mode switches immediately to the track mode with a narrow aperture window equal to or wider than the guard time between bursts, and the detection threshold is increased to ϵ. This aperture window for the detected unique word is centered one frame period after its detection. If the unique word is missed by the aperture window, the center of the next aperture window will be located one frame period afterward. For a TDMA system of 90 Mbps, a Doppler shift of 40 ns/s is equivalent to 3.6 bps. If the aperture window is taken to be ± 32 bits centered around the expected unique word position, the time a burst takes to drift out of its aperture is approximately 9.8 s.

6.5.1 Receive Frame Acquisition and Receive Frame Synchronization

In order for a traffic station to enter the TDMA operation, it must first perform the receive frame acquisition. This is a procedure in which the station detects the unique words of the reference bursts to establish the receive frame timing. This procedure may be applied to both reference bursts RB1 and RB2 in the frame in parallel or sequentially and employs the unique word search mode. When a reference burst unique word is detected, the procedure switches to the unique word track mode. Normally the reference burst is declared "acquired" if its unique word is detected in three additional consecutive frames. If the unique word is missed in any of the three additional consecutive frames following the first detection, the procedure is restarted for the reference burst. In practice the receive frame acquisition process takes about 1 s or less.

The receive frame synchronization procedure begins when the reference burst is declared acquired. It uses the unique word track mode to periodically track the detected reference burst using a narrow aperture whose center is located one frame later than detection of the reference burst unique word in the current frame. The reference burst may be declared "synchronized" if its unique word is detected N times during M consecutive frames where $M > N$. The receive frame timing is the instant of occurrence of the last bit of the reference burst unique word.

To speed up receive frame acquisition, that is, to acquire the reference burst when it is declared "not acquired" by the RFA procedure or when it is declared "not synchronized" by the RFS procedure, the *assisted receive frame acquisition* (ARFA) procedure can be applied to the reference burst if the receive frame timing from the other reference burst in the frame is available, that is, if the other reference burst is synchronized. Since the position of reference burst RB1 relative to reference burst RB2 is known from the receive burst time plan stored at the traffic station, the unique word of one reference burst can be detected in a predicted window using the receive frame timing from the other reference burst. The ARFA procedure is therefore used with the unique word track mode whose aperture window is centered at the predicted time derived from the receive frame timing of the other reference burst and the fixed offset between the two reference bursts. Normally the reference burst is declared acquired when its unique word is detected for four consecutive frames.

When a reference burst is declared *synchronized*, the receive frame timing can be obtained. Since there are two reference bursts in a frame, namely, RB1 and RB2, one of these two reference bursts carries the status code as a primary reference burst and the other as a secondary reference burst. The receive frame timing may be derived from either the PRB or the SRB according to the following criteria:

1. When both the PRB and the SRB have been declared *synchronized,* the receive frame timing should be derived from the PRB. When the PRB unique word has been missed, the timing is derived from internal timing for a predetermined number of frames.
2. When the PRB has not yet been declared *synchronized* (meaning that it has not yet been acquired) or when the PRB has been declared *not synchronized* and the SRB has been declared *synchronized*, the receive frame timing may be derived from the SRB. When the SRB unique word has been missed, the timing is derived from internal timing for a predetermined number of frames.
3. When the PRB has not been declared *synchronized* or when the PRB has been declared *not synchronized*, and the SRB has been declared *not synchronized*, the receive frame timing may be derived from internal timing for a number of frames. Afterward, the station has to start the RFA procedure again.

A flowchart for the receive frame timing derivation procedure is shown in Fig. 6.13.

Once the receive frame timing has been established and maintained by the receive frame synchronization procedure, the receive burst timing of each traffic burst may be derived from the unique word detection of that burst by employing the unique word track mode with a narrow aperture window centered on the assigned position of the traffic burst. The assigned position of the traffic burst can be determined by adding a fixed offset to the receive frame timing according to the receive burst time plan. Since there are two reference bursts in the frame—the primary reference burst and the secondary reference burst—the receive frame timing is obtained from the primary reference burst. When the primary reference burst is declared *not synchronized* or *not acquired*, the traffic stations should take the receive frame timing from the secondary reference burst. The offset between the two receive frame timings—one from the primary reference burst and one from the secondary reference burst—are contained in the receive burst time plan, thereby enabling the traffic stations

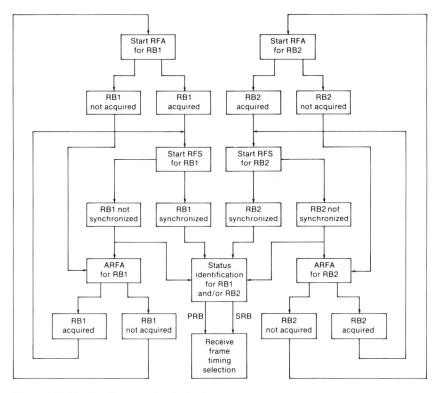

Figure 6.13 Receive frame timing derivation.

to establish their transmit burst timing from either receive frame timing. In summary, the primary reference burst transmitted by the primary reference station forms the basis for receive frame acquisition and synchronization of the entire TDMA network. As long as the primary reference burst is in the frame, the instant of detection of the last bit or symbol of its unique word (the occurrence of the unique word detection pulse) marks the receive frame timing.

6.5.2 Transmit Frame and/or Burst Acquisition and Transmit Frame and/or Burst Synchronization

Establishment of the transmit frame timing and/or transmit burst timing depends on whether the traffic stations operate in a single beam such as a global beam or regional beam where they can receive their own transmitted bursts, as shown in Fig. 6.14, or whether they operate in multiple beams, as shown in Fig. 6.15, where they may not be able to receive their own traffic bursts. Traffic station A in Fig. 6.14 transmits a traffic burst

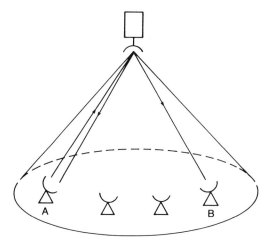

Figure 6.14 Operation in a global beam.

to station B in the same beam. It also receives its own traffic burst, since station A is located in the same beam as station B. On the other hand, traffic station A in the east beam in Fig. 6.15 transmits a traffic burst to traffic station B in the west beam. Since station A is located in the east beam, it cannot receive its own traffic burst which is retransmitted by the satellite in the west beam.

For the TDMA network in Fig. 6.14, the traffic station does not need to establish the transmit frame timing. Instead, it may establish the trans-

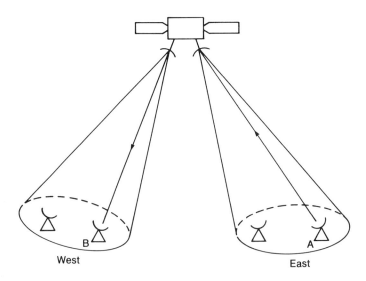

Figure 6.15 Operation in multiple spot beams.

mit burst timing of its traffic bursts directly from the receive frame timing by observation of the traffic burst position error ϵ relative to the fixed offset B_N (contained in the transmit burst time plan) between the receive burst timing of the traffic burst and the receive frame timing of the primary reference burst (or secondary reference burst), as seen in Fig. 6.16. This is called *loopback control* because acquisition and synchronization are achieved by using the station's own traffic burst. Two procedures are involved in the loopback control method: transmit burst acquisition (TBA) and transmit burst synchronization (TBS).

To start transmit burst acquisition, traffic station N transmits a short burst (SB) which is the preamble only of a traffic burst at an estimated offset Δ_N relative to the receive frame timing established by the primary reference burst (or the secondary reference burst). The estimated offset Δ_N, called the transmit burst delay can be obtained in a number of ways. One way to obtain Δ_N is to estimate the range between the earth station and the satellite with a knowledge of their coordinates, and this will be discussed in a later section. Another way to obtain Δ_N is for the traffic station to transmit a low-power short burst to search for the allocated position in the frame. The peak power of the short burst is normally maintained at 20 or 25 dB below the nominal traffic burst power to avoid interference with other traffic bursts in the frame. The low-power short burst is swept through the frame by a different value of the transmit burst delay Δ_N relative to the receive frame timing. Once the low-power short burst is observed to fall into the traffic burst's assigned time slot, the corresponding value of Δ_N is employed to start the transmission of a full-power short burst. Any error ϵ introduced by Δ_N is corrected by observing the position of the short burst relative to the receive frame timing one satellite round-trip propagation later using the unique word track mode.

Once the traffic station has successfully positioned the short burst within the assigned traffic burst time slot, it can start transmit burst synchronization whose purpose is to maintain the traffic burst position to within an error of $\epsilon = \pm 2$ bits of the assigned position by measuring the offset B_N between the receive burst timing of the traffic burst and the receive frame timing. Whenever the error ϵ is $+2$ or -2 bits for a number of consecutive frames within one satellite roundtrip delay, the station may adjust its transmit burst timing Δ_N to $\Delta_N \pm \epsilon$ so that the offset B_N remains correct. At most, the correction is made once per satellite roundtrip delay, since a new error measurement may not be taken until the result of the previous correction has been received. Figure 6.16 shows the error correction for maintaining burst synchronization, and Figs. 6.17 and 6.18 show transmit burst acquisition and synchronization in loopback control.

In a multiple-beam TDMA network, as shown in Fig. 6.15, the traffic station might not be able to receive its own traffic burst, hence loopback control might not be applicable. In this type of network, the reference sta-

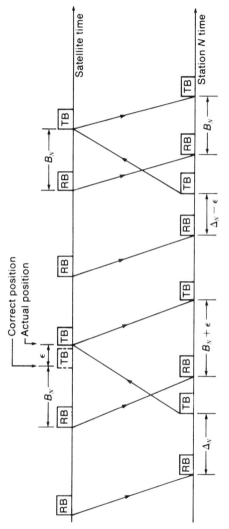

Figure 6.16 Correction of burst position in a loopback control network.

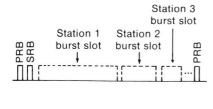

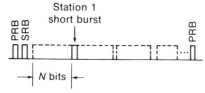

During acquisition, station 1 transmits a short burst (preamble only) offset by N bits from its nominal position in the burst slot using transmit burst delay Δ_N.

After acquisition of its own short burst, station 1 repositions the short burst at the beginning of its burst slot by advancing the transmit burst timing N bits.

After successful repositioning its short burst to the beginning of the burst slot, station 1 adds traffic data to the short burst and begins to transmit its complete traffic burst. The remaining TDMA bursts are acquired in a similar manner.

Figure 6.17 Transmit burst acquisition.

tions supply the control information to the traffic stations for transmit acquisition and synchronization. This information is contained in the transmit timing channel of the primary reference burst (and secondary reference burst). The traffic station may decode this information on reception of the reference burst to establish the transmit frame timing and the transmit burst timing of the traffic bursts. This is called *feedback control* because transmit acquisition and synchronization are achieved based on control information supplied by the reference station. One type of control information sent to traffic station N through the reference burst's transmit timing channel is the *transmit frame delay* D_N. Traffic station N establishes its transmit frame timing at time D_N after its receive frame timing and a number of frames after reception of the reference burst to allow for processing time:

$$\text{TFT} = \text{RFT} + D_N + kT_f \qquad (6.11)$$

where kT_f is an integer number of frames.

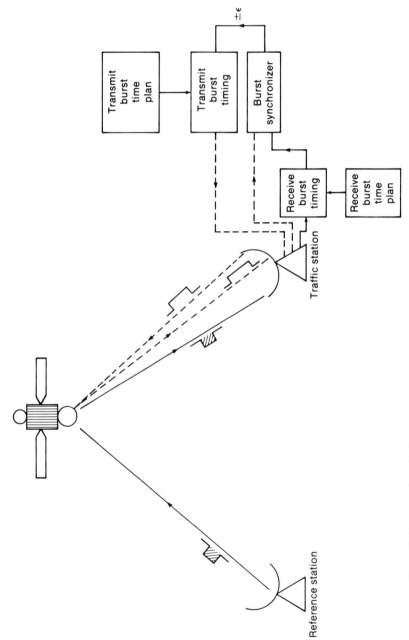

Figure 6.18 Transmit burst synchronization.

The calculation of D_N is shown in Fig. 6.19. The reference station transmits the reference burst to the satellite to establish a TDMA frame with period T_f. All the traffic bursts at the satellite transponder occupy allocated positions in the frame relative to the reference burst. The traffic burst for station N, must arrive at a time B_N after the reference burst according to the transmit burst time plan. Because station N is at a distance d_N from the satellite, there is a time shift in the TDMA frame established at the satellite transponder and the TDMA frame returned to the satellite by station N transmission. This time shift exists because the roundtrip propagation delay between the satellite and station N, $2d_N/c$, is normally not a multiple of the TDMA frame. Therefore a transmit frame delay D_N must be introduced to mark transmit frame timing of station N:

$$D_N = MT_f - \frac{2d_N}{c} \qquad (6.12)$$

where d_N represents the distance between the satellite and station N and M is the smallest integer chosen such that $D_N \geq 0$ for all d_N (for a geostationary satellite, $\max_N 2d_N/c = 278$ ms for a $0°$ elevation angle). If station N transmits a burst at its transmit frame timing, this burst will arrive at the satellite at the same time as the reference burst. If a station N traffic burst is assigned a position B_N relative to the reference burst in a frame, it will be transmitted at time $t = \text{TFT} + B_N$ during each frame. Note that the transmit frame delay D_N can be related to the transmit burst delay Δ_N involved in loopback control:

$$\Delta_N = D_N + B_N \qquad (6.13)$$

For example, let $\max_N 2d_N/c = 278$ ms and let the TDMA frame length be $T_f = 15$ ms; then the smallest integer M is 19, which yields $MT_f = 285$ ms and $D_N = 285 - 278 = 7$ ms for any earth station with $2d_N/c = 278$ ms. From the expression of D_N in (6.12) it is seen that D_N can be determined in terms of the satellite–earth station distance d_N. Knowing the coordinates of the satellite and earth station N, the reference station can compute d_N from this information and thus D_N for acquisition.

Two procedures are involved in the feedback control method: transmit frame acquisition and transmit frame synchronization. For a traffic station that is not yet in operation, the reference station continuously supplies, via the transmit timing channel of its reference burst, an acquisition code TA (for transmit acquisition) together with a transmit frame delay D_N for acquisition. Once the traffic station has obtained the receive frame timing from the receive frame synchronization, it can start transmit frame acquisition. Upon decoding the value D_N and the acquisition code TA, the traffic station starts transmitting a short burst (the preamble of the traffic burst) at a designated position in the frame (in the middle of the traf-

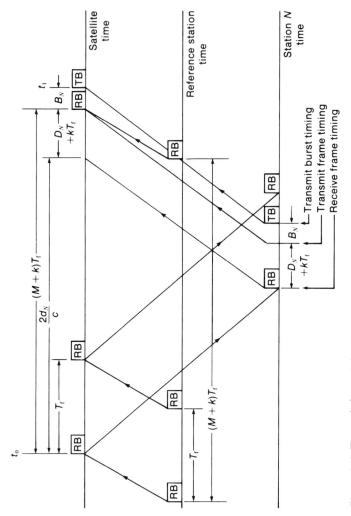

Figure 6.19 Transmit frame delay D_N.

256

fic burst time slot, for example) using the transmit frame timing derived from the receive frame timing according to (6.11). The reference station monitors the position of the short burst relative to its target position and updates the value of D_N. The reference station corrects the value of D_N by observing the position error of the short burst relative to its allocated position in the frame in a manner identical to the loopback method used by the traffic station to determine the burst position error (Fig. 6.16). Note that the position of any traffic burst in the frame is measured relative to the reference burst. Quantitatively speaking, this position is the number of bit or symbol counts between detection of the unique word of the reference burst and of the traffic burst. The traffic station will continue transmitting the short burst until it receives the synchronization code TS (for transmit synchronization) together with the transmit frame delay D_N for synchronization from the reference station. When the TS code is received, the traffic station will start the transmission of its traffic bursts using the corresponding value D_N and the values B_N that determine the positions of the traffic bursts according to the transmit burst time plan. The traffic station will then start transmit frame synchronization to maintain the positions of its traffic bursts within the guard time. Because of variations in the satellite position with time, the value of D_N will be corrected on a periodic basis by the reference station. Again the reference station corrects the value of D_N by observing the position error of the traffic burst in the frame. If the Doppler shift is 40 ns/s or 3.6 bps for a TDMA bit rate of 90 Mbps and the guard time is 64 bits, the time required for the traffic burst, at its correct position, to drift out half the guard time is 8.9 s (assuming the adjacent burst drifts in the opposite direction). In a TDMA network, the reference station normally sends a new D_N to each station every NT_f, where T_f is the frame length. For $N = 40$ and $T_f = 15$ ms, the correction is received every 600 ms, which is short enough to ensure that the traffic bursts stay within their allocated positions. During transmit frame synchronization, if D_N is lost once, the traffic station will use the last received D_N. This is called the *flywheel period*. If D_N is lost more than once, the station will take the receive frame timing from the secondary reference burst and the value of D_N contained in its transmit timing channel to generate the transmit frame timing.

In a TDMA network where both loopback control and feedback control can be employed, a trade-off must be carefully considered. To use loopback control the traffic station must be equipped with a burst synchronizer (in practice two burst synchronizers are normally employed for reliability) to measure traffic burst positions relative to the reference burst (correct Δ_N), thus increasing equipment costs as compared to the feedback control method where a burst synchronizer is not needed. The advantage of loopback control over feedback control is that the guard time for bursts in a loopback TDMA system is about 12 to 16 bits as compared to 32 to 64 bits for a feedback control system. Thus the frame efficiency is increased by using loopback control.

6.6 SATELLITE POSITION DETERMINATION

In a TDMA network where feedback control is employed, the transmit frame delay D_N must be calculated by the reference station and sent to traffic station N for transmit frame acquisition. In order to calculate D_N, the distance d_N from the satellite to traffic station N must be known, because D_N and d_N are related by

$$D_N = MT_f - \frac{2d_N}{c}$$

where M is the smallest integer chosen such that $D_N \geq 0$ for all d_N. The distance d_N can be calculated from a knowledge of the satellite coordinates and the coordinates of earth station N in real time. In the following discussion two methods of satellite position determination are presented: the *single-station ranging* method which is suitable when the reference station can receive its own bursts, and the *three-station ranging* method which is suitable when the reference station cannot receive its own burst.

6.6.1 Single-Station Ranging

The satellite position is normally determined in a geocentric Cartesian coordinate system as shown in Fig. 6.20, where

The X axis is the intersection of the equatorial plane (0° latitude) and the Greenwich meridian plane (0° longitude) and is oriented from the center of the earth.

The Z axis is the polar axis oriented from south to north.

The Y axis completes the right-handed Cartesian coordinate system.

To calculate the satellite coordinates (X_S, Y_S, Z_S) we introduce two additional Cartesian coordinate systems as shown in Fig. 6.21. The X', Y', Z' coordinate system is geocentric, with the X' axis defined as the intersection of the equatorial plane and the θ_L meridian plane where θ_L is the longitude of the reference station. The Z' axis is the polar axis oriented from south to north, and the Y' axis completes the right-handed Cartesian coordinate system. The x, y, z coordinate system has the origin R at the position of the reference station. The z axis is the vector **OR**. The y axis is the vector with origin R, parallel to the Y' axis and oriented in the same direction. The x axis completes the right-handed Cartesian coordinate system.

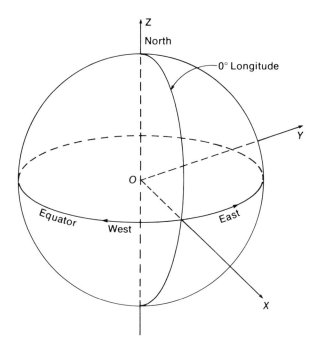

Figure 6.20 Geocentric Cartesian coordinate system X, Y, Z.

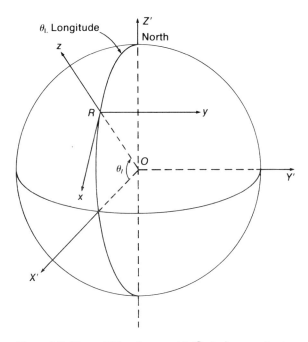

Figure 6.21 Two additional geocentric Cartesian coordinate systems x, y, z and X', Y', Z'.

The transformations between these three coordinate systems are

$$
\begin{bmatrix} X' \\ Y' \\ Z' \end{bmatrix} = \begin{bmatrix} \sin \theta_1 & 0 & \cos \theta_1 \\ 0 & 1 & 0 \\ -\cos \theta_1 & 0 & \sin \theta_1 \end{bmatrix} \begin{bmatrix} x \\ y \\ z \end{bmatrix} + \begin{bmatrix} R_e \cos \theta_1 \\ 0 \\ R_e \sin \theta_1 \end{bmatrix} \tag{6.14}
$$

$$
\begin{bmatrix} X \\ Y \\ Z \end{bmatrix} = \begin{bmatrix} \cos \theta_L & \sin \theta_L & 0 \\ -\sin \theta_L & \cos \theta_L & 0 \\ 0 & 0 & 1 \end{bmatrix} \begin{bmatrix} X' \\ Y' \\ Z' \end{bmatrix} \tag{6.15}
$$

$$
\begin{bmatrix} X \\ Y \\ Z \end{bmatrix} = \begin{bmatrix} \cos \theta_L \sin \theta_1 & \sin \theta_L & \cos \theta_L \cos \theta_1 \\ -\sin \theta_L \sin \theta_1 & \cos \theta_L & -\sin \theta_L \cos \theta_1 \\ -\cos \theta_1 & 0 & \sin \theta_1 \end{bmatrix} \begin{bmatrix} x \\ y \\ z \end{bmatrix}
$$

$$
+ \begin{bmatrix} R_e \cos \theta_L \cos \theta_1 \\ -R_e \sin \theta_L \cos \theta_1 \\ R_e \sin \theta_1 \end{bmatrix} \tag{6.16}
$$

where θ_1 = reference station latitude; $\theta_1 > 0$ for north latitude and $\theta_1 < 0$ for south latitude

θ_L = reference station longitude; $\theta_L > 0$ for west longitude and $\theta_L < 0$ for east longitude

R_e = earth radius = 6378.155 km

Let d_R, E, and A denote the range from the reference station R to the satellite S, the elevation angle, and the azimuth angle, respectively, as shown in Fig. 6.22. Depending on the location of the reference station with respect to the satellite, the azimuth angle A is given by:

1. Northern Hemisphere
   Reference station is west of the satellite: $A = 180° - A'$
   Reference station is east of the satellite: $A = 180° + A'$
2. Southern Hemisphere
   Reference station is west of the satellite: $A = A'$
   Reference station is east of the satellite: $A = 360° - A'$

where $0° \leq A' \leq 90°$ is the angle between the x,z plane and the plane containing the satellite, the earth station, and the center of the earth, as shown in Fig. 6.22. It is seen that the x,y,z coordinates of the satellite are

$$
x_S = d_R \cos E \cos A'
$$

$$
y_S = d_R \cos E \sin A'
$$

$$
z_S = d_R \sin E \tag{6.17}
$$

Substituting (6.17) into (6.16) yields the X, Y, Z coordinates of the satellite as

$$\begin{bmatrix} X_S \\ Y_S \\ Z_S \end{bmatrix} = d_R \begin{bmatrix} \cos \theta_L \sin \theta_1 & \sin \theta_L & \cos \theta_L \cos \theta_1 \\ -\sin \theta_L \sin \theta_1 & \cos \theta_L & -\sin \theta_L \cos \theta_1 \\ -\cos \theta_1 & 0 & \sin \theta_1 \end{bmatrix} \begin{bmatrix} \cos E \cos A' \\ \cos E \sin A' \\ \sin E \end{bmatrix}$$

$$+ \begin{bmatrix} R_e \cos \theta_L \cos \theta_1 \\ -R_e \sin \theta_L \cos \theta_1 \\ R_e \sin \theta_1 \end{bmatrix} \tag{6.18}$$

From (6.18) it is seen that the reference station can determine the geocentric coordinates of the satellite from the knowledge of its range d_R, its elevation angle E, and its azimuth angle A. The range d_N from traffic station N to the satellite can be calculated according to the following expression

$$d_N = \sqrt{(X_S - X_N)^2 + (Y_S - Y_N)^2 + (Z_S - Z_N)^2} \tag{6.19}$$

X_N, Y_N, Z_N are the geocentric coordinates of traffic station N and are given by

$$X_N = R_e \cos \theta_{L,N} \cos \theta_{1,N}$$

$$Y_N = -R_e \sin \theta_{L,N} \cos \theta_{1,N}$$

$$Z_N = R_e \sin \theta_{1,N} \tag{6.20}$$

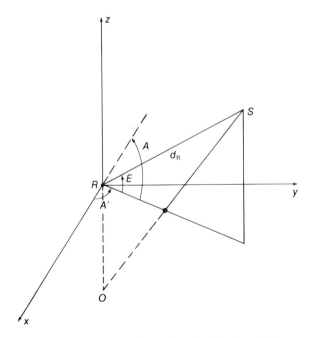

Figure 6.22 Range, elevation angle, and azimuth angle in an x, y, z coordinate system.

where $\theta_{1,N}$ = latitude of traffic station N ($\theta_{1,N} > 0$ for north latitude and $\theta_{1,N} < 0$ for south latitude) and $\theta_{L,N}$ = longitude of traffic station N ($\theta_{L,N} > 0$ for west longitude and $\theta_{L,N} < 0$ for east longitude).

Example 6.1 Consider a reference station R situated at longitude $\theta_1 = 120°W$ and latitude $\theta_1 = 40°N$. The measured azimuth angle $A = 150.48°$, the elevation angle $E = 39.32°$, and the range $d_R = 37,832.96$ km. Find the satellite coordinates.
(a) The coordinates of reference station R can be found from (6.16) by setting $x = y = z = 0$: $X_R = -2442.9751$ km, $Y_R = -4231.3570$ km, and $Z_R = 4099.799$ km.
(b) The coordinates of the satellite are given by (6.18): $X_S = -7321.4346$ km, $Y_S = -41,523.6794$ km, $Z_S = -0.9552$ km.

From the above analysis it is seen that three key parameters, d_R, E, and A, must be measured correctly in order to determine d_N and consequently the transmit frame delay D_N for traffic station N. To obtain the elevation angle E and the azimuth angle A the reference station antenna must be equipped with an automatic tracking system with a resolution of 0.001°. The value of d_R can be obtained by single-hop measurements where the reference station transmits a reference burst at its transmit frame timing and establishes the corresponding receive frame timing. The satellite roundtrip delay is defined as the difference between TFT and its corresponding RFT. Consequently, d_R is calculated as

$$d_R = \frac{c}{2}(\text{RFT} - \text{TFT} - S_R) \tag{6.21}$$

where S_R = sum of internal propagation delays of transmit and receive sides of reference station and satellite transponder.

To resolve frame ambiguity, a special identification number may be included in the reference bursts transmitted for single-hop measurement. Alternatively, d_R can be derived according to the following equation:

$$d_R = \frac{c}{2}(KT_f + \text{RFT} - \text{TFT*} - S_R) \tag{6.22}$$

where KT_f = an integer number of frames
   RFT = receive frame timing
   TFT* = transmit frame timing immediately preceding RFT

The frame ambiguity can be resolved by using the value of d_R obtained from the nominal satellite position.

6.6.2 Three-Station Ranging

Satellite positions can also be determined based on the distances and coordinates of three earth stations according to

$$d_R^2 = (X_S - X_R)^2 + (Y_S - Y_R)^2 + (Z_S - Z_R)^2 \tag{6.23}$$

$$d_K^2 = (X_S - X_K)^2 + (Y_S - Y_K)^2 + (Z_S - Z_K)^2 \qquad (6.24)$$

$$d_M^2 = (X_S - X_M)^2 + (Y_S - Y_M)^2 + (Z_S - Z_M)^2 \qquad (6.25)$$

where d_R, d_K, and d_M are the ranges of reference station R and the ranging stations K and M to the satellite, respectively, and (X_P, Y_P, Z_P), $P = $ S,R,K,M are the geocentric coordinates of the satellite, the reference station, and the ranging stations K and M, respectively.

From the above three equations the geocentric coordinates of the satellite can be calculated.

Because the satellite position is calculated at the reference station, the distances d_K and d_M must be known at the reference station. If the reference station can receive its own burst, the distance d_R can be calculated as in the single-station ranging method. To calculate d_K and d_M, double-hop ranging may be employed. Double-hop range measurements require measurement of the roundtrip propagation delay time of a burst transmitted by the reference station and transponded back by a ranging station using its own burst. The double-hop range between the reference station and ranging station K is given by

$$2(d_R + d_K) = c(KT_f + RFT - TFT^* - D_K - S_R - S_K - B_K)$$
$$(6.26)$$

where $KT_f = $ an integer number of frames that can be obtained by using values of d_R and d_K obtained from nominal satellite position

$RFT = $ receive frame timing at reference station

$TFT^* = $ transmit frame timing immediately preceding RFT

$D_K = $ transmit frame delay employed by ranging station K; this value of D_K is sent back to the reference station in the transponded burst

$S_R = $ sum of internal propagation delays of reference station and satellite transponder accessed by reference burst

$S_K = $ sum of internal propagation delays of ranging station K and satellite transponder accessed by transponded burst from ranging station K

$B_K = $ transponded burst position relative to reference burst

If d_R and $2(d_R + d_K)$ are known, the value of d_K can be calculated. The distance d_M for ranging station M can be found similarly by measuring $2(d_R + d_M)$.

If the reference station cannot receive its own burst, an initial estimate of d_R based on the nominal satellite position can be used to calculate d_K, d_M, and (X_S, Y_S, Z_S). A new value of d_R is then calculated based on the just obtained (X_S, Y_S, Z_S), and the procedure is repeated until convergence is achieved, that is, until the error in (X_S, Y_S, Z_S) is within a predetermined value.

Once the satellite position has been determined, the reference station can generate the transmit frame delay D_N for acquisition of traffic station N (feedback control open-loop acquisition). If such a method is also used for synchronization, the reference station must calculate D_N periodically based on knowledge of the satellite coordinates, and the process is called *feedback control open-loop synchronization*. In practice synchronization is normally achieved by measuring the position error of the traffic station's burst relative to the reference burst if the reference station can receive its own burst, or relative to another reference burst that accesses the same transponder as the traffic burst if the reference station cannot receive its own burst, and adjusting the value of D_N accordingly. (The reference station observes the position error by counting the number of bits or symbols between detection of the unique word of its own reference burst and of the unique word of the traffic burst and compares it to the stored count in the memory.) This process is called *feedback control closed-loop synchronization*. Feedback control open-loop acquisition is commonly employed, even in a loopback control TDMA network. In Sec. 6.7 feedback control closed-loop synchronization, that is, the adjustment of D_N, will be presented for two typical TDMA networks.

6.6.3 Satellite Position Error: Single-Station Ranging

It has been shown that, if the satellite coordinates are determined by the reference station ranging, they are functions of three parameters, namely, the reference station range d_R, the elevation angle E, and the azimuth angle A. All these parameters have measurement errors, and the resulting satellite position may not be exactly at the calculated coordinates. To estimate the position errors, recall that the satellite coordinates (X_S, Y_S, Z_S) are given by (6.18); thus their errors can be expressed as

$$
\begin{bmatrix} \Delta X_S \\ \Delta Y_S \\ \Delta Z_S \end{bmatrix} \cong \begin{bmatrix} \dfrac{\partial X_S}{\partial d_R} & \dfrac{\partial X_S}{\partial E} & \dfrac{\partial X_S}{\partial A'} \\ \dfrac{\partial Y_S}{\partial d_R} & \dfrac{\partial Y_S}{\partial E} & \dfrac{\partial Y_S}{\partial A'} \\ \dfrac{\partial Z_S}{\partial d_R} & \dfrac{\partial Z_S}{\partial E} & \dfrac{\partial Z_S}{\partial A'} \end{bmatrix} \begin{bmatrix} \Delta d_R \\ \Delta E \\ \Delta A' \end{bmatrix} \tag{6.27}
$$

By using (6.18) we get

$$
\begin{bmatrix} \Delta X_S \\ \Delta Y_S \\ \Delta Z_S \end{bmatrix} \cong \begin{bmatrix} \cos \theta_L \sin \theta_1 & \sin \theta_L & \cos \theta_L \cos \theta_1 \\ -\sin \theta_L \sin \theta_1 & \cos \theta_L & -\sin \theta_L \cos \theta_1 \\ -\cos \theta_1 & 0 & \sin \theta_1 \end{bmatrix} \times
$$

$$\begin{bmatrix} \cos E \cos A' & -d_R \sin E \cos A' & -d_R \cos E \sin A' \\ \cos E \sin A' & -d_R \sin E \sin A' & d_R \cos E \cos A' \\ \sin E & d_R \cos E & 0 \end{bmatrix} \begin{bmatrix} \Delta d_R \\ \Delta E \\ \Delta A' \end{bmatrix}$$

$$(6.28)$$

where ΔE and $\Delta A'$ are given in radians. The errors ΔX_S, ΔY_S, and ΔZ_S will result in errors in the transmit frame delay D_N, hence will determine the width of the window allocated for acquisition. (If feedback control open-loop synchronization is employed, the error in D_N will determine the guard time between traffic bursts.) The error in D_N will be discussed in a later section.

6.6.4 Satellite Position Error: Three-Station Ranging

When the reference station and two additional ranging stations are employed to determine the satellite position, the satellite coordinates are explicit functions of the distances from the satellite to the ranging stations and are given by (6.23–6.25). The errors ΔX_S, ΔY_S, and ΔZ_S can be expressed in terms of the range measurement errors Δd_R, Δd_K, and Δd_M by the matrix equation

$$\begin{bmatrix} \Delta X_S \\ \Delta Y_S \\ \Delta Z_S \end{bmatrix} \cong \begin{bmatrix} \dfrac{\partial X_S}{\partial d_R} & \dfrac{\partial X_S}{\partial d_K} & \dfrac{\partial X_S}{\partial d_M} \\ \dfrac{\partial Y_S}{\partial d_R} & \dfrac{\partial Y_S}{\partial d_K} & \dfrac{\partial Y_S}{\partial d_M} \\ \dfrac{\partial Z_S}{\partial d_R} & \dfrac{\partial Z_S}{\partial d_K} & \dfrac{\partial Z_S}{\partial d_M} \end{bmatrix} \begin{bmatrix} \Delta d_R \\ \Delta d_K \\ \Delta d_M \end{bmatrix} \qquad (6.29)$$

It is noted that the satellite coordinate errors are strongly influenced by the coordinates of the three ranging stations. In other words, it can be seen that widely separated ranging stations are necessary to reduce errors.

6.6.5 Errors in D_N

Recall that, in a feedback control TDMA network, the reference station has to send the transmit frame delay D_N to the traffic station for open-loop acquisition, where

$$D_N = MT_f - \frac{2d_N}{c}$$

and d_N can be obtained in terms of the satellite coordinates (X_S, Y_S, Z_S)

and station N coordinates (X_N, Y_N, Z_N) as

$$d_N = \sqrt{(X_S - X_N)^2 + (Y_S - Y_N)^2 + (Z_S - Z_N)^2}$$

If open-loop synchronization is also employed, the reference station has to calculate D_N periodically (in addition to the initial value of D_N for open-loop acquisition) and sends D_N to station N. If closed-loop synchronization is employed, the reference station needs only to correct the value of D_N accordingly by burst position control and periodically sends the updated D_N to station N without calculating D_N and performing the satellite position determination periodically.

It is seen from the above two equations that the errors in the satellite coordinates will result in an error in D_N. For open-loop acquisition, the error in D_N can be compensated for by allocating an acquisition window wide enough to make certain that the short burst (only the preamble of a traffic burst of station N) falls within the window to avoid burst overlapping. The situation is more critical for open-loop synchronization.

Recall that the guard time between bursts must be at least as long as the aperture window used to detect the reference burst unique word to establish the receive frame timing. For open-loop synchronization the guard time must take into account error in D_N.

There are two error sources in the transmit frame delay D_N: (1) measurement errors in the parameters d_R, A, and E(single-station ranging) or d_R, d_K, and d_M (three-station ranging) which result in an erroneously calculated value of D_N transmitted to traffic station N, and (2) the implementation error introduced at station N when it executes D_N. Since TDMA systems normally operate with QPSK modulation, the clock at the earth station runs at the TDMA symbol rate (one QPSK symbol duration is twice the bit duration, and thus the symbol rate is half the bit rate).

Satellite–earth station range measurement errors may consist of five types:

1. The satellite range is determined by the difference between an appropriate transmit frame timing and an appropriate receive frame timing at the reference station. Since the transmit frame timing is generated by the reference station clock while the receive frame timing is generated by the received ranging burst, a reclocking error results which is distributed evenly between 0 to 1. Since the satellite range is half of the roundtrip measurement, the range measurement error ϵ_r is

$$-0.5 \leq \epsilon_r \leq 0.5 \text{ symbol}$$

2. The roundtrip measurement in symbols is not necessarily an even number, therefore, a truncation error ϵ_t occurs:

$$-0.5 \leq \epsilon_t \leq 0.5 \text{ symbol}$$

3. As a result of the Doppler shift induced by satellite motion, the receive frame rate varies during a roundtrip measurement which is about 274 ms maximum (at a 5° elevation angle). For an orbit with a given inclination, the Doppler shift is $\pm\delta$ nanoseconds per second. The error induced over a 274-ms interval is $\pm0.274\delta$ nanosecond $=\pm0.274\delta R_s$ symbol, where R_s is the TDMA symbol rate. The range error due to the Doppler shift is then

$$-0.274\delta R_s \leq \epsilon_\delta \leq 0.274\delta R_s \text{ (symbol)}$$

For a Doppler shift of ±40 ns/s the resulting error is ±0.5 symbol when the TDMA symbol rate is 45 megasymbols per second (Msps). Note that this error does not occur when the ranging station is also the reference station.

4. The propagation velocity fluctuations about a nominal value resulting from atmospheric effects at a low elevation angle below 10° can be ±10 ns, which yields an error of $\epsilon_p = \pm10^{-8}R_s$, where R_s is the TDMA burst symbol rate. For $R_s = 45$ Msps, the error is ±0.5 symbol.

5. The location of a ranging station may be calibrated within ±20 m of its longitude and latitude. This results in an error of about ±2 symbols for a 20° elevation angle and an 45-Msps system. In addition the calibration error of the equipment can be as high as ±2 symbols. The overall calibration error ϵ_c can be about ±4 symbols.

Based on the above discussion, the worse-case range measurement error at a ranging station R is

$$\Delta d_R = \epsilon_r + \epsilon_t + \epsilon_\delta + \epsilon_p + \epsilon_c \tag{6.30}$$

For a 45-Msps TDMA system, the total range measurement error can be estimated to be about ±6 symbols.

For systems that employ the single-station ranging method, the azimuth and elevation angle data can be obtained from the antenna encoders of the autotrack system. The azimuth and elevation angles can be measured to within an error of $\pm0.001°$ or $\pm0.175 \times 10^{-4}$ rad.

The transmit frame delay D_N is calculated from the equation

$$D_N = MT_f - \frac{2d_N}{c}$$

in which there are three possible errors, namely, the error in the satellite range d_N, the assumption of a constant propagation velocity c, and the assumption of a constant frame length T_f at the satellite transponder without considering the Doppler effect. However, the second and third errors are very small compared to the first; therefore the error in the calculated D_N

can be explicitly expressed in terms of the error in D_N as

$$\Delta D_N = -\frac{2\ \Delta d_N}{c} \tag{6.31}$$

Upon using the equation

$$d_N = \sqrt{(X_S - X_N)^2 + (Y_S - Y_N)^2 + (Z_S - Z_N)^2}$$

we obtain

$$\Delta D_N = -\frac{2}{c}\left[\frac{(X_S - X_N)\ \Delta X_S + (Y_S - Y_N)\ \Delta Y_S + (Z_S - Z_N)\ \Delta Z_S}{\sqrt{(X_S - X_N)^2 + (Y_S - Y_N)^2 + (Z_S - Z_N)^2}}\right] \tag{6.32}$$

Besides Δ_N, which is the error in the calculated value of D_N sent to traffic station N by the reference station, errors occur when traffic station N implements the transmit delay D_N. The following errors can occur.

1. From the time t the reference station calculates D_N to the time t' the traffic burst at station N arrives at the satellite on using this D_N, the satellite has changed position. Since the Doppler shift affects both the reference and traffic stations, the burst position error, in the worse case, can be the sum of the two Doppler shifts. For a Doppler shift of ± 40 ns/s and a time interval of $t' - t = 2$ s, the error can be ± 80 ns or ± 3.6 symbols for a 45-Msps TDMA system.

2. The clock at the traffic station must be stable enough to be used for establishing frame timing in case D_N is lost for K frames, so that clocking errors are not significant. A clock of stability of 1 part in 10^8 introduces an error of $10^{-8}KT_f$ seconds or $10^{-8}R_SKT_f$ symbols, where R_S is the TDMA symbol rate. For $KT_f = 2$ s and $R_S = 45$ Msps, the clocking error is 0.9 symbol.

3. As discussed in the case of the ranging station, the error due to the location uncertainty of the traffic station can be ± 2 symbols. In addition, the calibration error of the equipment can add another ± 2 symbols. The overall calibration error is about ± 4 symbols.

The implementation error ϵ is the sum of the above three errors, in the worse case, and can be up to ± 8.5 symbols for a TDMA symbol rate of 45 Msps. The total burst position error in open-loop synchronization is $\pm(\Delta D_N + \epsilon)$.

Example 6.2 Let us consider the errors in single-station ranging method using the parameters given in Example 6.1. Assume that $\Delta d_R = \pm 40$ m, $\Delta E = \pm 0.175 \times 10^{-4}$ rad, and $\Delta A' = \pm 0.175 \times 10^{-4}$ rad. There are eight combinations of the error set $(\Delta d_R, \Delta E, \Delta A')$; therefore there are also eight combinations of the errors $(\Delta X_S, \Delta Y_S, \Delta Z_S)$ given in meters.

To find the error in the transmit frame delay D_N, consider a traffic station N located at longitude $\theta_{L,N} = 80°W$ and latitude $\theta_{l,N} = 35°N$. The coordinates of station N

can be calculated from (6.20): $X_N = 907.2559$ km, $Y_N = -5,145.3041$ km, and $Z_N = 3,658.3594$ km.

Using the eight combinations of $(\Delta X_S, \Delta Y_S, \Delta Z_S)$ we can evaluate ΔD_N from (6.32): -738 ns $\leq \Delta D_N \leq 738$ ns. The procedure can be repeated for all stations $N = 1,2,\dots,N_0$ in the network. Finally, the maximum error in the transmit frame delay of the network is

$$\Delta D_{N,\max} = \max_{1 \leq N \leq N_0} \{\Delta D_N\}$$

6.7 BURST TIME PLAN

As discussed before, in a TDMA network a traffic station transmits its bursts on time to their allocated positions in the frame at the satellite transponder according to a *transmit burst time plan* and receives bursts in the frame returned by the satellite transponder according to a receive burst time plan. The burst time plan is thus a map that indicates the position and length of bursts in the frame and also the position and length of information subbursts within a burst. Since bursts and subbursts carry traffic (voice, data, video) between stations, the burst time plan is simply the traffic assignment within a frame. If the total traffic of a TDMA network exceeds the capacity of one transponder, the network has to operate with more than one transponder. This means that a traffic station might transmit bursts to more than one transponder and might be required to receive bursts from more than one transponder (transponder hopping). In such a multiple-transponder operation, the burst time plan is the assignment of traffic to transponders and time ordering of the assigned traffic within a frame.

A typical burst time plan format is shown below:

Message data
  Burst time plan identification
  Traffic station identification
  Number of transmit bursts
  Number of receive bursts
Burst data
  Burst identification
  Transmit-receive flag
  Transponder identification
  Burst position
  Number of transmit-receive subbursts
Subburst data
  Transmit receive subburst identification
  Subburst position
  Subburst length

6.8 CONTROL AND COORDINATION BY THE REFERENCE STATION

In a TDMA network, the reference station performs the following main functions:

1. Determines the satellite position and generates the transmit frame delay D_N for transmit frame acquisition of the traffic station, as shown in Fig. 6.23, or for open-loop synchronization.
2. Provides and updates D_N on a periodic basis for transmit frame synchronization (closed-loop synchronization), as in Fig. 6.24.
3. Coordinates traffic stations when there is a burst time plan change to avoid disruption of traffic.
4. Performs frame reconfiguration, that is, changes the traffic burst position and length on a demand-assigned basis (to be discussed in Chap. 7).

The first two control functions depend very much on whether the network operates in a single- or multiple-beam environment. Modern TDMAs are designed to operate with multiple-beam satellites and, in addition, a traffic station might be required to transmit nonoverlapping bursts (if possible) to more than one transponder and/or to receive nonoverlapping bursts (if possible) from more than one transponder. This technique is called transponder hopping (Chap. 3 discussed transponder hopping equipment) and is employed to achieve flexibility in traffic arrangement for a number of transponders and to minimize traffic station equipment costs. Since the traffic station establishes its transmit frame timing and receive frame timing from the reference burst in one transponder, it can hop to other transponders (i.e., transmit traffic bursts to other transponders) if and only if the reference bursts of these transponders are synchronized with a common reference burst in a designated transponder. Only then can the transmit frame timing and receive frame timing from the traffic stations that access these transponders be synchronized with a common timing reference and only then can these transponders be employed as a common pool. In the following section we will discuss burst position control for feedback closed-loop synchronization in two typical satellite networks.

6.8.1 Burst Position Control

Consider the dual-polarized satellite communication network shown in Fig. 2.19 where the continental United States, Alaska, and Hawaii are illuminated by four downlink satellite beams: the east beam, the west beam, the CONUS beam, and the global beam over the frequency band 11.7 to 12.2 GHz; and by two uplink satellite beams: the CONUS beam

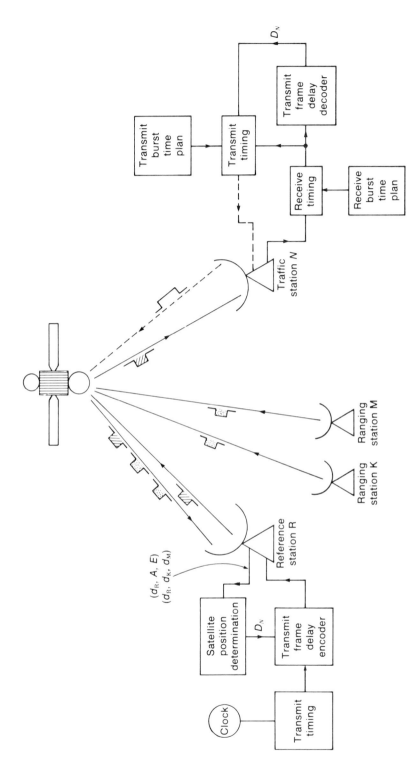

Figure 6.23 Satellite position determination for transmit frame acquisition.

271

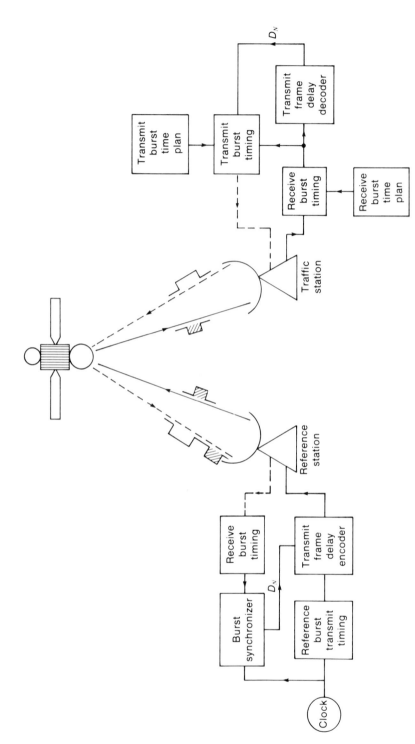

Figure 6.24 Correction of burst position for transmit frame synchronization.

and the global beam over the frequency band 14 to 14.5 GHz. The satellite transponder based on the downlink beams may be grouped as:

East transponders
West transponders
CONUS transponders
Global transponders

In order to control the burst positions and to update D_N for a traffic station in this type of TDMA network, two reference stations are required: a primary reference station (PRS) in the east beam and a secondary reference station (SRS) in the west beam. The roles of the PRS and SRS can be interchanged.

The PRS and SRS transmit primary reference bursts and secondary reference bursts, respectively, to the satellite transponders using the uplink global beam on the vertical polarization (two transponders), the uplink CONUS beam on the vertical polarization (six transponders), and the uplink CONUS beam on the horizontal polarization (eight transponders). All the PRBs are staggered in time by a fixed offset relative to a PRB in a transponder designated as a timing and reference transponder (TRT). The SRB in a transponder is separated by a fixed offset relative to its corresponding PRB. This arrangement permits all 16 transponders of the satellite to be synchronized for transponder hopping. When feedback control is employed, it is administered by the PRS and SRS independently in the following manner:

1. For traffic stations that transmit their traffic bursts into east transponders.
   a. Control is administered at the PRS by observing the traffic burst position relative to the PRB to correct D_N.
   b. For the SRS to be able to administer control, each traffic station must be capable of transmitting a short burst (preamble only) to a designated west transponder or, alternatively, to a designated CONUS or global transponder. An acquisition time slot must also be allocated in this designated transponder for sequential acquisition of the traffic stations. Control is administered at the SRS by observing the short burst position relative to the SRB to correct D_N.
2. For traffic stations that transmit traffic bursts to the west transponders,
   a. Each must be capable of transmitting a short burst to a designated east transponder or, alternatively, to a CONUS or a global transponder. An acquisition time slot must be allocated in the designated transponder for sequential acquisition of the traffic stations.

Control is administered at the PRS by observing the short burst position relative to the PRB to correct D_N.

b. The SRS administers control by observing the traffic burst position relative to the SRB to correct D_N.

3. For traffic stations that transmit traffic bursts to both east and west transponders

a. Control is administered at the PRS by observing the traffic burst positions relative to the PRB in the east transponders to correct D_N

b. Control is administered at the SRS by observing the traffic burst positions relative to the SRB in the west transponders to correct D_N.

The transmission of short bursts by traffic stations whenever necessary enables the PRS and SRS to initiate feedback control acquisition and synchronization for all the traffic stations in the network independently. If the PRS fails, the SRS can take over control immediately. In this type of network, the satellite may consist of many local synchronized communities of transponders (LSCTs), each serving a number of traffic stations, and these may constitute subnetworks whose control and coordination are independent of each other. The reference burst in each transponder of a LSCT may contain identical information in its signaling channel. Each traffic station in a LSCT obeys the control and coordination exerted by the reference burst in a designated transponder. All the traffic bursts from a traffic station that access a LSCT contain identical information in their signaling channel.

Another way to control the burst positions of traffic stations in this type of network, besides the above-mentioned feedback control method involving the PRS and SRS, is the loopback control method. This is possible because a traffic station in the east beam (west beam) that does not transmit its traffic bursts to the east transponders (west transponders) always has the ability to transmit a short burst to a designated east transponder (west transponder) so that it can receive its own short burst and perform loopback synchronization. Other traffic stations that transmit their traffic bursts to their own beam can receive their own traffic bursts and need not transmit short bursts. In this way loopback synchronization can be performed by each individual traffic station employing a burst synchronizer. Acquisition may be carried out using feedback control from the PRS and SRS.

In the above discussion we have described a TDMA network where only two reference stations are needed, a PRS and a backup SRS. This is possible because the PRS and SRS can transmit their reference bursts to all transponders in the satellite using the uplink CONUS beam or the global beam. For a network that operates with downlink spot beams and

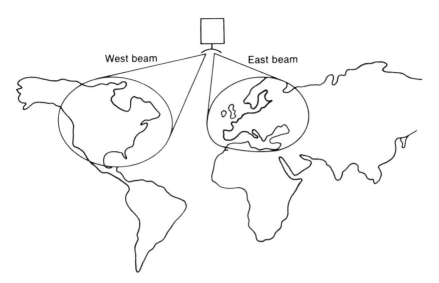

Figure 6.25 Spot beam satellite network.

also with uplink spot beams, as shown in Fig. 6.25, the situation is much different. The satellite beams illuminate two geographic regions with two downlink beams—the east and west beams over the frequency band 3.7 to 4.2 GHz. The satellite receives traffic from two uplink beams—the east and west beams over the frequency band 5.9 to 6.4 GHz. The satellite transponders are arranged according to their beam connectivity as

East-west transponders
West-east transponders
East-east transponders
West-west transponders

Earth stations in the east (west) beam can transmit their bursts only to east-west (west-east) transponders and east-east (west-west) transponders and can receive bursts only from west-east (east-west) transponders and east-east (west-west) transponders. This type of beam connectivity necessitates the use of a pair of reference stations in each beam, namely, an east primary reference station (EPRS) and a backup east secondary reference station (ESRS) in the east beam, and a west primary reference station (WPRS) and a backup west secondary reference station (WSRS) in the west beam.

Let EPRB (WPRB) and ESRB (WSRB) denote the east primary reference burst (west primary reference burst) and the east secondary reference burst (west secondary reference burst) transmitted by the EPRS

(WPRS) and ESRS (WSRS), respectively. The EPRS (WPRS) and ESRS (WSRS) transmit the EPRB (WPRB) and ESRB (WSRB) to east-west (west-east) and east-east (west-west) transponders, respectively. All the EPRBs (WPRBs) are staggered in time by a fixed offset relative to an EPRB (WPRB) in a transponder designated as the east (west) timing and reference transponder. The ESRB (WSRB) in a transponder is separated by a fixed offset relative to its corresponding EPRB (WPRB). Since the east-west (west-east) and east-east (west-west) transponders are synchronized with a single set of EPRBs (WPRBs) and ESRBs (WSRBs), transmit transponder hopping between them is possible. In order to achieve receive transponder hopping between west-east (east-west) transponders using the WPRB (EPRB) and WSRB (ESRB) as timing references and east-east (west-west) transponders using the EPRB (WPRB) and ESRB (WSRB) as timing references, the WPRB must be synchronized with the EPRB to provide a common timing reference for all the transponders. This can be carried out by either the EPRS or the WPRS using feedback control. Let the EPRS be selected to perform reference burst synchronization. Since the EPRS can receive its own EPRB from the east-east transponder, it can observe the position of the WPRB received from the west-east transponders relative to its EPRB and corrects the transmit frame delay D_{WPRS} that it sends to the WPRS for synchronization. In this manner all the WPRBs and EPRBs can be synchronized to a common timing reference (the EPRB in the east timing and reference transponder) and both transmit and receive transponder hopping is possible.

When feedback control is employed, it is administered by the reference stations as follows:

1. For traffic stations that transmit their traffic bursts to east-west (west-east) transponders, control is administered by the WPRS (EPRS) by observing the traffic burst positions relative to the EPRB (WPRB) to correct D_N. The WPRS (EPRS) then sends the updated D_N to the traffic stations in its WPRB (EPRB) via west-east (east-west) transponders.

2. For traffic stations that transmit their traffic bursts to east-east (west-west) transponders, control is administered by the EPRS (WPRS) by observing the traffic burst positions relative to the EPRB (WPRB) to correct D_N. The EPRS (WPRS) then sends the updated D_N to the traffic stations in its EPRB (WPRB) via east-east (west-west) transponders.

3. For traffic stations that transmit their traffic bursts to both east-west (west-east) and east-east (west-west) transponders, control is administered in the same way as in case 1. This is possible because the WPRB and EPRB are synchronized.

6.8.2 Traffic Coordination: Burst Time Plan Change

Operation of the traffic stations in a TDMA network, that is, the transmission and reception of traffic bursts, is defined by the burst time plan. When the network grows or when the network traffic changes, the burst time plan has to be modified accordingly. The new burst time plan is sent by the reference station to individual traffic stations in the network which then store it in background memory. A change in the burst time plan is implemented by causing the traffic stations to transmit and receive bursts as directed by the new plan. The reference station coordinates the change in such a way that all traffic bursts transmitted according to the new plan arrive at the satellite for the first time in their allocated positions in the same frame to avoid burst overlapping. This means that all the traffic stations in the network must implement their transmit burst time plan from the transmit frame timing derived from the receive frame timing obtained from the same reference burst in the same frame. Let K be a special codeword contained in the primary reference burst broadcast to all stations for the burst time plan change. All stations receive the PRB that contains the special codeword K and establish the receive frame timing RFT_K of the same TDMA frame transmitted to them by the satellite. The new transmit burst time plan is implemented by station N at the transmit frame timing as

$$TFT = RFT_K + D_N + kT_f \tag{6.33}$$

where kT_f = integer number of frames allowed for processing D_N. Let t_0 be the time the satellite retransmits the PRB to all the stations. Let $t_{1,N}$ be the time a station N traffic burst, assigned a position B_N relative to the reference burst according to the new transmit burst time plan, arrives at the satellite as shown in Fig. 6.26. The time $(t_{1,N} - TBT) + (RFT_K - t_0) = 2d_N/c$ represents the satellite roundtrip delay for station N, where TBT is the transmit burst timing of station N. Note that $TBT = TFT + B_N$; therefore upon using (6.33) we have

$$\frac{2d_N}{c} = t_{1,N} - TFT - B_N + RFT_K - t_0$$

$$= t_{1,N} - t_0 - B_N - (TFT - RFT_K)$$

$$= t_{1,N} - t_0 - B_N - D_N - kT_f \tag{6.34}$$

Hence from (6.12) we get

$$t_{1,N} - t_0 - B_N = \frac{2d_N}{c} + D_N + kT_f = (M + k)T_f$$

$$= LT_f \tag{6.35}$$

Since LT_f is a constant, it is seen that all the traffic bursts transmitted at

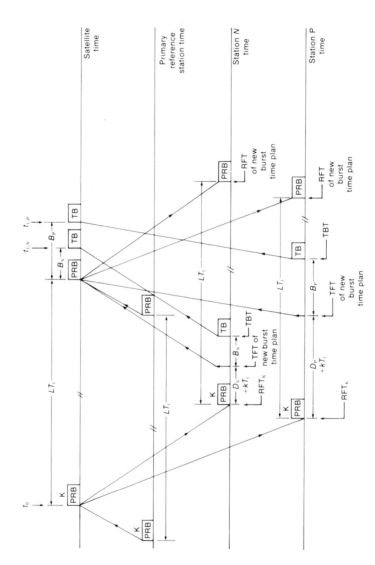

Figure 6.26 Burst time plan change.

the transmit frame timing defined in (6.33) will arrive at the satellite in their respective positions in the same frame; that is,

$$t_{1,1} - t_0 - B_1 = t_{1,2} - t_0 - B_2 = \cdots = t_{1,N} - t_0 - B_N = LT_f \quad (6.36)$$

for stations 1, 2,....,N. From Fig. 6.26, it is obvious that station N should implement the receive burst time plan at the receive frame timing defined as

$$RFT = RFT_K + LT_f \quad (6.37)$$

In practice the burst time plan change may be carried out by a countdown procedure where the special codeword K is actually a countdown sequence, say, $K, K-1, \ldots, 1, 0$. Each traffic station in the network may be required to correctly decode n out of K countdown numbers ($n \leq K$) and to lock their internal countdown to the countdown sequence and implement the burst time plan change at countdown 0. This is a safeguard for the situation where some of the countdown numbers are lost as a result of link error.

6.9 TDMA TIMING

In a TDMA network, the frame timing is established at the satellite by the reference station which transmits the reference bursts at a constant period equal to the frame length T_f. The frame period T_f is derived from a highly stable transmit symbol clock with frequency f_0 equal to the TDMA symbol rate R_s. For example, if $R_s = 44.776$ Msps, then $f_0 = 44.776$ MHz. The clock frequency offset is normally less than 1 part in 10^{11} from f_0. The frame period T_f is derived from f_0 by counting the number of symbols N that must be transmitted per frame; that is,

$$T_f = \frac{N}{f_0} \quad (6.38)$$

For example, if $f_0 = 44.776$ MHz, and $N = 671,640$ symbols, then $T_f = 15$ ms. In this case, the reference station establishes the frame period $T_f = 15$ ms by counting 671,640 symbol clock periods using the clock of frequency 44.776 MHz.

A geostationary satellite is not perfectly stable in its orbit because of the effects of the moon and the sun, hence its motion induces a different frame period at the satellite and at the traffic stations. Consider two stations in a TDMA network as shown in Fig. 6.27. The reference station R establishes the frame period $T_f = N/f_0$. We will show that the frame length at the satellite and at traffic station N varies with time and how to remedy this problem.

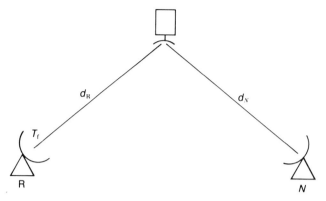

Figure 6.27 Reference station and traffic station.

Consider the two reference bursts transmitted by the reference station at time t and $t + T_f$ as shown in Fig. 6.28. These two reference bursts are received by the satellite at time $t + d_R(t)/c$ and $t + T_f + d_R(t + T_f)/c$, respectively. Because of satellite motion, the distance d_R between the satellite and the reference station varies with time, and therefore the average frame period $T_{f,s}$ at the satellite is

$$T_{f,s} = T_f + \frac{d_R(t + T_f)}{c} - \frac{d_R(t)}{c}$$

$$= T_f + \Delta T_R \tag{6.39}$$

where the frame period variation $\Delta T_R = [d_R(t + T_f) - d_R(t)]/c$ can be positive or negative.

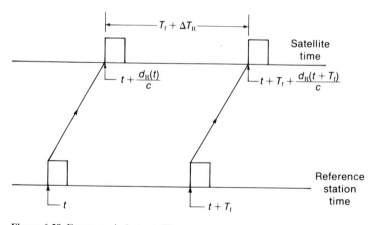

Figure 6.28 Frame period at satellite.

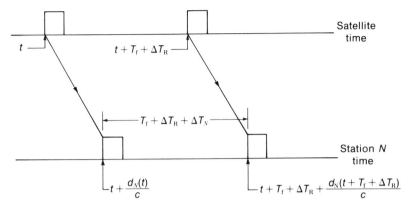

Figure 6.29 Receive frame period at a traffic station.

To establish the receive frame period at traffic station N, consider the two reference bursts that appear at the satellite at time t and $t + T_{f,s} = t + T_f + \Delta T_R$ as shown in Fig. 6.29. These two reference bursts are received at traffic station N at time $t + d_N(t)/c$ and $t + T_f + \Delta T_R + d_N(t + T_f + \Delta T_R)/c$, respectively. Since the distance d_N between the satellite and station N may vary with time, the average receive frame period $T_{f,r}$ at station N is

$$T_{f,r} = T_f + \Delta T_R + \frac{d_N(t + T_f + \Delta T_R)}{c} - \frac{d_N(t)}{c}$$

$$= T_f + \Delta T_R + \Delta T_N = T_{f,s} + \Delta T_N \qquad (6.40)$$

where the frame period variation $\Delta T_N = [d_N(t + T_f + \Delta T_R) - d_N(t)]/c$ can be positive or negative.

To derive the average transmit frame period $T_{f,t}$ at station N, we note that the transmit frame timing is derived from the receive frame timing; therefore the average frame period established at the satellite by station N must be equal to the average frame period established at the satellite by the reference station. Thus

$$T_{f,s} = \frac{T_{f,t} + T_{f,r}}{2} \qquad (6.41)$$

and therefore

$$T_{f,t} = 2T_{f,s} - T_{f,r}$$

$$= 2(T_f + \Delta T_R) - (T_f + \Delta T_R + \Delta T_N)$$

$$= T_f + \Delta T_R - \Delta T_N$$

$$= T_{f,s} - \Delta T_N \qquad (6.42)$$

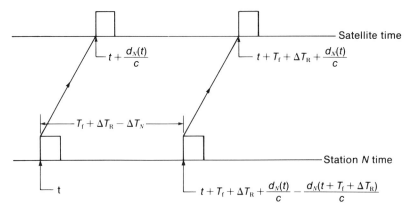

Figure 6.30 Transmit frame period at a traffic station.

The transmit frame period at station N is illustrated in Fig. 6.30.

Variations in the transmit and receive frame periods introduce variations in the transmitted and received data rate which must be eliminated to provide an interface with a terrestrial network having a constant data rate.

In a digital satellite network, the terrestrial switches may be mutually interconnected through TDMA links such as those shown in Fig. 6.31, where the switches i and j operate at clock frequencies f_i and f_j, respectively, and stations A and B operate with clock frequency f_0. It is obvious that there must be an overall network synchronization which encompasses the following problems:

1. The relationship between the clock frequencies f_i and f_j of the terrestrial switches and the TDMA clock frequency f_0.
2. Compensation for frame period variations due to satellite motion.

6.9.1 Slip Rate in Digital Terrestrial Network

In a digital terrestrial network, digital switches are interconnected by digital transmission links. A transmit digital signal leaves the originating switch at a rate determined by the switch's clock. At the terminating switch, the signal is read into a buffer. The read-in rate is determined by the originating switch. A buffer is provided for each arrival signal to accommodate the variations in arrival time of data from different switches. It serves as a reservoir in which data is stored until it is read out. The signal is then read out of the buffer at a rate controlled by the terminating switch's clock. If the read-out rate is slower than the read-in rate, the buffer will eventually become full and data will be overwritten before

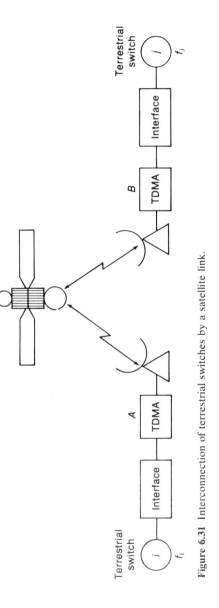

Figure 6.31 Interconnection of terrestrial switches by a satellite link.

being read out; that is, bits of data will be lost. If the read-out rate is faster than the read-in rate, the buffer will be emptied and scanned in succession, hence bits of data will be repeated. Either one of these situations, loss or duplication of information due to the difference between the number of bits transmitted and accepted per unit of time, is referred to as a *slip*. It is therefore necessary to keep the clock frequencies of the switches nearly identical on the average; clocks at different switches should be kept within acceptable tolerances of one another. For a PCM network operating with a frame period of 125 μs (Fig. 1.6), sufficient buffering must be provided to ensure that, when a slip occurs, an entire PCM frame is either deleted or repeated so that the PCM stream will not lose frame synchronization. Thus a slip rate objective of 1 in 20 h leads to a clock accuracy objective of 125 μs/20 h = 1.74×10^{-9}, that is, 1.74 parts in 10^9 for the average relative rate difference between clocks.

There are a number of approaches to maintaining timing accuracy between switches. They include pulse stuffing operation, plesiochronous (nearly synchronous) operation, and synchronous operation.

In a terrestrial network, different bit streams with different origins are normally asynchronous, and a pulse stuffing technique may be used to multiplex them together; that is, dummy pulses can be inserted into multiple asynchronous bit streams to produce a common bit rate slightly higher than the bit rate for time division multiplexing. Dummy pulses are removed at reception, and each stream is returned to its original asynchronous bit rate. In a digital network, channels in bit streams with different origins that arrive at the switch before switching may be packed into the same bit stream after switching. Unless all these channnels are synchronous, pulse stuffing and destuffing must be done on an individual channel basis at every switch in the network. For a complete digital network, this is not cost-effective, since the cost of pulse stuffing equipment can no longer be shared by many users.

In a plesiochronous operation, each switch in the network is controlled by a highly stable clock. Since the clocks are independent, they differ slightly in frequency over a period of time. For clock frequencies f_i and f_j, the relative clock rate difference $\Delta = |f_i - f_j|/f_0$, where f_0 is the desired operating frequency, may be 2×10^{-8} (2 parts in 10^8) for a good crystal oscillator and perhaps as stable as 2 parts in 10^{11} for a cesium beam clock. The slip rate objective for a plesiochronous PCM network (125 μs frame length) with $\Delta = 2 \times 10^{-11}$ is one slip in 72 days. The amount of buffer storage required for each bit stream interchanged between two switches is determined by the relative clock rate difference Δ, the bit rate R, and the time interval τ over which Δ is allowed to accumulate. The number of bits the buffer must absorb during the time interval τ is

$$N = \tau R \Delta \qquad (6.43)$$

If the required time interval is $\tau = 72$ days for two switches exchanging data streams at 44.736 Mbps, using clocks with a rate difference of $\Delta = 2 \times 10^{-11}$, then the required buffer size should be

$$N = (72 \times 24 \times 3600) \, (44.736 \times 10^6) \, (2 \times 10^{-11})$$

$$= 5566 \text{ bits}$$

Thus in a plesiochronous network, the slip rate objective performance is determined by the relative accuracy of the switch clocks. A less accurate clock will cost less but will incur additional costs with larger buffer sizes and require more frequent adjustments to maintain a specified slip rate objective.

In a synchronous operation, all the switch's clocks are automatically controlled by a frequency standard. One of the most simple and commonly used synchronous operations is a hierarchial master-slave structure in which all the switch clocks are slaved either directly or indirectly through intermediate switches to a master clock by phase-locking to the master clock reference timing signal. The concept of hierarchial master-slave clock synchronization is shown in Fig. 6.32. The direction of the arrows indicates the flow of the reference timing signal. The clock at the head of the arrow is phased-locked to the clock at the tail of the arrow. If the reference timing signal from the master clock is lost, the switch's clock will either select the most suitable incoming route to use as its reference timing signal or use its own independent clock as a timing source until the master timing signal is restored. It is necessary that the switch clocks at the hierarchial levels be stable and accurate and that they can free-run for a period of time (days) before overflowing the switch buffers. One problem in master-slave operations is the transmission delay occurring in the path between the switch clock and the master clock. Changes in the transmission delay can also change the buffer reading. For example, con-

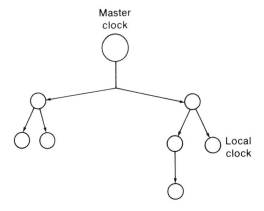

Figure 6.32 Master-slave clock synchronization.

sider two interconnected switches in a master-slave hierarchy. If the transmission delay increases in both directions, more bits will be stored in the transmission lines and the buffer reading will decrease at both switches. This also causes perturbations when switching from one path to another for timing, unless large enough buffers are provided to accommodate changes in transmission delays.

6.9.2 TDMA System and Terrestrial Network Interconnection

From the above discussion, it is seen that the interconnection of digital terrestrial networks through satellite TDMA links must involve some form of compensation for the difference in the clock frequencies used in the networks and the satellite Doppler effect which causes variations in the receive and transmit frame rates at the earth station. These problems can be solved by appropriate design of the digital terrestrial interfaces. Figure 6.33 illustrates a digital terrestrial network connected to a satellite TDMA link. If the TDMA clock and the terrestrial clock are independent but are derived from highly stable clocks with a stability of 1 part in 10^{11}, the terrestrial interfaces are said to be *plesiochronous*. If the terrestrial clock is derived from the TDMA clock, the terrestrial interfaces are *synchronous*. If the TDMA clock and the terrestrial clock are of the same nominal frequency but unrelated, the terrestrial interfaces are *asynchronous*.

Clock synchronization between the TDMA system and the digital terrestrial network alone does not correct for the Doppler effect due to satellite motion. Each terrestrial interface accepts a continuous data stream of constant bit rate R_k from the digital terrestrial network and also delivers a continuous data stream of constant bit rate R_k to the digital terrestrial network. Since at the same time the interface must accept or deliver data with a variable burst rate from or to the TDMA system, because of the variations in the frame period due to the Doppler effect, there must be some means to accommodate the differences between the bit rates. This can be accomplished at the terrestrial interface by using a transmit Doppler buffer and a receive Doppler buffer. To understand how

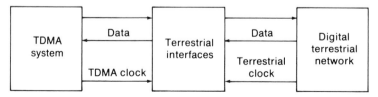

Figure 6.33 Digital terrestrial network connected to a TDMA link.

these Doppler buffers operate, recall that the terrestrial interface delivers or accepts continuous data to or from the digital terrestrial network while it delivers or accepts data at the burst rate to or from the TDMA system on a frame period basis. Assume for the time being that there is no Doppler effect and that the TDMA system at the traffic station of a TDMA network operates with a transmit and receive frame period T_f. Then each terrestrial interface must provide a buffer for storing the continuous data stream at constant bit rate R_k that it accepts from the digital terrestrial network for one frame length T_f and then deliver the stored data to the TDMA at the TDMA burst rate on a frame period basis. This buffer is called a *compression buffer* and possesses a capacity of μ:

$$\mu = R_k T_f \text{ (bits)} \tag{6.44}$$

Similarly, each terrestrial interface must provide a buffer for storing the subburst that it accepts from the TDMA system at the TDMA burst rate every frame period T_f and then converting the stored data to a data stream at a constant bit rate R_k that it must deliver to the digital terrestrial network. This buffer is called an *expansion buffer* and also possesses a capacity of $\mu = R_k T_f$.

Now if the Doppler effect is taken into account at the traffic station, the TDMA system must operate at the receive frame period $T_{f,r} = T_f + \Delta T_R + \Delta T_N$ and the transmit frame period $T_{f,t} = T_f + \Delta T_R - \Delta T_N$, where ΔT_R and ΔT_N are the frame period variations due to satellite motion relative to the reference station and the traffic station, respectively. Thus we have

$$T_f - |\Delta T_R| - |\Delta T_N| \leq T_{f,r} \leq T_f + |\Delta T_R| + |\Delta T_N|$$

$$T_f - |\Delta T_R| - |\Delta T_N| \leq T_{f,t} \leq T_f + |\Delta T_R| + |\Delta T_N| \tag{6.45}$$

where the upper and lower limits of $T_{f,r}$ and $T_{f,t}$ represent the worst case. This means that the terrestrial interface must now store an extra ΔR_k bits per frame length when it accepts the burst data from the TDMA system to deliver to the digital terrestrial network. It must also store an extra ΔR_k bits per frame length when it accepts the continuous data stream from the digital terrestrial network to deliver to the TDMA system. The extra ΔR_k bits are given by

$$\Delta R_k = R_k (|\Delta T_R| + |\Delta T_N|) \tag{6.46}$$

The effect of frame period variations on the storage is cumulative; that is, the terrestrial interface must store ΔR_k bits on the receive side and ΔR_k bits on the transmit side every frame period for the duration t in which the satellite moves from its nominal position to the position where the maximum path length deviation occurs. The Doppler shift $(|\Delta T_R| + |\Delta T_N|)/T_f$ is about 40 ns/s, and $t = 8$ h. Thus the maximum capacity of the transmit or receive Doppler buffer is

$$B = \frac{t(\Delta R_k)}{T_f} = \frac{tR_k}{T_f} \left(|\Delta T_R| + |\Delta T_N| \right)$$

$$= 1.15 \times 10^{-3} R_k \text{ (bits)} \qquad (6.47)$$

Hence the transmit or receive Doppler buffer at each terrestrial interface must be capable of accepting 1.15 ms of data at a rate of R_k bits per second. Since at the start-up of a TDMA system the satellite's position is not known, a Doppler buffer of capacity $2B$ may be used so that by starting at the center of the buffer, overflow and underflow can be prevented.

6.9.3 Plesiochronous Interfaces

The mutual connection of digital terrestrial networks through TDMA links requires appropriate terrestrial interfaces to interconnect the terrestrial path and the satellite path. For a *plesiochronous* (nearly synchronous) terrestrial network interconnection, system synchronization via TDMA links is possible using highly stable clocks at the reference station and at the terrestrial network. The clock accuracy is 1 part in 10^{11}; that is, the relative clock rate difference is $\Delta = 2 \times 10^{-11}$. As previously discussed, frame slips can occur at the interfaces of two plesiochronous networks. The introduction of a TDMA link between these networks will also cause variations in the data rate because of satellite motion. If the interfaces between the plesiochronous networks and the TDMA link are properly designed, the Doppler effect can be completely corrected. The net effect of the introduction of a TDMA link between two plesiochronous networks is the same as that for two plesiochronous interfaces. Figure 6.34 shows plesiochronous interfaces where network synchronization is obtained by employing, on both the transmit and receive sides, frame aligners to introduce frame slips and Doppler buffers to absorb data rate variations induced by satellite motion.

The TDMA local clock is phase-locked to the receive frame timing provided by the reference burst and thus synchronizes the local frame timing with the network frame timing. This avoids the requirement for a highly stable clock at the traffic station, since the local clock tracks, except for a slight Doppler offset, a highly stable reference station clock frequency (1 part in 10^{11}) derived from the received reference burst unique word detection or from the recovered symbol clock of the reference burst. The local clock stability relative to the network clock is chosen such that it does not introduce considerable error when being used temporarily as internal timing during an interval of time when the reference bursts are lost for some reason. If the TDMA synchronization procedure allows K reference bursts to be lost before declaring that synchronization has been lost, the local clock stability must be such that

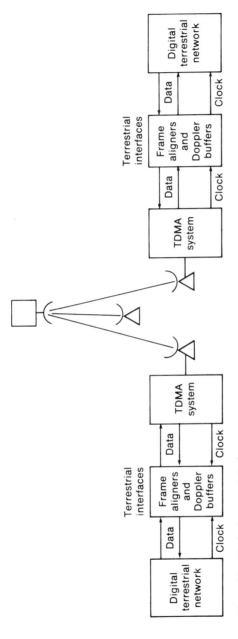

Figure 6.34 Plesiochronous interfaces.

clock error introduced during the interval KT_f seconds, where T_f is the frame length, does not exceed the design goal. Suppose the TDMA symbol rate is $R_s = 44.776$ Msps, $KT_f = 2$ s, and the design objective is a local clock error of 1 symbol per KT_f seconds when the clock is not phase-locked to the reference station clock; then the local clock stability must be $1/R_s KT_f = 1.1 \times 10^{-8}$ or 1.1 part in 10^8 as compared to the reference station clock stability of 1 part in 10^{11}.

As discussed previously, if the clock at the digital terrestrial network is accurate to 1 part in 10^{11}, a frame slip will occur once in 72 days. A frame aligner introduces a slip of one full PCM frame of 125 μs to avoid a loss of PCM frame synchronization. A frame slip repeats or eliminates one PCM sample in each voice channel affected and in practice is inaudible. For data transmission, the effect of slips can be much more serious, and a slip can impair the operation of some data sets for several seconds because of retransmission of blocks of data. The frame aligner contains a buffer storage of at least two PCM frames and must be able to recognize the PCM frame alignment signal. Consider a plesiochronous interface designed to handle a T1 carrier of 1.544 Mbps or 24 PCM channels. Each PCM frame consists of 193 bits (Fig. 1.6); thus the buffer capacity required for the transmit or the receive frame aligner is 386 bits. The capacity of the transmit or the receive Doppler buffer is given by (6.47) as $B = (1.15 \times 10^{-3})(1.544 \times 10^6) = 1776$ bits. The total storage capacity for both frame aligners and Doppler buffers is $2 \times 386 + 2 \times 1776 = 4324$ bits (starting at the center of the buffers).

6.9.4 Asynchronous Interfaces

In this type of connection no special frequency relationship between the TDMA system and the terrestrial networks is assumed. Slip-free operation can be achieved through the use of pulse stuffing on the transmit side to account for the difference between the TDMA clock and the terrestrial clock, and pulse destuffing on the receive side to restore the original data rate. Doppler buffers must again be employed on both the transmit and receive sides to account for the effects of satellite motion. Figure 6.35 illustrates *asynchronous* interfaces. Each of these interfaces contains a stuffer and a destuffer as shown in Fig. 6.36.

On the transmit side, the incoming terrestrial data stream is asynchronous; that is, its bit rate fluctuates around the nominal bit rate R_k. The clock frequency is recovered by the clock recovery circuit. To make the data stream synchronous with the TDMA clock at frequency f_0 (or a submultiple of f_0, namely f_0/m, where m is an integer), an elastic buffer is employed. Asynchronous data is written into the elastic buffer by the write clock at a frequency of $f_k = R_k$, and synchronous data is read out of the elastic buffer by means of the TDMA clock. Since data must be

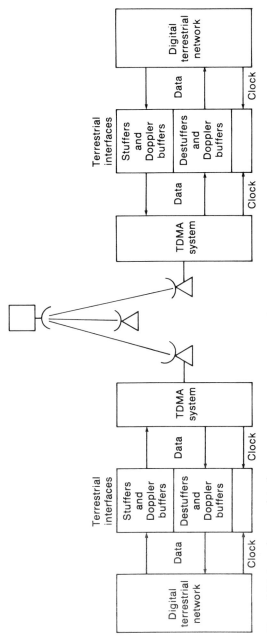

Figure 6.35 Asynchronous interfaces.

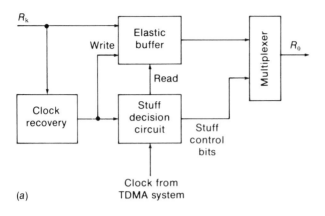

(a)

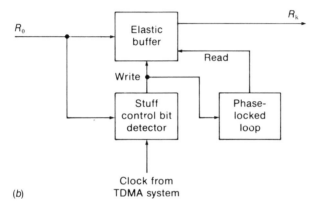

(b)

Figure 6.36 (*a*) Stuffing. (*b*) Destuffing.

written into the elastic buffer before it can be read out, the read clock must operate at a faster rate than the write clock; that is, the TDMA clock rate f_0 is set higher than the data clock rate f_k. The frequency difference $f_0 - f_k$ is the stuffing rate. The read clock cannot be allowed to overtake the write clock to the extent that it attempts to read a bit that is not yet in the buffer. To prevent this situation, the phase difference between the write and read clocks is determined by a phase comparator. When the phase difference between the clocks reaches a certain threshold, the stuff decision circuit generates a stuff request to make the read clock dwell for one additional time slot. As a result, one bit in the elastic buffer is read twice or, in effect, a dummy bit is stuffed into the data stream. Also, the stuff control bits are multiplexed to the data stream at precise intervals according to a framing scheme which allows identification of the stuffing bits on the receive side so that they can be removed from the data stream. The output of the stuffer is a data stream operated at the TDMA clock rate f_0 or at a bit rate of $R_0 = f_0$ bits per second.

On the receive side, the synchronous data stream at bit rate R_0 is received by the destuffer. By detecting the stuff control bits, the stuffed bits can be removed from the data stream. Because of removal of the stuffed bits the write clock is jittered. By generating a read clock at the average frequency of the write clock, the effect of jitter is smoothed and the data is read out of the elastic buffer with the same clock frequency f_k with which it enters the stuffer on the transmit side. Frequency control of the read clock is achieved by a phase-locked loop. The output of the destuffer is an asynchronous data stream at a nominal bit rate of $R_k = f_k$ bits per second operated at the nominal read clock frequency f_k.

6.9.5 Synchronous Interfaces

In a *synchronous* operation, the digital terrestrial network clock is synchronized with the TDMA local clock. This local clock is phase-locked to the reference station clock recovered from detection of the reference burst's unique word. Each terrestrial interface accepts an incoming data stream from the digital terrestrial network at a rate of R_k bits per second and delivers an outgoing data stream at a rate of R_k bits per second to the terrestrial network. Slip-free operation, that is, no loss of data, is achieved by synchronizing the outgoing data stream with a clock of $f_k = R_k$ derived from the TDMA clock at frequency f_0. Assume that each terrestrial interface is designed to accommodate a T_1 carrier at a bit rate of $R_k = 1.544$ Mbps and that the TDMA local clock frequency is $f_0 = 44.776$ MHz; then the terrestrial interface clock can be obtained as $f_k = f_0/29 = 1.544$ MHz. The terrestrial interface multiplexes the data stream and the clock and transmits them to the digital terrestrial network where the clock f_k is extracted from the data stream and used to generate the incoming data stream (from the terrestrial network to the TDMA system). Thus the incoming data stream is also synchronized with the TDMA clock. Because the outgoing data stream at the terrestrial interface is synchronized with the TDMA local clock—phase-locked to the reference station clock—no Doppler buffer is needed on the receive side of the terrestrial interface. However, because of periodic modifications of the transmit frame using the transmit frame delay D_N, Doppler buffers are needed on the transmit side. Figure 6.37 illustrates synchronous interfaces for a TDMA system. Note that in this mode of operation the PCM frame (125 μs) in the digital terrestrial network and the TDMA frame may not be synchronized, although the data clocks are synchronized with the receive TDMA clock. That is why Doppler buffers are employed on the transmit side.

If the terrestrial network operates with analog signals and when the timing for the PCM encoding of the analog signal is derived from the

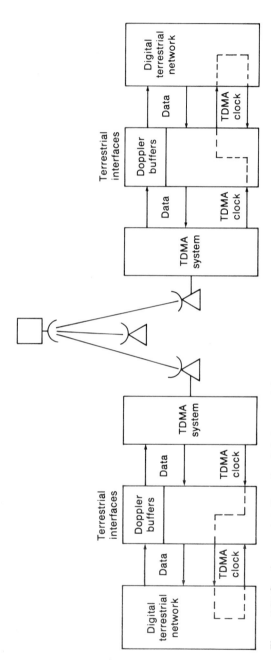

Figure 6.37 Synchronous interfaces for digital terrestrial networks.

transmit frame timing and the timing for the PCM decoding is derived from the receive frame timing, no Doppler buffer is required. This is because the PCM frames in the terrestrial network and the TDMA frame are fully synchronized. Furthermore, the data clocks are synchronized with the TDMA clocks, and therefore slip-free operation is achieved. Figure 6.38 illustrates *fully synchronous* interfaces between a TDMA system and analog terrestrial networks.

6.10 TDMA EQUIPMENT

In Secs. 6.1 to 6.9 we concentrated on the system aspects of a TDMA network. In this section the equipment functions of a TDMA station are described. A TDMA traffic terminal generally consists of

A modulator and demodulator (QPSK is the common form of modulation)
A TDMA processor
A terrestrial interface module (TIM)
TDMA monitoring and control

as shown in Fig. 6.39. The TDMA terminal delivers or accepts IF signals to or from the RF terminal (described in Chap. 3) and delivers or accepts continuous data streams to or from the terrestrial network. The IF carrier is normally at a frequency of 70 or 140 MHz. The QPSK modulator and demodulator will be described in detail in Chaps. 9 and 10. In this section we focus out attention on the TDMA processor, the terrestrial interface module, and TDMA monitoring and control.

6.10.1 TDMA Processor

The *TDMA processor* serves as a data and signaling interface between the modem (modulator and demodulator) and the terrestrial interface modules. It performs the following functions:

Frame acquisition and synchronization
Phase-locking the local TDMA clock to the receive reference station clock
Receiving, analyzing, and implementing the messages contained in the management channel of the reference burst
Unique word detection and preamble generation
Multiplexing data subbursts from the TIM and transmitting traffic bursts in their allocated positions in the frame
Receiving the destination traffic bursts in the frame and demultiplexing subbursts to the TIM

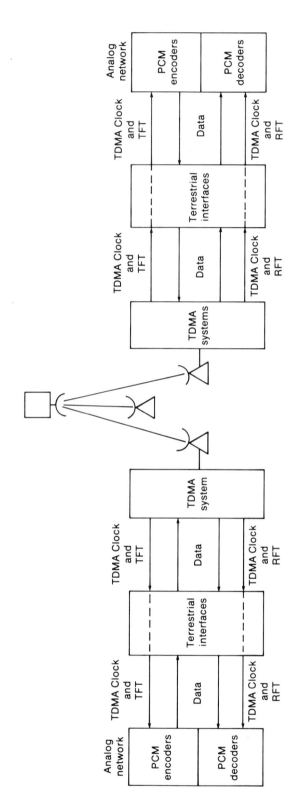

Figure 6.38 Synchronous interfaces for analog networks.

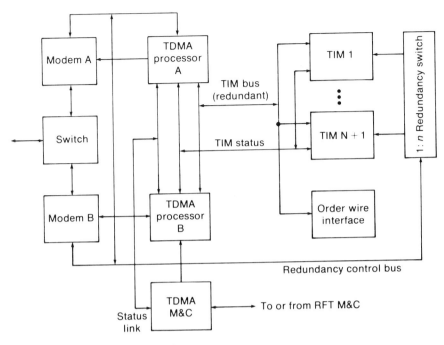

Figure 6.39 TDMA terminal equipment.

Transmitting the earth station status and fault conditions to the reference
    station

Demand assignment functions (which will be discussed in detail in
    Chap. 7)

The TDMA processor is always in a 1:1 redundancy configuration. Both
on-line and standby processors operate in parallel; that is, both receive
the signal from the satellite, but only the on-line processor transmits the
traffic to the satellite.

6.10.2 Terrestrial Interface Module

The *terrestrial interface module* serves as a data and signaling interface
between the TDMA processor and the terrestrial network and varies with
the application depending on whether the traffic is voice, data, facsimile,
or video signals. The TIM and the TDMA processors are interconnected
by a redundant TIM bus. Each bus may contain the address line, the control line, and separate transmit and receive 8-bit-wide data lines. The 8-
bit-data lines reduce the processing speed from R bits per second to $R/8$
bits per second, where R is the TDMA bit rate, thus relieving the critical
timing between the TDMA processor and the TIM.

The most commonly used terrestrial interface in TDMA systems is the T1 TIM which handles terrestrial T1 carriers (Sec. 1.2.3). The T1 TIM can be synchronous, asynchronous, or plesiochronous, as described in Sec. 6.9. Each interface includes compression and expansion buffers. Dual buffers are used for each type, so that the information on the terrestrial side of the buffers may be accessed on a continuous basis and the TDMA processor may simultaneously access the alternate buffer on a high-speed burst basis. As an example, consider a T1 TIM operating in the synchronous mode (Fig. 6.40). With this type of interface the data from the terrestrial T1 lines is first written into an elastic buffer at the rate of the recovered T1 clock and read out of the elastic buffer by a T1 clock derived from the local TDMA clock in order to eliminate jitter introduced by the terrestrial network. Serial data is converted to an 8-bit parallel form and stored in the compression buffer which is read out in synchronization with the TDMA transmit frame. As discussed in Sec. 6.9.5, the compression buffer must include a Doppler buffer to compensate for satellite motion. Twenty-four-channel blocks of data are read out of the compression buffer into the transmit bus by the transmit addresses and bus clock and transmitted in the allocated slots of the TDMA frame.

On the receive side, the T1 TIM stores burst data derived from the receive bus in appropriate locations in the expansion buffer. The data is read out of the expansion buffer on the basis of a T1 clock derived from the local TDMA clock, and therefore there is no need for a Doppler buffer.

For reliability and cost-saving purposes, terrestrial interfaces of the same type may be operated in a $1:n$ redundancy configuration ($n \geq 1$). The $1:n$ redundancy switch connects the selected terrestrial lines to selected interfaces and operates under control of the TDMA processor.

Terrestrial interfaces may also contain a voice or data order wire channel used by operators at different earth stations to communicate with each other. Normally, the transmitted signaling word of an order wire channel contains a source code corresponding to the originating station and a destination code corresponding to the receiving station when an order wire call is made. The received signaling words from the order wire channels of all received bursts are compared with the transmitted signaling word so that only the appropriate receive order wire is demodulated.

6.10.3 TDMA Monitoring and Control

The TDMA M&C system may be part of the TDMA processor or a separate minicomputer whose main tasks are to monitor the performance of the TDMA terminal, to diagnose failures, and to control switchover from on-line to standby equipment. It may also generate alarms for transmis-

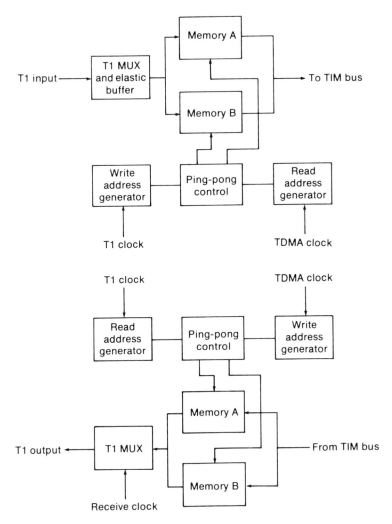

Figure 6.40 T1 terrestrial interface module.

sion to the network control center via the service channel in the traffic burst (Fig. 6.3) or superframe short burst (Fig. 6.12) and via an automatically dialed direct distance dialing (DDD) circuit (via a local telephone company) for backup purposes in case the communication via satellite is interrupted. The TDMA M&C may also be responsible for collecting the switching control commands from the NCC via the management channel of the reference burst and direct them to appropriate earth station subsystems for redundancy switchover (Fig. 3.37). Status display, switching of redundant equipment, and fault isolation routine setup and execution can be initiated via the TDMA M&C from a local portable terminal, as shown in Fig. 6.39 (and Fig. 3.37). Typical causes of TDMA switchover are

Loss of receive frame timing
Loss of transmit frame timing
Modem summary fault
TIM summary fault
Power supply fault
Remote NCC commands

To perform remote monitoring and control, the TDMA M&C system and the NCC may communicate using the following modes:

1. *Request mode.* When requested by the NCC through the management channel of the reference burst, the TDMA M&C system generates a status word which may include station identification, mode identification, event change, event priority, and subsystem identification and transmits the status word to the NCC via the service channel of the traffic burst or of the superframe short burst.
2. *Control mode.* This mode enables the NCC to control switchover functions of the TDMA terminal and the RF terminal of the traffic station and exercise remote tests. The control word may include station identification, mode identification, and control points and is transmitted in the management channel of the reference burst.
3. *Broadcast mode.* This mode enables the NCC to download messages to all traffic stations in the network. For example, the downloading of satellite ephemeris data to the TDMA M&C system to generate tracking informations for the antenna control unit.

6.11 ADVANCED TDMA SATELLITE SYSTEMS

So far we have studied basic TDMA satellite architecture where there are only a few spot beams and beam interconnections are static. To increase satellite capacity, spot beams must be employed so that the same frequency band can be spatially reused many times. Theoretically, if the United States is covered by N nonoverlapping spot beams, the satellite capacity will increase N-fold over that achieved using one beam. In addition, the use of a narrow antenna beam provides a high gain for the coverage area, hence permits power savings in both the uplink and downlink channels.

Satellite-switched TDMA (SS-TDMA) such as that planned for *IN-TELSAT VI* employs multiple spot beams. One inherent problem in multiple spot beam operation is the interconnectivity of upbeams (UBs) with downbeams (DBs). This is accomplished by dynamic satellite switching using a microwave switch matrix on-board the satellite. An illustration of the connectivity for three upbeams and three downbeams and the corresponding SS-TDMA frame is shown in Fig. 6.41. During a SS-TDMA frame the satellite switch is controlled by a sequence of switch states of various durations. The duration of a given switch state is selected to accommodate a segment of the total traffic between earth stations. The sum of the segments provided by the switch sequence is equal to the total traffic of the network. In essence, SS-TDMA operates as a set of parallel TDMA frames on the uplink, which are then switched by a sequence of switch states into a set of parallel TDMA frames on the downlink.

Another advanced TDMA satellite system is beamhopping TDMA which is very useful for serving areas in which traffic is spread out geographically and where traffic in no single area is sufficient to justify the use of a stationary spot beam. Beam hopping TDMA works as follows: The

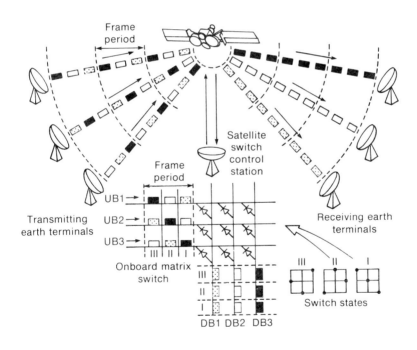

Frame slots of earth stations transmitting to DB1
Frame slots of earth stations transmitting to DB2
Frame slots of earth stations transmitting to DB3

Figure 6.41 Satellite-switched TDMA.

on-board phase-away antenna points a particular spot beam in the direction of a new burst and dwells for the duration of the burst plus guard time needed for burst position uncertainty. After the first burst in the frame has been received and stored in the uplink memory, the beam is steered in the direction of the second burst, the second burst is stored, and so on until all the bursts in the TDMA frame are stored; then the hopping sequence is repeated for a new TDMA frame. The stored uplink bursts are then processed by an on-board processor which performs demodulation of the uplink carriers followed by reconfiguration of the uplink bursts into new downlink bursts each for a particular downbeam dwell followed by remodulation on downlink carriers. On-board demodulation and remodulation decouple the uplink noise and interference from the downlink noise and interference to improve the bit error rate. Reconfiguration of bursts allows the grouping of all traffic into one burst destined for a particular region, thus avoiding the interburst interference associated with conventional TDMA. A combination of beam-hopping TDMA and SS-TDMA permits a mixture of low, medium and high bit rates. The planned NASA Advanced Communication Technology Satellite (ACTS) will employ on-board switching with beam hopping and may set a future trend for satellite communications.

REFERENCES

1. B. A. Pontano et al., "The INTELSAT TDMA/DSI System," *IEEE J. Selected Areas Commun.,* Vol. SAC-1, No. 1, Jan. 1983, pp. 165–173.
2. S. J. Campanella and R. J. Colby, "Network Control for Multibeam TDMA and SS/TDMA," *IEEE J. Selected Areas Commun.,* Vol. SAC-1, No. 1, Jan. 1983, pp. 174–187.
3. K. Inagaki et al. "International Connection of Pleisiochronous Networks via TDMA Satellite Link," *IEEE J. Selected Areas Commun.,* Vol. SAC-1, No. 1, Jan. 1983, pp. 188–198.
4. S. J. Campanella and D. Schaefer, "Time-Division Multiple-Access Systems (TDMA)," in K. Feher (ed.), *Digital Communications: Satellite/Earth Station Engineering.* Englewood Cliffs, N.J: Prentice-Hall, 1983.
5. H. L. Van Trees (ed.), *Satellite Communications,* Selected Reprint Series. New York: IEEE, 1979.
6. J. J. Spilker, Jr., *Digital Communications by Satellite.* Englewood Cliffs, N.J.: Prentice-Hall, 1977.
7. F. S. Leite and J. Albuquerque, "Heuristic Algorithms for TDMA Burst Assignment," 5th International Conference on Digital Satellite Communication, Genoa, Italy, 1981, pp. 387–392.
8. S. J. Campanella and J. V. Harrington, "Satellite Communications Networks," *Proc. IEEE,* Vol. 72, No. 11, pp. 1506–1519, Nov. 1984.
9. J. N. Sivo, "Advanced Communication Satellite Systems," *J. Selected Areas Commun.,* Vol. SAC-1, No. 4, pp. 580–588, Sept. 1983.

10. Special issues on satellite communications networks. *Proc. IEEE*, Vol. 72, No. 11, Nov. 1984.

PROBLEMS

6.1 Find the miss detection and false detection probabilities for a unique word with a length of $N = 24$ bits. The channel bit error rate is $p = 10^{-4}$, and the detection threshold is $\epsilon = 5$.

6.2 Find the approximate false detection probability for the unique word shown in Fig. 6.9 with $p = 10^{-3}$ and $\epsilon = 4$, assuming that the aperture window $W = 23$ bits and the occurrence of the unique word detection relative to the end of the aperture is $X = 12$ bits.

6.3 Consider the following unique word and carrier and clock recovery sequences:

UW: 011110001001
CCR: 010101010101

(a) Plot the correlation function as in Fig. 6.9 for an aperture window of $W = 13$ bits. The occurrence of the unique word detection relative to the end of the aperture is $X = 0$ bits.

(b) Find the false detection probability for $p = 10^{-3}$ and $\epsilon = 4$.

6.4 Repeat Prob. 6.3 for unique word and CCR sequences whose lengths are 24 bits each and which consists of the two consecutive 12-bit patterns given in Prob. 6.3, given that $W = 25$ bits, $X = 0$ bits, $p = 10^{-3}$, and $\epsilon = 4$.

6.5 Consider a TDMA frame with the following parameters.

TDMA frame length: 2 ms
TDMA burst bit rate: 120 Mbps
32 traffic bursts and 2 reference bursts
CCR sequence: 352 bits
UW sequence: 48 bits
Order wire channel: 64 bits
Management channel: 256 bits
Transmit timing channel: 320 bits
Service channel: 320 bits
Guard time: 48 bits

(a) Find the frame efficiency.

(b) The TDMA system carries delta modulation voice signals with a bit rate of 32 kbps and data signals with a bit rate of 9.6 kbps. Assume that two-thirds of the capacity is for voice and one-third is for data. Find the number of voice channels and data channels.

6.6 Consider the TDMA system in Prob. 6.5 and assume that there are eight stations in the network, each transmitting four traffic bursts. Design a superframe as shown in Figs. 6.11 and 6.12 where the reference station addresses each traffic station once per superframe and the service channel is put in a superframe short burst. Find the frame efficiency.

6.7 Consider a TDMA frame with the following parameters.

TDMA frame length: 16 ms
TDMA burst bit rate: 60 Mbps
32 traffic bursts and 2 reference bursts
CCR sequence: 256 bits
UW sequence: 20 bits
Order wire channel: 512 bits
Management channel: 256 bits
Transmit timing channel: 320 bits
Service channel: 256 bits
Guard time: 32 bits

(a) Find the frame efficiency.

(b) How many T1 carriers can the frame accommodate?

(c) How many 32-kbps voice channels can the frame accommodate?

6.8 Consider the TDMA system in Prob. 6.7 and assume that there are 32 traffic stations in the network, each transmitting one traffic burst.

(a) Design a superframe as shown in Figs. 6.11 and 6.12 where the reference station addresses each traffic station once per superframe, and service channel is put in a superframe short burst.

(b) Find the frame efficiency.

(c) Now assume that the system changes from feedback control to loopback control. The transmit timing channel is eliminated, and the guard time is reduced to 12 bits. Find the frame efficiency.

6.9 Consider a TDMA system where station N (farthest from the geostationary satellite) has an elevation angle of 10°. The nominal TDMA frame length is 9 ms.

(a) Find the nominal transmit frame delay D_N.

(b) Assume that the TDMA bit rate is 60 Mbps and that the type of carrier modulation is QPSK. A particular traffic burst is assigned a position of 90,000 symbols relative to the reference burst. What is the transmit burst delay if the station uses loopback control?

6.10 Verify (6.29).

6.11 Consider a reference station R situated at longitude $\theta_L = 120°W$ and latitude $\theta_l = 40°N$. At an instant in time the station measures the satellite coordinates as follows: $X_S = -7321.2154$ km, $Y_S = -41,523.5412$ km, and $Z_S = -0.9431$ km. Find the elevation angle and azimuth angle of the station at this instant in time, assuming the reference station is west of the satellite.

6.12 Consider the single-station ranging discussed in Examples 6.1 and 6.2 Assume that the range, elevation angle, and azimuth angle can be measured within an error of $\Delta d_R = \pm 100$ m, $\Delta E = \pm 0.05°$, and $\Delta A = \pm 0.05°$.

(a) Find the errors in the satellite coordinates.

(b) Find the maximum error in the transmit frame delay D_N for traffic station N.

6.13 Consider a station situated at longitude $\theta_L = 120°W$ and latitude $\theta_l = 35°N$. Assume that the earth station uses no tracking at all but instead predicts the satellite position based on the elevation and azimuth angles and the nominal satellite position at 105°W longitude. The angle errors are $\Delta E = \pm 0.05°$ and $\Delta A = \pm 0.05°$. Furthermore, the satellite–earth station distance has an uncertainty of ± 275 μs.

(a) Find the satellite position error.

(b) Find the transmit frame delay error.

6.14 Consider a TDMA network employing three-station ranging techniques. The locations of three ranging stations are

	R	K	M
Longitude	120°W	75°W	100°W
Latitude	45°N	43°N	32°N

One measurement shows that the geostationary satellite is at 105°W longitude and 0° latitude. Assume that the range measurement error is ± 40 m. Find the satellite position error.

6.15 For the TDMA network discussed in Prob. 6.5, find the cost of buffer memory (compression and expansion) plus the fractional transponder cost for frame overhead assuming a transponder costs $5 million and a memory bit costs 1¢.

6.16 Repeat Prob. 6.15 for the TDMA network discussed in Prob. 6.6.

6.17 A plesiochronous TDMA earth station carries 20 T1 carriers. Assume that the TDMA frame length is 15 ms and that the satellite Doppler shift is 20 ns/s. Find the total storage memory for the earth station.

6.18 A geostationary satellite has a peak-to-peak altitude variation of 0.2% of its orbit radius and an inclination of 0.05°. Four transponders of this satellite serve a plesiochronous TDMA network with a frame length of 16 ms. The maximum transponder efficiency is 97%, and the TDMA bit rate is 60 Mbps. Find the maximum storage memory of the plesiochronous TDMA network.

Chapter 7

Efficient Techniques: Demand Assignment Multiple Access and Digital Speech Interpolation

Consider a TDMA system carrying voice traffic as illustrated in Fig. 7.1. Long-distance telephone calls originating from earth station A are multiplexed into a 96-channel burst. Earth station A has preassigned 96 channels to four subbursts (24 channels each) to be received by four other earth stations, B, C, D, and E, which know the position of burst A in the TDMA frame and the positions of the subbursts in burst A according to a burst time plan. Earth station B receives the TDMA frame from the satellite and extracts burst A based on a knowledge of its position in the frame. Burst A is demultiplexed, and the subburst destined for station B is extracted. The telephone switch in city B then makes a connection with the called party via terrestrial facilities. The return connection from station B to station A is accomplished in a similar manner via burst B.

It is seen that in this operation, if all 24 channels in the subburst destined for earth station B are occupied, a new call entering the telephone switch will receive a busy signal even though the remaining 72 channels in burst A might not be occupied. This results in poor utilization of earth station A capacity from the occupancy point of view; that is, the subburst preassignment prevents reallocation of unoccupied channels in the burst. Furthermore, it is observed that, if all 96 channels in burst A are occupied, no additional calls can be transmitted from station A even though unoccupied satellite channels might be available in other bursts in the frame. This results in poor utilization of satellite transponder capacity;

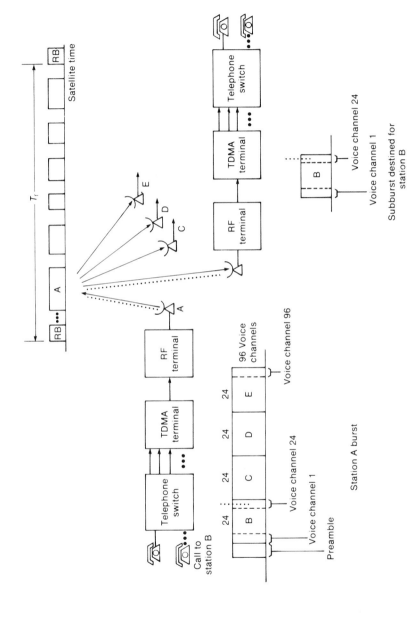

Figure 7.1 Preassigned TDMA, RB, reference burst; T_f, frame length.

that is, the fixed burst length prevents reallocation of unoccupied channels in the frame.

To improve efficiency, the earth station capacity should be assigned on demand when a new call arrives; that is, it should be possible to establish a call connection if there is an unoccupied channel in the station's burst. After the call is completed, the channel could be made available to other incoming calls regardless of their destination. In other words, the earth station capacity should be pooled. To further improve efficiency, the satellite transponder capacity or, equivalently, the frame capacity should also be assigned according to the station's traffic; that is, the earth station capacity should be allowed to vary adaptively with its traffic. This means that the burst length for each earth station in the network should be made to vary according to the station's needs. In other words, the frame capacity should also be pooled.

This efficient method of pooling capacities for use only when a demand arises is called *demand assignment multiple access* (DAMA). We distinguish between DAMA employed with TDMA and DAMA employed with FDMA by using the terms *demand assignment–time division multiple access* (DA-TDMA) and *demand assignment–frequency division multiple access* (DA-FDMA). Another efficient method for increasing satellite capacity, called *digital speech interpolation* (DSI), will also be discussed in this chapter. DSI interpolates an additional speech signal between pauses in a given speech signal and doubles the number of terrestrial channels carried by a satellite transponder.

7.1 THE ERLANG B FORMULA

Consider a communication capacity of n channels for voice traffic between two points. Assume that a call between these two points can use any of these channels if, when the call is placed, not all the channels are busy. When all n channels are busy, a new call arriving will find a busy signal and will be blocked. The call congestion, that is, the probability that an arriving call will be blocked on finding all channels busy, was studied by A. K. Erlang (1878–1929), a Danish mathematician who worked for the Copenhagen Telephone Company. The analysis is based on the following assumptions:

1. Call arrivals form a Poisson process. Let the random variable k be the number of call arrivals in an interval of length t and λ be the call arrival rate. We assume that the probability of a single arrival in an infinitesimal interval Δt is proportional to Δt, so that

$$\Pr\left[\text{one call arrival in } (t, t + \Delta t)\right] = \lambda \, \Delta t$$

We also assume that the probability of more than one call arrival in Δt is negligible, so that

$$\text{Pr [no call arrival in } (t,t + \Delta t)] = 1 - \lambda \, \Delta t$$

Furthermore, we assume that these arrivals are independent. From these assumptions it can be shown that the call arrivals are Poisson-distributed. Let $p_k(t)$ denote the probability of k arrivals in time interval t. The probability of k arrivals in time interval $t + \Delta t$ is equal to the product of $p_k(t)$ and the probability of no arrival in Δt plus the product of $p_{k-1}(t)$ and the probability of one arrival in Δt; that is,

$$p_k(t + \Delta t) = p_k(t) \, (1 - \lambda \, \Delta t) + p_{k-1}(t) \, \lambda \, \Delta t \qquad k \geq 1$$

or, equivalently, as $\Delta t \to 0$,

$$\frac{dp_k(t)}{dt} = \lambda[p_{k-1}(t) - p_k(t)]$$

Solution of this differential equation yields

$$p_k(t) = \frac{(\lambda t)^k}{k!} \exp(-\lambda t) \tag{7.1}$$

2. The call service time or holding time is independent and exponentially distributed with the parameter μ; that is, the service time distribution function is $1 - \exp(-\mu t)$, where μ is the mean hang-up rate. Note that the probability of a call being terminated in Δt is $\mu \, \Delta t$, and when there are j calls in progress, the probability that a call will end in Δt is $j\mu \, \Delta t$; that is, the rate μ_j at which the service ends is $\mu_j = j\mu$. Note that $1/\mu$ is the average service time or holding time.
3. Blocked calls are cleared. An arriving call that finds all n channels busy may leave the system and have no effect on it.

To derive the probability of a call being blocked, let p_k be the equilibrium probability that there are exactly k calls in progress in the system; that is, the system is said to be in state k. Transitions occur between states of the system. If the system is in state $k - 1$ and a new call arrives during the next interval Δt, the system will move to state k. Since the call arrival rate is λ, the transition rate from state $k - 1$ to state k is simply λp_{k-1}. Similarly, if the system is in state k, the transition rate from state k to state $k - 1$ is $k\mu p_k$ because the call hang-up rate is $k\mu$. In equilibrium the probability of finding the system in a given state is time-invariant; that is, the probability that there are more than k call arrivals in constant. The probability increases as the system moves from state $k - 1$ to state k and decreases as the system moves from state k to state $k - 1$. At any instant in time the net transition rate from $k - 1$ to k and from k to $k - 1$ must be zero to maintain equilibrium; therefore we must have

$$\lambda p_{k-1} = k\mu p_k \qquad k = 1, 2, \ldots, n \qquad (7.2)$$

or

$$p_k = \frac{\lambda}{k\mu} p_{k-1} \qquad k = 1, 2, \ldots, n \qquad (7.3)$$

Let

$$a = \frac{\lambda}{\mu} \qquad (7.4)$$

The parameter a is known as the *traffic intensity*. Thus

$$p_k = \frac{a}{k} p_{k-1} \qquad k = 1, 2, \ldots, n \qquad (7.5)$$

and

$$p_k = \frac{a^k}{k!} p_0 \qquad (7.6)$$

To eliminate p_0, note that the probabilities must sum to 1:

$$\sum_{k=0}^{n} p_k = p_0 \sum_{k=0}^{n} \frac{a^k}{k!} = 1 \qquad (7.7)$$

which yields

$$p_0 = \frac{1}{\displaystyle\sum_{k=0}^{n} a^k/k!} \qquad (7.8)$$

The probability that there are k calls in the system is

$$p_k = \frac{a^k/k!}{\displaystyle\sum_{k=0}^{n} a^k/k!} \qquad (7.9)$$

The probability that there are n calls in the system or the probability that an arriving call will be blocked by finding all n channels busy is

$$p_n = B(n,a) = \frac{a^n/n!}{\displaystyle\sum_{k=0}^{n} a^k/k!} \qquad (7.10)$$

$B(n,a)$ is called the *Erlang B formula* for an n-channel group with a *traffic intensity* of $a = \lambda/\mu$. Note that a is the product of the number of calls per second and the mean service time per call in seconds. The unit of a is the *erlang*. Equation (7.10) forms the cornerstone for evaluating the *grade of service* of a voice traffic network. The relationship between n, a, and B is given in Table 7.1.

Example 7.1 A traffic intensity of 1 erlang is offered to a group of three channels. The average call holding time is 2 min.

(a) The average number of call arrivals per hour can be calculated from (7.4):

$$\lambda = \frac{a}{1/\mu} = \frac{1}{2 \text{ min}} = \frac{1}{2[(1/60)] \text{ h}}$$

$$= 30 \text{ calls per hour}$$

(b) The probability that no call will arrive during a specified period of 2 min is given by (7.1):

$$p_0(2 \text{ min}) = 1/0! \exp[-30(2/60)]$$

$$= \exp(-1) = 0.368$$

(c) The probability that a call will be blocked is given by (7.10):

$$B(3,1) = 0.0625$$

Example 7.2 A traffic intensity of 15 erlangs is offered to a group of 25 channels.
(a) From Table 7.1, the call blocking probability is

$$B(25,15) = 0.005$$

(b) The probability that 1 channel is free is the same as the probability that 24 channels are busy and can be calculated from (7.9):

$$p_{24} = \frac{15^{24}/24!}{\sum\limits_{k=0}^{25} 15^k/k!} = \frac{25}{15} \times \frac{15^{25}/25!}{\sum\limits_{k=0}^{25} 15^k/k!}$$

$$= (25/15)[B(25,15)] = (25/15)(0.005) = 0.0083$$

(c) In the event that a 20% traffic overload or a 20% channel failure does not occur at the same time, the call blocking probability will not exceed 0.01. What is the minimum number of channels required for $B = 0.005$ during normal operation? From Table 7.1, the number of channels required for $B = 0.01$ *and* $a = 18$ erlangs is 28. The number of channels required for $B = 0.01$ and $a = 15$ is 24. So 29 channels must be provided for 5 channels out of order.

7.2 TYPES OF DEMAND ASSIGNMENTS

As an illustration, consider the case where earth station A communicates with earth stations B, C, D, and E. Assume that the traffic intensity for destination earth station i ($i = $ B,C,D,E) varies from a maximum value of $a_{max} = 15$ erlangs to a minimum value of $a_{min} = 1.8$ erlangs over 24 h. To achieve a typical grade of service of 1 blocked call per 100, that is, a call blocking probability of $B = 0.01$, it is required that 24 channels be preassigned to each destination earth station i to obtain the maximum traffic in-

Table 7.1 Erlang *B* table: Traffic intensity versus number of channels and blocking probability

n	One lost call in				n	One lost call in			
	50^a	100^b	200^c	1000^d		50^a	100^b	200^c	1000^d
1	0.020	0.010	0.005	0.001	51	41.2	38.8	36.8	33.4
2	0.22	0.15	0.105	0.046	52	42.1	39.7	37.6	34.2
3	0.60	0.45	0.35	0.19	53	43.1	40.6	38.5	35.0
4	1.1	0.9	0.7	0.44	54	44.0	41.5	39.4	35.8
5	1.7	1.4	1.1	0.8	55	45.0	42.4	40.3	36.7
6	2.3	1.9	1.6	1.1	56	45.9	43.3	41.2	37.5
7	2.9	2.5	2.2	1.6	57	46.9	44.2	42.1	38.3
8	3.6	3.2	2.7	2.1	58	47.8	45.1	43.0	39.1
9	4.3	3.8	3.3	2.6	59	48.7	46.0	43.9	40.0
10	5.1	4.5	4.0	3.1	60	49.7	46.9	44.7	40.8
11	5.8	5.2	4.6	3.6	61	50.6	47.9	45.6	41.6
12	6.6	5.9	5.3	4.2	62	51.6	48.8	46.5	42.5
13	7.4	6.6	6.0	4.8	63	52.5	49.7	47.4	43.4
14	8.2	7.4	6.6	5.4	64	53.4	50.6	48.3	44.1
15	9.0	8.1	7.4	6.1	65	54.4	51.5	49.2	45.0
16	9.8	8.9	8.1	6.7	66	55.3	52.4	50.1	45.8
17	10.7	9.6	8.8	7.4	67	56.3	53.3	51.0	46.6
18	11.5	10.4	9.6	8.0	68	57.2	54.2	51.9	47.5
19	12.3	11.2	10.3	8.7	69	58.2	55.1	52.8	48.3
20	13.2	12.0	11.1	9.4	70	59.1	56.0	53.7	49.2
21	14.0	12.8	11.9	10.1	71	60.1	57.0	54.6	50.1
22	14.9	13.7	12.6	10.8	72	61.0	58.0	55.5	50.9
23	15.7	14.5	13.4	11.5	73	62.0	58.9	56.4	51.8
24	16.6	15.3	14.2	12.2	74	62.9	59.8	57.3	52.6

n	0.02^a	0.01^b	0.005^c	0.001^d
25	17.5	16.1	15.0	13.0
26	18.4	16.9	15.8	13.7
27	19.3	17.7	16.6	14.4
28	20.2	18.6	17.4	15.2
29	21.1	19.5	18.2	15.9
30	22.0	20.4	19.0	16.7
31	22.9	21.2	19.8	17.4
32	23.8	22.1	20.6	18.2
33	24.7	23.0	21.4	18.9
34	25.6	23.8	22.3	19.7
35	26.5	24.6	23.1	20.5
36	27.4	25.5	23.9	21.3
37	28.3	26.4	24.8	22.1
38	29.3	27.3	25.6	22.9
39	30.1	28.2	26.5	23.7
40	31.0	29.0	27.3	24.5
41	32.0	29.9	28.2	25.3
42	32.9	30.8	29.0	26.1
43	33.8	31.7	29.9	26.9
44	34.7	32.6	30.8	27.7
45	35.6	33.4	31.6	28.5
46	36.6	34.3	32.5	29.3
47	37.5	35.2	33.3	30.1
48	38.4	36.1	34.2	30.9
49	39.4	37.0	35.1	31.7
50	40.3	37.9	35.9	32.5

n	0.02^a	0.01^b	0.005^c	0.001^d
75	63.9	60.7	58.2	53.5
76	64.8	61.7	59.1	54.3
77	65.8	62.6	60.0	55.2
78	66.7	63.6	60.9	56.1
79	67.7	64.5	61.8	56.9
80	68.6	65.4	62.7	57.8
81	69.6	66.3	63.6	60.3
82	70.5	67.2	64.5	59.5
83	71.5	68.1	65.4	60.4
84	72.4	69.1	66.3	61.3
85	73.4	70.1	67.2	62.1
86	74.4	71.0	68.1	63.0
87	75.4	71.9	69.0	63.9
88	76.3	72.8	69.9	64.8
89	77.2	73.7	70.8	65.6
90	78.2	74.7	71.8	66.6
91	79.2	75.6	72.7	67.4
92	80.1	76.6	73.6	68.3
93	81.0	77.5	74.3	69.1
94	81.9	78.4	75.4	70.0
95	82.9	79.3	76.3	70.9
96	83.8	80.3	77.2	71.8
97	84.8	81.2	78.2	72.6
98	85.7	82.2	79.1	73.5
99	86.7	83.2	80.0	74.4
100	87.6	84.0	80.9	75.3

[a] Probability of 0.02.
[b] Probability of 0.01.
[c] Probability of 0.005.
[d] Probability of 0.001.

tensity of 15 erlangs [the smallest integer value of n that makes the right side of (7.10) less than 0.01] and a total of 96 channels preassigned to burst A. This requires four *single-destination* (point-to-point) *terrestrial interface modules* (TIMs), each of which can handle 24 channels (a T1 carrier, for example). When the traffic intensity for each destination earth station i is at its minimum value of 1.8 erlangs, only 6 channels are required for each destination i to obtain a call blocking probability of 0.01, and a total of 24 satellite channels required for burst A. Thus, if the capacity of burst A could somehow be decreased from 96 channels to 24 channels, a saving of 72 channels could be obtained. This type of demand assignment yields a *gain* of 96/24 = 4. But it requires that the TDMA system have the capability to vary the burst length and possibly the burst position in a frame. It also requires *multidestination* (point-to-multipoint) *terrestrial interface modules*. In this example, four multidestination TIMs are required, each handling 24 channels and communicating with a maximum of four destinations. Note that, when the average traffic intensity is 5.7 erlangs per destination earth station, 12 channels are required for each destination, only two of the four multidestination TIMS are used (each communicating with two destinations), and the required capacity of burst A is 48 channels. This type of demand assignment is called a *variable-capacity demand assignment*.

As another example, consider the case where earth station 1 has to communicate with 30 other earth stations with an average traffic intensity of 0.21 erlang per destination earth station and where the traffic intensity remains fairly constant around an average value (say, a variation of ±0.04 erlang) over 24 h. To obtain a call blocking probability of 0.01 when the traffic intensity per destination is at its maximum value of 0.25 erlang, 3 channels per destination and a total of 90 channels for burst 1 are required. Four 24-channel TIMs are also required, each having the capability to communicate with $\lceil 30/4 \rceil = 8$ destinations ($\lceil x \rceil$ denotes the smallest integer equal to or greater than x). Note that 90 channels are also required for burst 1 when the traffic intensity per destination is at its minimum value of 0.17 erlang. In this case burst 1 carries four 24-channel subbursts (three 8-destination subbursts and one 6-destination subburst). It is seen that no demand assignment gain can be obtained by varying the capacity of burst 1 because burst 1 must always carry 90 channels in four preassigned subbursts.

Suppose that any channel in burst 1 can be assigned to any call; then the *total traffic* must be considered. The maximum total traffic intensity at earth station 1 is $30 \times 0.25 = 7.5$ erlangs. A call between earth stations 1 and 2 can be blocked under two conditions—when either all the channels in burst 1 or all the channels in burst 2 are occupied. This is in contrast to the first example where multidestination TIMs are employed. In this case a call can be blocked under only one condition—when all the channels (preassigned to destination earth station 2) in a multidestination TIM at earth station 1 are occupied. This happens because a duplex circuit (or two channels, one for the calling party in the transmit TIM and one for the called party in the receive TIM) is always established when a call arrives. Returning to the example under discussion, we see that the blocking probability for a call between earth stations 1 and 2 depends on the blocking probability B_1 at the originating station 1 and also on the blocking probability B_2 at the destination station 2. Note that the blocking probability at a station depends on the *total traffic* intensity at that station. The probability that a call will not be blocked at station 1 is $1 - B_1$, and at station 2 it is $1 - B_2$. If we assume that the traffic intensity between stations 1 and 2 is very low compared to their combined total traffic intensity (as in the case where stations 1 and 2 communicate with many other earth stations and each destination has a low traffic intensity), then the probability that a call will not be blocked at station 1 is independent of the probability that it will not be blocked at station 2. The blocking probability for any call between stations 1 and 2 is then given by

$$B = 1 - (1 - B_1)(1 - B_2)$$
$$= B_1 + B_2 - B_1B_2 \tag{7.11}$$

Assume that the *maximum total traffic* intensity at earth station 1 is equal to that at earth station 2, that is, 7.5 erlangs; then in order to obtain $B \leq 0.01$, it is sufficient that $B_1 = B_2 = 0.005$. In this case a total of 16 channels are required for burst 1, and the same number of channels are required for burst 2. This type of demand assignment yields a *gain* of $90/16 = 5.625$. Note that $B \approx B_1 + B_2$ for $B_1 < 0.1$ and $B_2 < 0.1$.

In this type of demand assignment one must calculate $N(N - 1)/2$ call blocking probability B values for a fully connected network of N stations based on N blocking probability B_i values, where B_i is the call blocking probability of station i ($i = 1, 2, \ldots, N$). To simplify the calculations, it is easier to require that $B_i = B/2$, $i = 1, 2, \ldots, N$, for a given grade of service B. For example, if $B = 0.01$, then $B_i = 0.005$ for $i = 1, 2, \ldots, N$. Now suppose that the total traffic intensity at station 1 decreases to its minimum value of $30 \times 0.17 = 5.1$ erlangs during a certain time of the day. Then in order to keep $B_1 = B/2 = 0.01/2 = 0.005$, it is necessary that burst 1 have a capacity of 12 channels instead of 16 channels. Again, this requires that the TDMA system have the capability to vary the burst length and possibly the burst position in the frame. This type of demand assignment is called a *per-call variable-capacity demand assignment*, since any channel in a station burst can be assigned on a per-call basis and a burst capacity (length) can be varied adaptively with the station's total traffic intensity.

From the above two examples, it is seen that a *variable-capacity de-*

mand assignment is a DA-TDMA suited to stations with a large range of slowly varying traffic intensity and with relatively few destinations. A *per-call variable-capacity demand assignment* is a DA-TDMA which best serves stations with fast fluctuations in the total traffic intensity and with many destinations, each of which has a low traffic intensity. When the fluctuations in the total traffic intensity at the station are small, say, less than 5%, variable-capacity techniques need not be implemented, and this type of DA-TDMA is called *per-call demand assignment*.

Variable-capacity demand assignment requires the reference station of the TDMA network to allocate the burst length and position, the sub-burst length and position, and the identification number of the subburst (i.e., the corresponding multidestination TIM) in the frame. This is exactly the burst time plan discussed in Chap. 6. Changing from an old burst time plan to a new one can take several minutes, and therefore variable-capacity demand assignment is suited to traffic with a known busy time during the day.

Per-call variable-capacity demand assignment only requires the reference station of the TDMA network to allocate the burst length and position. Since any channel in the burst can be assigned to any call, no subburst length, position, or identification number is involved. Instead, the position and length of the channel assigned to the calling party are communicated to the destination station via a special channel called a *common signaling channel*. Therefore allocation of the burst length and position in this case can be made very quickly, and this process is called *real-time frame reconfiguration* or *dynamic frame reconfiguration*. As just mentioned, the call connection in a per-call variable-capacity demand assignment can be established between stations via the common signaling channel without involving the reference station. A call connection involves exchange of the position and length of the channel, that is, of the subburst assigned to the call (if all the channels are of identical length, only the position of a channel is sufficient), and call supervisory signals such as address, busy, answer, and hangup.

Variable-capacity techniques can be easily used in DA-TDMA but are difficult to implement in DA-FDMA because the latter involves many carriers in a transponder. Changing the capacity and frequency of these carriers can vary the carrier-to-intermodulation ratio, hence the satellite TWTA output back-off. Therefore, only *per-call demand assignments* are employed in DA-FDMA where the satellite capacity for all the stations in the FDMA network is pooled and used on a single-channel-per-carrier basis; that is, for every channel assigned to a call, a carrier frequency is assigned to that channel. A pair of frequencies is assigned to a duplex circuit (two channels). Therefore a call can be blocked under only one condition—when no frequency pair is available from the pool of satellite capacity. This type of demand assignment is called *single-channel-per-carrier demand assignment multiple access* (SCPC-DAMA).

SCPC-DAMA is commonly employed in networks where there are many small stations and the total traffic intensity at each station is very low. The application in this case is a small settlement or work camp in a rural area that requires a single telephone line, or at most several lines, and which is mainly interested in communicating with a public switch network. The expected traffic intensity is about 0.1 to 0.2 erlang per station, and the network may involve more than 1000 small stations. As an example, consider the case of a geographically dispersed network of 2000 small stations each with a traffic intensity of 0.1 erlang. The total traffic intensity of the network is $2000 \times 0.1 = 200$ erlangs. Assume that a call blocking probability of 0.03 will suffice; then the network will require a satellite capacity of 208 channels to operate in the SCPC-DAMA mode. The advantage of SCPC-DAMA is that the station can be low in cost, since each station handles only a few channels at any one time and therefore does not require a large antenna. Instead, a small antenna and solid-state GaAs FET amplifiers for both HPAs and LNAs can be employed. The counterpart of SCPC-DAMA is *single-channel-per-burst demand assignment multiple access* (SCPB-DAMA) in which the TDMA station transmits one burst for every channel and the satellite capacity is made available to every station in the network.

Demand assignment types such as SCPC-DAMA and SCPB-DAMA where satellite capacity is pooled and shared by all earth stations in the network is also called *fully variable demand assignment*. It differs from per-call variable-capacity demand assignment discussed previously where only the capacity allocated to the earth station is pooled and shared by the terrestrial access circuits at that station.

A summary of the types of demand assignment multiple access follows:

1. *Variable-capacity demand assignment* is used for TDMA networks where each station has a large range of slowly varying traffic intensity and relatively few destinations. DAMA gain is reasonable.
2. *Per-call variable-capacity demand assignment* is used for TDMA networks where each station serves many destinations with low, rapidly varying traffic intensity. DAMA gain is close to maximum.
3. *Per-call demand assignment* is used for TDMA networks where each station serves many destinations with low traffic intensity. DAMA gain is reasonable.
4. *Fully variable demand assignment* is used for both FDMA and TDMA networks where each station serves many destinations with rapidly varying traffic intensity and total traffic at each station is low.

With FDMA it is called *single-channel-per-carrier demand assignment multiple access,* and with TDMA it is called *single-channel-per-burst demand assignment multiple access.* DAMA gain is maximum. The disadvantage of SCPB-DAMA is low frame efficiency, because each burst carries only one voice channel. Typical frame efficiency is about 80%, as compared to 98% for per-call variable-capacity demand assignment when both are employed in a TDMA network with feedback control for frame synchronization.

7.3 DAMA CHARACTERISTICS

In any DAMA system the most important function is to allocate the capacity required to set up a call. The method by which this function is handled illustrates the most significant difference between DAMA system concepts: centralized control versus distributed control.

In a centralized control DAMA system, a master control station assumes the responsibility of assigning the available duplex circuit (two channels) required to establish a call between two traffic stations. The master control station must first sense that a traffic station is initiating a call. Then it must locate the destination station to which the call is directed, determine the availability of channels at the satellite or at the two stations, assign a pair of channels, and let the two stations access each other through the satellite. The master control station also determines when a call is completed, so that it can release the duplex circuit; it then returns the pair of channels to the satellite pool or station pool, thus making them available to a new caller on demand. The master control station therefore must be able to keep track of actual satellite channel occupancy or channel availability at all the stations in the network. Figure 7.2 shows the communication paths required to establish a duplex circuit in a centralized control DAMA system.

In a distributed control DAMA system, there is no master control station. All traffic stations assume equal control status as far as normal network operation is concerned. Each traffic station must maintain an idle-busy table for the satellite channel pool or an idle-busy table for its own channel pool so that it may instantaneously select a free pair of satellite channels or a free channel from its own pool on detection of an off-hook signal (i.e., a call arrival). Note that a duplex circuit can be set up on call arrival if the traffic station maintains an idle-busy table for the satellite channel pool (or, equivalently, idle-busy tables for all other stations in the demand assignment network), and the call can be blocked under only one condition—when there is no free pair of satellite channels available. Only a simplex circuit (one channel) can be set up on call arrival if the traffic station maintains an idle-busy table only for its own channel pool. A

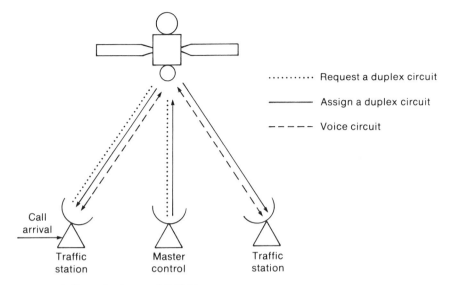

Figure 7.2 Centralized control DAMA.

duplex circuit can be connected only when the destination station has set up a return simplex circuit based on the idle-busy table for its own channel pool. Therefore the call can be blocked under two conditions—when there is no free channel at the originating station or no free channel at the destination station. Figures 7.3 and 7.4 show the communication paths necessary for the establishment of a duplex circuit in a distributed control DAMA system when the station maintains an idle-busy table for the satellite channel pool and when it maintains an idle-busy table only for its own channel pool, respectively.

Each method of control has advantages and disadvantages. The centralized control DAMA system is useful because it can maintain the status of the overall system. It can easily vary the capacity of each station according to traffic intensity and collect statistical data for traffic analysis and call logging data for off-line billing. It also has the advantage of reducing the amount of processing capability (and hence cost) required at the traffic station. Its major disadvantage is that an outage at the master control station (and its backup, if used) causes a total system failure. Furthermore, a large *capacity request channel* is required to connect all traffic stations to the master control station, and a *capacity assignment channel* for the master control station is needed to communicate with the traffic stations. Also, the size of the processing capability required at the master control station makes this type of control undesirable or a high-traffic DAMA system.

In a distributed control DAMA system, the stations make channel as-

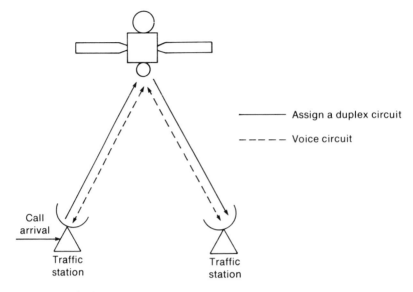

Figure 7.3 Distributed control DAMA: Station keeps idle-busy table of satellite channel pool.

signments by themselves via a *common signaling channel*. The failure of one station does not affect the other stations. The disadvantage is that a large processing capability is required at each station when it must keep an idle-busy table for the satellite channel pool. If the station had to keep

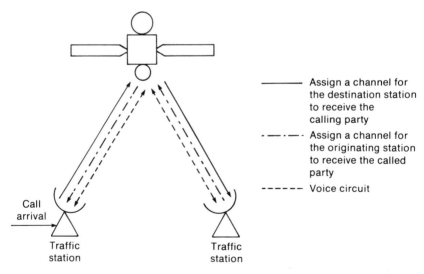

Figure 7.4 Distributed control DAMA: Station keeps idle-busy table of its channel pool.

an idle-busy table only for its own channel pool, the required processing capability would be greatly reduced. But then it would be difficult to vary the capacity assigned to the station according to its total traffic intensity unless this could be done with the assistance of a master control station. A mix of both centralized control and distributed control is normally employed for DA-TDMA, such as per-call variable-capacity demand assignment, where the reference station varies the capacity allocated to each station based on its requests and the stations make channel assignments by themselves. This hybrid system has the advantage that it requires a small processing capability at the reference station (since it neither keeps idle-busy tables nor makes call connections) and at each traffic station (since it has to keep only its own idle-busy table). The disadvantage is the requirement for a capacity request channel for the traffic stations to request capacity when needed and for a capacity assignment channel for the reference station to assign capacity to the traffic stations.

7.4 REAL-TIME FRAME RECONFIGURATION

Real-time frame reconfiguration or dynamic frame reconfiguration is the most important process in DA-TDMA employing per-call variable-capacity demand assignment. It is the ability of the reference station in a TDMA network to receive requests for capacity from traffic stations, reconfigure the frame format, distribute the new frame format to the traffic stations, and order implementation of the new frame format. All these steps must be executed in a matter of a few seconds so that the frame capacity can be allocated properly to traffic stations to meet the rapidly varying traffic. A large network might undergo frame reconfiguration as often as once per minute. Real-time frame reconfiguration is most effectively implemented in single-transponder-operation DA-TDMA.

7.4.1 Frame and Burst Structures for DA-TDMA

The frame structure for DA-TDMA is somewhat different from that for TDMA carry preassignment traffic shown in Fig. 6.3 because of the additional channels employed explicitly for demand assignment. A typical DA-TDMA frame structure is shown in Fig. 7.5. A *superframe* consists of N frames, where N is the number of stations transmitting in the frame. The superframe timing is marked by the timing of frame 1. The reference station transmits *capacity assignment messages* to the traffic stations via the *capacity assignment channel* of the reference burst. Each traffic station can be addressed once per frame, for example, station 1 is addressed in frame 1, station 2 in frame 2, . . . , and station N in frame N. Or it can be addressed once per superframe; that is, the message is broken into N segments and sent sequentially in N frames, hence the complete message can

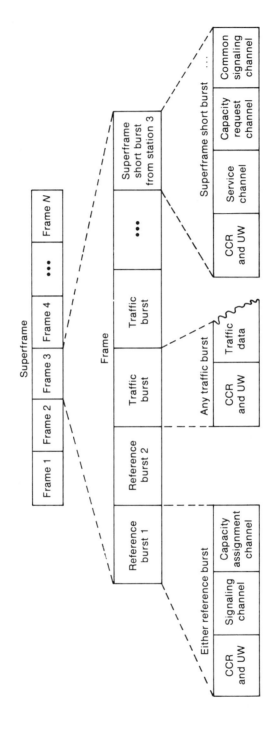

Figure 7.5 DA-TDMA frame structure.

322

be sent in one superframe. The former approach requires one superframe to address N stations, while the latter approach requires N superframes to address N stations, hence reduces the bit rate of the capacity assignment channel by N times although the frame efficiency is increased because of smaller overhead in the reference burst. Each traffic station also transmits a *superframe short burst* once per superframe: The SSB in frame 1 belongs to station 1, that in frame 2 belongs to station 2, and so on. Normally the SSB is at the end of the frame or contiguous with the RB2 for ease in carrying out frame reconfiguration. In such cases, the RB1 and RB2 and the SSB are always fixed in the frame and unaffected by frame reconfiguration. The SSB normally carries the *capacity request channel* by which the traffic station communicates with the reference station and requests additional capacity to meet its traffic load; the *common signaling channel* is employed between traffic stations only to implement distributed call processing and call assignment. *Call signaling messages* are transferred between originating and destination stations on this channel. The call signaling messages may include dual-tone multiple-frequency (DTMF) digits, the position of the allocated time slot for the call, call initiation and termination requests, and alarm condition and acknowledgment messages.

Because of the critical control information on the capacity assignment channel, the capacity request channel, and the common signaling channel, an 8:1 bit-by-bit redundancy coding scheme is normally employed to protect data integrity, where the information is transmitted in a broadcast mode eight times. (8:1 instead of a 7:1 bit-by-bit redundancy coding is used because the TDMA system processes 8 bits in parallel.) An 8:1 bit-by-bit redundancy coding scheme ensures an error probability equal to

$$P_e \approx \binom{8}{4} P_b^4$$

where P_b is the channel bit error rate. For TDMA systems the threshold bit error rate is normally 10^{-3}, hence the corresponding error probability is

$$P_e \approx 70(10^{-3})^4 = 7 \times 10^{-11}$$

When the traffic at a station exceeds its allocated capacity in the frame, the station sends a *capacity request message* to the reference station asking for additional capacity (a certain number of voice channels, for example) via the *capacity request channel*. The station continues to send this message once per superframe until it receives an acknowledgment from the reference station and then stops transmission of the message. This is to prevent the message from being lost at the reference station because of the unique word miss detection of the station's superframe short burst.

When the reference station finishes the frame reconfiguration, it sends *capacity assignment messages* to all the stations affected by the reconfiguration via the *capacity assignment channel*. Each station is addressed once per frame or once per superframe sequentially depending on the channel rate. The capacity assignment message contains information on the burst length (in voice channels, for example), burst position (in symbols, relative to the last symbol of the unique word of the primary reference burst), and possibly the transponder. Although the channel is almost error-free as a result of using an 8:1 bit-by-bit redundancy coding scheme, an error still can occur (e.g., during rain fades at the reference station). Since an error in the *capacity assignment message* can cause the station to transmit to the wrong time slot in the frame and cause burst overlapping, an acknowledgment from the affected station is necessary and is normally the same *capacity assignment message* echoed back to the reference station via the *capacity request channel*. The reference station may compare the *capacity assignment message* that it sent to the echoed-back message from the affected station and check for possible errors. If an error occurs, the same message will be retransmitted (even when the original message is received correctly by the affected station and echoed back with an error) with an identification number specifying the first, second,-..., n retransmissions. Frame reconfiguration is not implemented until the reference station verifies that all the *capacity assignment messages* have been correctly received by the affected stations.

The reference station starts the frame reconfiguration by broadcasting *countdown codes* to the affected stations in the capacity assignment channel: for example, countdown code $N - 1$ in frame 1 of the superframe, countdown code $N - 2$ in frame 2 of the same superframe,..., and countdown code 0 in frame N of the same superframe. The affected stations may lock a local countdown to the received countdown. If k or more countdown codes are received ($k \leq N$), the countdown is declared complete at countdown code 0. From (6.35) it is seen that the reconfigured transmit frame can be implemented for the first time at the *transmit frame timing* occurring at time (assuming feedback control is employed for the DA-TDMA network)

$$\text{TFT} = \text{RFT}_0 + D_N + kT_f \tag{7.12}$$

where $\text{RFT}_0 = $ *receive frame timing* of frame containing countdown code 0 and $kT_f = $ an integer number of frames allowing for processing *transmit frame delay* D_N. Also, from (6.39) the reconfigured receive frame is implemented at the *receive frame timing* occurring at time

$$\text{RFT} = \text{RFT}_0 + (M + k)T_f \tag{7.13}$$

where M is an integer number defined in (6.12).

Tables 7.2 and 7.3 show typical messages for the *capacity request*

Table 7.2 Messages of capacity request channel

Types of messages	Data blocks
Capacity request	
Message identification	First X bits
Requested capacity (in voice channels)	Next Y bits
Acknowledgment of capacity assignment	
Message identification	First X bits
Burst length (in voice channels)	Next Y bits
Burst position (in symbols)	Next Z bits
Transponder	Next W bits
Capacity release	
Message identification	First X bits
Release capacity (in voice channels)	Next Y bits

channel and the *capacity assignment channel,* respectively. Table 7.2 includes the *capacity release message* transmitted by the traffic station to the reference station, releasing the excess capacity that it does not need. (By monitoring the number of calls being blocked, the traffic station can calculate the required capacity, hence release excess capacity or request needed capacity.) The station continues to send this message once per superframe until it receives an acknowledgment from the reference station; it then stops transmission of the message to prevent the message from being lost at the reference station because of the unique word miss detection of the station's *superframe short burst*. Acknowledgment of the *capacity release message* is included in Table 7.3. Capacity release messages enable the reference station to keep track of the excess capacity in each burst in the frame at any instant in time. The reference station may also do this by monitoring the *common signaling channel* and determine

Table 7.3 Messages of capacity assignment channel

Types of messages	Data blocks
Acknowledgment of capacity request	
Message identification	First X bits
Capacity assignment	
Message and retransmission identification	First X bits
Burst length (in voice channels)	Next Y bits
Burst position (in symbols)	Next Z bits
Transponder	Next W bits
Frame reconfiguration	
Countdown codes	First X bits
Acknowledgment of capacity release	
Message identification	First X bits

the *idle-busy status* of the traffic stations. From the idle-busy status, the reference station can determine the excess capacity in the burst (but not the needed capacity).

7.4.2 Capacity Search for DA-TDMA

The most important process in real-time frame reconfiguration is the ability of the reference station to quickly search the frame to find available capacity and then assign it to the traffic stations that request it. In doing this the reference station might have to change the length and position of other station bursts in the frame. Since the position of a call subburst in a burst is measured relative to the last bit or symbol of the unique word of the burst, changing the position of a burst does not affect the position of ongoing calls. However, reducing the length of a burst may affect the position of on-going calls. Therefore, when the burst position is changed, the traffic station must inform other traffic stations via the *common signaling channel* of the new position of its burst in the frame before frame reconfiguration occurs. On the other hand, when the burst length is reduced and its position is changed, the traffic station must *repack* its burst (to a shorter burst and at a new position) via the *common signaling channel* by informing other stations of the new position of its repacked burst in the frame and the new positions of the on-going calls in the repacked burst before frame reconfiguration occurs.

There are many ways to assign the capacity requested by stations by taking channels from the excess capacity of other bursts in the frame. One way to do this is to take one channel from each burst, starting with the one at the beginning of the frame and continuing to the one at the end of the frame (excluding the bursts that request capacity, of course) and repeating the process until the number of channels requested by a station is met, and then to proceed in the same fashion for the next station if more than one station has requested additional capacity.

Consider the frame shown in Fig. 7.6 and assume that there are N traffic bursts, each from a traffic station, numbered in ascending order from the start to the end of the frame. Let burst i be the burst that needs additional capacity; then the position of any burst j ($1 \leq j \leq i-1$) on the left of burst i will be decreased (relative to the last symbol of the unique

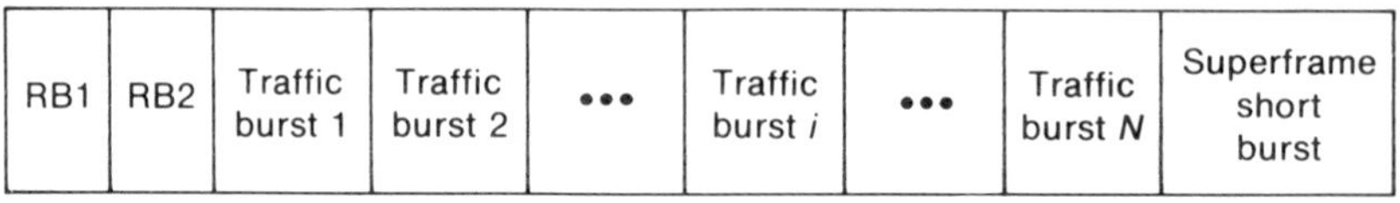

RB1	RB2	Traffic burst 1	Traffic burst 2	•••	Traffic burst i	•••	Traffic burst N	Superframe short burst

Figure 7.6 A DA-TDMA frame with N traffic bursts.

word of reference burst RB1 or RB2) by the number of voice channels (in symbols) taken away from the bursts that precede burst j. Similarly the position of any burst $j(i + 1 \leq j \leq N)$ on the right of burst i will be increased (relative to the last symbol of the unique word of reference burst RB1 or RB2) by the number of voice channels (in symbols taken away from the bursts that follow burst j. The length of a repacked burst is reduced only by the number of voice channels taken away from it.

The *frame reconfiguration algorithm* that includes the capacity search and assignment of burst position and burst length in the frame is shown in Fig. 7.7. The length $L(j)$, the position $P(j)$, and the excess capacity $\Delta L(j)$ of a burst j $(j = 1, 2, \ldots, N)$ from the previous frame reconfiguration are stored in a first-in first-out buffer and are used as inputs to the present frame reconfiguration. The algorithm can handle M $(1 \leq M \leq N)$ bursts that request additional capacity. These reconfigured bursts are designated as $i(m)$, $m = 1, 2, \ldots, M$, in ascending order $1 \leq i(1) < i(2) < \cdots < i(M) < N$, and they are among the N bursts, namely, burst 1, burst 2, $\ldots$, burst N, in the frame. The requested capacity for burst $i(m)$ is designated as $\Delta C(m)$. Whenever one capacity unit, which can be one voice channel or a group of voice channels, is found in the excess capacity $\Delta L(j)$ of a burst j, $\Delta L(j)$ and $L(j)$ are decreased by one capacity unit and the requested capacity of the reconfigured burst under consideration $\Delta C_{i(m)}$ and its length $L[i(m)]$ are increased by one capacity unit. If burst j is on the left of burst $i(m)$, the positions $P(n)$ of bursts $n = j + 1, j + 2, \ldots, i(m)$ decrease by one capacity unit. (Since burst positions are normally measured in symbols, the capacity unit here must also be expressed in symbols instead of voice channels.) If burst j is on the right of burst $i(m)$, the positions $P(n)$ of bursts $n = i(m) + 1$, $i(m) + 2, \ldots, j$ must increase by one capacity unit. The process is repeated until $\Delta C_{i(m)} = \Delta C(m)$ for each reconfigured burst $i(m)$, $m = 1, 2, \ldots, M$, or until there is no more excess capacity available in the frame. The final length $L(j)$, position $P(j)$, and excess capacity $\Delta L(j)$ for burst j, $j = 1, 2, \ldots, N$, are then stored in a first-in first-out buffer to be used as input for the next frame reconfiguration. Once $L(j)$ and $P(j)$ are obtained, the reference station just sends them in *capacity assignment messages* via the *capacity assignment channel* to the affected traffic stations and waits for acknowledgment messages before sending the countdown for frame reconfiguration.

7.4.3 Repacking On-Going Calls

Before transmitting the acknowledgment of a *capacity assignment message* to the reference station, the traffic station must *repack* the on-going calls in its burst if the burst length is reduced by the frame reconfiguration as indicated in the *capacity assignment message*. Bursts that undergo

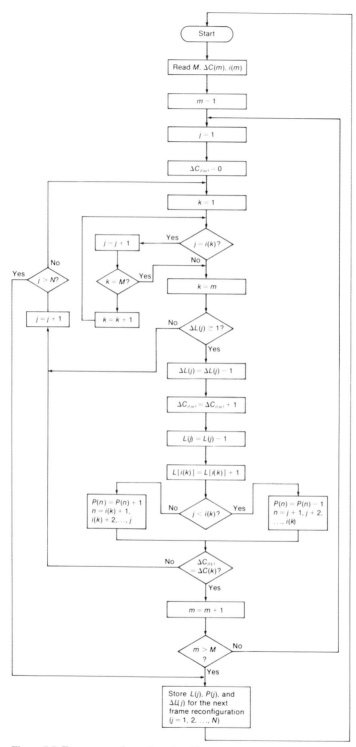

Figure 7.7 Frame reconfiguration algorithm.

only position change (no change in length) or both position change and length increase (bursts that request additional capacity) need not be repacked, because on-going calls will be transmitted in the same positions relative to the last bit or symbol of the unique word of the burst when the reconfigured transmit frame is implemented for the first time. To illustrate the on-going call *repacking process,* consider the example shown in Fig. 7.8*a* where a traffic burst has five on-going calls, namely, calls 1, 2, 4, 6, and 7, and three empty slots, namely, slots 3, 5, and 8 before frame reconfiguration. When the reference station informs the traffic station that its burst length will be reduced by one voice channel (one slot) from a capacity of eight voice channels to seven voice channels, the traffic station can repack its burst in three ways as shown in Fig. 7.8*b* to *d.* In Fig. 7.8*b*, empty slot 8 is taken away and the burst undergoes no repacking at all. In Fig. 7.8*c*, empty slot 5 is taken away and the burst has to repack original calls 6 and 7 to new calls 5 and 6. In Fig. 7.8*d*, empty slot 3 is taken away and the burst has to repack original calls 4, 6, and 7 to new calls 3, 5, and 6. It is seen that the case in Fig. 7.8*b* will involve the quickest repacking and that the case in Fig. 7.8 *d* will be the slowest. Since the repacking process involves sending the new positions of on-going calls in the *common signaling channel* of the *superframe short burst* once per superframe and waiting for an acknowledgment (the message echoed back) from the destination stations, the more calls to be repacked, the slower the repacking process, hence the slower the overall frame reconfiguration.

The *burst repacking algorithm* that reassigns the position of on-going calls before frame reconfiguration occurs is shown in Fig. 7.9. Assume that a burst is divided into K equal slots $S(1), S(2), \ldots, S(K)$ as shown in Fig. 7.10. Each on-going call occupies a slot that determines its position in the burst. An empty slot $S(k)$ is considered idle and is designated as $I(k)$. Even though the position of a call is determined in terms of the number of symbols from the last bit or symbol of the unique word of the burst, it is more efficient to send the positions of on-going calls in terms of the position of their occupied slots in the burst to reduce the length of the message and let the destination stations convert the positions of the slots to the number of bits or symbols from the last bit or symbol of the unique word of the burst. For example, a 32-kbps voice channel occupies a slot of length 240 symbols (1 symbol consists of 2 bits) in a DA-TDMA system with a frame length of 15 ms using QPSK modulation. Then the position of the on-going call in slot 16 will be 3840 symbols from the last symbol of the unique word of the burst. The *burst repacking algorithm* will begin if the burst length L_0 stored in the first-in first-out buffer of the traffic station is greater than the burst length L received from the *capacity assignment message.* If $\Delta R = L_0 - L > 0$, L will be read into the first-in first-out buffer to replace L_0, and the repacking process will begin by searching for the idle slot $I(k)$ starting at the end of the burst with slot

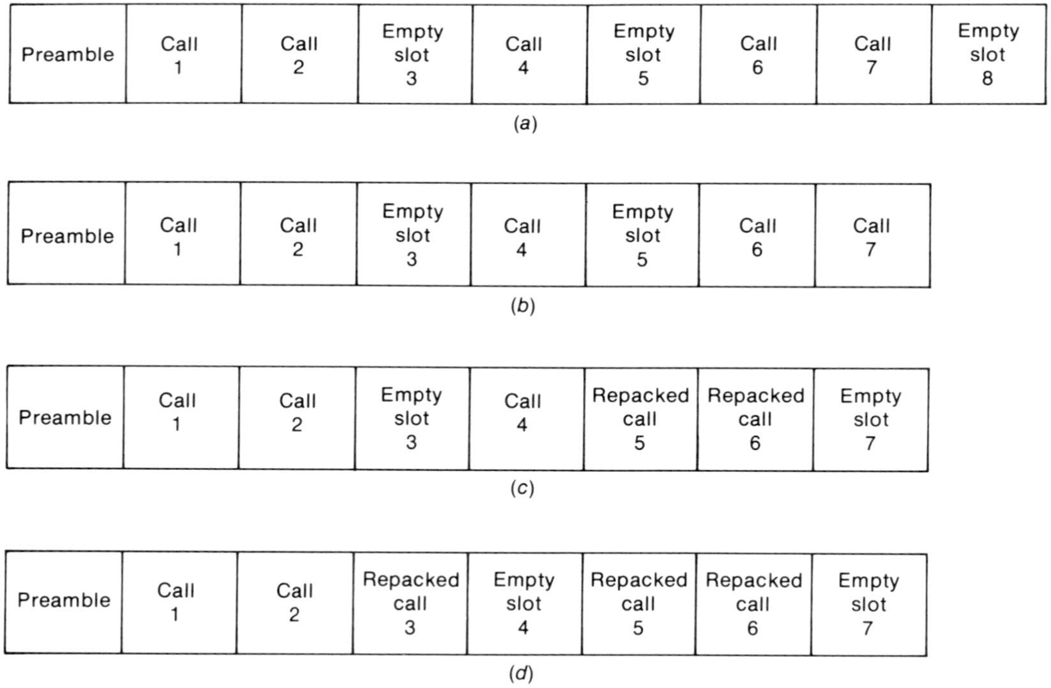

Preamble	Call 1	Call 2	Empty slot 3	Call 4	Empty slot 5	Call 6	Call 7	Empty slot 8

(a)

Preamble	Call 1	Call 2	Empty slot 3	Call 4	Empty slot 5	Call 6	Call 7

(b)

Preamble	Call 1	Call 2	Empty slot 3	Call 4	Repacked call 5	Repacked call 6	Empty slot 7

(c)

Preamble	Call 1	Call 2	Repacked call 3	Empty slot 4	Repacked call 5	Repacked call 6	Empty slot 7

(d)

Figure 7.8 Illustration of on-going call repacking process.

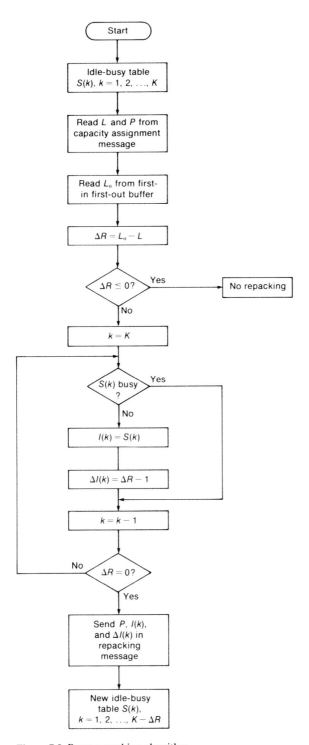

Figure 7.9 Burst repacking algorithm.

Figure 7.10 A burst with K equal voice channel slots.

$S(K)$. Whenever an idle slot $I(k)$ is found, the capacity repacking $\Delta I(k)$ associated with that idle slot will be computed.

For example, if $I(K-1)$ is the first idle slot found, then $\Delta I(K-1)$ will be equal to the total capacity reduction ΔR less one, that is, $\Delta I(K-1) = \Delta R - 1$. Next, if $I(K-4)$ is the second idle slot found, then $\Delta I(K-4) = \Delta R - 2$. Recall the burst in Fig. 7.8a and assume that $\Delta R = 3$; then $I(8)$ is the first idle slot with $\Delta I(8) = 2$, $I(5)$ is the second idle slot with $\Delta I(5) = 1$, and $I(3)$ is the third idle slot with $\Delta I(3) = 0$. When the total capacity reduction ΔR is reduced to 0, the repacking process is completed. The traffic station that repacks its burst then sends the new burst position P that it receives from the *capacity assignment message* together with the idle slots $I(k)$ and their associated capacity repacking $\Delta I(k)$. For example, let $\Delta R = 2$ and $I(k)$ and $I(k')$ be the two idle slots such that $1 \le k < k' \le K$. Then, all on-going calls $S(j)$ between $I(k)$ and $I(k')$, that is, $k < j < k'$, will be repacked to new slots $S[j - \Delta I(k) - 1] = S(j-1)$. All on-going calls $S(j)$ after $I(k')$, that is, $j > k'$, will be repacked to new slots $S[j - \Delta I(k') - 1] = S(j-2)$. All on-going calls $S(j)$ before $I(k)$, that is, $j < k$, will not be affected.

Table 7.4 shows typical *repacking messages* sent via the *common signaling channel*. The burst position P is expressed in symbols relative to the last symbol of the unique word of the reference burst. The idle slots $I(k)$ and their associated capacity repacking $\Delta I(k)$ are expressed in terms of slots (one slot for each voice channel). If there are K slots in the frame, $\log_2 K$ bits will be required to encode $I(k)$ and $\log_2 (\Delta R - 1)$ bits to encode $\Delta I(k)$. Since K and ΔR vary from one frame reconfiguration to another, a maximum value K_M for K and a maximum value ΔR_M for ΔR should be selected. K_M then represents the maximum capacity of a burst in the frame, and ΔR_M represents the maximum total capacity reduction of a burst in the frame. Note that $\Delta R_M \le K_M$. Also note that the contiguous idle slots $I(k), I(k+1), \ldots, I(k+x)$ where there are no on-going calls $S(j), j > k + x$, after them need not be sent, as seen in the case of the idle slot $I(8)$ in Fig. 7.8a. The *repacking message* in Table 7.4 will occupy a total of $L \ge X + Y + (\Delta R_M - 1) [\log_2 K_M + \log_2 (\Delta R_M = 1)]$ bits. For example, let $X = 1$ bit, $Y = 20$ bits, $\Delta R_M = 17$ slots, and $K_M = 24$ slots; then the repacking message will occupy a length of $L = 165$ bits. Acknowledgment of the repacking message will occupy a maximum length $L' \ge X + Y + K_M \log_2 M$ bits or 141 bits. Note that 1320 bits must be allocated to the repacking message if 8:1 redundancy coding is employed. If the message is to be transmitted in one superframe, the *common signaling channel* must have a minimum bit rate of $1320/NT_f$ bits per second, where N is the number of frames per superframe and T_f is the frame length.

7.4.4 How Fast Is Frame Reconfiguration?

The speed of frame reconfiguration depends mainly on the following time delays, assuming no retransmission due to message errors:

1. The satellite propagation delay t_1 in sending the capacity request message from the traffic station to the reference station
2. The processing time t_2 for the capacity request message at the reference station
3. The processing time t_3 for the frame reconfiguration algorithm at the reference station
4. The satellite propagation delay t_4 in sending the capacity assignment message from the reference station to the traffic station
5. The processing time t_5 for the capacity assignment message at the traffic station

Table 7.4 Repacking message

Types of messages	Data blocks
Repacking	
Message identification	First X bits
Burst position (symbols)	Next Y bits
First idle slot	
Position (slots)	Next $\log_2 K_M$ bits
Repacking (slots)	Next $\log_2(\Delta R_M - 1)$ bits
.	.
.	.
.	.
Last Idle slot	
Postion (slots)	Next $\log_2 K_M$ bits
Capacity repacking (slots)	Next $\log_2(\Delta R_M - 1)$ bits
Acknowledgment of repacking	
Message identification	First X bits
Burst position (symbols)	Next Y bits
New positions of on-going calls	
First call (slots)	Next $\log_2 K_M$ bits
.	.
.	.
.	.
Last call (slots)	Next $\log_2 K_M$ bits

6. The processing time t_6 for the burst repacking algorithm at the traffic station
7. The satellite propagation delay t_7 in sending the repacking message from one traffic station to another
8. The processing time t_8 for the repacking message at the traffic station
9. The satellite propagation delay t_9 in sending an acknowledgment of the repacking message at the traffic station
10. The processing time t_{10} for acknowledgment of the repacking message at the traffic station
11. The satellite roundtrip delay t_{11} in sending an acknowledgment of the capacity assignment message from the traffic station to the reference station
12. The processing time t_{12} for acknowledgment of the capacity assignment message at the reference station
13. The satellite propagation delay t_{13} in sending the frame reconfiguration countdown codes from the reference station to the traffic stations
14. One superframe time t_{14} for completing the countdown codes.

Thus the minimum delay t_D in implementing a frame reconfiguration is

$$t_D = \sum_{i=1}^{14} t_i$$

For illustration, let $t_1 = t_4 = t_7 = t_9 = t_{11} = t_{13} = 250$ ms, $t_2 = t_5 = t_8 = t_{10} = t_{12} = 30$ ms, $t_3 = t_6 = 1000$ ms, and $t_{14} = 300$ ms; then $t_D = 3.95$ s. Therefore the frame reconfiguration can be performed several times a minute if the network experiences rapid fluctuations in its traffic load at various nodes such that the reference station continuously receives *capacity request messages* from the traffic stations. Any capacity request received after the reference station starts the *frame reconfiguration algorithm* will be queued and processed during the next frame reconfiguration started right after the *countdown code* 0 of the present frame reconfiguration has been sent.

In summary, real-time frame reconfiguration can provide near optimum performance for a *per-call-variable-capacity demand assignment* network that operates via one common transponder. Since the burst length can be varied continuously to adapt to the traffic load at each station, the satellite channel pool is in effect shared by all stations. The above real-time frame reconfiguration can be used as well for a variable-capacity demand assignment network employing multidestination TIMs as discussed in Sec. 7.2. (In this case only the algorithm for the capacity search in Fig. 7.7 is sufficient.) In this type of network no distributed call-by-call processing between traffic stations via a *common signaling channel* is performed. Note that when multidestination TIMs are used, TIM *identification* for each subburst in the traffic burst is necessary, otherwise the

TDMA terminal may not know to which TIM the received subburst should be routed. In other words, the reference station must keep a TIM *interconnectivity map* of the individual stations and update it whenever there is a frame reconfiguration. The frame reconfiguration in this case is the implementation of a new burst time plan as discussed in Sec. 6.7 where the transmit or receive subburst identification in the burst time plan format is the transmit or receive TIM identification.

7.5 DAMA INTERFACES

As discussed in Sec. 7.2, a significant reduction in satellite transponder capacity can be realized by employing multidestination TIMs at the traffic station. Since traffic between the earth station and the terrestrial network is carried by T1 carriers (Sec. 1.2.3), the most commonly used terrestrial interfaces for variable-capacity demand assignment networks are multidestination T1—TIM where a single T1 carrier delivered to the station can have (1) channels that are destined for different stations and transmitted in different subbursts and (2) channels that are received from different bursts and repacked and delivered by the TIM to the terrestrial network in a single T1 carrier. When there is a requirement for less than 24 PCM channels between two earth stations, the transponder capacity would be wasted if the entire T1 line were allocated between them. Thus by splitting the T1 line between stations only the required number of channels will be transmitted via the satellite. Figure 7.11 illustrates this concept where station A is required to transmit 8 channels to station B and 16 channels to station C in a single T1 carrier. On the receive side,

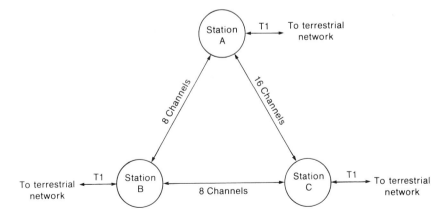

Figure 7.11 Multidestination T1 lines.

station A receives 8 channels from one of station B's bursts and 16 channels from one of station's C bursts and repacks these 24 channels into a single T1 carrier for delivery to the terrestrial network.

One problem involved in splitting a T1 carrier is alignment of the PCM frames. Recall from Sec. 1.2.3 that the signaling bits in every sixth frame of the T1 carrier (normally called the A and B signaling bits) carry the telephone signaling information; therefore the frames must be aligned properly. In order to do this, the data in the compression buffer of the TIM must be aligned before transmission to the TDMA frame; that is, the first transmitted byte in any transmitted channel must originate from a specific T1 frame (e.g., the first frame) in the T1 multiframe. Thus it is necessary that the TDMA frame length be an integer multiple of the T1 multiframe (1.5 ms). This allows the A and B signaling bits from different sources to be correctly positioned without additional processing by the receive side of the TIM, hence received channels from various originating stations can be combined into a single outgoing T1 carrier at the destination station. The PCM frames in these outgoing T1 carriers are always aligned. The above alignment process completely eliminates the requirement that A and B signaling bits be extracted and reinserted at the destination station.

For *per-call variable-capacity demand assignment* operation, channel interconnections are carried out on a single-channel basis via the *common signaling channel* of the *superframe short burst*, as shown in Fig. 7.5. Therefore it is necessary that the A and B signaling bits be extracted or inserted from or into the incoming or outgoing T1 streams. Figure 7.12 shows a DAMA T1 TIM that can operate both as a multidestination T1 TIM and a fully variable demand-assigned T1 TIM.

On the transmit side the DAMA T1 TIM receives data on the basis of the local TDMA clock turned around by the terrestrial network (Figs. 6.37–6.38). The data is written into an *elastic buffer* at the rate of the recovered T1 clock and read out of the elastic buffer by a T1 clock derived from the local TDMA clock to eliminate jitter introduced by the terrestrial equipment. The T1 serial data is applied to the *frame extractor* and *channel timing* which extract the *framing bits* and *channel clock* and present them to the compression buffer control circuitry. The serial T1 data is converted to an 8-bit parallel form and stored in the *compression buffer*. The compression buffer must include a *Doppler buffer* that compensates for the effect of satellite motion. The data in the compression buffer is aligned before transmission to the TDMA frame to allow multidestination operation. The data is also inputted to the *signaling processor* to detect the *dual-tone multiple frequency* and extract A and B *signaling bits*. Blocks of data from selected channels are read out of the compression buffer on the basis of the transmit TIM address and TIM bus clock and transmitted during the preassigned time slots in the TDMA frame. On the receive side, data derived from the receive TIM bus is

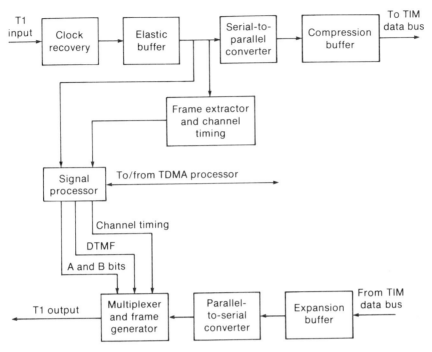

Figure 7.12 Schematic diagram of a DAMA T1 TIM.

stored in appropriate locations of the *expansion buffer* and read out on the basis of the channel clock and T1 frame synchronization. The 8-bit parallel data is then converted to the serial T1 format. DTMF tones and A and B signaling bits are then inserted by the signaling processor into the outgoing T1 stream.

The signaling processor collects the signaling information and forwards it to the TDMA processor. The TDMA processor then allocates capacity, signals it over the *common signaling channel* to other stations, and maps the T1 channels onto appropriate locations in the TDMA burst. The TDMA processor also receives messages over the common signaling channel from other stations and sends them to the signaling processor for reformatting to the A and B signaling bits.

7.6 SCPC-DAMA

As discussed in Sec. 7.2, SCPC-DAMA is best suited to a network with many small, remote stations (100 or more), each with a low traffic intensity. For such a system, *centralized control* is normally employed to reduce the cost of the stations. A *master control station* provides an interface between the small stations and the switch network. A communication circuit between small stations may be established only via the master control station using a *request channel* and an *assignment channel*. The request channel handles *frequency request* and *end-of-call message* sent by the small stations to the master control station. This can be done in a variety of ways, the most typical of which is the response to a poll from the master control. This procedure is attractive when the number of stations is small (less than 50) but incurs a large delay in requests when the number of stations is large. In the latter case, the request channel can be accessed by all the small stations in a random manner; that is, the small stations send request messages whenever calls arrive. There will be collisions, of course, and the request messages will be destroyed (contain many errors) and not received by the master control station. The small station may expect a response to a request or end-of-call message within a preset time. Should the response from the master control station not arrive within this time, the small station assumes that the previous message has been involved in a collision and retransmits the message after a random delay. This type of random multiple access is called *Aloha* and gives a good probability of uncollided message and short delay when the number of call attempts per small station is very low and is best suited to network with a large number of small stations (in the hundreds). We will study the *Aloha* mode of access in detail in Chap. 8. The *assignment channel* is used to communicate frequency assignments, end-of-call messages, and verification messages from the master control station to the small stations. The *assignment channel* is thus a broadcast channel and can be implemented using *time division multiplexing* where small stations are addressed sequentially.

Associated with each new call arrival is a new *request message*, an end-of-call message, a fraction of a new request retransmitted as a result of call blocking, and a call busy. Let λ be the network call arrival rate, B the network call blocking probability, α the network call busy probability, N_1 and N_2 the request and end-of-call message lengths, and R the bit rate of the request channel. Then the throughput of the request channel in messages per message time is

$$S = \frac{\lambda N_1/R}{(1-B)(1-\alpha)} + \frac{\lambda N_2}{R} \qquad (7.14)$$

In Chap. 8 we will show that the maximum throughput of the request channel operating in the *Aloha* random access mode is

$$S_{max} = \frac{1}{2e} = 0.184 \qquad (7.15)$$

This is the case because messages transmitted at random will collide and have to be retransmitted. If we let a be the network traffic intensity and M

be the number of receivers in the network, then the probability that a called party will be busy is

$$\alpha = \frac{a}{M} \tag{7.16}$$

As an example consider the case where $a = 200$ erlangs, the mean call holding time $= 1/\mu = 3$ min, $M = 1000$, and $B = 0.1$. $N_1 = N_2 = N$. Then

$$\lambda = a\mu = 200/[(3)(60)] = 1.11 \text{ calls per second}$$

$$\alpha = 200/1000 = 0.2$$

$$\frac{N}{R} = \frac{S}{2.65} < \frac{0.184}{2.65} = 0.069$$

Assume that a message consists of 120 bits; then the bit rate of the request channel must be

$$R > 1739 \text{ bps}$$

If it is desired to operate the request channel at a nominal throughput $S_0 = 0.05$ (to reduce delay and avoid instability), then

$$R = 6360 \text{ bps}$$

7.7 SPADE

The *single-channel-per-carrier pulse code modulation multiple access demand assignment equipment* (SPADE) system is a SCPC-DAMA satellite network using *distributed control*. The system is *fully variable*, allowing all circuits (each circuit consists of two channels) to be selected by any terminal on demand. Neither end of a circuit is permanently associated with any terminal but is assigned from the satellite channel pool as required and released again to the pool when no longer needed. The system does not require a master control station, as in the case of SCPC-DAMA as described in Sec. 7.6. Instead, it uses a computer-controlled *demand assignment signaling* and *switching* (DASS) system at each station for self-assignment of circuits using a network idle-busy table based on continually updated circuit allocation status data exchanged between stations via a *common signaling channel* (CSC). The SPADE network has sufficient capacity to accommodate 48 stations, each of which has a maximum capacity of 60 terrestrial circuits.

PCM is used for digitized voice circuits and QPSK is the carrier modulation scheme. Each channel of a voice circuit occupies a 45-kHz bandwidth, and the paired RF carriers of these circuits are separated by 18.045 MHz. PSK is used as the carrier modulation for the *common sig-*

naling channel which is accessed by the stations in the SPADE network using a TDMA mode. The *common signaling channel* occupies 160 (±80) kHz of RF bandwidth with the carrier offset from the center of the DAMA RF band toward the lower end of the band by 18.045 MHz. The DAMA RF band occupies 36 MHz of bandwidth with a reference pilot (used by the receiving station for automatic frequency control) located at the band's center. The CSC carrier is located 18.045 MHz below the reference pilot. The DAMA RF bandwidth accommodates 800 RF channel carriers (45 kHz each). The SPADE frequency plan is shown in Fig. 7.13. Carriers 1, 2, 400, 401, 402, and 403 are not used because of their proximity to the CSC carrier and the *reference pilot* (interference would occur). Thus only 397 pairs of carriers are usable.

The SPADE system (Fig. 7.14) provides for voice circuit switching and call signaling as required to systematically initiate, supervise, and terminate all calls. When a call request is received, the DASS automatically selects a frequency pair from the pool of available frequencies and alerts the destination station to the incoming call and the frequency assignment for response. Any number of DASS systems from 2 to 48 utilize the signaling information disseminated by the CSC to update an idle-busy table such that the frequencies just assigned become unavailable for new calls. The frequency selected is provided to the channel unit by means of a *frequency synthesizer* capable of generating any of the 800 discrete frequencies required using digital codes provided by the DASS system. When the channel QPSK modem is turned on, the DASS system conducts a two-way circuit continuity check. Once the call has been established, the voice signal received by the channel unit is sent to a PCM *coder* and *decoder* (codec) which transforms the analog voice signal to a digital signal for outgoing transmission and transforms a digital signal to an analog signal for returning signals. The content of the voice channel coming from the terrestrial network is detected by a voice detector which is used to gate the channel carrier on or off. This conserves the satellite power as a function of talker activity so that 800 voice channels can be supported. The digital bit stream in and out of the PCM codec is synchronized by the *channel synchronizer*, where timing, buffering, and framing functions are performed. The QPSK modem modulates the assigned carrier frequency with the outgoing bit stream and coherently demodulates the incoming bursts by recovering carrier and bit timing associated with the received signals. The modulated carriers, both outgoing and incoming, are passed through a common IF subsystem which interfaces with the earth station upconverters and downconverters at 70 MHz intermediate frequency. When the call is completed, a control signal from the terrestrial network allows the DASS system to return the circuit to the frequency pool for reassignment. This information is passed to all stations via the CSC. All interunit signaling between the

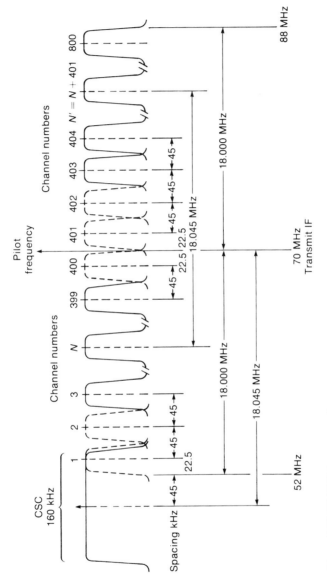

Figure 7.13 SPADE frequency plan.

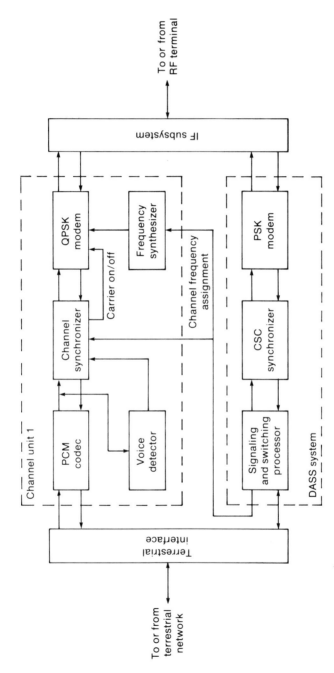

Figure 7.14 Block diagram of SPADE terminal

demand-assigned stations is conducted via the CSC which is routed through the satellite.

The CSC is shared on a time division basis by all the stations; that is, each operating station transmits one short local data burst of signaling information within each time frame such that these local data bursts are received consecutively without overlap by the satellite and then broadcast back to all the SPADE stations. Thus each station, while transmitting only one local data burst at a prescribed time within the TDMA frame, receives all the other stations' local data bursts. The CSC's TDMA frame length is 50 ms, and it uses a bit rate of 128 kbps. Each frame has fifty 1-ms time slots. One time slot is assigned to the reference burst and is used to synchronize the network; time slot 1, subsequent and adjacent to the reference burst, is reserved for test operations. This leaves 48 time slots to be assigned to SPADE stations in the network for transmission of their local data bursts. One SPADE station in the network is selected to assume the reference role and originate the CSC reference burst. Provisions are made for SPADE stations to assume this reference role in a sequential manner in the event that the previously designated reference station fails to originate the reference burst. The CSC reference burst and local data burst are shown in Fig. 7.15.

7.8 DIGITAL SPEECH INTERPOLATION

In Sec. 7.1 to 7.7 we studied demand assignment as an efficient technique for increasing satellite capacity. In this section we will investigate another efficient technique—*digital speech interpolation* (DSI). In a normal two-way telephone conversation, one participant listens while the other speaks. Thus on the average speech signals are present for less than 50% of the conversation time. In addition, even during the time one talker is speaking, pauses occur between sentences, words, and even syllables, and there are times when the circuit is idle for thinking and so forth. Actual measurements show that speech signals rarely exceed 40% of a typical two-way telephone conversation time. Since each circuit uses two channels for transmission, each channel is free 60% of the time on the average. To improve utilization of the circuit, speech interpolation may be used, with a transmission channel assigned to the telephone channel during speech spurts only. In this process, N conversations are carried on M transmission channels, where $M < N$. The ratio N/M is referred to as the *speech interpolation gain.*

Speech interpolation has been used on transatlantic submarine cable channels since the mid-1960s to double the capacity of analog FDM voice circuits using the well-known *time assignment speech interpolation*

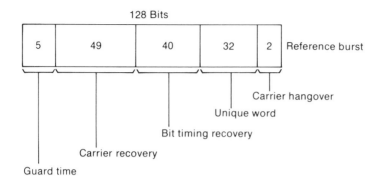

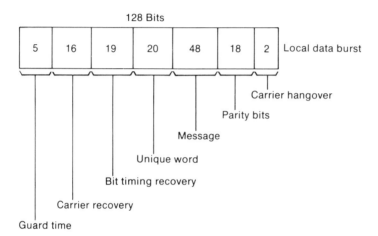

Figure 7.15 Reference burst and local data burst of the SPADE common signaling channel.

(TASI). With the advent of digital satellite communications using PCM for voice encoding and TDMA for satellite access, DSI was employed instead of TASI and shown to have many advantages over analog TASI.

DSI operates by assigning a satellite channel only when a speech spurt is present. Hence DSI becomes more efficient as the number of terrestrial channels increases. During periods of heavy speech activity there may not be enough satellite channels to accommodate all terrestrial channels. An incoming speech channel which cannot be assigned a satellite channel for transmission is said to be frozen-out, which results in a form of clipping of the initial portion of a speech spurt, or *competitive clipping.* Speech clips longer than 50 ms are not acceptable for high quality transmission. Acceptable performance has been demonstrated when the probability that speech clipping exceeding 50 ms will occur is less than 0.02 or

2%. (Such clipping would produce only minor articulation damage and would happen on the average once every 1.5 min during conversation to one of the participants assuming an average speech spurt duration of 1.5 s and a speech activity factor of 0.4 or 40%).

The probability that the number of simultaneous speech spurts on N terrestrial channels with an activity probability of α will equal or exceed M satellite channels $(M < N)$ is given by the cumulative binomial distribution [1]

$$B(M,N,\alpha) = \sum_{x=M}^{N} \frac{N!}{x!\,(N-x)!}\, \alpha^{x}(1-\alpha)^{N-x} \qquad (7.17)$$

The probability that a speech spurt will be clipped for more than T seconds is given by

$$B(M,N,\theta) = B\left[M,N,\alpha \exp\left(\frac{-T}{L}\right)\right] \qquad (7.18a)$$

where

$$\theta = \alpha \exp\left(\frac{-T}{L}\right) \qquad (7.18b)$$

represents an exponential speech spurt distribution and L is the mean speech spurt duration. Thus $B(M,N,\theta)$ expresses the probability of encountering M or more simultaneous speech spurts after T seconds, which is the same as the probability of a speech spurt being clipped for T seconds or more. Figure 7.16 plots the probability of encountering a speech

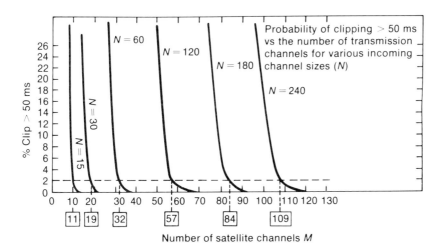

Figure 7.16 Percent of clipping greater than 50 ms.

clip greater than $T = 50$ ms with $L = 1.5$ s versus the number of satellite channels M for various numbers of terrestrial channels N. It also indicates the number of satellite channels needed to accommodate a given number of terrestrial channels N for 2% chance of encountering a speech clip greater than 50 ms. The ratio N/M is called the DSI *gain* and is plotted in Fig. 7.17 as a function of the number of terrestrial channels N.

The performance of DSI can also be analyzed based on the fraction of speech lost as a result of the freeze-out of all N terrestrial channels. New speech spurts beginning at times when all satellite channels are busy are frozen out and must wait on a first-come first-served basis for an available channel. Once a satellite channel is assigned to a particular speech spurt, the channel is held until the spurt ends. As described above, a freeze-out typically causes the initial part of a speech spurt to be clipped. A speech spurt can be lost entirely if the waiting time is longer than the spurt duration.

The fraction of time that exactly $x < N$ terrestrial channels are active with an activitiy probability of α is given by

$$b(x,N,\alpha) = \frac{N!}{x!(N-x)!}\, \alpha^x (1-\alpha)^{N-x} \tag{7.19}$$

and $x - M$ of these terrestrial channels will incur speech loss whenever $x > M$. Thus over a long period of time, the ratio of total speech loss time for all terrestrial channels to their total active time is

$$\Phi(M,N,\alpha) = \frac{\displaystyle\sum_{x=M+1}^{N} (x-M)b(x,N,\alpha)}{\displaystyle\sum_{x=1}^{N} xb(x,N,\alpha)}$$

$$= \frac{1}{N\alpha} \sum_{x=M+1}^{N} (x-M)b(x,N,\alpha)$$

$$= \frac{1}{N\alpha} \sum_{x=M+1}^{N} \frac{N!\,(x-M)}{x!(N-x)!}\, \alpha^x(1-\alpha)^{N-x} \tag{7.20}$$

Table 7.5 gives the values of Φ as a function of the number of terrestrial channels N for $\alpha = 0.4$ and the number of satellite channels $M = 36$. Experiments have found that the performance is acceptable if the freeze-out fraction is less than 0.5%. This threshold is almost equivalent to the 2% threshold for speech clipping exceeding 50 ms discussed previously, and the DSI gain N/M with $\Phi = 0.005$ is almost coincident with that in Fig. 7.17.

DSI is employed in TDMA systems as a data compression process which uses the speech activity factor to compress the number of satellite

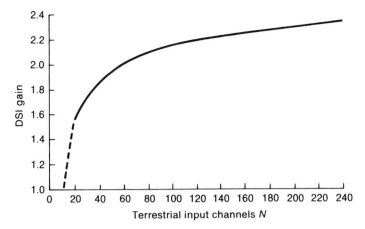

Figure 7.17 DSI gain. This curve is based on the assumption that less than 2% of the voice spurts will experience clipping of 50 ms or longer.

channels required to transmit a number of terrestrial channels. A speech detector is used to detect speech in each terrestrial channel input to the DSI system. Channels where speech is present are assigned a satellite channel for transmission to the receiving station. The DSI system includes an *assignment channel*(AC) which is used to transmit satellite channel assignments to the receiving station for proper reconnection to outgoing terrestrial channels. Satellite channel assignments are transmitted by means of assignment messages (AMs) which consist of the terrestrial channel number and an associated satellite channel number. As in the case of the single-destination and multidestination T1 terrestrial interface

Table 7.5 Freeze-out fraction versus terrestrial channels for $M=36$ satellite channels and speech activity probability $\alpha = 0.4$

N	Φ
60	3.79×10^{-5}
70	1.48×10^{-3}
75	5.14×10^{-3}
80	1.35×10^{-2}
85	2.87×10^{-2}
90	5.14×10^{-2}
100	1.14×10^{-1}

modules discussed in Secs. 6.10.2 and 7.6, a DSI-TIM may communicate with another DSI-TIM or with several DSI-TIMs. The first mode of operation is referred to as single-destination (point-to-point) DSI, while the second is called multidestination (point-to-multipoint) DSI. A single-destination DSI-TIM transmits and receives one subburst per frame, while a multidestination DSI-TIM transmits one subburst per frame but receives more than one subburst from different traffic bursts and recombines them. In practice single-destination DSI-TIMs carry larger traffic links, utilizing the maximum DSI system capacity. A multidestination DSI-TIM on the other hand is used to combine smaller traffic links and channels left over from larger links into one multidestination unit. However, the connectivity of terrestrial channels for a multidestination DSI-TIM is more complicated.

REFERENCES

1. K. Bullington and J. M. Fraser, "Engineering Aspects of TASI," *Bell Syst. Tech. J.,* Vol. 38, No. 2, Mar. 1959, pp. 353–364.
2. J. G. Puente and A. M. Werth, "Demand-Assigned Service for the Intelsat Global Network," *IEEE Spectrum,* Vol. 8, No. 1, Jan. 1971, pp. 59–69.
3. G. Frenkel, "The Grade of Service in Multiple-Access Satellite Communications Systems with Demand Assignments," *IEEE Trans. Commun. Tech.,* Vol. COM-22, No. 10, Oct. 1974, pp. 1681–1685.
4. S. J. Campanella, "Digital Speech Interpolation," *COMSAT Tech. Rev.,* Vol. 6, Spring 1976, pp. 127–158.
5. J. M. Keelty and S. Hatzigeorgiou, "Alternate Architectures and Technologies for INTELSAT Type DSI Design," *IEEE J. Selected Areas Commun.,* Vol. SAC-1, No. 1, Jan. 1983, pp. 214–222.

PROBLEMS

7.1 Consider a satellite system in which an earth station is required to communicate with 40 destinations. The traffic to each destination is 0.5 erlang, and the required blocking probability is 0.01.

    (a) Find the required number of satellite channels for the earth station when the channels are preassigned to each destination.

    (b) If any channel can be demand-assigned to any destination, how many channels will be needed?

    (c) Find the demand assignment gain.

7.2 Determine the minimum number of channels required to handle an offered traffic intensity of (a) 0.3 erlang, (b) 11 erlangs, and (c) 70 erlangs at a blocking probability of 0.005 with the condition that the blocking probability shall not exceed 0.01 in the event that either a 10% traffic overload or a single channel failure does not occur at the same time.

7.3 A group of four demand-assigned terrestrial interface modules is observed at intervals of

1 min during 10 busy hours. The number of occasions when various numbers of TIM were observed to be engaged simultaneously are listed in the following table.

Number of TIMs engaged	Number of observations
0	89
1	164
2	173
3	114
4	60

It is observed that the total number of blocked calls is 34. Assume that time and call congestion are equal, which is true with Poisson-distributed arrivals (Call congestion is the proportion of blocked calls in the long run, while time congestion is the proportion of time during which all channels are busy during the busy hours.)

    (a) Estimate the average call duration.

    (b) Estimate the average traffic intensity without using Erlang's formula.

7.4 Consider an earth station that communicates with many destinations. The total traffic at the earth station is 1151 erlangs, and the required blocking probability for any destination is 0.01. Assume that the earth station employs per-call demand assignment. Find the required number of channels.

7.5 Molina's "blocked calls held" formula is based on the same assumption as the Erlang B "blocked call cleared" formula, except that assumption 3 in Sec. 7.1 is replaced by the following assumption.

    When a call finds all channels busy, it continues to demand service for a period equal to the holding time it would have had if successful. If during this period a channel becomes free, it is seized and rendered unavailable for the unelapsed part of the holding time, but the call is still regarded as lost.

    This is an unrealistic assumption, but the result yields a slightly lower traffic capacity for a given blocking probability and is regarded as a useful safety margin in North America.

    Show that Molina's blocked calls held formula is

$$B_M = \sum_{k=n}^{\infty} e^{-a} \frac{a^k}{k!}$$

7.6 Express the Erlang B formula in terms of the Molina formula in Prob. 7.5.

7.7 The Erlang B formula is based on the assumption that calls occur in accordance with the Poisson distribution. This implies an infinite number of calling sources. When the number of calling sources is relatively small, the probability of call arrival may depend on the number of calling sources already being served, and the Poisson distribution is not applicable. Let M be the number of sources and assume that blocked calls are cleared.

    (a) Show that the probability that k channels are busy at a random instant is

$$p_k = \frac{\binom{M}{k} \rho^k}{\sum_{k=0}^{n} \binom{M}{k} \rho^k}$$

where ρ = probability that a new call will arise from a free source (or calling rate per free source) and n = number of channels. The time congestion probability is p_n which is the probability that all n channels will be busy.

(b) Show that the probability that all n channels will be busy when a call arrives is

$$B_c = \frac{\binom{M-1}{n} \rho^n}{\sum_{k=0}^{n} \binom{M-1}{k} \rho^k}$$

The Engset formula B_c is called the call congestion probability and differs from the time congestion probability p_n. (Note that the call congestion probability is the same as the time congestion probability for the Erlang B formula.) The call congestion probability B_c should be used when the number of call sources is not significantly larger than the number of channels (a finite calling source).

(c) Express ρ in terms of the traffic intensity a and the number of calling sources M.

7.8 Using the result in Prob. 7.7, find the call congestion probability B_c when a traffic of 12 erlangs is offered to 20 channels by various numbers of sources:
   (a) $M = 20$.
   (b) $M = 200$.
   (c) $M = \infty$.
Compare the result in (c) with that obtained from the Erlang B formula.

7.9 The traffic intensity in erlangs between earth stations A, B, and C is

	A	B	C
A	0	$13 \leq a \leq 36$	$36 \leq a \leq 71$
B	$13 \leq a \leq 36$	0	$10 \leq a \leq 28$
C	$33 \leq a \leq 71$	$10 \leq a \leq 28$	0

(a) How many voice circuits should be preassigned to these three earth stations for a blocking probability of 0.01? How many point-to-point T1 TIMs are needed?

(b) If variable-capacity demand assignment is employed at each station, what is the maximum demand-assignment gain for A, B, and C?

(c) Assume that the busy and nonbusy hours of A, B, and C are the same. What is the maximum demand assignment gain for a network consisting of these three earth stations?

(d) Find the number of multidestination T1 TIMs required for variable-capacity demand assignment.

7.10 A fully connected TDMA network consists of 30 earth stations. There are 400 one-way links with 0.4 erlang, and the rest have 0.2 erlang.

(a) Find the number of preassigned channels for the network required to provide a 0.01 grade of service.

(b) If the network employs SCPB-DAMA, how many channels will be needed?

7.11 Consider a TDMA network employing per-call variable-capacity demand assignment with the following network parameters.

TDMA frame length: 16 ms
TDMA burst bit rate: 43 Mbps
Number of bursts: 32 traffic bursts, 2 reference bursts, and 1 superframe short burst
CCR sequence: 256 bits
UW sequence: 20 bits
Order wire channel (reference bursts only): 512 bits
Management channel: 256 bits
Transmit timing channel: 320 bits
Service channel (superframe short burst only): 256 bits

Demand assignment channel (superframe short burst only): 1600 bits

Guard time: 32 bits

Voice channel bit rate: 32 kbps

Because of the finite time it takes to reconfigure the frame to obtain the required blocking probability, 5% of the total frame capacity available to the network is reserved as a safety margin. In other words, each traffic burst is assigned an extra 5% capacity all the time.

(a) Find the usable capacity of the frame.

(b) Assume that the blocking probabilities at earth stations are mutually independent and equal 0.01. Find the total network traffic intensity.

7.12 Consider a TDMA network employing SCPB-DAMA, where each voice channel is transmitted by a separate burst. The network parameters are

TDMA frame length: 16 ms

TDMA burst bit rate: 43 Mbps

CCR sequence: 256 bits

UW sequence: 20 bits

Order wire channel (reference bursts only): 512 bits

Management channel: 256 bits

Transmit timing channel: 320 bits

Service channel (superframe short burst only): 256 bits

Demand assignment channel (superframe short burst only): 1600 bits

Station identification (traffic burst only): 40 bits

Guard time: 32 bits

Voice channel bit rate: 32 kbps

Number of bursts: 2 reference bursts, 1 superframe short burst, and N traffic bursts

(a) Find the maximum number of traffic bursts N_{max}.

(b) Assume that each earth station maintains an idle-busy table for the frame. Find the total network traffic intensity so that the blocking probability at each station is 0.01.

7.13 Compare the TDMA networks in Probs. 7.11 and 7.12 in terms of usable satellite capacity.

7.14 A SCPC-DAMA network employs Aloha multiple access for its request channel. The network uses a capacity of 500 satellite channels to obtain a blocking probability of 0.1. Its request message is 120 bits and the end-of-call message from the traffic station to the master control station is 32 bits. Assume that the number of receivers in the network is 1500 and that the mean call holding time is 5 min. Find the bit rate of the request channel if it is designed to work at a throughput of 0.1.

7.15 Digital speech interpolation is used in the TDMA system in Prob. 6.5. where all the traffic is PCM voice at 64 kbps. Each DSI-TIM in the network accommodates 240 terrestrial channels and interpolates them into 127 satellite channels with one DSI-AC channel at 64 kbps. How many terrestrial channels can the satellite transponder accommodate?

Chapter 8

Satellite Packet Communications

The majority of satellite communication systems have been designed for voice and data traffic with fixed or demand assignment using a multiple access protocol such as FDMA or TDMA. Such systems thus work as circuit switching networks (a voice telephone network is an example of a circuit switching network) where a complete physical path is established from the sender to the receiver that remains in effect for the duration of the connection. The process of selecting a path or circuit establishment may take on the order of seconds for a complex network. Once the circuit is established, data transfer is continuous through the network, and no delays are added by the switches. End-to-end transmission time through the network is limited only by the propagation time of the circuit medium employed, which is dominated by the satellite propagation delay.

Circuit switching systems are efficient for voice calls or data with long messages compared to the time required to make new circuit allocations. Data traffic, however, has more diverse characteristics than voice traffic. In particular, data traffic generated in many data processing applications has great variability in its transmission requirements. The length of the message ranges from a single character to thousands of bytes. One such message is often made available instantly by a control signal and must be transported to the source within specific delay constraints. As such, data traffic in which a given data source duty cycle is low is often characterized as "bursty" (having a large peak-to-average ratio of the data rate). That is, if one were to observe the user's transmission for a period of time, one would see that he requires the communication resource infrequently, but when he does, he requires a rapid response. If fixed-assignment capacity allocation of the communication resource is employed, then each user must be assigned enough capacity to meet his peak transmission rate, with the consequence that the resulting channel utilization is low because of the large peak-to-average ratio of the data rate. To efficiently transmit bursty data traffic where a fast response is required, the data is formatted into one or more fixed-length *packets* which are routed through a *shared* communication resource by a sequence of node switches. Packet switching makes no attempt to store packets for a prolonged period of time while attempting delivery. Rather, packets are discarded if difficulties are encountered in their delivery, in which case they must be retransmitted by senders. Packet switching systems are designed to rapidly forward packets to their destination with the only delay in the node because of a finite transmission capacity.

The use of satellite packet switching for data traffic can have great economic advantages over conventional satellite circuit switching, especially when there are a large number of geographically distributed users. A shared broadcast satellite channel offers full connectivity between users within the satellite global beam, thus eliminating routing and node switches. Furthermore, each user can listen to her own transmission and thus receive automatic acknowledgment. This allows the implementation of special multiple access protocols for dynamic allocation of satellite capacity to all users to achieve statistical averaging of traffic loads. The key performance of a multiple access protocol for satellite packet communications is that of the satellite channel throughput versus the average packet delay characteristic. The *throughput* of a satellite channel is defined as the rate at which packets are successfully transmitted.

8.1 PRELIMINARIES

To study packet communications, the following traffic model is assumed:

1. Each user generates messages according to a *Poisson process* with an average arrival rate, equal to λ messages per unit time.
2. The message consists of one or more packets and has an average length of $1/\mu$ time units. Each packet carries a destination address so that, when it is transmitted over the satellite channel with no interference from other users, it will be received by the proper addressee.

It is known that telephone traffic can often be modeled as a Poisson process (Sec. 7.1). This too has been verified for data traffic. A message input to a system is characterized by its average arrival rate λ and its average length or service time $1/\mu$ (μ being the average service rate). The average arrival rate multiplied by the $1/\mu$ service time is called the *traffic intensity* and represents the average load to the system:

$$\rho = \frac{\lambda}{\mu} \tag{8.1}$$

The probability of the arrival of exactly k messages during an interval of length t is given by the Poisson law:

$$\Pr[k] = \frac{(\lambda t)^k}{k!} \exp(-\lambda t) \tag{8.2}$$

Queueing systems are commonly used to model processes in which messages arrive, wait in a buffer for service, are serviced by servers, and leave. Examples of queues are theater ticket lines and supermarket checkout cashier lines. Queueing systems are characterized by the arrival process (interarrival time probability density function, message length probability density function), service discipline (priority scheme), number of servers (outgoing trunks), and buffer size (finite, infinite). In this chapter we will concentrate on an infinite buffer and a single server using a first-come first-served discipline. Queueing systems are usually symbolized by the notation $A/B/C$, where A is the interarrival time, B is the message length distribution or service time distribution, and C is the number of servers. The distributions A and B can be of the following three types:

1. "M" stands for "Markov" and is used for Poisson arrival or the equivalent exponential distribution. (A Markov process is a stochastic process whose past has no influence on the future if its present is specified.) Note that Poisson arrivals generate an exponential probability density.
2. "D" stands for "deterministic" and is used for a constant service time.
3. "G" stands for "general" and is used for arbitrary distributions.

Thus an M/M/1 queue has a Poisson arrival, an exponential service distribution, and one server. An M/G/1 queue has a Poisson arrival, a general service distribution, and one server.

8.2 MESSAGE TRANSMISSION BY FDMA: THE M/G/1 QUEUE

In this section we analyze the *average message delay* versus the satellite channel *throughput* performance using fixed-assignment FDMA as a multiple access protocol. The analysis is based on the M/G/1 queue as shown schematically in Fig. 8.1, where messages arrive, according to a Poisson process, at the input of the buffer with infinite capacity. For gen-

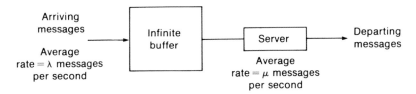

Figure 8.1 A M/G/1 queue.

erality we assume the message length or service time has a general distribution; that is, each is of a randomly varying length. The server works on one message at a time until completion, and then service begins on the next message (first-come first-served or first-in first-out basis).

Assume that the system is already in its steady state, that n messages exist in the buffer at the departure time t_0, and that one of n messages served at time t_0 departs after an interval t. Also, let k be the number of messages arriving during this interval t; then the number of messages existing in the buffer at the end of the interval t is

$$n' = \max \ (n - 1,0) + k$$
$$= n - 1 + \delta_n + k \tag{8.3a}$$

where

$$\delta_n = \begin{cases} 1 & n = 0 \\ 0 & n > 0 \end{cases} \tag{8.3b}$$

The expected (average) value of n' is expressed as

$$E\{n'\} = E\{n\} + E\{\delta_n\} + E\{k\} - 1 \tag{8.4}$$

By using (8.2), the expected value of k for a given interval t is

$$E\{k|t\} = \sum_{k=0}^{\infty} k \ \mathrm{Pr}[k]$$

$$= \sum_{k=0}^{\infty} k \ \frac{(\lambda t)^k}{k!} \ \exp(-\lambda t)$$

$$= \lambda t \tag{8.5}$$

Hence

$$E\{k\} = \int_0^{\infty} E\{k|t\} g(t) \ dt$$

$$= \int_0^{\infty} \lambda t g(t) \ dt = \frac{\lambda}{\mu} = \rho \tag{8.6}$$

where $g(t)$ = arbitrary probability density function of service time with mean (average) $1/\mu$ and variance σ^2.

Note that $\rho < 1$, otherwise the buffer will build up indefinitely and the system will become unstable. Using the steady-state condition, that is, $E\{n'\} = E\{n\}$, we have

$$E\{\delta_n\} = 1 - E\{k\} = 1 - \rho \tag{8.7}$$

When we square both sides of (8.3a), take the expectation, and rearrange terms, we obtain

$$E\{(k-1)^2\} + E\{\delta_n\} + 2E\{n(k-1)\} + 2E\{\delta_n(k-1)\} = 0 \tag{8.8}$$

taking into account the steady-state condition, that is,

$$E\{n'^2\} = E\{n^2\} \tag{8.9}$$

and that

$$\delta_n^2 = \delta_n$$

$$n\delta_n = 0$$

Also, we note that the messages arrive with a Poisson distribution; that is, k is independent of n or δ_n, hence

$$E\{n(k-1)\} = E\{n\}\ E\{k-1\} \tag{8.10}$$

$$E\{\delta_n(k-1)\} = E\{\delta_n\}\ E\{k-1\} \tag{8.11}$$

Substituting (8.7), (8.10), and (8.11) into (8.8) yields

$$E\{n\} = \rho + \frac{E\{k^2\} - \rho}{2(1-\rho)} \tag{8.12}$$

To evaluate the mean square value of k, that is, $E\{k^2\}$, we note that

$$E\{k^2\} = \int_0^\infty E\{k^2|t\}g(t)\ dt \tag{8.13}$$

and

$$E\{k^2|t\} = \sum_{k=0}^\infty k^2\ \frac{(\lambda t)^k}{k!}\ \exp(-\lambda t)$$

$$= \lambda t + (\lambda t)^2 \tag{8.14}$$

Therefore

$$E\{k^2\} = \lambda \int_0^\infty tg(t)\ dt + \lambda^2 \int_0^\infty t^2 g(t)\ dt$$

$$= \frac{\lambda}{\mu} + \lambda^2\left(\frac{1}{\mu^2} + \sigma^2\right) = \rho + \rho^2 + \lambda^2\sigma^2 \tag{8.15}$$

Hence

$$E\{n\} = \rho + \frac{\rho^2 + \lambda^2\sigma^2}{2(1 - \rho)} \qquad (8.16)$$

The result in (8.16) is known as the *Pollaczek-Khinchine* equation and represents the average number of messages waiting in the buffer including the one being served; it is often called the *average M/G/1 queue length* or the *average buffer occupancy.*

The average message delay (a message delay is defined as the time elapsing between the arrival of a message at the buffer and the departure of the complete message) can be found using Little's [1] result:

$$T = \frac{E\{n\}}{\lambda} = \frac{1}{\mu} + \frac{\rho^2 + \lambda^2\sigma^2}{2\lambda(1 - \rho)}$$

$$= \frac{1}{\mu} + \frac{\lambda(1/\mu^2 + \sigma^2)}{2(1 - \rho)} \qquad (8.17)$$

The average time spent in a queue waiting to be served or the waiting time of messages is simply the average message delay less the average service time; that is,

$$W = T - \frac{1}{\mu} = \frac{\rho^2 + \lambda^2\sigma^2}{2\lambda(1 - \rho)} = \frac{\lambda(1/\mu^2 + \sigma^2)}{2(1 - \rho)}$$

When the service time is exponentially distributed, that is, when $\sigma^2 = 1/\mu^2,$

$$T = \frac{1}{\mu(1 - \rho)} = \frac{1}{\mu - \lambda} \qquad (8.18a)$$

When the service time is constant, that is, when $\sigma^2 = 0,$

$$T = \frac{2 - \rho}{2\mu(1 - \rho)} = \frac{2 - \lambda/\mu}{2(\mu - \lambda)} \qquad (8.18b)$$

From (8.18) it is seen that the message delay increases quickly as ρ approaches 1.

Now consider a satellite channel of capacity R bits per second used in the FDMA mode by N users. Each user is assigned a channel of capacity R/N bits per second. Assume that the average message length is b bits per message; then the average service rate of the FDMA channel is

$$\mu = \frac{R}{Nb} \qquad (8.19)$$

Let λ be the average message arrival rate for each user. Then the traffic intensity for each channel is $\rho = \lambda/\mu.$

The message delay for the FDMA channel including the satellite roundtrip delay T_R may thus take on one of two models.

1. *An exponentially distributed message length:*

$$T_{\text{FDMA}} = \frac{1}{(R/Nb)(1-\rho)} + T_R$$

$$= \frac{1}{R/Nb - \lambda} + T_R \qquad (8.20)$$

2. *A constant message length:*

$$T_{\text{FDMA}} = \frac{2-\rho}{2(R/Nb)(1-\rho)} + T_R$$

$$= \frac{2 - \lambda/(R/Nb)}{2(R/Nb - \lambda)} + T_R \qquad (8.21)$$

Figure 8.2 plots the average message delay $T_{\text{FDMA}} - T_R$ versus traffic intensity ρ for a constant message length as functions of $Nb/R = 1/\mu$.

Example 8.1 Consider a FDMA system of 200 users sharing a satellite channel with a capacity of $R = 12.8$ Mbps. Assume that each user generates a constant-length message of 10^4 bits according to a Poisson process at a rate of three messages per second. (a) Each user is assigned a subchannel of capacity

$$\frac{R}{N} = \frac{12.8 \text{ Mbps}}{200} = 64 \text{ kbps}$$

(b) The average service rate of each FDMA channel is

$$\mu = \frac{R}{Nb} = \frac{64 \times 10^3}{10^4} = 6.4 \text{ messages per second}$$

(c) The average message delay for each user including a 270-ms satellite roundtrip delay is

$$T_{\text{FDMA}} = 495 \text{ ms}$$

8.3 MESSAGE TRANSMISSION BY TDMA

In this section we study the average message delay versus the traffic intensity for message transmission over a shared satellite channel using a TDMA protocol. For each user we assume that messages arrive according to a Poisson process with an average arrival rate of λ messages per second. Each message may consist of one or more fixed-length packets according to a general probability distribution. The TDMA frame length T_f is divided into N slots indexed from 1 to N as shown in Fig. 8.3. Time slots with the same index in consecutive frames form a TDMA channel. The satellite channel thus consists of N TDMA channels. A TDMA

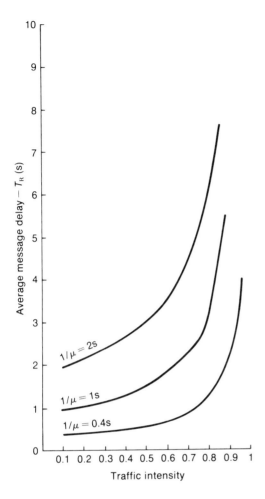

Figure 8.2 Average message delay of a FDMA channel (less the satellite roundtrip propagation delay T_R).

channel user transmits one packet to a time slot of T_f/N seconds (if the buffer is not empty) at a TDMA burst rate of R bits per second. The user then becomes idle for the next $N - 1$ time slots or $(N - 1)T_f/N$ seconds before it can transmit another packet. It can be easily seen that the data rate of a TDMA channel is R/N bits per second, even though its burst

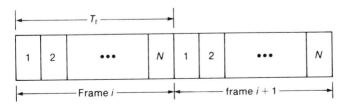

Figure 8.3 A TDMA frame with N slots.

bit rate is R bits per second. If a packet is chosen to have b bits, then its length is simply b/R seconds, which is equal to one time slot in the frame:

$$\frac{T_f}{N} = \frac{b}{R}$$

which yields

$$T_f = \frac{Nb}{R} \tag{8.22}$$

Thus given the TDMA channel capacity, the user population, and the number of bits per packet, the TDMA frame length can be determined. For example, let $R = 250$ kbps, $N = 100$, and $b = 1000$ bits; then $T_f = 0.4$ s. Note the similarity between a TDMA channel and a FDMA channel with the same channel capacity R, user population N, and number of bits per packet b. That is, they have the same channel capacity of R/N bits per second. Also, the average message length $1/\mu$ (8.19) of a FDMA channel is exactly equal to the frame length T_f of a TDMA channel.

To analyze the message delay in the TDMA channel assume that an arbitrary message consisting of k_j packets arrives at the station buffer t seconds before the end of a frame, where k_j is a random variable. It will take $t + (k_j - 1)T_f + T_f/N$ seconds to transmit this entire message. In addition to this delay, the newly arrived message also experiences a delay due to the transmission of previously arrived messages which consist of q packets held over from previous frame periods and L messages arriving during the present cycle in the time interval $T_f - t$.

It then takes qT_f seconds to transmit all the heldover packets from the previous frame periods, and $T_f \sum_{i=1}^{L} k_i$ seconds to transmit all L messages arriving in the interval $T_f - t$, where k_i is the number of packets in the ith of these L messages. The total delay of a newly arrived message including the satellite roundtrip delay T_R is

$$T = t + (k_j - 1)T_f + qT_f + T_f \sum_{i=1}^{L} k_i + \frac{T_f}{N} + T_R \tag{8.23}$$

Note that L and t are mutually dependent random variables; thus we have three independent random variables in the above expression, namely, $(k_j - 1)T_f$, qT_f, and $t + T_f\sum_{i=1}^{L} k_i$. The average message delay (Prob. 8.7) is given by

$$T_{\text{TDMA}} = \frac{1}{\mu} + \frac{\lambda(1/\mu^2 + \sigma^2)}{2(1 - \rho)} - \frac{T_f}{2} + \frac{T_f}{N} + T_R \tag{8.24}$$

where σ^2 = variance of message length and $\rho = \lambda/\mu$.

By comparing (8.24) and (8.17) we conclude that the average message delay for a TDMA channel can be expressed in terms of the average message delay for a FDMA channel of the same data rate R/N:

$$T_{\text{TDMA}} = T_{\text{FDMA}} - \frac{T_f}{2} + \frac{T_f}{N} \qquad (8.25)$$

In other words, in regard to message delay TDMA performs $T_f/2 - T_f/N$ seconds better than FDMA. Figure 8.4 shows the message delay (less the satellite roundtrip delay) versus the traffic intensity for various frame lengths.

So far we have considered conventional channel allocation schemes.

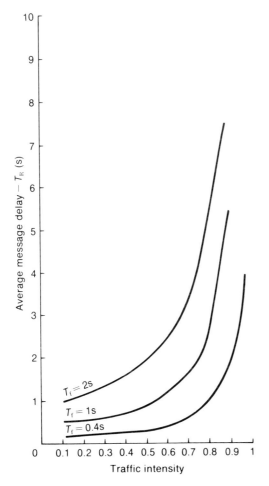

Figure 8.4 Average message delay of a TDMA channel (less the satellite roundtrip propagation delay).

It is seen that, when a user does not transmit data, the capacity is just wasted because no one else is allowed to use the channel. Furthermore, in computer systems interactive data traffic is very bursty (a peak-traffic-to-average-traffic ratio of 1000:1 is very common), and consequently the channels in FDMA or TDMA will be idle most of the time. In subsequent sections we will study random multiple access schemes that can resolve these problems.

8.4 PURE ALOHA: SATELLITE PACKET SWITCHING

The *Aloha* protocol is a random access scheme pioneered at the University of Hawaii for interconnecting terminals and computers via radio and satellites. In the Aloha system, a satellite channel of capacity R bits per second is shared by a *large population* of M users whose traffic is very bursty; that is, it has a high peak-to-average ratio and a low delay constraint. Each user station transmits packets "randomly" at the channel bit rate R whenever its buffer contains one packet. Each packet contains parity bits for error detection. Assume that the satellite channel has a *broadcast* capability (i.e., all stations are within its downlink antenna beam); then a station can receive its own transmitted packet on the downlink after a satellite roundtrip delay. If the previously transmitted packet is received correctly, assuming that the satellite link has a low error rate, the transmit station can assume that the packet has been received correctly at the destination station and consider the transmission successful. In the situation where packets from different stations overlap at the satellite channel (called *packet collision*) the transmission error can be detected at the transmit stations on the downlink. The stations then retransmit the packets until they are free from overlap. If two packets from two transmit stations collide at the satellite, they will surely collide again if they are retransmitted after a fixed time-out. To avoid repeated collisions, the interval of packet retransmission is randomized for each station. The Aloha protocol is shown schematically in Fig. 8.5.

To analyze the average packet delay versus the satellite channel throughput in the Aloha system, we assume an infinitely large number of user stations that collectively generate packets according to a Poisson process with rate λ_c packets per second. Also, we let the packet length be τ seconds; then the average channel input rate or channel throughput is

$$S = \lambda_c \tau \text{ (packets per packet length)} \qquad (8.26)$$

It is seen that $0 \leq S < 1$ because, if $S > 1$, the user population will be generating packets at a rate higher than that which the channel can handle and nearly every packet will collide. An infinite population assumption is necessary to ensure that S does not decrease as users wait to find out

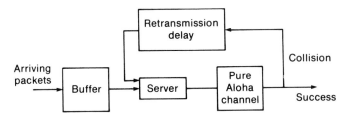

Figure 8.5 Representation of an Aloha multiple access protocol.

whether their packets have been successfully transmitted. In addition to the newly arrived packets, the satellite channel also contains retransmitted packets. Let G denote the average satellite channel traffic (newly arrived and retransmitted packets) in packets per packet length and assume that this traffic is also a Poisson process with mean G (this is true if the randomized retransmission delay is sufficiently large). Then the probability that k packets will arrive at the satellite channel during any interval of t packet lengths is

$$\Pr [k,t] = \frac{(Gt)^k}{k!} \exp(-Gt) \tag{8.27}$$

Assume that even a partial overlap may cause a collision, as shown in Fig. 8.6; then the probability that no collision will occur when a packet is transmitted is exactly the probability that no other packet will be generated during an interval of two packet lengths. Thus from (8.27) we have

Pr[newly generated packet is successfully transmitted]

$$= \Pr [k = 0, t = 2] = \exp(-2G) \tag{8.28}$$

Since the channel throughput S is just the channel traffic G times the probability that a newly generated packet will not suffer a collision, we

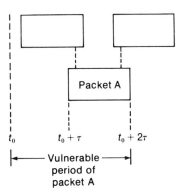

Vulnerable period of packet A

Figure 8.6 Vulnerable period (two packet lengths) of an Aloha protocol.

have

$$S = G \exp(-2G) \tag{8.29}$$

Based on (8.29) it is noted that the *maximum throughput* occurs at a channel traffic of $G = 0.5$:

$$S_{\max} = \frac{1}{2e} = 0.184 \tag{8.30}$$

This shows that the channel throughput of the Aloha system is very poor, but this is expected since every user is allowed to transmit at will. But, as will be seen later, the Aloha protocol is more appropriate for serving a large number of earth stations whose traffic is very bursty and when satellite channel capacity is limited. In these situations, the average packet delay of the Aloha system is much better than that of a TDMA or FDMA system. A plot of the Aloha channel throughput versus the channel traffic is shown in Fig. 8.7.

From (8.29) it is seen that $G = f(S)$ is a double-valued function; for a given value of S, there are two values of G, namely, G and $G' > 0.5 > G$, such that $S = Ge^{-2G} = G'e^{-2G'}$. This indicates that, as the channel traffic increases past 0.5, the throughput drops because the number of packets that suffer a collision increases (which means more packet retransmission). Hence there is a further increase in channel traffic and consequently a decrease in channel throughput, creating a runaway effect. This instability is an inherent characteristic of the Aloha channel and can be prevented only by operating it well below the maximum throughput with enough margin for peak traffic or with some sort of control. The latter method will be studied in the next section when we deal with the slotted Aloha channel.

The average packet delay in an Aloha channel consists of the service time τ (packet length), the average retransmission delay $E\{T\}$, and the satellite propagation delay T_R:

$$T_{\text{Aloha}} = T_R + \tau + E\{T\} \tag{8.31}$$

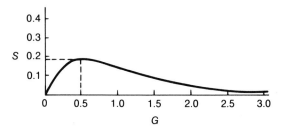

Figure 8.7 Throughput versus channel traffic for an Aloha channel.

Since τ and T_R are known, it remains to evaluate the average retransmission delay $E\{T\}$. As mentioned previously, if two packets collide at the satellite channel, each station involved must initiate a retransmission. If the timing of each retransmission is the same, then the collision will persist. Thus some strategy must be used to avoid persistent collisions. One strategy is to assign each station a fixed time-out delay. This approach has the advantage that it completely avoids persistent collisions. It has the disadvantage that some stations will experience a large delay. The second strategy is to use a randomized retransmission approach, where two interfering stations select a retransmission interval from a random sequence of retransmission delays. If each station has a different sequence, then there will be a low probability of persistent collisions. This approach has an advantage over the fixed time-out approach in that it shortens the retransmission delay, but it has the disadvantage that there is a nonzero probability of repeated collisions.

Consider the randomized retransmission strategy where the random time delay introduced is uniformly distributed over 1 to K intervals of τ seconds each (i.e., the number of intervals between the first and second transmissions may be $T_R/\tau + 1, T_R/\tau + 2, \ldots, T_R/\tau + K$, each with probability $1/K$). The average delay before retransmission is $(K + 1)\tau/2$, and the retransmission delay of a packet after r retransmission is

$$T = r\left[T_R + \frac{(K + 1)\tau}{2}\right] \tag{8.32}$$

Let Q_r be the probability of a successful retransmission after r retries; then the average number of retransmissions of a packet is

$$E\{r\} = \sum_{r=1}^{\infty} rQ_r \tag{8.33}$$

and the average retransmission delay of a packet is

$$E\{T\} = E\{r\}\left[T_R + \frac{(K + 1)\tau}{2}\right] \tag{8.34}$$

Now let q be the probability of a successful transmission given that a new packet has been generated and q' be the probability of a successful transmission given that a retransmitted packet is transmitted. Then it is seen that, for $r \geq 1$,

$$Q_r = (1 - q)(1 - q')^{r-1}q' \tag{8.35}$$

Substituting (8.35) into (8.33) yields

$$E\{r\} = \frac{1 - q}{q'} \tag{8.36}$$

If we assume that the probability of success is the same on any try, that is,

$q = q'$, which is reasonably correct for sufficiently large K, then

$$E\{r\} \approx \frac{1 - q}{q} \qquad (8.37)$$

From (8.28) we have, for $K \gg 1$,

$$q \approx \exp(-2G) \qquad (8.38)$$

Therefore

$$E\{r\} \approx \exp(2G) - 1 \qquad (8.39)$$

Thus the average retransmission delay of a packet is

$$E\{T\} \approx [\exp(2G) - 1]\left[T_R + \frac{(K + 1)\tau}{2}\right]$$

Putting all these results together we obtain the average packet delay as

$$T_{\text{Aloha}} \approx T_R + \tau + [\exp(2G) - 1]\left[T_R + \frac{(K + 1)\tau}{2}\right] \qquad K \gg 1 \quad (8.40)$$

Note that, since the channel traffic G is a double-valued function of the channel throughput S, there are two average packet delays T_{Aloha} and $T'_{\text{Aloha}} > T_{\text{Aloha}}$ corresponding to channel traffic $G < 0.5$ and $G' > 0.5$. Note that T_{Aloha} can become arbitrarily large when the Aloha channel traffic exceeds 0.5, creating a runaway effect which further increases G if the peak load persists.

A plot of the average packet delay versus the channel throughput of an Aloha system is shown in Fig. 8.8, with K as a parameter, for $R = 250$ kbps, $b = 1000$ bits, $\tau = b/R = 4$ ms, $T_R = 250$ ms, and user population $N = \infty$. Note that the average packet delay does not change for values of K between 10 and 50, which means that the actual value of K selected is not critical. The average packet delays using FDMA and TDMA protocols are also plotted for comparison using $N = 1500$. Note the clear advantages of the Aloha channel in terms of the packet delay.

8.5 SLOTTED ALOHA

The advantage of the pure Aloha protocol is its simplicity, which can result in low-cost user stations since no synchronization is required between stations in the system. Each station transmits a packet whenever its buffer has one. The disadvantage is somewhat inefficient channel utilization; that is, the maximum channel throughput is only 18.4% of the available capacity. One strategy for improving pure Aloha channel throughput is to coordinate transmissions between stations by synchronizing the

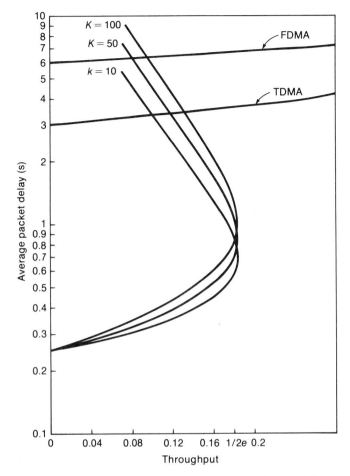

Figure 8.8 Average packet delay versus throughput for an Aloha channel ($R = 250$ kbps, $b = 1000$ bits, $T_R = 250$ ms).

transmit and retransmit timing as in TDMA, in effect providing time slots of the same duration as a packet length. A station can transmit a packet only at the start of a slot. Thus when two packets collide at the satellite channel, they will overlap completely as shown in Fig. 8.9, where the vulnerable period of a packet is one packet length τ. Partial overlaps like those occurring in a pure Aloha channel will never occur. This modified Aloha scheme is called *slotted Aloha* and is a synchronized random access method.

If we assume that the user population N is infinitely large, as in the pure Aloha analysis it can be shown that the slotted Aloha throughput is

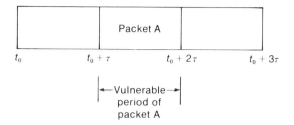

Figure 8.9 Vulnerable period (one packet length) of a slotted Aloha protocol.

related to the channel traffic by

$$S = G \exp(-G) \qquad (8.41)$$

This is because the channel throughput is just the channel traffic G times the probability that a newly generated packet will not suffer a collision (is successfully transmitted) where, as is seen from (8.27),

Pr[newly generated packet is successfully transmitted]

$$= \Pr[k = 0, t = 1]$$

$$= \exp(-G) \qquad (8.42)$$

The slotted Aloha *maximum throughput* occurs at $G = 1$:

$$S_{\max} = \frac{1}{e} = 0.368 \qquad (8.43)$$

which is twice the maximum capacity of the pure Aloha channel. A plot of the channel traffic G versus the throughput S is shown in Fig. 8.10.

The slotted Aloha channel can also be analyzed for a finite earth station population N. Let G_i be the probability that a station will transmit a packet in a given slot. The average traffic due to the ith station is therefore

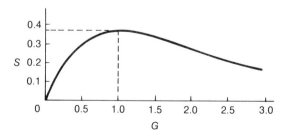

Figure 8.10 Throughput versus channel traffic for a slotted Aloha channel.

G_i. Hence the average channel traffic is

$$G = \sum_{i=1}^{N} G_i \tag{8.44}$$

Let S_i be the probability that a newly generated packet will be successfully transmitted. Then the average throughput due to the ith station is S_i, and the channel throughput is

$$S = \sum_{i=1}^{N} S_i \tag{8.45}$$

The probability S_i that a packet from station i will be successfully transmitted is simply

$$S_i = G_i \prod_{\substack{j=1 \\ j \neq i}}^{N} (1 - G_j) \tag{8.46}$$

When all the stations are statistically identical, that is, $S_i = S/N$ and $G_i = G/N$, we have

$$S = \sum_{i=1}^{N} S_i = \sum_{i=1}^{N} G_i \prod_{\substack{j=1 \\ j \neq i}}^{N} (1 - G_j)$$

$$= G\left(1 - \frac{G}{N}\right)^{N-1} > G \exp(-G) \tag{8.47}$$

Note that, as $N \to \infty$, $(1 - G/N)^{N-1} \to \exp(-G)$ and $S = G \exp(-G)$, which is the case of an infinite population.

To analyze the average packet delay of the slotted Aloha channel, we note that in the pure Aloha operation a packet arriving at an empty buffer is transmitted immediately; however, in the slotted channel, this situation occurs with a probability of zero; that is, a newly arrived packet must await the beginning of a slot to be transmitted. Given that a packet arrival occurring in a time slot is uniformly distributed over the interval $(0,\tau)$, where τ is the duration of the slot or packet length, then the additional average waiting time is simply $\tau/2$. This also applies to any retransmitted packet. Thus the average packet delay of the slotted Aloha channel for an infinitely large number of users can be expressed similarly to that for the pure Aloha channel:

$$T_{\text{s-Aloha}} = T_R + \tau + \frac{\tau}{2} + \frac{1-q}{q'}\left[T_R + \frac{(K+1)\tau}{2} + \frac{\tau}{2}\right] \tag{8.48}$$

For $K \gg 1$, $T_{\text{s-Aloha}}$ can be approximated as

$$T_{\text{s-Aloha}} = T_R + \frac{3\tau}{2} + [\exp(G) - 1]\left[T_R + \frac{(K+2)\tau}{2}\right] \qquad K \gg 1 \tag{8.49}$$

In reference to (8.49), it is noted that there are two possible delays for a given S and K. They correspond to a small delay $T_{s-Aloha}$ and a large delay $T'_{s-Aloha}$. We will refer to the equilibrium point given by S_0, K, and $T_{s-Aloha}$ as the *channel operating point,* since this is the desired channel performance given S_0 and K. The existence of two channel equilibrium points suggests that the slotted Aloha channel (or pure Aloha channel) is inherently *unstable.* Starting with an initially empty system (no traffic) the channel moves toward equilibrium. It stays at the channel operating point for a finite period until stochastic fluctuations gives rise to high channel traffic and push the channel throughput toward its maximum value, consequently reducing it; this in turn increases the channel traffic, hence the packet delay. As the fluctuations continue, the channel becomes overloaded with retransmitted packets, the throughput vanishes to zero, and channel saturation occurs. Thus the delay-versus-throughput performance at equilibrium as studied above is not sufficient to characterize the dynamic behavior of the slotted Aloha channel. An accurate measure of the channel's performance must include a stability study of the delay-throughput trade-off. This is what we will consider in the following discussion.

To study the *delay-throughput-stability* performance we will adopt the following model. Consider N earth stations, each of which can be in one of two states: *thinking* or *blocked.* In the thinking state, a station transmits newly generated packets according to a *Poisson distribution* with a mean of $\alpha << 1$ packet per slot (one slot is a τ-second interval where τ is the packet length); that is, α can be thought of as the probability that a given station will transmit a newly generated packet in a given slot. If the *mean think time* (interarrival time) of a packet is t_a, then $\alpha = \tau/t_a$. From the time the station transmits a newly generated packet to the time the packet is successfully received, which includes retransmission time if the packet suffers a collision, the station is blocked in the sense that it cannot transmit any newly generated packet. A packet which suffers a collision at the channel and is waiting for retransmission is said to be *backlogged.* It is this waiting time that determines the performance of the channel. As in the previous study of delay-throughput performance, the waiting time consists of K slots where the retransmission time is uniformly distributed. Including the satellite roundtrip delay of T_R/τ slots, the number of slots between the first and second retransmissions may be $T_R/\tau + 1$, $T_R/\tau + 2, \ldots,$ $T_R/\tau + K$, each with probability $1/K$. If K is small, the chance of recollision is large, but the average retransmission delay is small; for example, if two stations have collided packets and each must wait either 0 or 10 slots with equal probability, then the chance that the two packets will collide a second time is 0.1. On the other hand, if the retransmissions are spread out uniformly over the next 50 slots, the chance that the two packets will collide again in 0.02. Of course, the average retransmission delay in the latter case is larger than in the former.

This is the delay-throughput trade-off of the slotted Aloha (or pure Aloha) system.

Unfortunately, the randomization retransmission strategy is difficult to analyze because of the inclusion of the satellite roundtrip delay T_R/τ slots, which necessitates a model with memory consisting of a channel history of at least T_R/τ consecutive slots. Instead, a probabilistic model called a *Markov model* is adopted, in which a backlogged packet is retransmitted with probability p in a given slot following the original transmission until it is successfully received. The average delay before retransmission is now assumed to be geometrically distributed with the probability of an n-slot delay given by $p(1-p)^{n-1}$. The average delay before retransmission attempts is

$$\sum_{n=1}^{\infty} np(1-p)^{n-1} = \frac{1}{p} \tag{8.50}$$

Note that the average retransmission delay (the time the station is blocked), which is denoted by $E\{T\}$ and henceforth will be called the average backlog time, is different from $1/p$ since a packet can be retransmitted many times.

The assumption that the delay before retransmission attempts has a memoryless geometric distribution ($T_R = 0$) permits an analytical study using a Markov model. Simulation has shown that the slotted Aloha channel performance in terms of average delay and throughput is dependent primarily on the average delay between retransmission attempts and is quite insensitive to the exact probability distribution considered. In order to use the analytic result of the Markov model to analyze the delay-throughput performance with $T_R \neq 0$ to approximate the slotted Aloha channel with randomization retransmission uniformly distributed over K slots, it is necessary to equate the average delay $1/p$ with the average randomization retransmission delay; that is, we must let (in the subsequent discussion we will use a slot instead of a second as the unit of time)

$$\frac{1}{p} = \frac{T_R}{\tau} + \frac{1}{2} + \frac{K+1}{2}$$

$$= \frac{T_R}{\tau} + \frac{K+2}{2} \tag{8.51}$$

Thus, if the value of p is determined for a channel operating point, the corresponding value of K can be derived, and vice-versa.

The slotted Aloha channel state at any time t can be described by the total number of backlogged packets. In state n there are n packets backlogged with retransmission probability p in the current slot, yielding an average retransmission traffic of np packets per slot. Besides the n blocked stations (each with one backlogged packet), there are $N - n$ thinking stations which are busy transmitting newly generated packets at a collective rate of $(N - n)\alpha$ packets per slot. The quantity $S = (N - n)\alpha$ is called the *channel input rate* and must be equal to the average channel throughput (the fraction of slots that contain exactly one packet) at equilibrium.

Now consider the behavior of the Markov process. Let r be the number of backlogged packets retransmitted from i blocked stations in a current slot, with $0 \le r \le i$. Also, let m be the number of newly generated packets transmitted by $N - i$ thinking stations during the same slot, with $0 \le m \le N - i$. The transition probability p_{ij} that the channel in state i will move to state j in the next slot is given by

$$p_{ij} = \begin{cases} 0 & j \le i - 2 \\ \Pr[m = 0]\ \Pr[r = 1] & j = i - 1 \\ \Pr[m = 0]\ \Pr[r \ne 1] + \Pr[m = 1]\ \Pr[r = 0] & j = i \\ \Pr[m = 1]\ \Pr[r \ge 1] & j = i + 1 \\ \Pr[m = j - i] & j \ge i + 2 \end{cases}$$

(8.52)

where $\Pr[m = x]$, $x = 0, 1, \ldots, N - i$, is the probability that x newly arrived packets will be transmitted in the current slot and $\Pr[r = 0]$, $\Pr[r = 1]$, $\Pr[r \ge 1]$, and $\Pr[r \ne 1]$ represent the probabilities that none, one, one or more than one, and none or more than one blocked stations will attempt a retransmission in the current slot. These probabilities are given below for the case of finite N and α and for the case of $N \to \infty$ and $\alpha \to 0$ such that $N\alpha = S < \infty$. (In fact, $N\alpha < 1$ since the total rate of newly generated packets must always be less than one packet per slot.)

Finite case		Infinite case
$\Pr[m = 0]$	$= (1 - \alpha)^{N-i}$	$\exp(-S)$
$\Pr[m = 1]$	$= (N - i)\,\alpha\,(1 - \alpha)^{N-i-1}$	$S\exp(-S)$
$\Pr[m = j - i]$	$= \binom{N - i}{j - i} \alpha^{j-i}(1 - \alpha)^{N-j}$	$\dfrac{S^{j-i}}{(j - i)!}\exp(-S)$
$\Pr[r = 0]$	$= (1 - p)^i$	$(1 - p)^i$
$\Pr[r = 1)$	$= ip(1 - p)^{i-1}$	$ip(1 - p)^{i-1}$
$\Pr[r \ge 1]$	$= 1 - (1 - p)^i$	$1 - (1 - p)^i$
$\Pr[r \ne 1]$	$= 1 - ip(1 - p)^{i-1}$	$1 - ip(1 - p)^{i-1}$

Assume that N and α are both time-invariant; the Markov process that represents the number of blocked stations n can be described by a stationary probability distribution P_n, that is, the probability of finding the system in equilibrium state n (time-invariant). P_n can be found from a set of linear $N + 1$ dependent equations:

$$P_n = \sum_{i=0}^{N} P_i\, p_{in} \qquad n = 0, 1, \ldots, N$$

subject to the constraint

$$\sum_{i=0}^{N} P_i = 1$$

which means that the probability that the system will move from its current state to some state is unity.

Given the equilibrium state probabilities, the average number of blocked stations (or the average backlogged packets) can be obtained from

$$E\{n\} = \sum_{n=0}^{N} nP_n$$

The throughput of the channel in any state n, $S(n)$, is the probability that a packet will be successfully sent, given that n stations are blocked and $N - n$ stations are thinking. A successful transmission occurs when either exactly one thinking station transmits a packet and none of the blocked stations attempts a retransmission or exactly one blocked station attempts a retransmission and none of the thinking stations transmits a newly arrived packet. Therefore $S(n)$ can be expressed as

$$S(n) = \Pr[r = 0]\ \Pr[m = 1] + \Pr[r = 1]\ \Pr[m = 0] \qquad (8.53)$$

For the finite case, $N < \infty$, $\alpha > 0$,

$$S(n) = (1 - p)^n (N - n)\alpha(1 - \alpha)^{N-n-1} + np(1 - p)^{n-1}(1 - \alpha)^{N-n} \tag{8.54}$$

For the infinite case $N \to \infty$, $\alpha \to 0$, $(N - n)\alpha \to N\alpha = S$,

$$S(n) = (1 - p)^n S \exp(-S) + np(1 - p)^{n-1} \exp(-S) \qquad (8.55)$$

The steady-state channel throughput S is simply the sum of the throughput of each state weighted by the equilibrium state probability; that is,

$$S = \sum_{n=0}^{N} P_n S(n)$$

By using Little's result, the average backlog time (average retransmission time) in slots is given by

$$\frac{E\{n\}}{S} = \frac{\displaystyle\sum_{n=0}^{N} nP_n}{\displaystyle\sum_{n=0}^{N} P_n S(n)}$$

Using the previous notation, the average backlog time $E\{T\}$ in seconds is

$$E\{T\} = \tau \frac{E\{n\}}{S} = \tau \frac{\displaystyle\sum_{n=0}^{N} nP_n}{\displaystyle\sum_{n=0}^{N} P_n S(n)} \tag{8.56}$$

Consequently, the average packet delay $T_{\text{s-Aloha}}$ can be expressed as

$$T_{\text{s-Aloha}} = T_R + \frac{3\tau}{2} + E\{T\}$$

$$= T_R + \frac{3\tau}{2} + \tau \frac{\displaystyle\sum_{n=0}^{N} nP_n}{\displaystyle\sum_{n=0}^{N} P_n S(n)} \tag{8.57}$$

Normally the computation of the equilibrium state probabilities P_n, $n = 0, 1, \ldots, N$, is quite involved. Simulation has shown that S and $E\{n\}$ are closely approximated by the throughput S_0 and backlog n_0 at the channel operating point for a stable system, which we will discuss next.

As shown above, the channel throughput $S(n)$ at any state n, which represents the number of blocked stations, varies with the retransmission probability p. Assuming that N and α are time-invariant, a plot of $S(n)$ in packets per slot versus the backlog n with K (hence p) as a parameter is shown in Fig 8.11 for $T_R = 0.250s$, $\tau = 4ms$ ($T_R/\tau = 62.5$ slots), $N = 500$, and $\alpha = 7.36 \times 10^{-4}$. Also plotted in Fig. 8.11 is the *channel load line*, representing the channel input rate versus the backlog, $S = (N - n)\alpha$. The channel load line intersects the n axis at N and has a slope of $-1/\alpha$. At equilibrium the channel throughput must be equal to the channel input rate. Thus when the S axis and the $S(n)$ axis have the same unit scale, the intersections of the curve $S(n)$ versus n and the channel load line represent the equilibrium points. Four cases are considered in Fig. 8.12. In Fig. 8.12a the channel load line intersects the throughput-backlog curve at exactly one point. The channel is said to be stable, and the channel operating point is said to be a globally stable equilibrium point. The arrows on the channel load line point in the direction of decreasing backlog size, since the channel throughput is always greater than the channel input rate. If N is finite, a stable channel can always be achieved by using a sufficiently small p or, equivalently, a sufficiently large K. Of course, a large K implies that the equilibrium backlog n_0 is large, and so is the average packet delay. In Fig. 8.11 the globally stable equilibrium point is obtained for $K = 100$ at

$$n_0 \approx 37$$

$$S_0 \approx 0.342$$

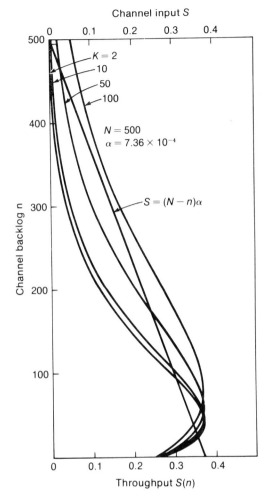

Figure 8.11 Channel backlog versus throughput for a slotted Aloha channel ($T_R = 62.5$ slots, $N = 500$, $\alpha = 7.36 \times 10^{-4}$).

which yields the average backlog time (average retransmission time) in seconds as

$$E\{T\} = \tau \frac{n_0}{S_0} = 0.004 \frac{37}{0.342} = 0.433 \text{ s}$$

hence the average packet delay time is

$$T_{\text{s-Aloha}} = T_R + \frac{3\tau}{2} + E\{T\} = 0.250 + 0.006 + 0.433$$

$$= 0.689 \text{ s}$$

A necessary condition for the stable channel in Fig. 8.12*a* is that the max-

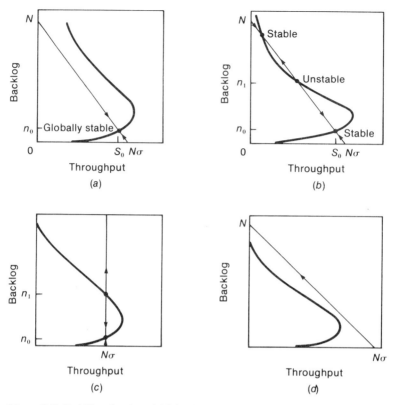

Figure 8.12 Stability of a slotted Aloha channel. (a) Globally stable. (b) Bistable. (c) Unstable. (d) Overloaded.

imum channel input rate be below the maximum channel throughput, that is,

$$S_{max} = N\alpha \leq \left(1 - \frac{1}{N}\right)^{N-1} \approx 0.368$$

Thus for a given α (the probability that a newly generated packet will be transmitted in a time slot) the upper bound on the number of stations is given by

$$N \leq \frac{S_{max}}{\alpha} \approx \frac{0.368}{\alpha}$$

For example, if a packet contains $b = 1000$ bits and the channel capacity is $R = 250$ kbps, then each time slot (one packet length) is 4 ms long. For an average thinking time (packet interarrival time) $t_a = 10s$, $\alpha = \tau/t_a = 4 \times 10^{-4}$, and $N \leq 0.368/4 \times 10^{-4} = 920$.

In Fig. 8.12b, the channel load line intersects the throughput-backlog curve at three points which can be considered equilibrium points. But only two of them are stable equilibrium points which we refer to as local stable equilibrium points. One stable equilibrium point corresponds to a low backlog (hence a short delay) and is the desired channel operating point; the other corresponds to a huge backlog where the channel is almost saturated. The third equilibrium point is unstable in the sense that the backlog at this point will remain there for a finite but short period of time. If the channel backlog is close to n_0, the channel will remain at its operating point. Since n is a random process the channel cannot maintain the local stable equilibrium (S_0, n_0) for an infinite time. Fluctuations in the input rate will drive the channel toward the unstable equilibrium point. As soon as the backlog surpasses n_1, the channel input rate will exceed the channel throughput, the backlog will keep building up, and the channel will settle at the high-backlog local stable equilibrium point for a finite but probably very long period of time. For this case (Fig. 8.12b) the channel is said to be bistable, as shown in Fig. 8.11 for $K = 2,10,50$.

The case in Fig. 8.12c corresponds to a channel with $N \to \infty$, $\alpha \to 0$, and $N\alpha \leq S_{max} = 0.368$. This is an unstable channel, since it will certainly fail as soon as n surpasses n_1 and the backlog will grow without bound.

Figure 8.12d shows the case of an overload channel where the input rate is always greater than the throughput. This system certainly will fail as soon as it starts up, and the only way to correct it is to reduce N.

In designing a stable slotted Aloha channel for a given average retransmission delay, that is, for a fixed K or p, the designer has no choice but to limit the number of stations N such that the channel load line intersects the throughput-backlog curve at only one point, assuming that $N\alpha$ will never exceed its maximum value of 0.368. This results in inefficient utilization of the channel capacity per station. As an example, for $K = 50$ in Fig. 8.11, the number of earth stations in the network must be reduced to about $N = 425$ to obtain a stable channel as compared to the maximum attainable $N = 500$ which can be achieved only if K is raised to about 100 (giving some safety margin). The probelm can be solved by some sort of adaptive control. Since K or, equivalently, the retransmission probability p actually determines the stability of the channel, it would be ideal if K could be increased (or p could be decreased) to make the channel input rate less than the maximum throughput when the channel backlog is high so as to reduce it, and if K could be decreased (or p could be increased) when the channel backlog is low to achieve a shorter delay.

One way to control the channel is to estimate the channel traffic G. Recall that the channel throughput is maximum at $G = 1$. When $G > 1$, the channel throughput drops and can fall below the channel input rate, creating a higher backlog; thus K should be increased by the stations. When $G < 1$ the channel throughput exceeds the channel input rate and

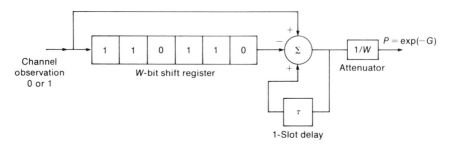

Figure 8.13 Controlled slotted Aloha protocol.

K should be decreased to reduce the delay. To estimate G, note that the probability that a packet will be successfully transmitted is $P_0 = \exp(-G)$ or $G = -\ln P_0$. P_0 can be estimated by maintaining the channel history for W slots and recording only the fraction of empty slots (no collisions or transmissions). The size of W for each station is very important for successful estimation. If W is too large, the channel behavior may change when action is taken. If W is too small, large errors may occur in approximating the probability of zero channel traffic based on the fraction of empty slots. A good initial estimate is that W should be larger than the satellite roundtrip propagation delay T_R/τ slots. Since it is only necessary to record the empty slots, a W-bit shift register as shown in Fig. 8.13 would suffice. The stored bit string represents the channel history in a window of W slots. An empty channel slot is represented by 1, while a nonempty slot is represented by 0. The value of p can be selected as $p = e^{-G}/(T_R/\tau + 1)$, which decreased ($K$ increases) when $G > 1$ and increases (K decreases) when $G < 1$. This gives $K = 2(e^G - 1)(T_R/\tau + 1)$ as seen from (8.51).

8.6 PACKET RESERVATION

We have studied in detail two main types of satellite multiple access schemes for packet transmission: channel fixed assignments such as FDMA and TDMA which are appropriate for a small user population with high traffic density but perform poorly for a large population of bursty users, and random access systems such as pure Aloha and slotted Aloha which are independent of the user population but yield poor throughput.

To overcome the drawbacks of the above two schemes, a number of protocols called *packet reservation* have been proposed that work like the slotted Aloha when the channel throughput is low and move gradually to

TDMA when the channel throughput exceeds $1/e$. Packet reservation, like random access, is intended for use when a satellite channel is shared by a large number of earth stations. The satellite capacity is demand-assigned to individual packets or messages from earth stations. The objective of reservation is to avoid the collisions of packets that occur with random access, and the price paid is an increase in the packet or message delay of at least twice the satellite roundtrip delay excluding the service time (one roundtrip delay due to making a reservation if it is successful). To coordinate all the stations, a global queue is maintained for satellite channel access. When it has a packet in the buffer, each station generates a request to reserve a place in the queue. A fraction of the satellite capacity is used to accommodate reservation request traffic. Since the earth stations are geographically distributed, the multiple access problem in the reservation request channel still exists, along with the problem of maintaining information on the status of the global queue to make decisions on when to access the channel. Readers may refer to [2–5] for more information.

8.7 TREE ALGORITHM

Recall that pure Aloha and slotted Aloha are random multiple access protocols where the conflict (packet collision) is resolved by randomized retransmissions. In this section we present an alternative algorithm called a *tree algorithm* for resolving the conflict. As in slotted Aloha, the channel we consider is divided into slots of equal length. The length of each slot equals the packet length. Also, we consider the satellite to have a broadcast capability, hence each station can detect a packet collision in any slot. Also, we assume that the user station population is infinitely large and that packets arrive at the satellite channel according to a Poisson process with an average arrival rate of S packets per slot. Thus S is the channel input rate. Since the number of users is infinite but the channel input rate is finite, the probability that any station will have more than one packet arrival in a finite interval is zero. Also, if a second packet arrives at a station, it will not be transmitted until the first packet has been successfully transmitted.

The tree algorithm for conflict resolution is based on the observation that the contention among active stations (stations with one packet to transmit) can be resolved if all the stations can be divided into groups such that each group contains only one active station. The tree algorithm can be best described by an example.

First consider a binary tree where each node consists of only two branches and all the stations in the network are arranged like the leaves of the tree, as shown in Fig. 8.14 for eight stations $(X_0, X_1, X_2, \ldots, X_7)$. The

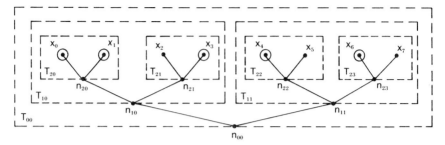

Figure 8.14 A binary tree.

satellite channel time is slotted as shown in Fig. 8.15 and the slots are paired. Each slot pair SL_{ij} is separated by T_R/τ slots, where T_R is the satellite roundtrip delay and τ is the slot length and both are in seconds. Note that $\tau = b/R$, where R is the channel capacity in bits per second and b is the packet length in bits. In the binary tree algorithm, the stations in the tree T_{ij} transmit in the slot pair SL_{ij}. Now assume that no collision exists until time t_0, the start of slot pair SL_{00} where stations X_0, X_1, X_3, X_4, and X_6 become active. Then the binary tree algorithm takes the following steps in the indicated slot pairs:

SL_{00}: All the active stations in tree T_{00} (i.e., X_0, X_1, X_3, X_4, and X_6) transmit is SL_{00}. Active stations in tree T_{10} (i.e., X_0, X_1, and X_3) transmit in the first slot, and active stations in tree T_{11} (i.e., X_4 and X_6) transmit in the second slot. This results in two collisions, one between X_0, X_1, and X_3 and the other between X_4 and X_6. Since there is at least one collision in SL_{00}, new packets generated at these active stations will not be transmitted until the conflict is resolved.

SL_{10}: Since there is a collision in tree T_{10}, T_{10} transmits in the slot pair SL_{10}. Active stations in tree T_{20} (i.e., X_0 and X_1) transmit in the first slot, causing a collision, while the only active station

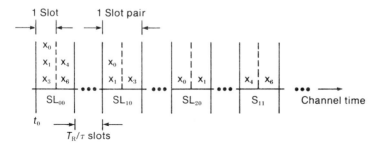

Figure 8.15 Slotted pairs in the tree algorithm.

X_3 in tree T_{21} transmits in the second slot, resulting in a successful transmission.

SL$_{20}$: A collision causes tree T_{20} to transmit in slot pair SL$_{20}$ so that active stations X_0 and X_1 transmit in the first and second slots, respectively, resulting in two successful transmissions.

SL$_{11}$: Since there is a collision in tree T_{11}, trees T_{22} and T_{23} transmit in the first and second slots of SL$_{11}$, resulting in two successful transmissions by stations X_4 and X_6. The process is repeated for newly generated packets.

For the binary tree algorithm it has been shown that the average *throughput* (the fraction of slots that contain exactly one packet each when the channel is at equilibrium) S of the satellite channel can achieve a maximum of 0.347 packet per slot [6]. Note that at equilibrium the average throughput is equal to the average channel input rate. The average packet delay T versus the average channel input rate S for the tree algorithm is normally expressed in terms of the number of algorithm steps (an algorithm step is equal to a satellite roundtrip delay plus two slots). There is no closed-form expression for T, however, T is characterized by its upper and lower bounds as given in Fig. 8.16.

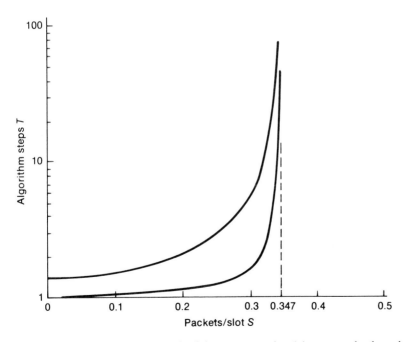

Figure 8.16 Upper and lower bounds of the average packet delay versus the throughput for the tree algorithm.

REFERENCES

1. D. Little, "A Proof for the Queueing Formula: $L = \lambda W$," *Oper. Res.*, Vol. 9, May 1961, pp. 383–387.
2. W. Crowther et al., "A System for Broadcast Communication: Reservation-Aloha," *Proc. Sixth Hawaii Int. Conf. Syst. Sci.*, 1973, pp. 371–374.
3. L. Roberts, "Dynamic Allocation of Satellite Capacity through Packet Reservation," *Proc. NCC, 1973, pp. 711–716.*
4. R. Binder, "A Dynamic Packet Switching System for Satellite Broadcast Channels," *Proc. ICC*, 1975, pp. 41.1–41.5a.
5. I. M. Jacobs et al., "General Purpose Packet Satellite Networks," *Proc. IEEE*, Vol. 66, No. 11, Nov. 1978, pp. 1448–1467.
6. J. I. Capetanakis, "Tree Algorithms for Packet Broadcast Channels," *IEEE Trans. Inform. Theory*, Vol. IT-25, No. 5, Sept. 1979, pp. 505–515.
7. J. I. Capetanakis, "Generalized TDMA: The Multi-accessing Tree Protocol," *IEEE Trans. Commun.*, Vol. COM-27, No. 10, Oct. 1979, pp. 1476–1484.
8. S. S. Lam, "Packet Switching in a Multi-access Broadcast Channel with Application to Satellite Communication in a Computer Network," Ph.D. Dissertation, Computer Science Department, University of California, Los Angeles, Mar. 1974.
9. M. Schwartz, Computer *Communication Network Design and Analysis*. Englewood Cliffs, N.J.: Prentice-Hall, 1977.
10. A. Tanenbaum, *Computer Networks*. Englewood Cliffs, N.J.: Prentice-Hall, 1981.

PROBLEMS

8.1 The mean time delay for a satellite channel of capacity R bits per second when λ packets per second arrive is T. Each packet has a length drawn from an exponential probability density function with mean $1/\mu$ bits per packet. Suppose the satellite channel is partitioned into N independent subchannels each with equal capacity. Find the mean packet delay for this FDMA system and compare this delay with T.

8.2 Consider a satellite channel of capacity 2 Mbps serving an earth station population of 100 using a FDMA protocol. Each message consists of a single packet with a length of 8000 bits. Find the average message delay for a station located at longitude 100°W and latitude 40°N assuming the geostationary satellite longitude is 105°W.

8.3 Consider a satellite channel of capacity 2 Mbps serving an earth station population of 100 using a FDMA protocol. Let the messages which arrive at an earth station consist of a single packet and/or 10 packets, each packet length being 8000 bits. Assume the traffic intensity is 0.8 and the satellite roundtrip delay is 250 ms. Find the average message delay for the following cases:
    (a) Single-packet messages only.
    (b) Single-packet and 10-packet messages at an equal packet rate.
    (c) Single-packet and 10-packet messages at equal message rates.
    (d) Ten-packet messages only.

8.4 Priority queueing: In this problem we extend the analysis of a M/G/1 queue to cover multiple class users, each of different priority. Suppose the messages in a queueing system can be classified into r distinct priorities labeled $p = 1, 2, \ldots, r$ in order of decreasing importance, so that priority $p = 1$ has the highest priority and priority $p = r$ the lowest. As soon as the server finishes with a message, it goes on to serve the next message in the queue having the highest priority. If several messages in the queue have the same priority, they are served on a first-come first-served basis. A message may then move ahead of lower-priority messages on the queue but cannot preempt a lower-priority message undergoing service. Assume that a message of priority k arrives according to a Poisson process with average rate λ_k and that its length has mean $1/\mu_k$ and variance σ_k^2. Let T_0 be the service time of a message of priority p at instant t, T_k be the service time of m_k messages of priority $k = 1, 2, \ldots, p$ already waiting in line at instant t, and T_k' be the service time of m_k' messages of priority $k = 1$, $2, \ldots, p - 1$ that may arrive during the waiting time and be put ahead of the message under consideration. Show that the average delay time of a priority p message is

$$E\{T_p\} = \frac{\sum\limits_{k=1}^{r} \lambda_k \, (1/\mu_k^2 + \sigma_k^2)}{2(1 - \sum\limits_{k=1}^{p-1} \rho_k)(1 - \sum\limits_{k=1}^{p} \rho_k)} + \frac{1}{\mu_p}$$

where $\rho_k = \lambda_k/\mu_k$.

8.5 Consider a satellite channel with a capacity of 2 Mbps serving an earth station population of 100. Let the messages which arrive at the earth station consist of a single packet or 10 packets at equal packet rates. The message with a single packet has priority 1 (higher priority), and that with 10 packets has priority 2 (lower priority). This is a shortest-message-first queue. Assume that each packet consists of 8000 bits. Find the average delay time for both single-packet and 10-packet messages in a FDMA protocol.

8.6 Consider a TDMA channel with a capacity of 250 kbps serving an earth station population of 100. Assume that messages arriving at the earth station are of two types: single-packet and 10-packet. A packet has a fixed length of 1000 bits. Find the average message delay when the traffic intensity is 0.8.

8.7 Verify (8.24) by showing that

$$E\left\{ t + T_f \sum_{i=1}^{L} k_i \right\} = \frac{T_f}{2} + \frac{\lambda T_f^2 \, E\{k\}}{2}$$

$$E\{q\} = \frac{\lambda T_f E\{k^2\} + \rho^2 - \rho}{2(1 - \rho)}$$

8.8 Find the average packet delay for a 50-kbps Aloha satellite channel operating at a throughput of 8 kbps with 1000-bit packets. The average satellite roundtrip delay is 13 packets, and the randomized retransmission interval is 10 packet lengths. If the average user throughput over time is 1 packet/2 min, estimate the number of users the channel could support.

8.9 An Aloha satellite channel serves a community of 4000 users, each with an average throughput of 2 bps. The channel capacity is 50 kbps, and the packet length is 1000 bits. Find the average packet delay if the randomized retransmission interval is 20 packet lengths. The average satellite roundtrip delay is 12.5 packets.

8.10 A community of N earth stations shares a 64-kbps Aloha satellite channel. Each earth station sends out a 1000-bit packet on an average of once every 100 s. What is the maximum value of N?

8.11 Find the channel backlog in packets for a slotted Aloha satellite channel with an infinite population and at equilibrium. The channel input rate is 0.346, the randomized retransmission interval is 60 packet lengths, and the satellite roundtrip propagation delay is taken to be 12 packet slots. What is the average packet delay in packet slots?

8.12 The average packet delay in an infinite-population slotted Aloha satellite channel is 211.68 packet slots at equilibrium. The channel is operating with a channel throughput of

0.275 and a randomized retransmission delay of 40 packet lengths. Find the average channel backlog in packets.

8.13 Consider a slotted Aloha satellite channel serving a user population M. The satellite propagation delay is 12 packet slots. Assume that the mean user think time is 615 slots and the randomized retransmission interval is 10 packet lengths.

    (a) Is the channel stable, bistable, or unstable if $M = 100$? Find the throughput of the channel and the average packet delay.

    (b) Find the channel throughput when $M = 150$. Is the channel unstable toward a sudden increase in channel input rate equal to 40% of the channel throughput at equilibrium?

8.14 A slotted Aloha satellite channel with a bit rate of 250 kbps serves M users. The packet length is 1000 bits, and the satellite roundtrip propagation delay is 250 ms. It is desired to operate the channel at a throughput of 0.32 and maintain stability at all times for $M = 300$ users. Assuming that the randomized retransmission interval is 10 packet lengths, what is the mean user thinking time?

Chapter *9*

Digital Modulation

Digital modulation is a technique used to transmit *baseband digital information* over a *bandpass channel*. The digital information is assumed to be binary, that is, 0 and 1, and occurs at a rate of 1 bit per T_b seconds. Two signals $s_1(t)$ and $s_2(t)$ are needed to represent the binary digits 0 and 1, respectively. This type of transmission is called *binary signaling*. Alternatively, the binary digits can be segmented into blocks where each block consists of k bits. Since there are $M = 2^k$ distinct blocks, M different signals are required to represent the k-bit blocks unambiguously. Each k-bit block is called a *symbol*. It is obvious that a symbol duration is $T_s = kT_b$ seconds. This type of transmission is called *M-ary signaling*. The most commonly used digital modulation in satellite communications is *M-ary phase-shift keying* (*M*-ary PSK) where the signal amplitude is constant and M different phases are used to represent M distinct symbols of a binary sequence. In this chapter we will study the *modulation* and *demodulation* of digital information in an additive white Gaussian noise channel with emphasis on the derivation of the average probability of bit error. In Chap. 10 we will look at the design and performance of an *M*-ary PSK demodulator operating in the *TDMA burst mode*.

9.1 OPTIMUM COHERENT DEMODULATION

In this section we will discuss the structure of an *optimum demodulator* for detecting one out of M possible transmitted signals in an *M*-ary sig-

385

naling scheme with a minimum probability of error. We will assume that the demodulator is synchronized with the transmitter. In other words, we assume that the initial phase of the transmitted signal is known perfectly at the receiver. This is called *coherent demodulation*. For example, let the signals $s_1(t)$ and $s_2(t)$ represent the binary digits 1 and 0:

$$s_1(t) = A \cos(\omega_c t + \theta) \qquad 0 \le t \le T_b$$

$$s_2(t) = -A \cos(\omega_c t + \theta)$$

$$= A \cos(\omega_c t + \pi + \theta) \qquad 0 \le t \le T_b$$

where ω_c = carrier frequency
θ = initial phase of transmitted signals
A = signal amplitude
T_b = bit duration

This type of binary signaling is called phase-shift keying where the modulation is achieved by changing the phase ϕ of the carrier $A \cos(\omega_c t + \phi + \theta)$ between 0 and π depending on whether the transmitted binary digit is 1 or 0. Coherent detection requires the demodulator to generate the initial phase θ in order to extract the modulating information. A detailed discussion of the carrier phase recovery will be postponed until Chap. 10.

Now consider an M-ary signaling scheme where the finite-energy signals $s_1(t)$, $s_2(t)$, $\ldots$, $s_M(t)$ represent the M distinct k-bit symbols H_1, $H_2, \ldots, H_M$ with symbol duration T_s, respectively. By the Gram-Schmidt orthogonalization process (Appendix 9A), the signal set $\{s_m(t)\}$, $m = 1, 2, \ldots, M$, can be expressed as linear combinations of N orthogonal basis functions where $N \le M$ with equality holds when $s_m(t)$, $m = 1, 2, \ldots, M$, is linearly independent [1]:

$$s_m(t) = \sum_{n=1}^{N} s_{mn} u_n(t) \qquad m = 1, 2, \ldots, M \tag{9.1}$$

where for each m and n the coefficients of the expansion are defined by

$$s_{mn} = \int_0^{T_s} s_m(t) u_n(t) \, dt \tag{9.2}$$

The basis functions $u_1(t)$, $u_2(t)$, $\ldots$, $u_N(t)$ are orthonormal, which means that

$$\int_0^{T_s} u_k(t) u_n(t) \, dt = \begin{cases} 1 & k = n \\ 0 & k \ne n \end{cases}$$

Thus each signal $s_m(t)$, $m = 1, 2, \ldots, M$, is completely characterized by the signal vector of its coefficients:

$$\mathbf{s}_m = \begin{bmatrix} s_{m1} \\ s_{m2} \\ \cdot \\ \cdot \\ \cdot \\ s_{mN} \end{bmatrix} \qquad m = 1,2,\ldots,M$$

The set of signal vectors $\{\mathbf{s}_m\}$, $m = 1,2,\ldots,M$, defines a corresponding set of M points in an N-dimensional Euclidean space called the *signal space* whose N mutually perpendicular axes are the orthonormal basis $\{u_n\}$, $n = 1,2,\ldots,N$. The reason for the geometric representation of a signal set is that one can replace the continuous-time operation with a discrete-time operation which is much simpler to deal with. For example, the energy of a signal $s_m(t)$ of duration T_s is equal to

$$E_m = \int_0^{T_s} s_m^2(t) \; dt \tag{9.3}$$

By using the corresponding signal vector $\mathbf{s}_m$ in (9.1) and the orthonormal condition of $\{u_n(t)\}$ we can write E_m in (9.3) as

$$
\begin{aligned}
E_m &= \int_0^{T_s} \left[\sum_{n=1}^{N} s_{mn} u_n(t) \right] \left[\sum_{k=1}^{N} s_{mk} u_k(t) \right] dt \\
&= \sum_{n=1}^{N} \sum_{k=1}^{N} s_{mn} s_{mk} \int_0^{T_s} u_n(t) u_k(t) \; dt \\
&= \sum_{n=1}^{N} s_{mn}^2
\end{aligned}
$$

Thus the energy of a signal $s_m(t)$ is equal to the squared length of its signal vector $\mathbf{s}_m$. Similarly the energy of the difference signal $s_m(t) - s_k(t)$ is equal to the squared length of the difference vector $\mathbf{s}_m - \mathbf{s}_k$:

$$\int_0^{T_s} [s_m(t) - s_k(t)]^2 \; dt = \sum_{n=1}^{N} (s_{mn} - s_{kn})^2$$

To derive the structure of the optimum demodulator we assume that the transmission channel is an *additive white Gaussian noise* (AWGN) channel, hence the received signal at the demodulator corresponding to a transmitted signal $s_m(t)$ can be characterized by

$$r(t) = s_m(t) + n(t) \qquad 0 \le t \le T_s \tag{9.4}$$

where $n(t) = $ AWGN of *zero mean* and *power spectral density* $N_0/2$. [The corresponding autocorrelation of $n(t)$ is $(N_0/2)\delta(\tau)$.] Now for each n, let

$$r_n = \int_0^{T_S} r(t) u_n(t) \, dt \tag{9.5}$$

Substituting (9.4) into (9.5) and using (9.2) yields

$$r_n = s_{mn} + n_n \qquad m = 1, 2, \ldots, M; \, n = 1, 2, \ldots, N \tag{9.6}$$

where, for each n,

$$n_n = \int_0^{T_S} n(t) u_n(t) \, dt \tag{9.7}$$

is a random variable with zero mean.

Note that $\sum_{n=1}^N r_n u_n(t)$ is not a representation of the Gaussian process $r(t)$. Instead, it can be shown that

$$r(t) - \sum_{n=1}^N r_n u_n(t) = s_m(t) + n(t) - \sum_{n=1}^N r_n u_n(t)$$

$$= \sum_{n=1}^N s_{mn} u_n(t) + n(t) - \sum_{n=1}^N (s_{mn} + n_n) u_n(t)$$

$$= n(t) - \sum_{n=1}^N n_n u_n(t)$$

$$\tag{9.8}$$

or, equivalently,

$$r(t) = \sum_{n=1}^N r_n u_n(t) + \hat{n}(t) \tag{9.9a}$$

where

$$\hat{n}(t) = n(t) - \sum_{n=1}^N n_n u_n(t) \tag{9.9b}$$

is also a zero mean Gaussian process.

We have thus characterized the random process $r(t)$ by a set of *observable variables* $r_1, r_2, \ldots, r_N$ and the process $\hat{n}(t)$. The question is whether these observables provide sufficient statistics to make any statistical decision regarding the transmitted symbols based on the a priori probabilities of the symbols. First we note that the observables $\{r_n\}$ are Gaussian variables because $r(t)$ is a Gaussian process. Furthermore, as seen from (9.6) and (9.7), the mean of r_n is $E\{r_n\} = s_{mn}$ and its variance can be obtained as

$$\text{var}\{r_n\} = E\{(r_n - s_{mn})^2\}$$

$$= E\{n_n^2\}$$

$$= E\left\{\int_0^{T_s}\int_0^{T_s} n(t)n(\tau)u_n(t)u_n(\tau)\ dt\ d\tau\right\}$$

$$= \int_0^{T_s}\int_0^{T_s} E\{n(t)n(\tau)\}u_n(t)u_n(\tau)\ dt\ d\tau$$

$$= \int_0^{T_s}\int_0^{T_s} \frac{N_0}{2}\,\delta(t-\tau)u_n(t)u_n(\tau)\ dt\ d\tau$$

$$= \frac{N_0}{2}\int_0^{T_s} u_n^2(t)\ dt = \frac{N_0}{2} \tag{9.10}$$

Also, the covariance of any two observables r_i and r_j, $i \neq j$, is

$$\mathrm{cov}\{r_i,r_j\} = E\{(r_i - s_{mi})(r_j - s_{mj})\}$$

$$= E\{n_in_j\}$$

$$= \frac{N_0}{2}\int_0^{T_s} u_i(t)u_j(t)\ dt = 0$$

by virtue of the orthogonality of the basis $\{u_n\}$. Therefore the observables $\{r_n\}$ are mutually uncorrelated. Since they are Gaussian variables, these observables are also statistically independent. Let $p(r_n|s_{mn})$ denote the conditional probability density function of the Gaussian variable r_n given s_{mn}; then

$$p(r_n|s_{mn}) = \frac{1}{\sqrt{\pi N_0}}\exp\left[\frac{-(r_n - s_{mn})^2}{N_0}\right] \tag{9.11}$$

Define the vector of N observables as

$$\mathbf{r} = \begin{bmatrix} r_1 \\ r_2 \\ \cdot \\ \cdot \\ \cdot \\ r_N \end{bmatrix}$$

Since these N observables are *statistically independent*, the conditional joint probability density function of $\mathbf{r}$ given that the symbol H_m was sent [or, equivalently, given that the signal $s_m(t)$ was sent] is

$$p(\mathbf{r}|H_m) = \prod_{n=1}^N p(r_n|s_{mn})$$

$$= \frac{1}{(\pi N_0)^{N/2}}\exp\left[-\sum_{n=1}^N \frac{(r_n - s_{mn})^2}{N_0}\right] \tag{9.12}$$

The function $p(\mathbf{r}|H_m)$ is also called the *likelihood function*. Recall from (9.9) that the Gaussian process $r(t)$ is characterized by the vector of observables $\mathbf{r}$ and the zero mean Gaussian process $\hat{n}(t)$ which depends only on the noise, as seen from (9.10). It is seen that $\hat{n}(t)$ is independent of any observable r_n, since for any j

$$E\{\hat{n}(t)r_j\} = E\{\hat{n}(t)\ (s_{mj} + n_j)\}$$

$$= E\left\{\left[n(t) - \sum_{n=1}^{N} n_n u_n(t)\right] (s_{mj} + n_j)\right\}$$

$$= E\left\{n(t) \int_0^{T_s} n(\tau)u_j(\tau)\ d\tau\right\} - \sum_{n=1}^{N} E\{n_n n_j\} u_n(t)$$

$$= \frac{N_0}{2} u_j(t) - \frac{N_0}{2} u_j(t) = 0$$

Therefore we conclude that the vector of observables $\mathbf{r}$ based on the received random process $r(t)$ provides sufficient statistics for a decision process.

Let $\Pr(H_m|\mathbf{r})$ be the *a posteriori* probability that the symbol H_m was sent, given the vector of observables $\mathbf{r}$. Based on the vector of observables $\mathbf{r}$, the demodulator makes the decision that the symbol H_m was sent. The probability of error in this decision is simply $1 - \Pr(H_m|\mathbf{r})$. Thus the optimum decision of the demodulator is

$$\Pr(H_m|\mathbf{r}) > \Pr(H_k|\mathbf{r}) \qquad \text{for all } k \neq m \qquad (9.13)$$

to minimize the probability of error. By applying *Bayes rule* to (9.13), we have

$$\frac{p(\mathbf{r}|H_m)\Pr(H_m)}{p(\mathbf{r})} > \frac{p(\mathbf{r}|H_k)\Pr(H_k)}{p|\mathbf{r})} \qquad \text{for all } k \neq m \qquad (9.14)$$

where $p(\mathbf{r})$ is the probability density function of $\mathbf{r}$ which is independent of which symbol was sent and can be ignored in (9.14) and $\Pr(H_m)$ is the *a priori* probability of the symbol H_m. Thus the *optimum decision* that minimizes the probability of error is

$$p(\mathbf{r}|H_m)\ \Pr(H_m) > p(\mathbf{r}|H_k)\ \Pr(H_k) \qquad \text{for all } k \neq m \qquad (9.15)$$

Substituting (9.12) into (9.15) and taking the natural logarithm of both sides, we have

$$\ln[\Pr(H_m)] - \frac{1}{N_0} \sum_{n=1}^{N} (r_n - s_{mn})^2 > \ln[\Pr(H_k)] - \frac{1}{N_0} \sum_{n=1}^{N} (r_n - s_{kn})^2$$

Simplifying and rearranging terms we obtain

$$\ln\left[\frac{\Pr(H_m)}{\Pr(H_k)}\right] + \frac{2}{N_0} \sum_{n=1}^{N} r_n(s_{mn} - s_{kn}) - \frac{E_m - E_k}{N_0} > 0 \qquad \text{for all } k \neq m$$

$$(9.16)$$

where

$$E_m = \sum_{n=1}^{N} s_{mn}^2 = \int_0^{T_s} s_m^2(t)\, dt$$

and

$$E_k = \sum_{n=1}^{N} s_{kn}^2 = \int_0^{T_s} s_k^2(t)\, dt$$

represent the signal energy of $s_m(t)$ and $s_k(t)$, respectively.

For the special case where the M signals $s_1(t), s_2(t), \ldots, s_M(t)$ have equal energy, that is, $E_m = E_s$ for all m, and all the symbols are equiprobable (i.e., they have equal a priori probabilities):

$$\Pr(H_m) = \frac{1}{M} \qquad \text{for all } m$$

then the optimum decision in (9.16) reduces to

$$z_{mk} = \sum_{n=1}^{N} r_n(s_{mn} - s_{kn}) > 0 \qquad \text{for all } k \neq m \qquad (9.17a)$$

or, equivalently,

$$z_{mk} = \int_0^{T_s} r(t)\,[s_m(t) - s_k(t)]\, dt > 0 \qquad \text{for all } k \neq m \qquad (9.17b)$$

This decision rule is referred to as *maximum likelihood*, and the demodulator that makes the decision for equiprobable symbols based on (9.15) or (9.17) is called the *maximum likelihood demodulator*. Based on (9.5) and (9.17a) it is seen that the maximum likelihood demodulator can be implemented by two structures. In the first structure (Fig. 9.1a) the received signal $r(t)$ is multiplied by each orthonormal basis signal $u_1(t)$, $u_2(t), \ldots, u_N(t)$, and each product is integrated over the signaling interval or symbol duration T_s to obtain the vector of observables $\mathbf{r}$. Next, M scalar products between $\mathbf{r}$ and the signal vector $\mathbf{s}_m$, $m = 1, 2, \ldots, M$, are formed:

$$S_m = \mathbf{r} \cdot \mathbf{s}_m = \sum_{n=1}^{N} r_n s_{mn} \qquad m = 1, 2, \ldots, M$$

and the decision is made in favor of the signal corresponding to the largest scalar product. Since the received signal $r(t)$ is correlated with each of the orthonormal basis signals, the demodulator is called a *correlation demodulator*.

A second method for generating the decision variable is to pass the received signal $r(t)$ through N parallel filters with impulse response $h_n(t) = u_n(T_s - t)$, $n = 1, 2, \ldots, N$, as shown in Fig. 9.1b. The output of

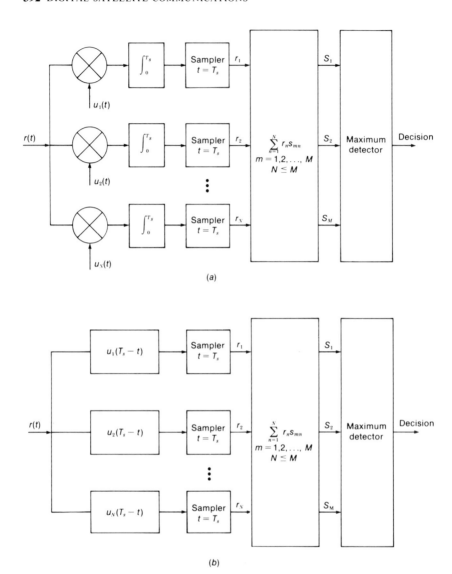

Figure 9.1 (*a*) Correlation demodulator. (*b*) Matched filter demodulator.

the *n*th filter at $t = T_s$ is

$$\int_0^{T_s} r(t) h_n(T_s - t) \ dt = \int_0^{T_s} r(t) u_n(t) \ dt = r_n$$

Thus the vector of observables **r** is obtained. Consequently, the remainder of the demodulator is the same as that of Fig. 9.1*a*. This is the *matched filter demodulator*, since the filters are matched to the waveforms $u_n(t)$, $n = 1, 2, \ldots, N$.

The maximum likelihood demodulators in Fig. 9.1a and b are obtained from the optimum decision rule in (9.17a). Similarly, when the optimum decision in (9.17b) is used, we obtain the two additional equivalent structures shown in Fig. 9.2a and b. This is because the decision rule based on (9.17b) is equivalent to computing M real numbers:

$$S_m = \int_0^{T_s} r(t) s_m(t) \, dt \qquad m = 1, 2, \ldots, M \qquad (9.18)$$

and decides in favor of the signal corresponding to the largest of the set $\{S_m\}$.

Both types of maximum likelihood demodulators (Figs. 9.1 and 9.2) perform identically. The choice depends on the types of waveforms in the M-ary signaling scheme. Note that the maximum likelihood demodulators in Figs. 9.1 and 9.2 are optimum demodulators for a linear AWGN channel with an infinite bandwidth. In a band-limited and/or nonlinear AWGN channel, these configurations are no longer optimum, and the performance will degrade. We thus have obtained the structure of an optimum coherent demodulator as characterized by (9.17), which we will use to analyze the performance of various signaling schemes in terms of the average probability of bit error in subsequent sections.

9.2 PHASE-SHIFT KEYING

As previously mentioned, *phase-shift keying* is a binary signaling scheme where the phase of the carrier changes between two values separated by π radians with each new binary digit. Thus two signals $s_1(t)$ and $s_2(t)$ are employed to represent the binary digits 1 and 0:

$$\begin{aligned} s_1(t) &= A \cos(\omega_c t + \theta) & 0 \leq t \leq T_b \\ s_2(t) &= A \cos(\omega_c t + \pi + \theta) \\ &= -A \cos(\omega_c t + \theta) & 0 \leq t \leq T_b \end{aligned} \qquad (9.19)$$

where ω_c = carrier frequency
θ = initial phase of carrier
A = signal amplitude
T_b = bit duration

Figure 9.3 is a block diagram of a PSK modulator. The carrier can be turned on and off by a carrier on/off signal to determine when the modulator will emit the PSK signal, as in the burst mode operation in a TDMA system described in Chap. 6. A delay adjuster is employed to synchronize the switching of the carrier on/off signal with the data stream. A delay ad-

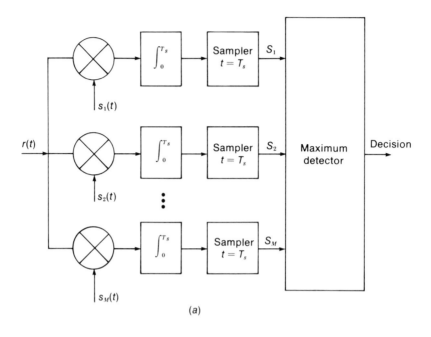

(a)

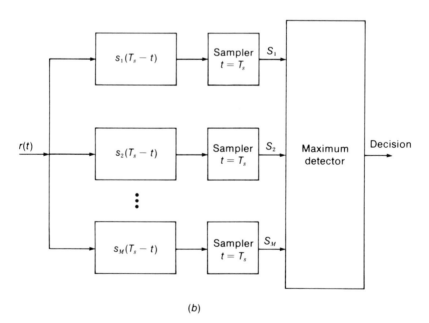

(b)

Figure 9.2 (*a*) Correlation demodulator. (*b*) Matched filter demodulator.

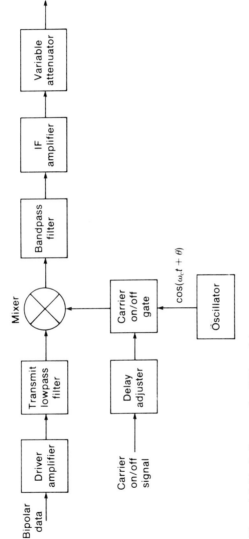

Figure 9.3 Block diagram of a PSK modulator.

justment is necessary to compensate for the delay when the data stream passes through the transmit lowpass filter which shapes the data spectrum. The bipolar information bits are applied to the driver amplifier which provides the proper signal level to the mixer. The driver amplifier is a dc amplifier with a temperature compensation circuit. The mixer uses the filtered information bit stream to modulate the carrier at frequency ω_c. The output of the mixer is a band-limited PSK signal which is passed through a bandpass filter centered at the carrier frequency ω_c. The bandwidth of the bandpass filter is wider than that of the PSK signal and has no effect on the signal spectrum. It is used to filter spurious signals generated by the mixer. (The bandpass filter may also be used to shape the data spectrum. In this case the transmit lowpass filter may not be needed.) The output of the bandpass filter is amplified to a required level by the postamplifier. A variable attenuator is normally provided to continuously adjust the output level if necessary.

The transmit lowpass filter in the PSK modulator is very important, especially in TDMA operations. This filter limits the bandwidth of the transmitted signal so that the out-of-band emission does not exceed a specified value. The power spectral density of a PSK signal with a bipolar modulating waveform of bit duration T_b is identical to the power spectral density of the modulating waveform shifted by the carrier frequency $f_c = \omega_c/2\pi$ and is given by (Appendix 9B)

$$S(f) = \frac{E_b}{2}\left\{\left[\frac{\sin\pi(f-f_c)T_b}{\pi(f-f_c)T_b}\right]^2 + \left[\frac{\sin\pi(f+f_c)T_b}{\pi(f+f_c)T_b}\right]^2\right\} \quad (9.20)$$

where E_b = signal energy of $s_1(t)$ or $s_2(t)$ and is also called the energy per bit. Figure 9.4 shows spectral nulls at $f = f_c \pm n/T_b$, and the first sidelobes are 13.5 dB down from the main lobe. In band-limited satellite channels, the sidelobes may interfere with adjacent channels, hence may require filtering by the transmit lowpass filter shown in Fig. 9.3. An example of a fil-

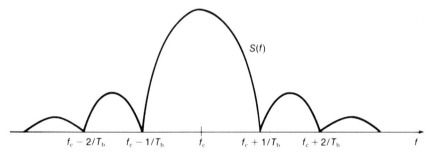

Figure 9.4 Normalized power spectrum of a PSK signal.

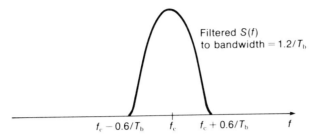

Figure 9.5 Narrowband filtered PSK spectrum.

tered PSK spectrum where the energy is confined to a narrow band is shown in Fig. 9.5. Filtering, however, changes the constant envelope of the PSK signal to a smooth time-varying envelope which goes to 0 at each phase change, as shown in Fig. 9.6. This results in amplitude modulation, introduces intersymbol interference, and increases the average probability of bit error. Furthermore, when a PSK signal with a time-varying envelope is transmitted through a nonlinear satellite channel, the amplitude nonlinearity of the earth station HPA and that of the satellite TWTA regenerate the sidelobes of the PSK spectrum. Therefore, in the design of the transmit lowpass filter in Fig. 9.3, a trade-off between the sharp roll-off of the filter amplitude response to confine the PSK spectrum within the bandwidth of the satellite channel to avoid adjacent channel interference (but introducing more intersymbol interference, hence increasing the average probability of bit error in the received PSK signal) and the smooth roll-off of the filter response to reduce intersymbol interference (but introducing more adjacent channel interference) must be carefully considered. We will discuss this subject further when we study the performance of a PSK signal over a nonlinear satellite channel in a later section. From (9.19) it can be easily seen that there is only one basis function of unit energy, namely,

$$u_1(t) = \sqrt{\frac{2}{T_b}} \cos(\omega_c t + \theta) \qquad 0 \le t \le T_b$$

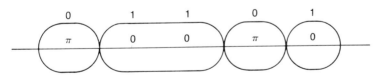

Figure 9.6 Time-varying envelope of a filtered PSK signal.

Note that the PSK signals $s_1(t)$ and $s_2(t)$ in (9.19) can be expressed in terms of $u_1(t)$:

$$s_1(t) = \sqrt{\frac{A^2 T_b}{2}}\, u_1(t)$$

$$s_2(t) = -\sqrt{\frac{A^2 T_b}{2}}\, u_1(t)$$

and therefore only one correlator is needed for optimum coherent demodulation of the PSK signal, as suggested by the general structure in Fig. 9.1a. A block diagram of a coherent PSK demodulator is shown in Fig. 9.7. The PSK signal is filtered by a bandpass filter which rejects undesirable adjacent channel interferences and limits the bandwidth of the thermal noise. The bandwidth of this bandpass filter is usually as wide as that of the received PSK signal from the satellite so as not to affect the spectrum shape. (This receive bandpass filter can also be designed to be compatible with the transmit bandpass filter in the modulator to reduce the intersymbol interference effect.) An automatic gain control (AGC) amplifier is used to provide carrier leveling over a wide dynamic range of the input signal for the circuit that follows. In a TDMA operation large burst-to-burst level differences arising from downlink fading due to rain attenuation can vary by as much as 12 dB over a long period. In the AGC circuit the peak power of the PSK signal is detected and fed back to the amplifier, changing its gain until the output level reaches a predetermined value. The response time of the AGC amplifier is several seconds in practice. Since a TDMA burst lasts only a few microseconds, the AGC amplifier cannot compensate for the burst-to-burst level differences over this short period of time, and these differences have to be taken into account in the design of the carrier phase and bit timing recovery circuits, which will be discussed in Chap. 10. The carrier phase recovery circuit extracts the carrier $\cos(\omega_c t + \theta)$ from the PSK signal and together with the mixer demodulates the received PSK signal to the information bit stream. The receive lowpass filter removes the high-frequency components and out-of-band noise and is designed to be compatible with the transmit lowpass filter in the modulator in Fig. 9.3 to reduce the intersymbol interference effect and to maximize the received signal-to-noise ratio.† The demodulated bit stream is regenerated at the decision circuit by using the bit timing clock recovered from the PSK signal by the bit timing recovery circuit.

---

† If the reduction of intersymbol interferences is accomplished by the bandpass filter, then the lowpass filter need only be designed to maximize the signal-to-noise ratio. In a non-band-limited channel, this lowpass filter is simply the integrator.

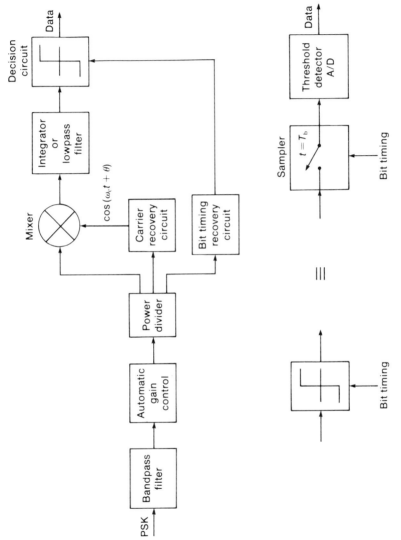

Figure 9.7 Block diagram of a coherent PSK demodulator.

9.2.1 Probability of Error

In this section we focus our attention on the performance of the optimum coherent demodulator for the PSK signal over an AWGN channel expressed in terms of the average probability of bit error. We assume that the channel has an infinite bandwidth. From (9.17) it is seen that the optimum decision rule is $z_{12} > 0$, given that the signal $s_1(t)$ which represents the bit H_1 is sent. Let $p(z_{12})$ be the probability density function of the decision variable z_{12}. Then the probability of error is simply

$$\Pr(z_{12} < 0 | H_1) = \int_{-\infty}^{0} p(z_{12}) \, dz_{12} \tag{9.21}$$

Since

$$z_{12} = \sum_{n=1}^{N} r_n(s_{1n} - s_{2n}) \tag{9.22}$$

is a linear combination of independent Gaussian variables r_n with mean s_{1n} and variance $N_0/2$, it must be itself a Gaussian variable. Its mean value is

$$\mu = E\{z_{12} | H_1\} = \sum_{n=1}^{N} (s_{1n} - s_{2n}) E\{r_n | s_{1n}\}$$

$$= \sum_{n=1}^{N} (s_{1n} - s_{2n}) s_{1n}$$

$$= \sum_{n=1}^{N} s_{1n}^2 - \sum_{n=1}^{N} s_{2n} s_{1n}$$

$$= 2E_b \tag{9.23}$$

where E_b is the signal energy of $s_1(t)$ or $s_2(t)$ and is also called the energy per bit. It is given by

$$E_b = \sum_{n=1}^{N} s_{1n}^2 = \int_0^{T_b} s_1^2(t) \, dt = \int_0^{T_b} A^2 \cos^2(\omega_c t + \theta) \, dt \approx \frac{A^2 T_b}{2}$$

$$= \sum_{n=1}^{N} s_{2n}^2 = \int_0^{T_b} s_2^2(t) \, dt = \int_0^{T_b} A^2 \cos^2(\omega_c t + \theta) \, dt \approx \frac{A^2 T_b}{2}$$

$$\tag{9.24}$$

since $\omega_c > 2\pi/T_b$. Also, by virtue of (9.19), we have

$$\sum_{n=1}^{N} s_{2n} s_{1n} = \int_0^{T_b} s_2(t) s_1(t) \, dt = \int_0^{T_b} -s_1^2(t) \, dt = -E_b \tag{9.25}$$

By using (9.10) the variance of z_{12} is obtained as

$$\sigma^2 = \text{var}\{z_{12}|H_1\} = \sum_{n=1}^{N} (s_{1n} - s_{2n})^2 \, \text{var}(r_n|s_{1n})$$

$$= \sum_{n=1}^{N} (s_{1n} - s_{2n})^2 \frac{N_0}{2}$$

$$= \frac{N_0}{2} \left[\sum_{n=1}^{N} s_{1n}^2 + \sum_{n=1}^{N} s_{2n}^2 - 2 \sum_{n=1}^{N} s_{2n}s_{1n} \right]$$

$$= 2E_b N_0 \qquad (9.26)$$

Thus $\Pr(z_{12} < 0|H_1)$ in (9.21) can be expressed by the integral

$$\Pr(z_{12} < 0|H_1) = \int_{-\infty}^{0} \frac{\exp[-(z_{12} - \mu)^2/2\sigma^2]}{\sqrt{2\pi}\sigma} \, dz_{12} \qquad (9.27)$$

By using the transformation $x = (z_{12} - \mu)/\sigma$, (9.27) can be written as

$$\Pr(z_{12} < 0|H_1) = \frac{1}{\sqrt{2\pi}} \int_{\mu/\sigma}^{\infty} \exp\left(\frac{-x^2}{2}\right) dx$$

$$= \frac{1}{\sqrt{2\pi}} \int_{\sqrt{2E_b/N_0}}^{\infty} \exp\left(\frac{-x^2}{2}\right) dx \qquad (9.28)$$

The integral in (9.28) is called the *Gaussian integral function* and is defined as

$$Q(y) = \frac{1}{\sqrt{2\pi}} \int_{y}^{\infty} \exp\left(\frac{-x^2}{2}\right) dx \qquad (9.29)$$

By symmetry we can show that $\Pr(z_{21} < 0|H_2) = \Pr(z_{12} < 0|H_1)$. Therefore the average probability of bit error for coherent PSK demodulation is

$$P_b = Q\left(\sqrt{\frac{2E_b}{N_0}}\right) \qquad (9.30)$$

Often the average probability of bit error is expressed in terms of the complement error function defined as

$$\text{erfc}(y) = \frac{2}{\sqrt{\pi}} \int_{y}^{\infty} \exp(-z^2) \, dz \qquad (9.31)$$

By using the transformation $x = \sqrt{2}z$, (9.31) can be expressed as

$$\text{erfc}(y) = \frac{2}{\sqrt{2\pi}} \int_{\sqrt{2}\,y}^{\infty} \exp\left(\frac{-x^2}{2}\right) dx$$

$$= 2Q(\sqrt{2}y) \qquad (9.32)$$

or, equivalently,

$$Q(y) = \frac{1}{2}\mathrm{erfc}\left(\frac{y}{\sqrt{2}}\right) \tag{9.33}$$

Thus the average bit error probability for coherent PSK can also be expressed as

$$P_b = \frac{1}{2}\mathrm{erfc}\left(\sqrt{\frac{E_b}{N_0}}\right) \tag{9.34}$$

Figure 9.8 shows the bit error rate performance of coherent PSK in terms of the ratio of the energy per bit to the noise density E_b/N_0.

We are now in a position to make the following observations. We note that the average power of the constant-amplitude PSK signal $s_m(t)$,

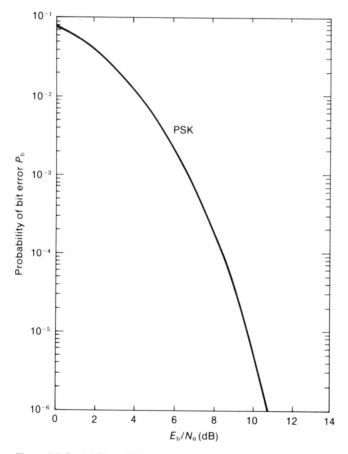

Figure 9.8 Probability of bit error for coherent PSK.

$m = 1,2$, is

$$C = \text{carrier power} = \frac{1}{T_b} \int_0^{T_b} s_m^2(t) \, dt = \frac{E_b}{T_b}$$

and that the noise power within received bandwidth B (Hz) is given by

$$N = \int_{-\infty}^{\infty} S_n(f) \, df$$

$$= \int_{-B/2-f_c}^{B/2-f_c} \frac{N_0}{2} \, df + \int_{-B/2+f_c}^{B/2+f_c} \frac{N_0}{2} \, df = N_0 B$$

where $f_c = \omega_c/2\pi$ and $S_n(f) = $ noise spectral density defined as

$$S_n(f) = \begin{cases} \dfrac{N_0}{2} & f_c - \dfrac{B}{2} \leq |f| \leq f_c + \dfrac{B}{2} \\ 0 & \text{elsewhere} \end{cases}$$

Thus the average probability of bit error for a PSK signal can be expressed is terms of the input carrier-to-noise ratio:

$$P_b = Q\left(\sqrt{\frac{2E_b}{N_0}}\right) = Q\left(\sqrt{2T_b B \left(\frac{C}{N}\right)}\right) \tag{9.35}$$

The bit duration–bandwidth product $T_b B$ is normally chosen to be between 1.1 and 1.4 in satellite communications.

9.2.2 Phase Ambiguity Resolution by Unique Word

Since PSK is a binary signaling technique where two waveforms $s_1(t)$ and $s_2(t)$ with a phase difference of π radians represent the binary digits 1 and 0, there is a twofold phase ambiguity. Unless the demodulator knows which waveform represents which binary digit, the data can be recovered in the correct state or in the inverted state depending on whether the recovered carrier is in-phase (0 rad) or out-of-phase (π radians) with respect to the transmitted signal phase. There are two methods for resolving this phase ambiguity. One method is to use a known code word sequence in the data stream such as the unique word (UW) in the TDMA burst discussed in Chap. 6. In principle, a UW detector and a complement-UW detector are employed at the demodulator. If the data is recovered in its correct state D, the UW detector will provide an indication that the recovered carrier is in-phase. If the data is recovered in the inverted state $\overline{D}$ (complement of D, e.g. $\overline{0} = 1$, $\overline{1} = 0$), the complement-UW detector will indicate that the recovered carrier is out-of-phase and the data will be inverted. The second method for resolving phase ambiguity is to transmit phase differences between consecutive bits instead of their absolute

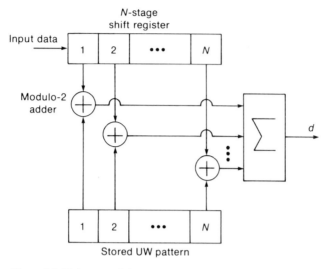

Figure 9.9 Unique word detector.

phases. This method is called differential encoding and will be discussed in the next subsection.

In practice phase ambiguity resolution by unique word is accomplished by making a decision based on the number of bit disagreements d between the demodulator bit stream and the known unique word pattern (d is also called the Hamming distance†) stored in the unique word detector discussed in Chap. 6 and shown again in Fig. 9.9. The relationship between the unique word detector output and the state of the decoded data is shown in Table 9.1a. Also, the decision regarding detection of the unique word is shown in Table 9.1b. Note that the threshold level ϵ must be less than $N/2$, where N is the unique word length in bits.

†The Hamming distance between the expected and received patterns is the number of bit disagreements between the two.

Table 9.1a Relationship between the Hamming distance and the state of the decoded data in PSK signaling

Decoded data	d
D	$d \leq N/2$
$\bar{D}$	$d > N/2$

Table 9.1*b* **Unique word detection decision**

UW detection	$d \leq N/2$	$d > N/2$
Yes	$d \leq \epsilon$	$N - d \leq \epsilon$
No	$d > \epsilon$	$N - d > \epsilon$

9.2.3 Phase Ambiguity Resolution by Differential Encoding

The second method for resolving phase ambiguity is to encode the binary digit not by the absolute phase of the carrier but by the phase change between successive bits. For example, the binary digit 0 is represented by a phase shift of π radians relative to the phase in the previous signaling interval or bit duration, and the binary digit 1 is represented by a zero phase shift relative to the phase in the previous bit duration, or vice versa. A block diagram of the differential encoder is shown in Fig. 9.10*a*. The encoding scheme given in Table 9.2 selects the next encoded bit d_k based on the input information bit b_k and the phase of the previous encoded bit d_{k-1} according to the logic function:

$$d_k = \overline{b_k \oplus d_{k-1}}$$

where $\oplus$ represents the exclusive-OR operation. For example, if the previous encoded bit is 0, then the input stream

$$0 \quad 0 \quad 1 \quad 0 \quad 1 \quad 1 \quad 0 \quad 1$$

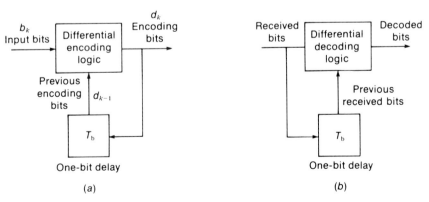

Figure 9.10(*a*) Block diagram of a PSK differential encoder. (*b*) Block diagram of a PSK differential decoder.

Table 9.2 Differential encoding scheme for PSK

Input bits b_k	Phase $\theta_k - \theta_{k-1}$	Previous encoded bits d_{k-1}	Phase θ_{k-1}	Encoded bits d_k	Phase θ_k
0	π	0	π	1	0
0	π	1	0	0	π
1	0	0	π	0	π
1	0	1	0	1	0

will be encoded as

$$1 \quad 0 \quad 0 \quad 1 \quad 1 \quad 1 \quad 0 \quad 0$$

and the corresponding transmitted phase will be

$$0 \quad \pi \quad \pi \quad 0 \quad 0 \quad 0 \quad \pi \quad \pi$$

From Table 9.2 it is seen that, if θ_k is the phase of the signal representing the encoded bit d_k and θ_{k-1} is the phase of the signal representing the previous encoded bit d_{k-1}, then the phase difference $\theta_k - \theta_{k-1}$ completely specifies the information bit b_k. Now suppose that the demodulator introduces a phase ambiguity β in the recovered carrier. Thus the phase of the received signal representing the encoded bit d_k is $\theta_k + \beta$ and the phase of the previously received signal representing the encoded bit d_{k-1} is $\theta_{k-1} + \beta$. The phase ambiguity can be perfectly resolved if the demodulator performs the decoding by making a decision based on the phase difference between d_k and d_{k-1}, which is $(\theta_k + \beta) - (\theta_{k-1} + \beta) = \theta_k - \theta_{k-1}$.

Figure 9.10b is a block diagram of a differential decoder with the decoding scheme shown in Table 9.3. For example, if the previously received bit is 0, the received data stream

$$1 \quad 0 \quad 0 \quad 1 \quad 1 \quad 1 \quad 0 \quad 0$$

Table 9.3 Differential decoding scheme for PSK

Received bits	Phase θ_k	Previous received bits	Phase θ_{k-1}	Decoded bits	Phase $\theta_k - \theta_{k-1}$
0	π	0	π	1	0
0	π	1	0	0	π
1	0	0	π	0	π
1	0	1	0	1	0

will be decoded as

$$0 \quad 0 \quad 1 \quad 0 \quad 1 \quad 1 \quad 0 \quad 1$$

The bit error rate for differential encoding PSK using coherent detection (DCPSK) as shown in Fig. 9.10 is about twice the bit error rate for coherent PSK, since the decision is made on the basis of the signals received in two consecutive signaling intervals, hence there is a tendency for bit errors to occur in pairs. The average probability of bit error for DCPSK [2] is:

$$P_{\mathrm{b}} = 2Q\left(\sqrt{\frac{2E_{\mathrm{b}}}{N_0}}\right)\left[1 - Q\left(\sqrt{\frac{2E_{\mathrm{b}}}{N_0}}\right)\right]$$

$$\approx 2Q\left(\sqrt{\frac{2E_{\mathrm{b}}}{N_0}}\right) = \mathrm{erfc}\left(\sqrt{\frac{E_{\mathrm{b}}}{N_0}}\right) \qquad (9.36)$$

at $P_{\mathrm{b}} < 10^{-2}$, where $E_{\mathrm{b}}/N_0 =$ ratio of energy per bit to noise density.

Differential encoding PSK can also be demodulated without a carrier reference, as shown in Fig. 9.11, by performing the differential decoding on the received signals instead of on the demodulated information stream. The received signal in the kth signaling interval is $A \cos(\omega_c t + \theta_k + \theta)$, where θ_k is the phase representing the encoded bit d_k and θ is the initial carrier phase. Similarly, the received signal in the $(k-1)$th signaling interval is $A \cos(\omega_c t + \theta_{k-1} + \theta)$. Multiplication of the two signals, followed by integration, which also lowpass-filters the second harmonic at $2\omega_c$, yields a decision variable proportional to $\cos(\theta_k - \theta_{k-1})$ which is either 1 or -1 depending on whether $\theta_k - \theta_{k-1} = 0$ or π radians. The information can be easily extracted from the decision variable.

Because the demodulation is performed by multiplying the received signals in consecutive bit durations, the AWGN that contaminates them is also being multiplied. This in effect doubles the variance of the AWGN and increases the bit error rate of differential encoding PSK with

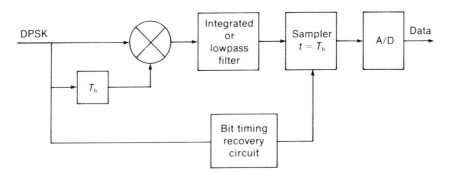

Figure 9.11 Block diagram of a DPSK demodulator.

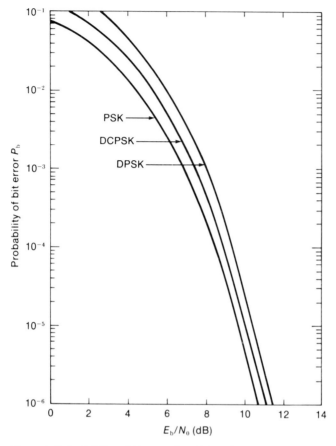

Figure 9.12 Probability of bit error for PSK, DCPSK, and DPSK.

noncoherent demodulation (DPSK). The average probability of bit error for DPSK [3] is given by

$$P_b = \frac{1}{2} \exp\left(-\frac{E_b}{N_0}\right) \tag{9.37}$$

where E_b/N_0 = ratio of energy per bit to noise density. The average probability of bit error for PSK, DCPSK, and DPSK is plotted in Fig. 9.12.

9.3 QUATERNARY PHASE-SHIFT KEYING

As we mentioned at the beginning of this chapter, binary digits can be transmitted using M-ary signaling, where $M = 2^k$ distinct k-bit symbols are represented by M signals. If T_b is the bit duration, then the symbol duration or signaling interval is simply $T_s = kT_b$. When $M = 4$, the four distinct symbols are 00, 01, 10, and 11. In *quaternary phase-shift keying* each of these four symbols is represented by a distinct phase of a constant-amplitude carrier:

$$s_m(t) = A \cos(\omega_c t + \theta_m + \theta) \qquad 0 \leq t \leq T_s \qquad (9.38)$$

where θ_m = modulation phase and θ = initial phase of carrier. The mapping of the 2-bit symbols 00, 01, 10, and 11 can be accomplished in a numbers of ways. The most preferred mapping is Gray encoding where two symbols which differ by only 1 bit are represented by adjacent phases, as shown in Fig. 9.13. *Gray encoding* is important in the demodulation of M-ary phase-shift keying because the errors caused by AWGN are likely to occur when the adjacent phase is selected for the transmitted phase. In such a case only one bit error occurs in a k-bit symbol. In practice the four symbols 00, 01, 10,11 are represented by the phase θ_m, $m = 1,2,3,4$, in Fig. 9.13, and the corresponding transmitted signals are

$$s_1(t) = s_{10}(t) = A \cos\left(\omega_c t + \frac{\pi}{4} + \theta\right)$$

$$s_2(t) = s_{00}(t) = A \cos\left(\omega_c t + \frac{3\pi}{4} + \theta\right) \qquad 0 \leq t \leq T_s$$

$$s_3(t) = s_{01}(t) = A \cos\left(\omega_c t + \frac{5\pi}{4} + \theta\right)$$

$$s_4(t) = s_{11}(t) = A \cos\left(\omega_c t + \frac{7\pi}{4} + \theta\right) \qquad (9.39)$$

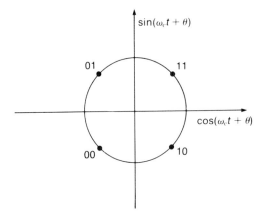

Figure 9.13 Example of Gray encoding for a QPSK signal.

or, equivalently,

$$s_1(t) = s_{10}(t) = \frac{A}{\sqrt{2}} \cos(\omega_c t + \theta) - \frac{A}{\sqrt{2}} \sin(\omega_c t + \theta)$$

$$s_2(t) = s_{00}(t) = -\frac{A}{\sqrt{2}} \cos(\omega_c t + \theta) - \frac{A}{\sqrt{2}} \sin(\omega_c t + \theta) \qquad 0 \le t \le T_s$$

$$s_3(t) = s_{01}(t) = -\frac{A}{\sqrt{2}} \cos(\omega_c t + \theta) + \frac{A}{\sqrt{2}} \sin(\omega_c t + \theta)$$

$$s_4(t) = s_{11}(t) = \frac{A}{\sqrt{2}} \cos(\omega_c t + \theta) + \frac{A}{\sqrt{2}} \sin(\omega_c t + \theta) \tag{9.40}$$

From (9.40) it is seen that the QPSK signal can be viewed as a linear combination of two PSK signals in phase quadrature. This suggests the structure of the QPSK modulator and demodulator in Figs. 9.14 and 9.15, respectively. Also, note that there are two orthonormal basis functions in the expansion of $s_m(t)$ with unit energy

$$u_1(t) = \sqrt{\frac{2}{T_s}} \cos(\omega_c t + \theta) \qquad 0 \le t \le T_s$$

$$u_2(t) = \sqrt{\frac{2}{T_s}} \sin(\omega_c t + \theta) \qquad 0 \le t \le T_s \tag{9.41}$$

hence $s_m(t)$ in (9.40) can be written as

$$s_1(t) = s_{10}(t) = \sqrt{\frac{A^2 T_s}{4}} u_1(t) - \sqrt{\frac{A^2 T_s}{4}} u_2(t)$$

$$s_2(t) = s_{00}(t) = -\sqrt{\frac{A^2 T_s}{4}} u_1(t) - \sqrt{\frac{A^2 T_s}{4}} u_2(t)$$

$$s_3(t) = s_{01}(t) = -\sqrt{\frac{A^2 T_s}{4}} u_1(t) + \sqrt{\frac{A^2 T_s}{4}} u_2(t)$$

$$s_4(t) = s_{11}(t) = \sqrt{\frac{A^2 T_s}{4}} u_1(t) + \sqrt{\frac{A^2 T_s}{4}} u_2(t) \tag{9.42}$$

In the modulator the bipolar bit stream enters the serial-to-parallel converter where it is converted to 2-bit streams, the in-phase P channel and the quadrature Q channel, each at one-half the original bit rate, as shown in Fig. 9.16. Thus a symbol is represented by a P digit and a Q digit. For example, the symbol 01 is represented by the P digit 0 and the Q digit 1. Note that each P digit or Q digit has a duration of $2T_b$, where T_b is the bit duration. Thus a symbol duration is $T_s = 2T_b$. The P stream is used to modulate the carrier $\cos(\omega_c t + \theta)$, and the Q stream is used to modulate the quadrature carrier $\sin(\omega_c t + \theta)$. The two modulated PSK signals are then added to form the QPSK signal by the power combiner.

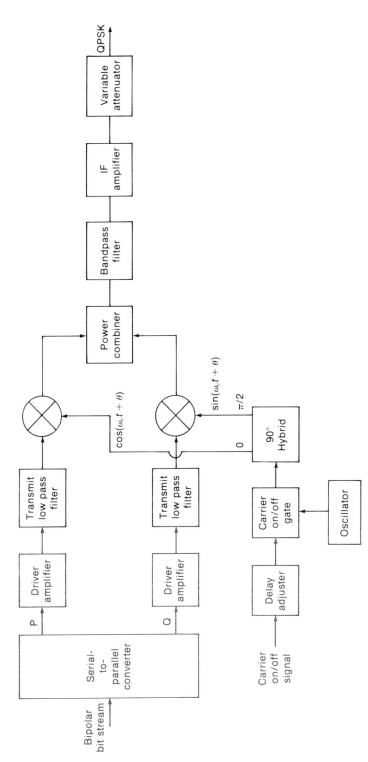

Figure 9.14 Block diagram of a QPSK modulator.

411

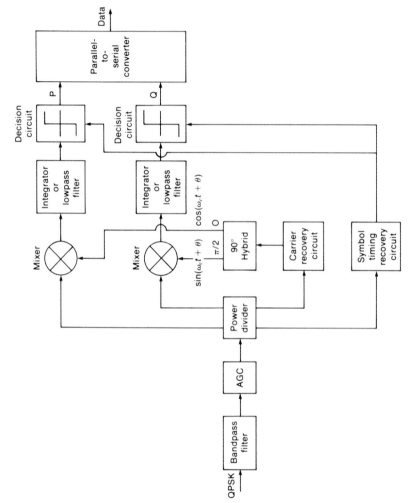

Figure 9.15 Block diagram of a coherent QPSK demodulator.

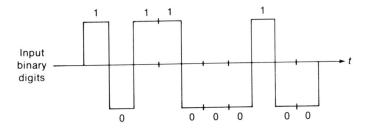

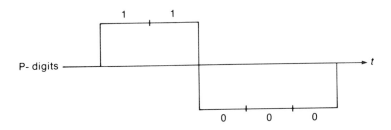

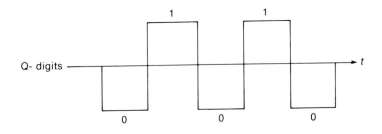

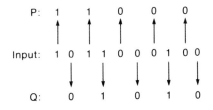

Figure 9.16 Serial-to-parallel conversion of a bipolar bit stream.

The function of other components in the QPSK modulator is similar to that of the PSK modulator in Fig. 9.3.

In the coherent QPSK demodulator, the QPSK signal is delivered to the P channel, the Q channel, the carrier recovery circuit, and the symbol timing recovery circuit. A 90° hybrid is used to obtain quadrature reference carriers to demodulate the PSK signals in the P and Q channels. The demodulated P and Q streams are regenerated at the decision circuits by the symbol clock recovered from the symbol timing recovery circuit and recombined by the parallel-to-serial converter to form a single-bit stream. The functions of other components of the QPSK demodulator are similar to that of the PSK demodulator in Fig. 9.7.

Since the QPSK signal is a linear combination of two PSK signals in phase quadrature whose signaling interval is T_s, the QPSK power spectral density is identical to the PSK power spectral density described in (9.20) with T_s replacing T_b:

$$S(f) = \frac{1}{2} E_s \left\{ \left[\frac{\sin \pi(f - f_c) T_s}{\pi(f - f_c) T_s} \right]^2 + \left[\frac{\sin \pi(f + f_c) T_s}{\pi(f + f_c) T_s} \right]^2 \right\} \quad (9.43)$$

where E_s = energy per symbol. Note that $T_s = 2T_b$, where T_b is the bit duration; thus the symbol rate of the QPSK signal is only half the bit rate. Thus a QPSK signaling scheme reduces the transmitted bandwidth by a factor of 2 compared to a PSK signaling scheme.

9.3.1 Probability of Error

To derive the average probability of error for a coherent QPSK signal in an AWGN channel with an infinite bandwidth we make the assumption that $\omega_c = 2\pi k/T_s$, where k is a positive integer. Thus the carrier frequency ω_c is a 2π multiple of the symbol rate. This simplifies the analysis, because the two signals $\cos(\omega_c t + \theta)$ and $\sin(\omega_c t + \theta)$ are orthogonal over any symbol duration since

$$\int_0^{T_s} \sin(\omega_c t + \theta) \cos(\omega_c t + \theta) \, dt = \frac{1}{2} \int_0^{T_s} \sin(2\omega_c t + 2\theta) \, dt$$

$$= -\frac{1}{2\omega_c} \cos(2\omega_c t + 2\theta) \Big|_0^{T_s} = 0$$

$$(9.44)$$

Now we return to the optimum decision rule of coherent demodulation in (9.17) and assume that the signal waveforms in (9.39) are used, which results in two reference carriers $\cos(\omega_c t + \theta)$ and $\sin(\omega_c t + \theta)$ at the demodulator. Suppose that the signal $s_1(t) = A \cos(\omega_c t + \pi/4 + \theta)$, representing the symbol H_1 (for 10), is transmitted:

$$s_1(t) - s_2(t) = \sqrt{2}\, A \cos(\omega_c t + \theta)$$

$$s_1(t) - s_4(t) = -\sqrt{2}\, A \sin(\omega_c t + \theta)$$

$$s_1(t) - s_3(t) = \sqrt{2}\, A \cos(\omega_c t + \theta) - \sqrt{2}\, A \sin(\omega_c t + \theta)$$

$$= [s_1(t) - s_2(t)] + [s_1(t) - s_4(t)]$$

We need to consider only two decision variables z_{12} and z_{14} in (9.17), because the decision variable $z_{13} = z_{12} + z_{14}$. With the orthogonality assumption made above it can be easily seen that $s_1(t)$ is orthogonal to $s_2(t)$ and $s_4(t)$ over a symbol duration. Using the same derivation as indicated in (9.21) to (9.26) we can show that both Gaussian variables z_{12} and z_{14} possess the mean μ and variance σ^2 given by

$$\mu = E_s \tag{9.45}$$

$$\sigma^2 = E_s N_0 \tag{9.46}$$

where E_s is the signal energy of $s_m(t)$, $m = 1,2,3,4$, in the signaling interval T_s, or simply the energy per symbol.

Furthermore, since $s_1(t) - s_2(t)$ and $s_1(t) - s_4(t)$ are orthogonal, we can show that the two Gaussian variables z_{12} and z_{14} are uncorrelated, hence statistically independent. Let

$$P_{e1} = \Pr(z_{12} < 0 | H_1) = Q\left(\frac{\mu}{\sigma}\right) = Q\left(\sqrt{\frac{E_s}{N_0}}\right)$$

$$P_{e2} = \Pr(z_{14} < 0 | H_1) = Q\left(\frac{\mu}{\sigma}\right) = Q\left(\sqrt{\frac{E_s}{N_0}}\right) \tag{9.47}$$

Since z_{12} and z_{14} are independent and $z_{13} = z_{12} + z_{14}$, the probability of error for the signal $s_1(t)$ is

$$P_e = \Pr(z_{12} < 0 \cup z_{14} < 0 \cup z_{13} < 0 | H_1)$$

$$= 1 - \Pr(z_{12} > 0 \cap z_{14} > 0 | H_1)$$

$$= 1 - \Pr(z_{12} > 0 | H_1)\Pr(z_{14} > 0 | H_1)$$

$$= 1 - (1 - P_{e1})(1 - P_{e2}) = 1 - (1 - P_{e1})^2$$

$$= 2P_{e1} - P_{e1}^2 \approx 2P_{e1} \tag{9.48}$$

for $P_{e1} < 10^{-2}$. By symmetry, we conclude that the average probability of symbol error for a coherent QPSK signal is

$$P_s \approx 2Q\left(\sqrt{\frac{E_s}{N_0}}\right) \tag{9.49}$$

where $E_s/N_0 =$ ratio of energy per symbol to noise density.

When *Gray encoding* is employed for symbol mapping and the adjacent phase is selected for the true phase, the symbol error contains only one bit error. Thus for Gray encoding the average probability of bit error for coherent QPSK is

$$P_b \approx \frac{1}{2} P_s \approx Q\left(\sqrt{\frac{E_s}{N_0}}\right)$$

$$\approx Q\left(\sqrt{\frac{2E_b}{N_0}}\right)$$

$$\approx \frac{1}{2} \text{erfc}\left(\sqrt{\frac{E_b}{N_0}}\right) \tag{9.50}$$

since $E_s = 2E_b$, where E_b is the energy per bit, because there are 2 bits per symbol in QPSK signaling.

Let C denote the QPSK carrier power and N denote the noise power within the received bandwidth B (Hz). Then

$$C = \frac{E_s}{T_s} = \frac{E_b}{T_b}$$

$$N = N_0 B$$

Therefore, in terms of the input carrier-to-noise ratio C/N, we have

$$\frac{E_s}{N_0} = \frac{2E_b}{N_0} = T_s B \left(\frac{C}{N}\right)$$

Thus the average probability of bit error for QPSK is

$$P_b \approx Q\left(\sqrt{\frac{E_s}{N_0}}\right) = Q\left(\sqrt{\frac{2E_b}{N_0}}\right) = Q\left[\sqrt{T_s B \left(\frac{C}{N}\right)}\right]$$

$$= Q\left[\sqrt{2T_b B \left(\frac{C}{N}\right)}\right] \tag{9.51}$$

In satellite communications the symbol duration–bandwidth product $T_s B$ is normally chosen between 1.1 and 1.4. Thus, for the same channel bandwidth, twice the bit rate can be achieved with QPSK as with PSK.

9.3.2 Phase Ambiguity Resolution by Unique Word

In the demodulation of the QPSK signal $s_m(t) = A \cos(\omega_c t + \theta_m + \theta)$, $0 \leq t \leq T_s$, given in (9.38) or (9.39), where $\theta_m = (2m-1)\pi/4$, $m = 1,2,3,4$, is the modulation phase and θ is the initial phase, the recovered carrier phase $\hat{\theta}$ can assume any one of the four possible values 0, $\pi/2$, π, or $3\pi/2$ referenced to the correct initial phase θ; that is, $\hat{\theta} = \theta + n\pi/2$, $n = 0,1,2,3$. This gives rise to a fourfold phase ambiguity. Of course there is no phase ambiguity if $\hat{\theta} = \theta$. For example, if

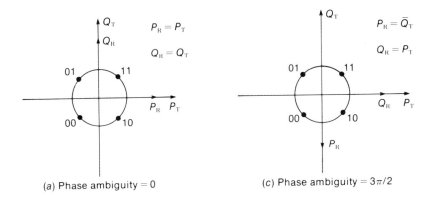

(a) Phase ambiguity = 0

(c) Phase ambiguity = $3\pi/2$

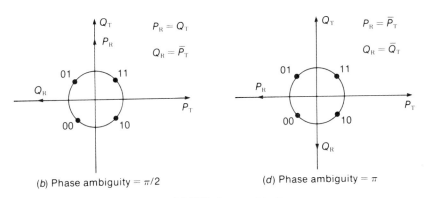

(b) Phase ambiguity = $\pi/2$

(d) Phase ambiguity = π

Figure 9.17 Four possible states of QPSK phase ambiguity.

$\hat{\theta} = \theta + \pi/2$, then the signal $s_1(t)$, $s_2(t)$, $s_3(t)$, and $s_4(t)$ in (9.39) would be incorrectly decoded as $s_2(t)$, $s_3(t)$, $s_4(t)$, and $s_1(t)$, respectively. Figure 9.17 shows the four possible states corresponding to the phase ambiguity of the recovered carrier and the relationship between the transmitted P digits (P_T) and Q digits (Q_T), and the received P digits (P_R) and Q digits (Q_R). For example, consider the state in Fig. 9.17b where the phase ambiguity is $\pi/2$. In this state, the received P channel is in-phase with the transmitted Q channel, and the received Q channel is π radians out-of-phase with the transmitted P channel. Therefore the correct P digits (P) and Q digits (Q) after phase ambiguity resolution as shown in Fig. 9.18 must be

$$P = P_T = \bar{Q}_R$$
$$Q = Q_T = P_R \tag{9.52}$$

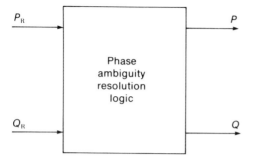

Figure 9.18 Phase ambiguity resolution of the received P_R digits and Q_R digits.

where the $\bar{Q}_R$ indicates the complement of Q_R (e.g., $\bar{0} = 1$ and $\bar{1} = 0$). This means that, when the received P digits (P_R) are found in the correct state and the received Q digits (Q_R) are found in the inverted state, the correct P digits (P) and Q digits (Q) should be assigned according to (9.52). The phase ambiguity resolution for P and Q digits in terms of P_R and Q_R for all four states is summarized in Table 9.4.

As in the case of coherent PSK demodulation, the phase ambiguity of coherent QPSK demodualtion can be resolved by using the known pattern of a unique word. It does not matter if identical unique word patterns or two distinct unique word patterns are used in the P and Q channels. Let d_P and d_Q denote the Hamming distances between the demodulator P and Q streams and the known unique word patterns stored in the P and Q channel unique word correlators, respectively, as shown in Fig. 9.9. The relationship between the Hamming distances d_P and d_Q and the state of the decoded P digits (P_R) and Q digits (Q_R) is shown in Tables 9.5a and 9.5b where N is the unique word length in P or Q digits. From Tables 9.4, 9.5a, and 9.5b the phase ambiguity can be resolved according to the decision given in Table 9.6. Also, the decisions for detection of the unique word from the Hamming distance d_P and d_Q are shown in Table 9.7 where ϵ is the threshold level and must be chosen to be less than $N/2$.

Table 9.4 Fourfold phase ambiguity resolution in terms of P_R and Q_R

Decision	Q_R	$\bar{Q}_R$
P_R	$P = P_R$ $Q = Q_R$	$P = \bar{Q}_R$ $Q = P_R$
$\bar{P}_R$	$P = Q_R$ $Q = \bar{P}_R$	$P = \bar{P}_R$ $Q = \bar{Q}_R$

Table 9.5a Relationship between the Hamming distance d_P and P_R

Decoded P digits	d_P
P_R	$d_P \leq N/2$
$\overline{P_R}$	$d_P > N/2$

Table 9.5b Relationship between the Hamming distance d_Q and Q_R

Decoded Q digits	d_Q
Q_R	$d_Q \leq N/2$
$\overline{Q_R}$	$d_Q > N/2$

Table 9.6 Phase ambiguity resolution for QPSK using a unique word

Decision	$d_Q \leq N/2$	$d_Q > N/2$
$d_P \leq N/2$	$P = P_R$ $Q = Q_R$	$P = \overline{Q}_R$ $Q = P_R$
$d_P > N/2$	$P = Q_R$ $Q = \overline{P}_R$	$P = \overline{P}_R$ $Q = \overline{Q}_R$

Table 9.7 Unique detection decision for QPSK

UW detection	$d_Q \leq N/2$	$d_Q > N/2$
$d_P \leq N/2$	$d_P + d_Q \overset{Yes}{\underset{No}{\lesseqgtr}} \epsilon$	$N - d_Q + d_P \overset{Yes}{\underset{No}{\lesseqgtr}} \epsilon$
$d_P > N/2$	$N - d_P + d_Q \overset{Yes}{\underset{No}{\lesseqgtr}} \epsilon$	$2N - (d_P + d_Q) \overset{Yes}{\underset{No}{\lesseqgtr}} \epsilon$

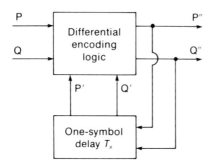

Figure 9.19 Block diagram of a QPSK differential encoder.

9.3.3 Phase Ambiguity Resolution by Differential Encoding

Another method of resolving fourfold phase ambiguity is to use a differential encoder as shown schematically in Fig. 9.19. An encoding scheme is given in Table 9.8 which selects the next encoded symbol based on the input symbol and the previous encoded symbol. For example, if the previous encoded symbol is 00, then the input sequence

$$01 \quad 10 \quad 00 \quad 11$$

will be encoded as

$$01 \quad 00 \quad 00 \quad 11$$

and the corresponding transmitted phase will be

$$5\pi/4 \quad 3\pi/4 \quad 3\pi/4 \quad 7\pi/4$$

In the demodulator the operation of the differential decoder shown in Fig. 9.20 is similar to that of the encoder in a reverse manner. The decoding scheme is shown in Table. 9.9. For example, if the previously received

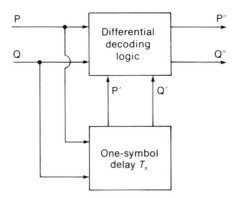

Figure 9.20 Block diagram of a QPSK differential decoder.

Table. 9.8 Differential encoding scheme for QPSK

Input symbol PQ	Phase	Previous encoded symbol $P'Q'$	Phase	Encoded symbol $P''Q''$	Phase
00	0	00	$3\pi/4$	00	$3\pi/4$
00	0	10	$\pi/4$	10	$\pi/4$
00	0	11	$7\pi/4$	11	$7\pi/4$
00	0	01	$5\pi/4$	01	$5\pi/4$
10	$3\pi/2$	00	$3\pi/4$	10	$\pi/4$
10	$3\pi/2$	10	$\pi/4$	11	$7\pi/4$
10	$3\pi/2$	11	$7\pi/4$	01	$5\pi/4$
10	$3\pi/2$	01	$5\pi/4$	00	$3\pi/4$
11	π	00	$3\pi/4$	11	$7\pi/4$
11	π	10	$\pi/4$	01	$5\pi/4$
11	π	11	$7\pi/4$	00	$3\pi/4$
11	π	01	$5\pi/4$	10	$\pi/4$
01	$\pi/2$	00	$3\pi/4$	01	$5\pi/4$
01	$\pi/2$	10	$\pi/4$	00	$3\pi/4$
01	$\pi/2$	11	$7\pi/4$	10	$\pi/4$
01	$\pi/2$	01	$5\pi/4$	11	$7\pi/4$

Table 9.9 Differential decoding scheme for QPSK

Receive symbol PQ	Phase	Previous received symbol $P'Q'$	Phase	Decoded symbol $P''Q''$	Phase
00	$3\pi/4$	00	$3\pi/4$	00	0
00	$3\pi/4$	10	$\pi/4$	01	$\pi/2$
00	$3\pi/4$	11	$7\pi/4$	11	π
00	$3\pi/4$	01	$5\pi/4$	10	$3\pi/2$
10	$\pi/4$	00	$3\pi/4$	10	$3\pi/2$
10	$\pi/4$	10	$\pi/4$	00	0
10	$\pi/4$	11	$7\pi/4$	01	$\pi/2$
10	$\pi/4$	01	$5\pi/4$	11	π
11	$7\pi/4$	00	$3\pi/4$	11	π
11	$7\pi/4$	10	$\pi/4$	10	$3\pi/2$
11	$7\pi/4$	11	$7\pi/4$	00	0
11	$7\pi/4$	01	$5\pi/4$	01	$\pi/2$
01	$5\pi/4$	00	$3\pi/4$	01	$\pi/2$
01	$5\pi/4$	10	$\pi/4$	11	π
01	$5\pi/4$	11	$7\pi/4$	10	$3\pi/2$
01	$5\pi/4$	01	$5\pi/4$	00	0

symbol is 00, then the received symbol sequence

$$01 \quad 00 \quad 00 \quad 11$$

will be decoded as

$$01 \quad 10 \quad 00 \quad 11$$

The average probability of symbol error for *differential encoding QPSK using coherent detection* (DCQPSK) is about twice the average probability of symbol error for coherent QPSK, since the decison is made on the basis of two consecutive symbols, hence there is a tendency for symbol errors to occur in pairs. The average probability of symbol error for DCQPSK [2] can be approximated as

$$P_s \approx 4Q\left(\sqrt{\frac{E_s}{N_0}}\right) \tag{9.53}$$

at $P_s < 10^{-2}$, where E_s/N_0 is the ratio of the energy per symbol to the noise density. When Gray encoding is employed, the average probability of bit error for DCQPSK is given by

$$P_b \approx \frac{1}{2} P_s = 2Q\left(\sqrt{\frac{E_s}{N_0}}\right)$$

$$\approx 2Q\left(\sqrt{\frac{2E_b}{N_0}}\right)$$

$$\approx \text{erfc}\left(\sqrt{\frac{E_b}{N_0}}\right) \tag{9.54}$$

where E_b/N_0 = ratio of energy per bit to noise density.

As in the case of DPSK, *differential encoding QPSK* signals with *noncoherent detection* (DQPSK) can also be achieved by performing the differential decoding on the received signals instead of on the demodulated P and Q streams, as shown in Fig. 9.21. The average probability of bit error for DQPSK with Gray encoding [3] is given by

$$P_b \approx Q\left(\sqrt{\frac{2E_b}{N_0}\left(4 \sin^2 \frac{\pi}{8}\right)}\right) \tag{9.55}$$

Thus DQPSK requires approximately 2.3 dB more in E_b/N_0 to achieve the same bit error rate performance as coherent QPSK, as indicated by (9.51).

9.4 M-ARY PHASE-SHIFT KEYING

In the previous two sections we analyzed the performance of PSK and QPSK. In this section we will consider *M-ary phase shift keying*, $M = 2^k$,

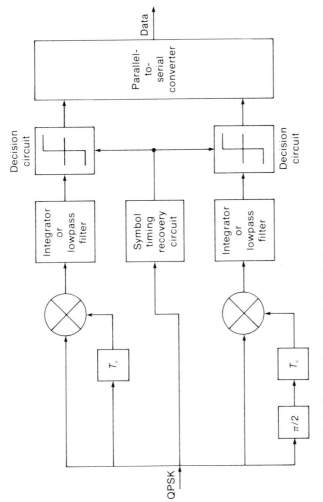

Figure 9.21 Block diagram of a DQPSK demodulator.

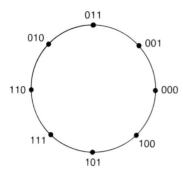

Figure 9.22 Example of Gray encoding for an 8-ary PSK signal.

and attempt to derive the average probability of bit error. An example of the symbol-phase mapping for 8-ary PSK using Gray encoding is shown in Fig. 9.22. The general waveform for M-ary PSK is given by

$$s_m(t) = A \cos(\omega_c t + \theta_m + \theta) \qquad 0 \le t \le T_s \qquad (9.56)$$

where θ_m, $m = 1, 2, \ldots, M$ = modulation phase
$$\theta = \text{initial phase}$$
$$\omega_c = \text{carrier frequency},$$
$$A = \text{signal amplitude}$$

The signaling interval or symbol $T_s = kT_b$, where T_b is the bit duration. From (9.56) we can express $s_m(t)$ as

$$s_m(t) = (A \cos \theta_m) \cos(\omega_c t + \theta) - (A \sin \theta_m) \sin(\omega_c t + \theta)$$

$$= A_c \cos(\omega_c t + \theta) - A_s \sin(\omega_c t + \theta) \qquad (9.57a)$$

where

$$A_c = A \cos \theta_m \qquad A_s = A \sin \theta_m \qquad (9.57b)$$

Equation (9.57) implies that the M-ary PSK signal may be generated by the modulator in Fig. 9.23, and the phase of the signal $s_m(t)$ contaminated by noise can be recovered by the coherent demodulator in Fig. 9.24 by deciding whether it is within $\pm \pi/M$ of θ_m. To derive the average probability of bit error for M-ary PSK we assume that $\omega_c = k2\pi/T_s$, where $k > 1$ is an integer, so that the in-phase channel is orthogonal to the quadrature channel.

Let $s_m(t)$ be the transmitted signal; then the received signal $r(t)$ can be expressed as

$$r(t) = s_m(t) + n(t)$$

where $n(t)$ = AWGN with zero mean and power spectral density $N_0/2$. Let r_c and r_s denote the observable variables at the output of the in-phase

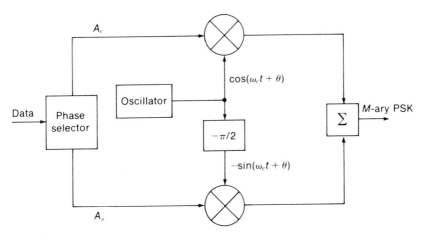

Figure 9.23 Block diagram of an *M*-ary PSK modulator.

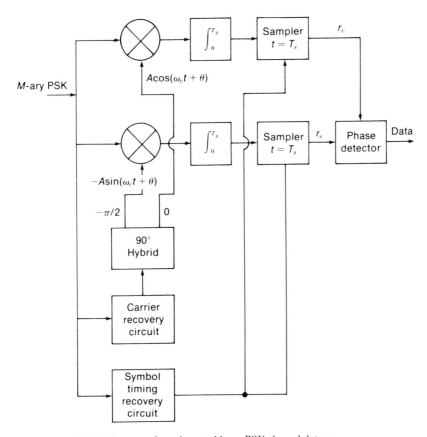

Figure 9.24 Block diagram of a coherent *M*-ary PSK demodulator.

and quadrature correlators, respectively; then

$$r_c = \int_0^{T_s} r(t) \, A \, \cos(\omega_c t + \theta) \, dt$$

$$r_s = -\int_0^{T_s} r(t) \, A \, \sin(\omega_c t + \theta) \, dt$$

hence

$$r_c = E_s \cos \theta_m + n_c$$

$$r_s = E_s \sin \theta_m + n_s \qquad (9.58)$$

where E_s is the signal energy

$$E_s = \int_0^{T_s} s_m^2 (t) \, dt = \frac{A^2 T_s}{2}$$

and n_c and n_s are zero mean Gaussian variables defined by

$$n_c = \int_0^{T_s} n(t) \, A \, \cos(\omega_c t + \theta) \, dt$$

$$n_s = -\int_0^{T_s} n(t) \, A \, \sin(\omega_c t + \theta) \, dt$$

With the assumption that $\omega_c = k 2\pi/T_s$ ($k > 1$ is an integer) we can easily show that n_c and n_s are uncorrelated and hence are independent Gaussian variables with equal variance $E_s N_0/2$.

The likelihood function when the signal $s_m(t)$ is sent is simply the joint probability density function for two Gaussian variables r_c and r_s with mean $E_s \cos \theta_m$ and $E_s \sin \theta_m$, respectively, and with equal variance $E_s N_0/2$:

$$p(r_c, r_s | s_m) = \frac{\exp\{-[(r_c - E_s \cos \theta_m)^2 + (r_s - E_s \sin \theta_m)^2]/E_s N_0\}}{\pi E_s N_0}$$

$$(9.59)$$

Since the demodulator makes a decision based on the phase

$$\hat{\theta}_m = \tan^{-1}\left(\frac{r_s}{r_c}\right)$$

we make the transformation

$$r_c = \hat{r} \cos \hat{\theta}_m$$

$$r_s = \hat{r} \sin \hat{\theta}_m$$

to obtain the joint probability density function

$$p(\hat{r}, \theta_m | s_m) = \frac{\hat{r}}{\pi E_s N_0} \exp\left\{-\frac{[\hat{r}^2 + E_s^2 - 2E_s \hat{r} \cos(\hat{\theta}_m - \theta_m)]}{E_s N_0}\right\}$$

$$(9.60)$$

Define $\phi = \hat{\theta}_m - \theta_m$; then (9.60) becomes

$$p(\hat{r},\phi|s_m) = \frac{\hat{r}}{\pi E_s N_0} \exp\left\{-\frac{[\hat{r}^2 + E_s^2 - 2E_s\hat{r}\cos\phi]}{E_s N_0}\right\} \quad (9.61)$$

A correct decision is made if $\hat{\theta}_m$ is within $\pm\pi/M$ of θ_m or, equivalently, if ϕ falls between $-\pi/M$ and π/M. Therefore, the probability of symbol error is given by

$$P_s = 1 - \int_{-\pi/M}^{\pi/M} \int_0^\infty p(\hat{r},\phi|s_m) \, d\hat{r} \, d\phi$$

$$= 1 - \int_{-\pi/M}^{\pi/M} p(\phi|s_m) \, d\phi \quad (9.62)$$

The probability density function of the phase ϕ (Prob. 9.17) is given by

$$p(\phi|s_m) = \frac{1}{2\pi} \exp\left(-\frac{E_s}{N_0}\right)\left\{1 + \left[\sqrt{2\pi}\,\sqrt{\frac{2E_s}{N_0}}\cos\phi\,\exp\left(\frac{E_s}{N_0}\cos^2\phi\right)\right]\right.$$
$$\left.\left[1 - Q\left(\sqrt{\frac{2E_s}{N_0}}\cos\phi\right)\right]\right\}$$

$$(9.63)$$

For the situation where $E_s/N_0 \gg 1$ we can approximate $Q(\sqrt{2E_s/N_0}\cos\phi)$ by

$$Q\left(\sqrt{\frac{2E_s}{N_0}}\cos\phi\right) \approx \frac{1}{\sqrt{2\pi}\sqrt{2E_s/N_0}\cos\phi}\,\exp\left(-\frac{E_s\cos^2\phi}{N_0}\right)$$

$$(9.64)$$

Substituting (9.64) into (9.63) yields

$$p(\phi|s_m) \approx \frac{1}{\sqrt{2\pi}}\sqrt{\frac{2E_s}{N_0}}\cos\phi\,\exp\left(-\frac{E_s}{N_0}\sin^2\phi\right) \quad (9.65)$$

By using (9.65) in (9.62) we obtain

$$P_s \approx 1 - \frac{1}{\sqrt{2\pi}}\int_{-\pi/M}^{\pi/M}\sqrt{\frac{2E_s}{N_0}}\cos\phi\,\exp\left(-\frac{E_s}{N_0}\sin^2\phi\right) d\phi$$

By transforming the variable ϕ to $u = \sqrt{2E_s/N_0}\sin\phi$ we obtain

$$P_s \approx 1 - \frac{1}{\sqrt{2\pi}}\int_{-\sqrt{2E_s/N_0}\sin(\pi/M)}^{\sqrt{2E_s/N_0}\sin(\pi/M)}\exp\left(-\frac{u^2}{2}\right) du$$

$$P_s \approx 1 - \frac{2}{\sqrt{2\pi}}\int_0^{\sqrt{2E_s/N_0}\sin(\pi/M)}\exp\left(-\frac{u^2}{2}\right) du$$

$$P_s \approx 2Q\left(\sqrt{\frac{2E_s}{N_0}}\sin\frac{\pi}{M}\right) \qquad M \geq 4 \quad (9.66)$$

When a *Gray encoding* scheme is employed, two symbols that correspond to adjacent phases differ by only 1 bit. Hence when the demodulator selects the adjacent phase for the true phase, only 1 bit in $k = \log_2 M$ bits is in error. Therefore, the average probability of bit error for M-ary PSK can be approximated by

$$P_b \approx \frac{P_s}{\log_2 M} \tag{9.67}$$

9.5 FREQUENCY-SHIFT KEYING

Besides PSK, a binary signaling scheme discussed in Sec. 9.2, there is another type of binary signaling called *frequency-shift keying* (FSK) that is commonly used in practice, especially in frequency hopping spread spectrum systems as discussed in Chap. 11.

9.5.1 Coherent FSK

In FSK the waveforms $s_1(t)$ and $s_2(t)$ used to convey the binary digits 0 and 1 are given by

$$s_1(t) = A \cos(\omega_1 t + \phi_1) \quad 0 \leq t \leq T_b$$
$$s_2(t) = A \cos(\omega_2 t + \phi_2) \quad 0 \leq t \leq T_b \tag{9.68}$$

where $\omega_1 \neq \omega_2$ and ϕ_1 and ϕ_2 are the initial phases of $s_1(t)$ and $s_2(t)$, respectively. The information in an FSK signal is carried by the carrier frequencies ω_1 and ω_2 and not by the phase as in PSK. In practice ω_1 and ω_2 may be selected such that $s_1(t)$ and $s_2(t)$ are orthogonal over a signaling interval. In this case FSK signals belong to a class of binary orthogonal waveforms. An example of frequency-orthogonal waveforms is

$$s_n(t) = \cos\left[\left(\omega_0 + \frac{\pi n}{T_b}\right)t + \phi_n\right] \quad 0 \leq t \leq T_b \tag{9.69}$$

where $\omega_0 = $ a multiple of π/T_b and $n = $ an integer.

The average probability of bit error for coherent orthogonal FSK can be found in a way similar to that in which the probability for coherent PSK is found using (9.21) to (9.28) and the orthogonality of $s_1(t)$ and $s_2(t)$:

$$\int_0^{T_b} s_1(t) s_2(t) \, dt = 0$$

Therefore the decision variable z_{12} in (9.17) is a Gaussian variable with mean $\mu = E_b$ and variance $\sigma^2 = E_b N_0$, hence the average probability of bit error is

$$P_b = Q\left(\frac{\mu}{\sigma}\right) = Q\left(\sqrt{\frac{E_b}{N_0}}\right) \tag{9.70}$$

It is seen that coherent orthogonal FSK requires 3 dB more in the ratio of energy per bit to noise density than PSK.

9.5.2 Noncoherent FSK

Most digital modulation schemes using orthogonal FSK employ noncoherent demodulation where knowledge of the initial phase of $s_1(t)$ or $s_2(t)$ is not required or cannot be accurately estimated. To study noncoherent FSK we will assume that the initial phase is a random variable with uniform probability density function $1/2\pi$. That is,

$$p(\phi_1) = \frac{1}{2\pi} \qquad 0 \le \phi_1 \le 2\pi$$

$$p(\phi_2) = \frac{1}{2\pi} \qquad 0 \le \phi_2 \le 2\pi \qquad (9.71)$$

This is a reasonable assumption since the uniform probability density function yields the maximum uncertainty of ϕ_1 or ϕ_2. Now consider the FSK signal $s_m(t)$, $m = 1,2$, and $\omega_1 \ne \omega_2$:

$$s_m(t) = A \cos(\omega_m t + \phi_m) \qquad 0 \le t \le T_b$$

$$= (A \cos \phi_m) \cos \omega_m t - (A \sin \phi_m) \sin \omega_m t \qquad (9.72)$$

With the orthogonality assumption that ω_m is some multiple of $2\pi/T_b$, (9.72) suggests the noncoherent demodulators for orthogonal FSK shown in Fig. 9.25 using four correlators, and in Fig. 9.26 using two filters matched to $s_1(t)$ and $s_2(t)$. Note that a square-law detector is employed after each integrate-and-dump filter in Fig. 9.25 or an envelope detector after each matched filter in Fig. 9.26 to remove the dependence on the phase.

To derive the average probability of bit error for noncoherent orthogonal FSK in an AWGN channel we will consider the demodulator in Fig. 9.25 and assume that the signal $s_1(t)$ was sent. The received signal is

$$r(t) = s_1(t) + n(t) \qquad (9.73)$$

where $n(t)$ = AWGN with zero mean and power spectral density $N_0/2$. The four observables are

$$r_{1c} = E_b \cos \phi_1 + n_{1c}$$

$$r_{1s} = E_b \sin \phi_1 + n_{1s}$$

$$r_{2c} = n_{2c}$$

$$r_{2s} = n_{2s} \qquad (9.74)$$

where E_b is the signal energy:

$$E_b = \int_0^{T_b} s_m^2(t)\ dt = \frac{A^2 T_b}{2} \qquad m = 1,2$$

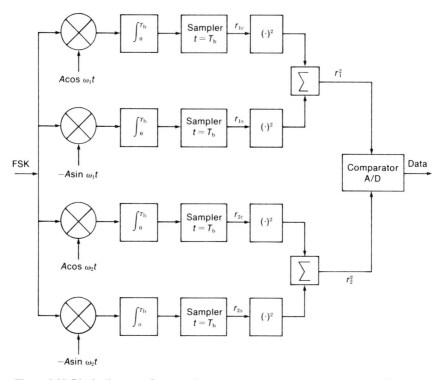

Figure 9.25 Block diagram of a noncoherent demodulator for an orthogonal FSK signal using correlators and square-law detectors.

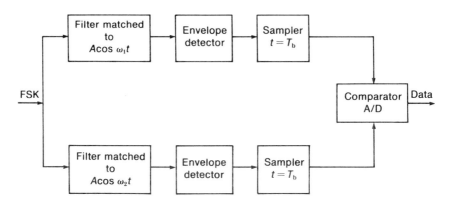

Figure 9.26 Block diagram of a noncoherent demodulator for an orthogonal FSK signal using matched filters and envelope detectors.

and n_{mc} and n_{ms}, $m = 1,2$, are zero mean Gaussian variables defined by

$$n_{mc} = \int_0^{T_b} n(t)A \cos \omega_m t \, dt$$

$$n_{ms} = -\int_0^{T_b} n(t)A \sin \omega_m t \, dt \qquad (9.75)$$

With the assumption that ω_m is a multiple of $2\pi/T_b$ and $\omega_1 \neq \omega_2$ (we leave it for the reader to show that the minimum separation between ω_1 and ω_2 is $2\pi/T_b$ for noncoherent demodulation instead of π/T_b as required for co-herent demodulation), the four variables in (9.75) are mutually indepen-dent Gaussian variables with zero mean and variance $E_b N_0/2$.

The likelihood function when the signal $s_1(t)$ is sent and the phase is ϕ_1 is simply the joint probability density function of the vector of the ob-servables $\mathbf{r} = [r_{1c}, r_{1s}, r_{2c}, r_{2s}]$:

$$p(\mathbf{r}|s_1, \phi_1) = p(r_{1c}, r_{1s}|s_1, \phi_1) \, p(r_{2c}, r_{2s}|s_1) \qquad (9.76a)$$

where

$$p(r_{1c}, r_{1s}|s_1, \phi_1) = \frac{\exp\{-[(r_{1c} - E_b \cos\phi_1)^2 + (r_{1s} - E_b \sin\phi_1)^2]/E_b N_0\}}{\pi E_b N_0} \qquad (9.76b)$$

$$p(r_{2c}, r_{2s}|s_1) = \frac{\exp[-(r_{2c}^2 + r_{2s}^2)/E_b N_0]}{\pi E_b N_0} \qquad (9.76c)$$

Since ϕ_1 is a uniformly distributed random variable between 0 and 2π, we can obtain the likelihood function $p(\mathbf{r}|s_1)$ by averaging over ϕ_1. To simplify the result we define the new variables

$$r_{mc} = r_m \cos \gamma \qquad (9.77a)$$

$$r_{ms} = r_m \sin \gamma \qquad (9.77b)$$

Then

$$p(r_{1c}, r_{1s}|s_1) = \frac{\exp[-(r_1^2 + E_b^2)/E_b N_0]}{\pi E_b N_0} \frac{1}{2\pi} \int_0^{2\pi}$$

$$\exp\left[\frac{2}{N_0}(r_1 \cos \gamma \cos \phi_1 + r_1 \sin \gamma \sin \phi_1)\right] d\phi_1$$

$$= \frac{\exp[-(r_1^2 + E_b^2)/E_b N_0]}{\pi E_b N_0} I_0\left(\frac{2}{N_0} r_1\right) \qquad (9.78a)$$

where $I_0(\cdot)$ is the modified Bessel function of order zero:

$$I_0\left(\frac{2}{N_0} r_1\right) = \frac{1}{2\pi} \int_0^{2\pi} \exp\left[\frac{2}{N_0} r_1 \cos(\phi_1 - \gamma)\right] d\phi_1$$

and

$$p(r_{2c}, r_{2s} | s_1) = \frac{\exp(-r_2^2 / E_b N_0)}{\pi E_b N_0} \tag{9.78b}$$

Thus

$$p(\mathbf{r} | s_1) = \frac{1}{(\pi E_b N_0)^2} \exp\left[-\frac{(r_1^2 + r_2^2 + E_b^2)}{E_b N_0}\right] I_0\left(\frac{2}{N_0} r_1\right) \tag{9.79}$$

By symmetry, the likelihood function $p(\mathbf{r} | s_2)$ is

$$p(\mathbf{r} | s_2) = \frac{1}{(\pi E_b N_0)^2} \exp\left[-\frac{(r_1^2 + r_2^2 + E_b^2)}{E_b N_0}\right] I_0\left(\frac{2}{N_0} r_2\right) \tag{9.80}$$

Returning to the optimum decision rule given in (9.15) with the assumption that the signals $s_1(t)$ and $s_2(t)$ have equal a priori probabilities, the decision rule for two signals will be in favor of $s_1(t)$ if

$$\ln p(\mathbf{r} | s_1) > \ln p(\mathbf{r} | s_2) \tag{9.81}$$

or, equivalently,

$$\ln I_0\left(\frac{2}{N_0} r_1\right) > \ln I_0\left(\frac{2}{N_0} r_2\right) \tag{9.82}$$

Since $I_0(x)$ is a monotonically increasing function of x, (9.82) is equivalent to

$$r_1 > r_2 \tag{9.83}$$

The probability of error given that the signal $s_1(t)$ was sent is simply

$$\Pr(r_2 > r_1 | s_1) = \int_0^\infty \int_{r_1}^\infty p(r_1, r_2 | s_1) \, dr_2 \, dr_1 \tag{9.84}$$

It remains to determine the joint probability function of r_1 and r_2 given $s_1(t)$. Since r_1 and r_2 are independent, we have

$$p(r_1, r_2 | s_1) = p(r_1 | s_1) \, p(r_2 | s_1) \tag{9.85}$$

Also, r_m, $m = 1, 2$, is related to r_{mc} and r_{ms} by the transformation given in (9.77); then the probability density functions $p(r_m | s_1)$, $m = 1, 2$, can be related to the joint probability density functions $p(r_{mc}, r_{ms} | s_1)$:

$$p(r_m | s_1) = 2\pi r_m \, p(r_{mc}, r_{ms} | s_1) \qquad m = 1, 2 \tag{9.86}$$

From (9.78a) and (9.78b) we conclude that

$$p(r_1 | s_1) = \frac{2r_1}{E_b N_0} \exp\left[\frac{-(r_1^2 + E_b^2)}{E_b N_0}\right] I_0\left(\frac{2}{N_0} r_1\right) \tag{9.87}$$

$$p(r_2 | s_1) = \frac{2r_2}{E_b N_0} \exp\left(\frac{-r_2^2}{E_b N_0}\right) \tag{9.88}$$

Substituting (9.85), (9.87), and (9.88) into (9.84) yields

$$\Pr(r_2 > r_1 | s_1) = \int_0^\infty p(r_1 | s_1) \int_{r_1}^\infty p(r_2 | s_1) \, dr_2 \, dr_1$$

$$= \int_0^\infty p(r_1 | s_1) \exp\left(-\frac{r_1^2}{E_b N_0}\right) dr_1$$

$$= \frac{1}{2} \exp\left(-\frac{E_b}{2N_0}\right) \int_0^\infty \frac{r_1}{E_b N_0/4} \exp\left(-\frac{r_1^2 + E_b^2/4}{E_b N_0/2}\right) I_0\left(\frac{2}{N_0} r_1\right) dr_1$$

$$= \frac{1}{2} \exp\left(-\frac{E_b}{2N_0}\right)$$

$$(9.89)$$

We have used the fact that the integral in the third equation in (9.89) is the integral of a Rician probability density function, hence must equal 1. By symmetry we have

$$\Pr(r_1 > r_2 | s_2) = \frac{1}{2} \exp\left(-\frac{E_b}{2N_0}\right) \qquad (9.90)$$

Thus the average probability of bit error is

$$P_b = \frac{1}{2} \exp\left(-\frac{E_b}{2N_0}\right) \qquad (9.91)$$

9.6 M-ary FREQUENCY-SHIFT KEYING

In this section we generalize the results of orthogonal binary FSK to orthogonal *M-ary frequency-shift keying* (*M*-ary FSK) with the purpose of conserving power at the expense of an increased bandwidth. In orthogonal *M*-ary FSK, the $M = 2^k$ distinct symbols are represented by *M* frequency-orthogonal waveforms:

$$s_m(t) = A \cos(\omega_m t + \phi_m) \qquad m = 1, 2, \ldots, M; \; 0 \leq t \leq T_s \quad (9.92)$$

and

$$\int_0^{T_s} s_i(t) s_j(t) \, dt = \begin{cases} E_s & i = j \\ 0 & i \neq j \end{cases} \qquad (9.93)$$

where ω_m = a multiple of π/T_s
ϕ_m = initial phase
A = signal amplitude
$T_s = kT_b$ = signaling interval
T_b = bit duration
$E_s = A^2 T_s/2$ = signal energy or energy per symbol

9.6.1 Coherent *M*-ary FSK

The demodulator for *coherent* M-ary FSK is exactly the one shown in Fig. 9.1. To derive the average probability of symbol error, we assume that the signal $s_1(t)$ was sent and the received signal is

$$r(t) = s_1(t) + n(t)$$

where $n(t) = $ AWGN with zero mean and power spectral density $N_0/2$. The M observable variables at the output of M correlators are the set S_m, $m = 1,2,\ldots,M$, given by (9.18), where

$$S_1 = E_s + n_1$$

$$S_m = n_m \qquad m = 2,3,\ldots,M \qquad (9.94)$$

where n_m, $m = 1,2,\ldots,M$, are independent Gaussian variables with zero mean and variance $E_s N_0/2$ defined by

$$n_m = \int_0^{T_s} n(t)\, s_m(t)\, dt$$

The probability density function of S_m, $m = 1,2,\ldots,M$, given that $s_1(t)$ was sent is ($\delta_{1m} = 1$ if $m = 1$, and 0 if $m \neq 1$)

$$p(S_m|s_1) = \frac{\exp\{-[(S_m - \delta_{1m}E_s)^2/E_s N_0]\}}{\sqrt{\pi E_s N_0}} \qquad (9.95)$$

Thus the probability that the demodulator will make an incorrect decision is simply

$$P_{e1} = 1 - \Pr(S_2 < S_1 \cap S_3 < S_1 \cap \cdots \cap S_M < S_1|s_1)$$

$$= 1 - \int_{-\infty}^{\infty} \int_{-\infty}^{S_1} \cdots \int_{-\infty}^{S_1} p(S_1,S_2,\ldots,S_M|s_1)\, dS_1\, dS_2 \cdots dS_M \qquad (9.96)$$

It remains to find the joint probability density function $p(S_1,S_2,\ldots,S_M|s_1)$. Since S_m, $m = 1,2,\ldots,M$, are independent, we have

$$p(S_1,S_2,\ldots,S_M|s_1) = \prod_{m=1}^{M} p(S_m|s_1) \qquad (9.97)$$

Substituting (9.97) into (9.96) and using the fact that

$$\int_{-\infty}^{S_1} p(S_m|s_1)\, dS_m, \quad m = 2,3,\ldots,M$$

are identical, we have

$$P_{e1} = 1 - \int_{-\infty}^{\infty} \left[\int_{-\infty}^{S_1} p(S_m|s_1)\, dS_m \right]^{M-1} p(S_1|s_1)\, dS_1 \qquad (9.98)$$

Note that

$$\int_{-\infty}^{S_1} p(S_m|s_1)\, dS_m = 1 - \int_{S_1}^{\infty} \frac{\exp(-S_m^2/E_sN_0)}{\sqrt{\pi E_sN_0}}\, dS_m \qquad (9.99)$$

By using the change in variable $x = \sqrt{2/E_sN_0}\, S_m$ we can write (9.99) as

$$\int_{-\infty}^{S_1} p(S_m|s_1)\, dS_m = 1 - \int_{\sqrt{2/E_sN_0}S_1}^{\infty} \frac{\exp(-x^2/2)}{\sqrt{2\pi}}\, dx$$

$$= 1 - Q\left(\sqrt{\frac{2}{E_sN_0}}\, S_1\right) \qquad (9.100)$$

where $Q(\cdot)$ is the familiar Gaussian integral function given in (9.29). Substituting (9.95), with $m = 1$, and (9.100) into (9.98) yields

$$P_{e1} = 1 - \int_{-\infty}^{\infty} \left[1 - Q\left(\sqrt{\frac{2}{E_sN_0}}\, S_1\right)\right]^{M-1} \frac{\exp[-(S_1 - E_s)^2/E_sN_0]}{\sqrt{\pi E_sN_0}}\, dS_1$$

$$(9.101)$$

Now let $y = \sqrt{2/E_sN_0}\,(S_1 - E_s)$; then P_{e1} becomes

$$P_{e1} = 1 - \frac{1}{\sqrt{2\pi}} \int_{-\infty}^{\infty} \left[1 - Q\left(y + \sqrt{\frac{2E_s}{N_0}}\right)\right]^{M-1} \exp\left(\frac{-y^2}{2}\right) dy$$

$$= 1 - \frac{1}{\sqrt{2\pi}} \int_{-\infty}^{\infty} \left[1 - Q\left(y + \sqrt{\frac{2E_b \log_2 M}{N_0}}\right)\right]^{M-1} \exp\left(\frac{-y^2}{2}\right) dy$$

$$(9.102)$$

where E_b is the energy per bit.

Since all the signaling waveforms are equally likely, the average probability of symbol error is simply

$$P_s = P_{e1} \qquad (9.103)$$

To derive the average probability of bit error we note that all symbol errors occur with equal probability:

$$\frac{P_s}{M-1} = \frac{P_s}{2^k - 1} \qquad (9.104)$$

Since there are $\binom{k}{h}$ ways in which h out of k bits in a symbol may be in error, the average number of bit errors per k-bit symbol is

$$\sum_{h=1}^{k} h\binom{k}{h} \frac{P_s}{2^k - 1} = k\, \frac{2^{k-1}}{2^k - 1}\, P_s$$

Because there are k bits in a symbol, the average probabilities of bit error is simply

$$P_b = \frac{2^{k-1}}{2^k - 1}\, P_s \qquad (9.105)$$

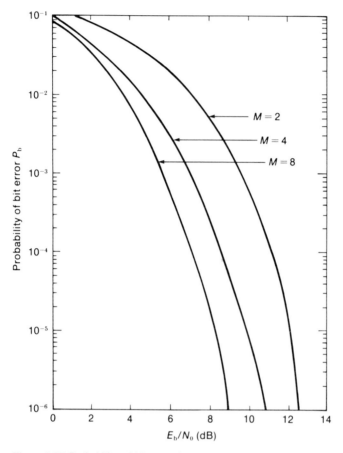

Figure 9.27 Probability of bit error for coherent orthogonal M-ary FSK.

The average probability of bit error for coherent orthogonal M-ary FSK is plotted in Fig. 9.27. It is seen that, by increasing the number of waveforms, one can reduce the ratio of energy per bit to noise density but at the expense of the bandwidth. Note that the minimum frequency separation for coherent orthogonal M-ary FSK is π/T_s rad/s.

9.6.2 Noncoherent M-ary FSK

The demodulator for *noncoherent* orthogonal M-ary FSK consists of a bank of $2M$ correlators similar to the one shown in Fig. 9.25. We leave it for the reader to show that the minimum frequency spacing for noncoherent orthogonal M-ary FSK is $2\pi/T_s$ rad/s. Assuming that the signal $s_1(t)$ with uniformly distributed random phase ϕ_1 was sent, then the received signal is

$$r(t) = s_1(t) + n(t)$$

where $n(t) = $ AWGN with zero mean and power spectral density $N_0/2$. The vector of $2M$ observable variables is $\mathbf{r} = [r_{1c}, r_{1s}, r_{2c}, r_{2s}, \ldots, r_{Mc}, r_{Ms}]$ where

$$r_{1c} = E_s \cos \phi_1 + n_c$$

$$r_{1s} = E_s \sin \phi_1 + n_{1s}$$

$$\left.\begin{matrix} r_{mc} = n_{mc} \\ r_{ms} = n_{ms} \end{matrix}\right\} \quad m = 2,3,\ldots,M \qquad (9.106)$$

where $E_s = $ signal energy or energy per symbol as defined in (9.93) and n_{mc} and n_{ms}, $m = 1,2,\ldots,M$, are independent Gaussian variables with zero mean and variance $E_s N_0/2$ defined as

$$n_{mc} = \int_0^{T_s} n(t)\, A \cos \omega_m t\, dt$$

$$n_{ms} = -\int_0^{T_s} n(t)\, A \sin \omega_m t\, dt$$

Using the analysis in Sec. 9.5 we can easily show that the decision rule will favor $s_1(t)$ if

$$r_1 > r_m \qquad m = 2,3,\ldots,M \qquad (9.107)$$

where

$$r_m^2 = r_{mc}^2 + r_{ms}^2 \qquad m = 1,2,\ldots,M \qquad (9.108)$$

Thus the probability of error given that the signal $s_1(t)$ is sent is given by

$$P_{e1} = 1 - \Pr(r_2 < r_1 \cap r_3 < r_1 \cap \cdots \cap r_m < r_1 | s_1)$$

$$= 1 - \int_0^\infty \int_0^{r_1} \cdots \int_0^{r_1} p(r_1, r_2, \ldots, r_M | s_1)\, dr_1\, dr_2 \cdots dr_M$$

$$(9.109)$$

Since r_m, $m = 1,2,\ldots,M$, are independent variables, we have

$$p(r_1, r_2, \ldots, r_M | s_1) = \prod_{m=1}^M p(r_m | s_1) \qquad (9.110)$$

Note that $p(r_1 | s_1)$ is given by (9.87) and $p(r_m | s_1)$, $m = 2,3,\ldots,M$, are given by (9.88) with r_m replacing r_2. Furthermore,

$$\int_0^{r_1} p(r_m | s_1)\, dr_m = 1 - \exp\left(\frac{-r_1^2}{E_s N_0}\right) \qquad m = 2,3,\ldots,M$$

$$(9.111)$$

Substituting (9.110) and (9.111) into (9.109) yields

$$P_{e1} = 1 - \int_0^\infty \left[\int_0^{r_1} p(r_m|s_1) \, dr_m \right]^{M-1} p(r_1|s_1) \, dr_1$$

$$= 1 - \int_0^\infty \left[1 - \exp\left(\frac{-r_1^2}{E_s N_0}\right) \right]^{M-1} p(r_1|s_1) \, dr_1 \qquad (9.112)$$

To evaluate the integral on the right-hand side of (9.112) we note the equivalence

$$P_{e1} = \int_0^\infty \left\{ 1 - \left[1 - \exp\left(\frac{-r_1^2}{E_s N_0}\right) \right]^{M-1} \right\} p(r_1|s_1) \, dr_1 \qquad (9.113)$$

Note that

$$1 - \left[1 - \exp\left[\frac{-r_1^2}{E_s N_0}\right] \right]^{M-1} = 1 - \sum_{j=0}^{M-1} (-1)^j \binom{M-1}{j} \exp\left(\frac{-jr_1^2}{E_s N_0}\right)$$

$$= \sum_{j=1}^{M-1} (-1)^{j+1} \binom{M-1}{j} \exp\left(\frac{-jr_1^2}{E_s N_0}\right)$$

$$(9.114)$$

Substituting (9.87) and (9.114) into (9.113) yields

$$P_{e1} = \sum_{j=1}^{M-1} (-1)^{j+1} \frac{1}{j+1} \binom{M-1}{j} \exp\left[-\frac{j}{j+1} \left(\frac{E_s}{N_0}\right) \right] \qquad (9.115)$$

Since all signal waveforms are equally likely, by symmetry the average probability of symbol error is

$$P_s = P_{e1} \qquad (9.116)$$

Based on (9.105), the average probability of bit error is

$$P_b = \frac{2^{k-1}}{2^k - 1} P_s \qquad (9.117)$$

and is plotted in Fig. 9.28. Note that $E_b = E_s/\log_2 M$.

9.7 OQPSK AND MSK

In Sec. 9.3 we investigated a bandwidth- and power-efficient quadrature carrier modulation scheme, namely, QPSK. In this section we examine two additional quadrature carrier modulation schemes that perform similarly to QPSK. These are *offset QPSK* (also called *staggered QPSK*) and *minimum-shift keying* (MSK).

Offset QPSK is a modification of QPSK where the Q digits are

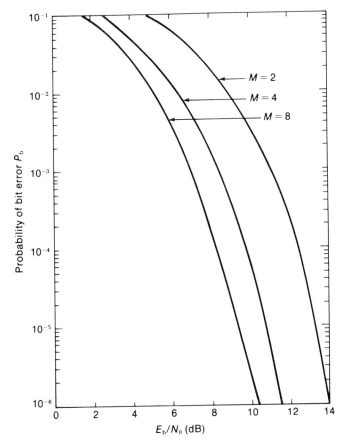

Figure 9.28 Probability of bit error for noncoherent orthogonal M-ary FSK.

delayed by 1 bit or T_b seconds with respect to the corresponding P digits. Figure 9.29 illustrates the modulation and demodulation of OQPSK. The difference in time alignment between P- and Q-channel data streams does not change the power spectral density, hence OQPSK possesses the same power spectral density as QPSK. Furthermore, since the orthogonality between the P and Q channels is preserved, the average probability of bit error for OQPSK is the same as that of QPSK.

MSK can be thought of as a modification of OQPSK where the rectangular symbol pulse is replaced by a half-cycle sinusoidal symbol pulse. Figure 9.30 illustrates the modulation and demodulation of MSK. If we let $x_c(t)$ and $x_s(t)$ denote the in-phase and T_b-delayed quadrature symbol

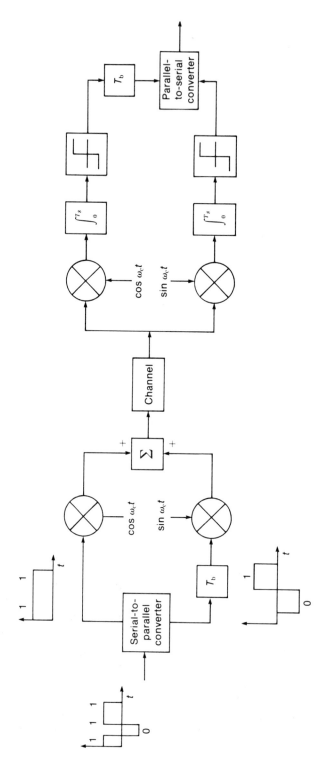

Figure 9.29 OQPSK modulator and demodulator.

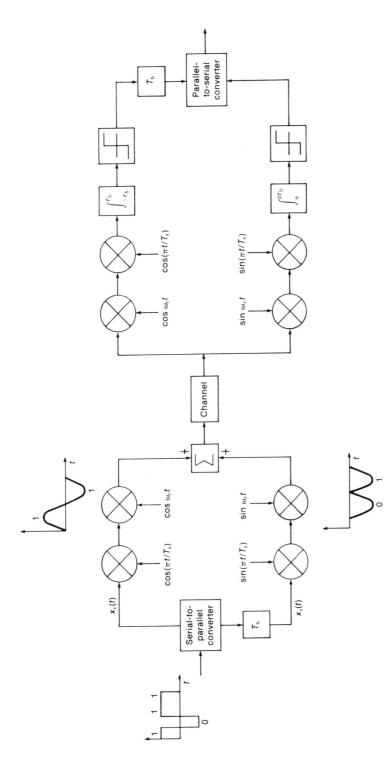

Figure 9.30 MSK modulator and demodulator.

streams represented by

$$x_c(t) = \sum_{k=-\infty}^{\infty} a_k g(t - kT_s) \qquad a_k = \pm 1$$

$$x_s(t) = \sum_{k=-\infty}^{\infty} b_k g(t - kT_s + T_b)$$

$$= \sum_{k=-\infty}^{\infty} b_k g[t - (k - \tfrac{1}{2}) T_s] \qquad b_k = \pm 1$$

where $g(t) =$ a rectangular pulse of unit amplitude and duration T_s. Then the MSK signal at the output of the modulator can be written as

$$s(t) = \frac{A}{\sqrt{2}} x_c(t) \cos\left(\frac{\pi t}{T_s}\right) \cos \omega_c t$$

$$+ \frac{A}{\sqrt{2}} x_s(t) \sin\left(\frac{\pi t}{T_s}\right) \sin \omega_c t \qquad (9.118)$$

Using a well-known trigonometric identity we can rewrite (9.118) as

$$s(t) = \frac{A}{\sqrt{2}} \cos\left[\omega_c t - x_c(t) x_s(t) \frac{\pi t}{T_s} + \theta\right] \qquad (9.119a)$$

where

$$\theta = \begin{cases} 0 & x_c(t) = 1 \\ \pi & x_c(t) = -1 \end{cases} \qquad (9.119b)$$

Equation (9.119) shows that MSK is constant-envelope continuous-phase FSK (the phase is continuous at the bit transition instants when the carrier frequency $f_c = \omega_c/2\pi$ is chosen to be an integral multiple of $1/4T_b = 1/2T_s$) with signaling frequencies

$$f_1 = f_c + \frac{1}{2T_s}$$

$$f_2 = f_c - \frac{1}{2T_s}$$

Hence the minimum frequency spacing equals the symbol rate or half the bit rate:

$$\Delta f = f_1 - f_2 = \frac{1}{T_s} = \frac{1}{2T_b}$$

This is the minimum frequency spacing which allows the two FSK signals to be orthogonal, hence the name "minimum-shift" keying. The power spectral density of an MSK signal (Appendix 9B) is given by

$$S(f) = \frac{8E_b}{\pi^2} \left\{ \left[\frac{\cos \pi (f - f_c) T_s}{4(f - f_c)^2 T_s^2 - 1} \right]^2 + \left[\frac{\cos \pi (f + f_c) T_s}{4(f + f_c)^2 T_s^2 - 1} \right]^2 \right\}$$

$$(9.120)$$

$S(f)$ is plotted in Fig. 9.31 together with the power spectral densities of QPSK and OQPSK signals. It is observed that MSK has a wider main lobe (the first null is at $1.5/T_s$) than QPSK or OQPSK (the first null is at $1/T_s$) but lower sidelobes.

The average probability of bit error for MSK is the same as that for QPSK or OQPSK, since the orthogonality of the P and Q channels is preserved over a symbol duration. This is verified by observing the MSK orthonormal basis functions derived from (9.118):

$$u_1(t) = 2\sqrt{\frac{1}{T_s}} \cos\left(\frac{\pi t}{T_s}\right) \cos \omega_c t \qquad 0 \le t \le T_s$$

$$u_2(t) = 2\sqrt{\frac{1}{T_s}} \sin\left(\frac{\pi t}{T_s}\right) \sin \omega_c t \qquad 0 \le t \le T_s$$

9.8 BAND-LIMITED NONLINEAR SATELLITE CHANNEL

In the previous sections we studied the performance of digitally modulated signals in a bandpass channel with no constraints imposed on it. In a digital satellite system, the signal is transmitted over a finite bandwidth and the channel is nonlinear; thus intuitively one would expect

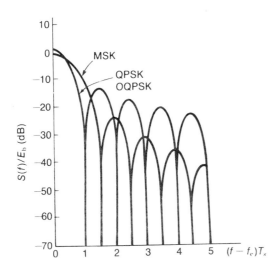

Figure 9.31 Power spectral densities of QPSK, OQPSK, and MSK.

the performance to degrade. The designer therefore must know how to analyze such a band-limited nonlinear channel and to select proper filters and a TWTA operating point to minimize performance degradation. This section is devoted to such a task. The first question concerns the bandwidth of the satellite channel. Given a channel symbol rate of $1/T_s$ symbols per second, where T_s is the symbol duration, we want to find the bandwidth required to support such a symbol rate. For M-ary signaling where $M = 2^k$ is the number of distinct symbols, the minimum bandwidth B is given by the dimensionality theorem [1] as

$$B \approx \frac{N}{2T_s} \qquad (9.121)$$

where N = number of orthonormal basis functions (Sec. 9.1). For example, the minimum bandwidth required for a QPSK signal ($N = 2$) is

$$B \approx \frac{1}{T_s} \qquad (9.122)$$

Now consider the QPSK or OQPSK power spectral density in Fig. 9.31. We see that the main lobe alone occupies a bandwidth $2/T_s$, centered around the carrier frequency f_c, which is twice the minimum required bandwidth of $1/T_s$. The question is whether we can shape the QPSK power spectral density such that it will occupy a bandwidth between $1/T_s$ and $2/T_s$ without serious performance degradation. For a linear satellite channel such optimum shaping is possible. When the channel is nonlinear, however, there is only suboptimum shaping. Shaping of the power spectral density of a digital modulation signal is accomplished by the transmit filter (lowpass or bandpass) in the modulator (Fig. 9.14) and the receive filter (lowpass or bandpass) in the demodulator (Fig. 9.15). Recall that the power spectral density plotted in Fig. 9.31 results from a bipolar baseband signal (the modulating waveform) which is a rectangular pulse train of amplitude $+1$ or -1 and pulse duration T_s. Thus, shaping this spectrum means reshaping the rectangular pulse that represents the M-ary symbols. The question now is, What type of pulse shape should we employ? We know that, when rectangular pulses pass through a filter, the pulse for each symbol spreads in time and causes interference with the pulses representing the adjacent symbols. This effect is called *intersymbol interference* and is illustrated in Fig. 9.32. If the intersymbol interference effect is too strong at the sampling time, an erroneous decision may occur. For example, a 0 transmitted may be decoded as a 1 if the tails of the adjacent pulses add up to a higher value that exceeds the decision threshold

Is it possible to minimize intersymbol interference? The answer is affirmative for a linear channel. Let us consider the satellite channel illustrated in Fig. 9.33. The transmit filter in the modulator, the receive filter in

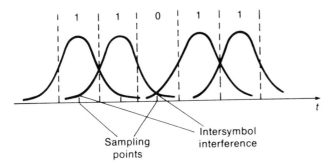

Sampling
points

Intersymbol
interference

Figure 9.32 Concept of intersymbol interference.

the demodulator, and the satellite input and output multiplexers are described by their amplitude and group delay characteristics. The nonlinear satellite TWTA and earth station HPA are described by their input/output amplitude and phase transfer characteristics (Fig. 5.2). Now assume that both the TWTA and the HPA are operated in the linear region, hence the overall channel can be considered a band-limited channel consisting of a cascade of four filters. The equivalent lowpass transfer characteristics of the overall linear channel is

$$\hat{H}(f) = \hat{H}_1(f) \; \hat{H}_2(f) \; \hat{H}_3(f) \; \hat{H}_4(f) \qquad (9.123)$$

where $\hat{H}(f)$, $i = 1,2,3,4$, represents the equivalent lowpass transfer characteristic of the filters in Fig. 9.33. By using the equivalent lowpass transfer characteristic we can consider the modulating baseband signal instead of the digitally modulated signal. Now consider the pulse train $x(t)$ applied at the input of the channel:

$$x(t) = \sum_{n=-\infty}^{\infty} a_n g(t - nT_s) = \left[\sum_{n=-\infty}^{\infty} a_n \, \delta(t - nT_s) \right] \circledast g(t)$$

$$(9.124)$$

where a_n may take on any of the allowed M multilevels [e.g., $a_n = \pm 1$ for PSK, $a_n = (\pm 1 \pm j)/\sqrt{2}$ for QPSK] and $g(t)$ is a rectangular pulse of unit amplitude and duration T_s. The output of the linear channel is

$$y(t) = \left[\sum_{n=-\infty}^{\infty} a_n \, \delta(t - nT_s) \right] \circledast p(t) \qquad (9.125)$$

where

$$p(t) = g(t) \circledast \hat{h}(t) \qquad (9.126)$$

and $\hat{h}(t)$ is the inverse Fourier transform of $\hat{H}(f)$ in (9.123). The function $p(t)$ is the pulse shape at the channel output.

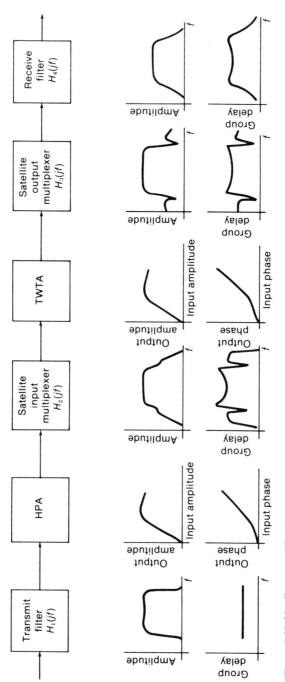

Figure 9.33 Nonlinear satellite channel.

The purpose now is to select a particular pulse shape $p(t)$ for minimizing intersymbol interference at the sampling time $t = nT_s$. One of the pulse shapes is the $\sin x/x$ type, where $p(t) = (1/T_s) \sin(\pi t/T_s)/\pi t/T_s$. The spectrum of this pulse—the Fourier transform of $p(t)$—is $P(f) = 1$ for $|f| < 1/2T_s$ and 0 for $|f| > 0$. It is obvious that the required equivalent lowpass bandwidth of the channel is simply $B_L = 1/2T_s$, and therefore the required bandpass bandwidth centered around the carrier frequency must be $B = 1/T_s$ (the symbol duration–bandwidth product being $T_s B = 1$). Thus the $\sin x/x$ type of overall pulse shape produces a minimum bandwidth and zero intersymbol interference at the sampling time $t = kT_s$, since

$$y(t = kT_s) = \left\{ \sum_{n=-\infty}^{\infty} a_n \delta[(k-n)T_s] \right\} \circledast p(kT_s) = a_k p(0)$$

for $k = n$ and 0 when $k \neq n$. However, two practical difficulties make the $\sin x/x$ pulse shape undesirable:

1. The ideal shape of $P(f)$ is physically unrealizable.
2. The slow decay rate of $p(t)$ as $1/t$ for large t requires precise synchronization of the symbol clock for sampling. A small timing error may cause intersymbol interference.

To overcome these difficulties and still achieve zero intersymbol interference for a linear channel, a raised cosine pulse is employed in practice. Such a pulse shape is given in both time domain and frequency domain as

$$p(t) = 2f_0 \left(\frac{\sin 2\pi f_0 t}{2\pi f_0 t} \right) \left[\frac{\cos 2\pi r f_0 t}{1 - (4rf_0 t)^2} \right] \tag{9.127}$$

$$P(f) = \begin{cases} 1, & |f| < (1-r)f_0 \\ \frac{1}{2}\left\{1 + \cos\left[\frac{\pi(|f| - (1-r)f_0)}{2rf_0}\right]\right\} & (1-r)f_0 < |f| < (1+r)f_0 \\ 0, & |f| > (1+r)f_0 \end{cases}$$
$$\tag{9.128}$$

The parameter r in (9.127) and (9.128) is called the *roll-off factor*. Plots of $P(f)$ and $p(t)$ are shown in Fig. 9.34 for a roll-off factor of $r = 0$ (0%), $r = 0.5$ (50%), and $r = 1$ (100%). The case $r = 0$ corresponds to the minimum equivalent lowpass bandwidth $B_L = f_0$. If f_0 is chosen to be $f_0 = 1/2T_s$, then $B_L = 1/2T_s$. The corresponding minimum bandpass bandwidth centered around the carrier frequency f_c is of course $B = 2f_0 = 1/T_s$ (the symbol duration–bandwidth product being $T_s B = 1$). The pulse shape for $r = 0$ is simply the $\sin x/x$ type discussed previously. As the roll-off factor increases, as in the case of $r = 0.5$ (50%) or $r = 1$ (100%), the absolute bandpass bandwidth also increases from $B = 2f_0 = 1/T_s$ to $B = 3f_0 = 1.5/T_s$ or $B = 4f_0 = 2/T_s$, respectively. Also,

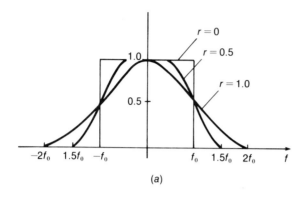

(a)

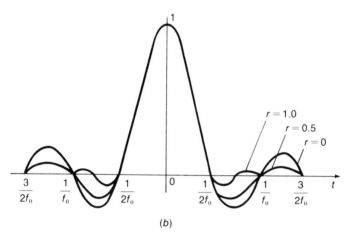

(b)

Figure 9.34 (a) Raised-cosine filter. (b) Raised-cosine pulse.

the synchronization timing requirement can be relaxed since $p(t)$ decays faster then $1/t$ for large t. In practical satellite channels, the roll-off r is normally chosen between 0.4 (40%) and 0.5 (50%). The remaining problem is how to design the various filters in the satellite channel shown in Fig. 9.33. From (9.126) we conclude that

$$P(f) = G(f)\hat{H}(f) \quad (9.129)$$

Substituting (9.123) into (9.129) yields

$$P(f) = G(f)\hat{H}_1(f)\hat{H}_2(f)\hat{H}_3(f)\hat{H}_4(f) \quad (9.130)$$

Given the satellite input and output multiplexer characteristics $\hat{H}_2(f)$ and $\hat{H}_3(f)$, the transmit and receive filter characteristics can be selected as

$$\hat{H}_1(f) = \frac{\sqrt{P(f)}}{G(f)\ \sqrt{\hat{H}_2(f)\hat{H}_3(f)}} \tag{9.131a}$$

$$\hat{H}_4(f) = \frac{\sqrt{P(f)}}{\sqrt{\hat{H}_2(f)\hat{H}_3(f)}} \tag{9.131b}$$

When the baseband signal is a bipolar waveform, $g(t)$ is simply a rectangular pulse of unit amplitude and duration T_s. Therefore $G(f) = T_s \sin(\pi f T_s)/\pi f T_s$, which is of the form $\sin x/x$. A compensation filter of the form $x/\sin x$ can be built and put in cascade with the filter $\sqrt{P(f)}/\sqrt{\hat{H}_2(f)\hat{H}_3(f)}$ to realize $\hat{H}_1(f)$ in (9.131a). In this case $\hat{H}_1(f)$ and $\hat{H}_4(f)$ can be made identical and take the form $\sqrt{P(f)}/\sqrt{\hat{H}_2(f)\hat{H}_3(f)}$. Furthermore, if the amplitude characteristics of the satellite input and output multiplexers are approximately flat over the channel bandwidth, and their phase characteristics are approximately linear (the group delay is approximately constant), then $\hat{H}_1(f)$ and $\hat{H}_4(f)$ can be designed as linear phase filters with amplitude response $\sqrt{P(f)}$; that is, each transmit and receive filter accomplishes half the desired pulse shaping (in satellite terminology $\sqrt{P(f)}$ is called $\sqrt{100r\%}$ raised-cosine filter).

As shown in Sec. 9.1, the effect of channel noise is minimized by using a correlator or matched filter as the receive filter. But if a matched filter is used as $\hat{H}_4(f)$ in (9.131b), the overall pulse shape will not necessarily satisfy the raised cosine characteristic in (9.127) for minimum intersymbol interference. However, if the linear satellite channel is downlink-limited (i.e., the uplink noise contributes negligibly), for a flat downlink noise spectrum, the transmit and receive filters with characteristics given in (9.131) minimize the effect of both intersymbol interference and downlink noise [4,5].

In a digital satellite channel such as the one used in a TDMA operation, the requirement for a high level of transmit power to reduce the noise effect gives rise to the need to operate the satellite TWTA at or near saturation, and to a lesser extent to operate the earth station HPA in the nonlinear region. Therefore, the satellite channel can be highly nonlinear, and the above pulse shaping criterion for minimizing intersymbol interference in a linear channel no longer holds. Furthermore, nonlinear operation of the HPA and TWTA produces spectrum spreading of the filtered modulated signal (Fig. 4.15). Spectrum spreading causes both uplink and downlink adjacent channel interference (ACI). As the name implies, *adjacent channel interference* is caused by a spillover of energy from adjacent channel(s) into the desired channel.

For a narrowband satellite channel ($BT_s < 1.4$), uplink adjacent channel interference is caused mainly by spectrum spreading at the output of the earth station HPA. It is almost independent of the selection of the transmit filter and can be reduced by backing off the HPA operating point or by placing a filter at the output of the HPA. The latter solution is not practical because of the insertion loss of the filter. Furthermore, the filter prevents the use of transponder hopping in a TDMA operation. Downlink adjacent channel interference is caused by spectrum spreading at the output of the satellite TWTA. It can be reduced by tightening the out-of-band attenuation of the satellite output multiplexer. The receive filter also helps to reduce downlink adjacent channel interference.

A much more devastating interference due to spectrum spreading is the *multipath effect*, where spillover energy from the desired channel falls into adjacent channels and combines with the desired signal at the transponder output to form a composite signal whose amplitude and phase differ from those of the desired signal. This can cause severe distortion of the desired signal. Adjacent channel and multipath interference is illustrated in Fig. 9.35. The reason for multipath interference is that, when the main signals in the adjacent channels are not present, the adjacent channel TWTA amplifies the small multipath signal with a higher gain than the gain it gives the main signal because the TWTA is now operating in the linear region instead of at or near saturation. In this worst case, the multipath signal may experience an additional gain of 6 to 10 dB. Performance degradation caused by the worst-case multipath interference is much more severe than that caused by channel noise or other incoherent interference of the same power. At a bit error rate of 1 in 10^6, the degradation in E_b/N_0 can be as high as 2 dB.

An evaluation of adjacent channel interference power requires the power spectral density of the modulated signal at the output of the nonlinear HPA (for uplink ACI) and at the output of the nonlinear TWTA (for downlink ACI). The characteristic of these nonlinear amplifiers has been discussed in Chap. 5. Methods for evaluation of the power spectral density of QPSK, OQPSK, and MSK can be found in [6].

9.9 COMPUTER SIMULATION OF A SATELLITE CHANNEL

The theoretical analysis of a band-limited nonlinear satellite channel in terms of the average probability of bit error is extremely tedious if not impossible. This is the case because intersymbol interference now depends on the nonlinearity of the channel, the uplink noise, and the channel filter characteristics and cannot be eliminated by the pulse shaping technique discussed in Sec. 9.8. The problem is compounded by the addition of adjacent channel interference resulting from spectrum spreading of carriers in adjacent transponders and multipath interference caused by spectrum spreading of the main carrier itself (Fig. 9.35). Furthermore, the effect of

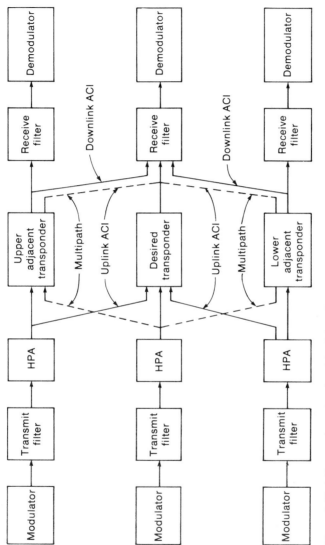

Figure 9.35 Adjacent channel and multipath interference.

451

cochannel interference must be taken into account if frequency reuse is employed. The complexity of a band-limited nonlinear satellite channel therefore necessitates the use of computer simulation to obtain the average probability of bit error.

For a given bit rate, the average probability of bit error P_b for a digitally modulated signal is influenced by the output back-offs of the earth station HPA and the satellite TWTA (the output back-off determines the degree of nonlinearity) and the roll-offs of the modem transmit and receive filters. These are the four parameters that the designer must adjust to obtain the lowest P_b (hopefully) for a given received E_b/N_0. The satellite input and output multiplexer characteristics are considered fixed. In many applications, the satellite TWTA is operated at or near saturation (0 to 0.5 dB output back-off) to provide the required downlink carrier-to-noise ratio, and thus the TWTA output back-off may also be fixed.

Computer simulation of a bandpass satellite channel is conveniently performed by its baseband (lowpass) equivalent model. This eliminates the need to translate the frequency band to radio frequency. A bandpass signal $x(t)$ can be represented by its complex envelope $\hat{x}(t)$ as

$$x(t) = \text{Re}[\hat{x}(t) \exp(j\omega_c t)] \tag{9.132}$$

where ω_c = carrier frequency. If the modulation is quadrature, such as QPSK, OQPSK, or MSK, the complex envelope $\hat{x}(t)$ is

$$\hat{x}(t) = x_c(t) + jx_s(t) \tag{9.133}$$

The information content of $x(t)$ is entirely contained within the in-phase and quadrature baseband signals $x_c(t)$ and $x_s(t)$. In this case $x(t)$ in (9.132) can also be written as

$$x(t) = x_c(t) \cos \omega_c t - x_s(t) \sin \omega_c t \tag{9.134}$$

It is seen that by using the complex envelope $\hat{x}(t)$ we can eliminate the need for simulating the high-frequency carrier $\cos \omega_c t$ and $\sin \omega_c t$.

Similarly, the impulse response $h(t)$ of a bandpass filter may also be represented by its complex envelope $\hat{h}(t)$ which is the complex impulse response of its lowpass equivalent:

$$h(t) = \text{Re}[2\hat{h}(t) \exp(j\omega_c t)] \tag{9.135}$$

The frequency domain version of (9.135) is obtained by taking the Fourier transform of both sides:

$$H(j\omega) = \hat{H}(j\omega - j\omega_c) + \hat{H}^*(-j\omega - j\omega_c) \tag{9.136}$$

The Fourier transform of $\hat{h}(t)$, which is $\hat{H}(j\omega)$, is simply the transfer function of the lowpass equivalent of $H(j\omega)$.

When the bandpass signal $x(t)$ is applied to the input of the bandpass filter $h(t)$, the desired output $y(t)$ is given by

$$y(t) = \text{Re}[\hat{y}(t) \, \exp(j\omega_c t)] \tag{9.137}$$

where the complex envelope $\hat{y}(t)$ of $y(t)$ is obtained by convolving $\hat{h}(t)$ with $\hat{x}(t)$:

$$\hat{y}(t) = \hat{h}(t) \circledast \hat{x}(t) \tag{9.138}$$

In summary, the digitally modulated signal and all the bandpass filters in a satellite channel can be simulated in complex baseband form. The input baseband signals $x_c(t)$ and $x_s(t)$ in (9.133) can be respresented by

$$x_c(t) = A \sum_{k=-\infty}^{\infty} a_k p(t - kT_s) \tag{9.139}$$

$$x_s(t) = A \sum_{k=-\infty}^{\infty} b_k q(t - kT_s) \tag{9.139b}$$

In (9.139), $a_k = \pm 1$, $b_k = \pm 1$ are equally likely independent random variables, and T_s is the symbol duration. For QPSK, $p(t)$ and $q(t) = p(t)$ are rectangular pulses of duration T_s, and for OQPSK, $q(t) = p(t - T_s/2)$. For MSK, $q(t) = p(t - T_s/2)$, and half-sine-wave pulse shaping filters must be employed in addition (see 9.118). In computer simulation the random variables a_k and b_k are generated by using maximal-length pseudo-random sequences of sufficient length (Chap 11).

To represent the nonlinear effects of the HPA or TWTA on the complex baseband input signal $\hat{x}(t) = x_c(t) + jx_s(t)$ we rewrite $x(t)$ as

$$x(t) = R(t) \cos[\omega_c t + \theta(t)] \tag{9.140a}$$

where

$$R(t) = \sqrt{x_c^2(t) + x_s^2(t)} \qquad \theta(t) = \tan^{-1}\left[\frac{x_s(t)}{x_c(t)}\right] \tag{9.140b}$$

$$x_c(t) = R(t) \cos \theta(t) \qquad x_s(t) = R(t) \sin \theta(t) \tag{9.140c}$$

The HPA or TWTA output may be expressed in the narrowband form (Chap. 5):

$$z(t) = G[R(t)] \cos\{\omega_c t + \theta(t) + F[R(t)]\} \tag{9.141}$$

where $G[\cdot]$ and $F[\cdot]$ represent the amplitude nonlinearity and phase nonlinearity, respectively. From (9.141) it is seen that $z(t)$ can be rewritten as

$$z(t) = \text{Re}[\hat{z}(t) \, \exp(j\omega_c t)]$$
$$= z_c(t) \cos\omega_c t - z_s(t) \sin\omega_c t \tag{9.142}$$

where

$$\hat{z}(t) = z_c(t) + jz_s(t) \tag{9.142b}$$

$$z_c(t) = G[R(t)] \cos\{\theta(t) + F[R(t)]\}$$

$$= \frac{G[R(t)] \cos F[R(t)]}{R(t)} x_c(t)$$

$$- \frac{G[R(t)] \sin F[R(t)]}{R(t)} x_s(t) \tag{9.142c}$$

$$z_s(t) = G[R(t)] \sin\{\theta(t) + F[R(t)]\}$$

$$= \frac{G[R(t)] \sin F[R(t)]}{R(t)} x_c(t)$$

$$+ \frac{G[R(t)] \cos F[R(t)]}{R(t)} x_s(t) \tag{9.142d}$$

The complex baseband output signal of the HPA or TWTA is simply $\hat{z}(t)$. The quadrature model of the HPA or TWTA described by the two terms

$$P[R(t)] = G[R(t)] \cos F[R(t)] \tag{9.143a}$$

$$Q[R(t)] = G[R(t)] \sin F[R(t)] \tag{9.143b}$$

can be represented by two-parameter models [see (5.45) and (5.46)] as

$$P[R] = \frac{\alpha_P R}{1 + \beta_P R^2} \tag{9.144a}$$

$$Q[R] = \frac{\alpha_Q R^3}{(1 + \beta_Q R^2)^2} \tag{9.144b}$$

The coefficients α_P, β_P, α_Q, and β_Q are obtained from the amplitude characteristic $G[R]$ and the phase characteristic $F[R]$ by least-squares fitting (see Appendix 5A).

One problem in computer simulation is to obtain the symbol timing at the receiver to determine the exact point at which the complex envelope at the receive filter output should be sampled. The symbol timing interval may be recovered by averaging the zero crossings of the complex envelope. The decision process is assumed to be based on instantaneous sampling at the midpoint of the symbol timing interval.

The remaining problems are to simulate noise and to calculate the probability of error. Noise can be simulated directly by adding samples of random noise to the samples of simulated signal and passing the resultant through the system. The average probability of symbol error is

$$P_s \approx \frac{L}{N}$$

for a simulation run of N transmitted symbols and L errors. This approximation becomes more accurate when N is large enough. In general at least $L = 100$ errors should be observed, requiring a run of $N = 100/P_s$. At $P_s = 10^{-4}$, the number of symbols N will have to be as large as 10^6, and the simulation can be very long. Furthermore, different simulation runs must be made for each value of E_s/N_0.

To avoid long simulation runs, noise can be simulated indirectly. First, the noise-free signal is simulated, and the amplitude $A_n + jB_n$, $n = 1,2,\ldots,N$, of all N symbols is determined at the sampling instants. Next, the noise variance σ^2 equal to the received noise power is calculated, and the conditional probability of symbol error is computed as

$$P_{s,n} = P_{i,n} + P_{q,n} - P_{i,n}\,P_{q,n} \qquad n = 1,2,\ldots,N$$

The subscripts i and q refer to the in-phase channel and quadrature channel, respectively. If we assume that the received noise is Gaussian with zero mean, then

$$P_{i,n} = Q\left(\frac{A_n}{\sigma}\right) = \frac{1}{2}\,\mathrm{erfc}\left(\frac{A_n}{\sqrt{2}\sigma}\right)$$

$$P_{q,n} = Q\left(\frac{B_n}{\sigma}\right) = \frac{1}{2}\,\mathrm{erfc}\left(\frac{B_n}{\sqrt{2}\sigma}\right)$$

The average probability of symbol error is obtained by averaging the conditional probability over N transmitted symbols:

$$P_s = \frac{1}{N}\sum_{n=1}^{N} P_{s,n}$$

The advantage of indirect simulation is that only short sequences of transmitted symbols are needed (1000 symbols typically). The only requirement is that the M symbols of the M-ary signal occur with approximately equal frequency. A limitation of indirect simulation is that only downlink noise can be considered because it can be modeled as Gaussian noise. Uplink noise that passes through a nonlinear channel is no longer Gaussian at the receiver input. Indirect simulation is applicable to downlink-limited situations. A block diagram of an indirect simulation is shown in Fig. 9.36.

9.10 DIGITAL MODULATION WITH ERROR-CORRECTION CODING

In satellite communications, it is desirable to minimize the average probability of bit error at the receiver subject to the constraints on received power and channel bandwidth. For example, the average probability of

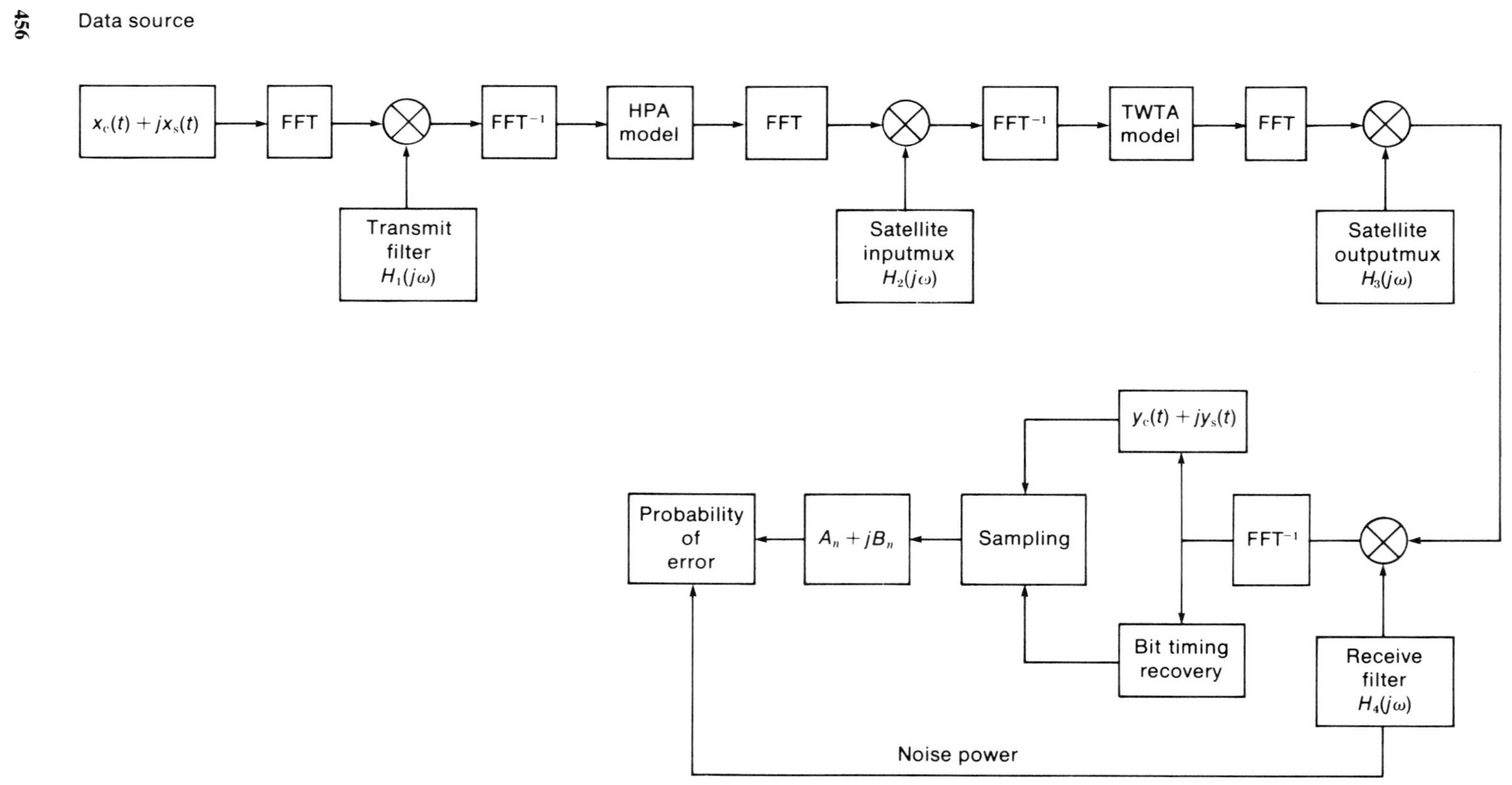

Data source

Figure 9.36 Indirect simulation of the digital modulation performance over a nonlinear satellite channel.

error in PSK modulation is

$$P_b = Q\left(\sqrt{\frac{2E_b}{N_0}}\right) = Q\left(\sqrt{\frac{2C}{R_b N_0}}\right)$$

and it decreases monotonically with increasing E_b/N_0. If the bit rate R_b is kept constant, we must increase the carrier power C, and this means an increase in the total power capability of the satellite itself and/or the transmitted power of the earth station. This might not be possible in many satellite systems which are power-limited. Another way to reduce the average probability of bit error is to use M-ary orthogonal signaling such as M-ary FSK, where for a fixed E_b/N_0 an increase in channel bandwidth (or M) reduces P_b, as is evident from Fig. 9.27. But satellite channels are bandwidth-limited, and it appears that increasing the bandwidth is not a solution to the problem either. An effective technique for reducing the average probability of error in satellite communications is the use of error-correction coding.

In 1948, Shannon demonstrated that, by proper encoding of information, errors introduced by AWGN in the channel can be reduced to any desired level without sacrificing the rate of information transmission provided that this rate R_b (bps) does not exceed the *channel capacity* $\mathscr{C}$ (bps) given by

$$\mathscr{C} = B \log_2\left(1 + \frac{S}{N}\right) \tag{9.145}$$

where B = channel bandwidth (Hz) and S/N = signal-to-noise ratio at input of receiver (in satellite communications, it is the familiar carrier-to-noise ratio). Since $S/N = R_b E_b/N_0 B$, (9.145) can be written as

$$\mathscr{C} = B \log_2\left(1 + \frac{R_b E_b}{B N_0}\right)$$

$$= R_b \frac{E_b}{N_0}\left[\frac{B N_0}{R_b E_b} \log_2\left(1 + \frac{R_b E_b}{B N_0}\right)\right] \tag{9.146}$$

Superficially (9.145) seems to indicate that $\mathscr{C} \to \infty$ as $B \to \infty$. This is not true, however; by taking the limit of $\mathscr{C}$ as $B \to \infty$ and by noting that

$$\lim_{x \to \infty} x \log_2\left(1 + \frac{1}{x}\right) = \log_2 e = \frac{1}{\ln 2} = 1.443$$

Hence

$$\lim_{B \to \infty} \mathscr{C} = 1.443 R_b \frac{E_b}{N_0} \tag{9.147}$$

Since R_b must always be less than $\lim_{B \to \infty} \mathscr{C}$, we have

$$R_b \le 1.443 R_b \frac{E_b}{N_0}$$

or

$$\frac{E_b}{N_0} \geq 0.693 = -1.6\,\text{dB} \tag{9.148}$$

Thus the channel capacity places an absolute limit on the minimum value of E_b/N_0 that is required to communicate with an error-free performance. In other words, there is a coding scheme which allows an arbitrarily low error rate. For example, PSK modulation requires $E_b/N_0 \approx 9.6$ dB to achieve $P_b = 10^{-5}$. According to (9.148), a margin of 11.2 dB is available for error-correction coding to provide the same $P_b = 10^{-5}$, but with E_b/N_0 smaller than 9.6 dB. This can be the case when the ratio E_b/N_0 is reduced by rain attenuation.

9.10.1 Concept of Error-Correction Coding

Error-correction coding is commonly used in digital satellite communications to improve the average probability of bit error. The two common features of all error-correction codes are structured redundancy and noise averaging. Structured redundancy is a method of inserting extra or redundant symbols into the information message. Noise averaging is obtained by making the redundant symbols depend on a span of several information symbols. The uniqueness of structured redundancy makes it possible to tolerate some fraction of the symbols in a block of several information symbols being in error without destroying the uniqueness of the information message it conveys, thereby causing a block error. Also, noise averaging indicates that the error rate decreases with increasing block length.

Two types of error-correction coding that we discuss in this section are linear *block coding* (or group coding) and linear *convolutional coding*. With block coding, the information data, which is usually binary symbols or bits (but may have been encoded in any alphabet of $q > 2$ symbols), is segmented into blocks of k information bits each, where k is called the *block length*. Each information block can represent any one of $M = 2^k$ distinct messages ($M = q^k$ if $q > 2$). The encoder then transforms each information block into a larger block of n bits ($n > k$) by adding $n - k$ redundant bits in a unique way. Each block of n bits from the encoder constitutes a *code word* contained in the set of $M = 2^k$ possible code words. The code words are then fed to the modulator which generates a set of finite time duration waveforms for transmission over the channel. A block encoder is a *memoryless* device because each n-bit code word depends only on a specific k-bit information block and on no others. It does not mean that the encoder does not contain memory elements. The ratio of information bits to total bits in a code word is called the *code rate R*. It is seen that

$$R = \frac{k}{n} \qquad (9.149)$$

If R_b is the information bit rate at the input of the encoder, the coded bit rate R_c at the output of the encoder is

$$R_c = \frac{R_b}{R} = \frac{nR_b}{k} \qquad (9.150)$$

Practical values for information block length k range from 3 to several hundreds and for R from $\frac{1}{4}$ to $\frac{7}{8}$. Error-correction coding can be achieved with (n,k) block coding because there are only $M = 2^k$ possible code words that could have been transmitted, and the number of possible received words 2^n is much greater than 2^k. The uniqueness of $n - k$ redundant bits added to a k-bit information block permits the decoder to identify the corresponding k-bit message.

In *convolutional coding*, the information data is passed through a linear shift register with K stages which shift k bits at a time. For every K information bits stored in the shift register, there are n linear logic circuits which operate on the shift register contents to produce n coded bits as the output of the encoder. The *code rate* R is therefore $R = k/n$. The parameter K is called the *constraint length* of the convolutional code. Because a particular information bit remains in the shift register for K/k shifts, it influences the value of nK/k coded bits. Thus, the convolutional encoder is a device with *memory*. Typical values for k and n are in the range of 1 to 8, for R in the range of $\frac{1}{4}$ to $\frac{7}{8}$, and for K in the range of 8 to 70. A convolutional encoder with $K = 3$, $k = 1$, and $n = 2$ is shown in Fig. 9.37.

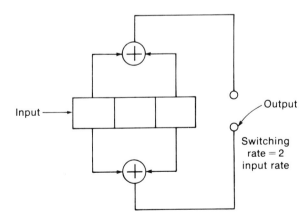

Figure 9.37 Convolutional encoder with $K = 3$, $k = 1$, and $n = 2$.

From the above discussion it is seen that error-correction coding requires more capacity, as is evident from (9.150). This can be in the form of a wider bandwidth in a FDMA channel or a longer subburst in a TDMA channel.

In the following discussion some more concepts involving a coded system will be presented: *hard-decision, soft-decision*, and *maximum likelihood decoding*. Consider a coded system operating on an AWGN waveform channel as shown in Fig. 9.38; from the viewpoint of the encoder and decoder, the *discrete memoryless channel* (DMC) is the most important. It is characterized by a set of M-ary input symbols, Q-ary output symbols, and *transition probabilities* $\Pr(j|i)$, $0 \leq i \leq M - 1$, $0 \leq j \leq Q - 1$ (where i is a modulator input symbol, j is a demodulator output symbol, and $P(j|i)$ is the probability of receiving j given that i was transmitted) that are time-invariant and independent from symbol to symbol. The most commonly encountered case of a DMC is a *binary symmetric channel* (BSC) where (1) binary modulation is used ($M = 2$), (2) the demodulator output is quantized to $Q = 2$ levels, and (3) $\Pr(0|1) = \Pr(1|0) = p$ and $\Pr(0|0) = \Pr(1|1) = 1 - p$. The *transition probability* p can be calculated from a knowledge of the waveform used, the probability density function of noise, and the output quantization threshold of the demodulator. For example, when coherent PSK is used with binary output quantization, the BSC transition probability is just the PSK average probability of bit error with equally likely transmitted symbols 0 and 1 given by $p = Q(\sqrt{2RE_b/N_0})$. The optimum threshold is 0, and the demodulator output is a 0 if the output voltage of the matched filter is neg-

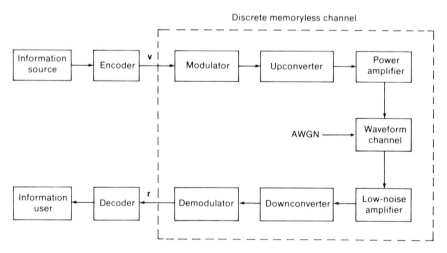

Figure 9.38 Coded satellite channel.

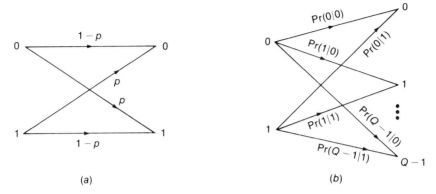

Figure 9.39 Transition probability diagram of a DMC. (*a*) BSC. (*b*) *Q*-ary channel.

ative. Otherwise, the output is a 1. Figure 9.39 shows the transition probability diagram for a DMC.

When binary demodulator output quantization is used ($Q = 2$), the decoder has only binary inputs. In this case, the demodulator is said to make hard decisions and the decoding process is termed *hard-decision decoding*. Many coded digital communications systems use binary encoding with hard-decision decoding because of its simplicity.

In some binary-encoded systems where Q-ary demodulator output quantization is used ($Q > 2$) or the output is left unquantized, the demodulator is said to make soft decisions. In this case the decoder accepts multilevel (or analog) inputs, and the decoding process is termed *soft-decision decoding*, as shown in Fig. 9.39*b*. A soft-decision decoder is more complex than a hard-decision decoder, as an automatic gain control is needed and $\log_2 Q$ bits have to be manipulated for every channel bit. But soft-decision decoding offers an additional coding gain of about 2 dB at realistic values of E_b/N_0 over hard-decision decoding. In practice eight-level quantization ($Q = 8$) is commonly used because there is only a small difference in performance between the eight-level quantization scheme and the unquantized case. The eight-level quantized outputs involve one decision threshold and three pairs of confidence thresholds, as shown in Fig. 9.40.

The concept of *maximum likelihood decoding* is similar to the concept of maximum likelihood demodulating presented in Sec. 9.1. It provides the best feasible decoding rule such that for a given code the probability of error is minimized. Let **r** represent the Q-ary received n-tuple at the output of the Q-level quantized demodulator in Fig. 9.38, where n represents the length of the code word. Also, let **v** be the potential transmitted code word as it would be received in the absence of

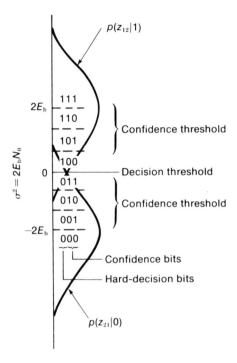

Figure 9.40 Eight-level quantization scheme.

noise. An optimum decoder would choose **v** which maximizes $\Pr(\mathbf{v}|\mathbf{r})$, the conditional probability that **v** is the code word actually transmitted given that **r** has been received. Applying Bayes' rule we write

$$\Pr(\mathbf{v}|\mathbf{r}) = \Pr(\mathbf{r}|\mathbf{v})\, \frac{\Pr(\mathbf{v})}{\Pr(\mathbf{r})} \qquad (9.151)$$

$\Pr(\mathbf{r})$ is independent of the decoding rule used, since **r** is produced before decoding. If all code words are equally likely [i.e., $\Pr(\mathbf{v})$ is the same for all **v**], then maximizing $\Pr(\mathbf{v}|\mathbf{r})$ is equivalent to maximizing $\Pr(\mathbf{r}|\mathbf{v})$. For a DMC,

$$\Pr(\mathbf{r}|\mathbf{v}) = \prod_{i=1}^{n} \Pr(r_i|v_i) \qquad (9.152)$$

since each received symbol depends only on the corresponding transmitted symbol. A decoder that chooses its estimate to maximize (9.152) is called a *maximum likelihood decoder* (MLD). Since $\log x$ is a monotonically increasing function of x, maximizing (9.152) is also equivalent to maximizing the log likelihood function:

$$\log \Pr(\mathbf{r}|\mathbf{v}) = \sum_{i=1}^{n} \log \Pr(r_i|v_i) \qquad (9.153)$$

Consider the case when a MLD is applied to a BSC. In this case $\mathbf{r}$ is a binary n-tuple which may differ from the transmitted code word $\mathbf{v}$ in some position because of channel noise. Let $\Pr(r_i|v_i) = p$ when $r_i \neq v_i$; then $\Pr(r_i|v_i) = 1 - p$ when $r_i = v_i$. Let d be the distance between $\mathbf{r}$ and $\mathbf{v}$ (i.e., the number of positions in which $\mathbf{r}$ and $\mathbf{v}$ differ). Then

$$\Pr(\mathbf{r}|\mathbf{v}) = p^d(1 - p)^{n-d}$$

or, equivalently,

$$\log \Pr(\mathbf{r}|\mathbf{v}) = d \log p + (n - d) \log (1 - p)$$

$$= d \log \frac{p}{1 - p} + n \log (1 - p)$$

Since $\log[p/(1 - p)] < 0$ for $p < \frac{1}{2}$ and $n \log (1 - p)$ is constant for all $\mathbf{v}$, the MLD rule for a BSC minimizes the distance d between $\mathbf{r}$ and $\mathbf{v}$; that is, it chooses the code word that differs from the received sequence in the fewest number of positions.

9.10.2 Linear Block Coding

In this subsection we discuss the performance of linear block coding applied to digital modulation. Linear binary codes form a linear vector space and have the property that two code words can be modulo-2-added to produce a third code word. An important parameter of block coding is the minimum distance that determines the random error-correction capabilities of a code. Let $\mathbf{v}$ be a binary n-tuple code word; the Hamming weight (or simply weight) of $\mathbf{v}$, denoted by $w(\mathbf{v})$, is the number of nonzero components of $\mathbf{v}$. The set of all Hamming weights in a code constitutes the Hamming weight structure of the code. For example, the Hamming weight of $\mathbf{v} = (110101)$ is 4. Suppose $\mathbf{v}$ and $\mathbf{u}$ are any two code words in an (n,k) block code. A measure of the difference between $\mathbf{v}$ and $\mathbf{u}$ is the number of positions in which they differ. This measure is called the *Hamming distance* (or simply the distance) between $\mathbf{v}$ and $\mathbf{u}$ and is denoted by $d(\mathbf{v},\mathbf{u})$. For example, if $\mathbf{v} = (110101)$ and $u = (111000)$, then $d(\mathbf{v},\mathbf{u}) = 3$. Clearly $d(\mathbf{v},\mathbf{u}) = w(\mathbf{v} \oplus \mathbf{u})$, where $v \oplus u = (001101)$ is a code word for linear block codes.

From the above discussion it is clear that the minimum distance $d_{\min}$ of a linear block code is equal to the minimum weight of its nonzero code words:

$$d_{\min} = w_{\min} = \min\{w(\mathbf{v}), \mathbf{v} \neq 0\} \qquad (9.154)$$

If a block code with *minimum distance* $d_{\min}$ is used for random error correction, one would like to know how many errors the code is able to correct. Let $\mathbf{v}$ and $\mathbf{r}$ be the transmitted code word and the received sequence, respectively. For any other code word $\mathbf{u}$ in this code, the Hamming distances between $\mathbf{v},\mathbf{r}$, and $\mathbf{u}$ satisfy the triangle inequality

$$d(\mathbf{v},\mathbf{r}) + d(\mathbf{u},\mathbf{r}) \geq d(\mathbf{v},\mathbf{u}) \geq d_{min} \qquad (9.155)$$

Suppose an error pattern of t errors occurs during the transmission of $\mathbf{v}$. Then $\mathbf{v}$ and $\mathbf{r}$ will differ in t places, and therefore $d(\mathbf{v},\mathbf{r}) = t$. Using (9.155) we obtain

$$d(\mathbf{u},\mathbf{r}) \geq d_{min} - t \qquad (9.156)$$

Based on the maximum likelihood decoding rule for a BSC, the received sequence $\mathbf{r}$ is correctly decoded as the transmitted code word $\mathbf{v}$ if the *Hamming distance* $d(\mathbf{v},\mathbf{r})$ is smaller than the *Hamming distance* $d(\mathbf{u},\mathbf{r})$. In other words,

$$d(\mathbf{u},\mathbf{r}) > t \qquad (9.157)$$

Both (9.156) and (9.157) indicate that the *minimum distance* d_{min} must be at least $2t + 1$. Therefore

$$d_{min} \geq 2t + 1 \qquad (9.158)$$

Summarizing the above results, a block code with minimum distance d_{min} guarantees the correction of all the error patterns of

$$t = \left\lfloor \frac{d_{min} - 1}{2} \right\rfloor \qquad (9.159)$$

or fewer errors, where $\lfloor x \rfloor$ denotes the largest integer no greater than x. The parameter $t = \lfloor (d_{min} - 1)/2 \rfloor$ is called the *random-error-correcting capability* of the code. The code is referred to as a *t-error-correcting* code and is usually capable of correcting many error patterns of $t + 1$ or more errors. Indeed, every (n,k) linear block code is capable of correcting 2^{n-k} error patterns including those with t or fewer errors. Codes that can correct all error patterns with t or fewer errors and no others are called *perfect codes*. However, only a small number of perfect codes have been found. For a *t*-error-correcting (n,k) code, the number of code words of Hamming distance less than or equal to t from a possible transmitted code word is

$$\sum_{i=0}^{t} \binom{n}{i}$$

Since there are 2^k possible transmitted code words, there are

$$2^k \sum_{i=0}^{t} \binom{n}{i}$$

such code words and they cannot exceed 2^n. Thus a *t*-error-correcting

(n,k) linear block code must satisfy the inequality

$$2^k \sum_{i=0}^{t} \binom{n}{i} \leq 2^n$$

or

$$\sum_{i=0}^{t} \binom{n}{i} \leq 2^{n-k} \tag{9.160}$$

This is the Hamming upper bound or the sphere-packing bound. The equality in (9.160) is achieved only for perfect codes.

The Hamming upper bound tends to provide the tightest bound for high-rate codes (i.e., large k/n). For low-rate codes the *Plotkin upper bound* [7,8] is tighter:

$$d_{\min} \leq \frac{n2^{k-1}}{2^k - 1} \tag{9.161}$$

Another bound is the *Varsharmov-Gilbert lower bound* [7,8] given as

$$\sum_{i=0}^{d_{\min} -2} \binom{n-1}{i} > 2^{n-k} \tag{9.162}$$

Example 9.1 Consider a $(63,k)$ linear block code with $d_{\min} = 5$. It is desired to find the range of the information block k. The Hamming upper bound gives

$$\sum_{i=0}^{2} \binom{63}{i} \leq 2^{n-k}$$

or

$$2017 \leq 2^{n-k} \tag{9.163}$$

Inequality (9.163) holds for $n - k \geq 11$. The Varsharmov-Gilbert lower bound states that

$$39{,}774 > 2^{n-k} \tag{9.164}$$

Inequality (9.164) holds for $n - k < 16$. Thus the range of k is

$$47 < k < 52$$

Important classes of block codes. In this subsection we shall describe some important classes of block codes: *Hamming codes, Golay codes, Bose-Chaudhuri-Hocquenghem* (BCH) *codes, Reed-Solomon codes,* and *maximal-length codes.*

Hamming codes have the following parameters.
Code length: $n = 2^m - 1$
Information block length: $k = 2^m - 1 - m = n - m$
Minimum distance: $d_{\min} = 3$
Error-correcting capability: $t = 1$

Hamming codes are examples of the few known *perfect codes*. Note that a perfect code must satisfy (9.160) with equality, which for Hamming codes is

$$1 + n = 2^{n-k}$$

Since $n = 2^{n-k} - 1$ for these codes, they are obviously perfect. Hamming codes comprise one of the few classes of codes for which the complete weight structure is known. The number of code words of weight i, A_i, is simply the coefficient of x^i in the expansion of the following weight enumerator polynomial:

$$A(x) = \frac{1}{n+1} [(1+x)^n + n(1-x)(1-x^2)^{(n-1)/2}]$$

For example, let $m = 3$, $n = 2^3 - 1 = 7$, $k = 7 - 3 = 4$; then the weight enumerator polynomial for the (7,4) Hamming code is

$$A(x) = \tfrac{1}{8}[(1+x)^7 + 7(1-x)(1-x^2)^3] = 1 + 7x^3 + 7x^4 + x^7$$

Therefore the weight structure for the (7,4) Hamming code is $A_0 = 1$, $A_3 = A_4 = 7$, and $A_7 = 1$.

Golay codes represent another class of perfect codes and have the following parameters.

Code length: $n = 23$
Information block length: $k = 12$
Minimum distance: $d_{min} = 7$
Error-correcting capability: $t = 3$.

The extended (24,12) Golay code is widely used with a minimum distance of 8 by adding an extra redundant bit and has the exact code rate $R_c = \frac{1}{2}$. The weight enumerator polynomial of the (23,12) Golay code is

$$A(x) = 1 + 253(x^7 + 2x^8 + 2x^{15} + x^{16}) + 1288(x^{11} + x^{12}) + x^{23}$$

The weight enumerator of the extended (24,12) Golay code is

$$A(x) = 1 + 759(x^8 + x^{16}) + 2576 x^{12} + x^{24}$$

BCH codes are a generalization of Hamming codes that allow multiple error correction. For any positive integer $m > 3$ and $t < 2^m - 1$ there exists a binary BCH code with the following parameters.

Code length: $n = 2^m - 1$
Information block length: $k \geq n - mt$
Minimum distance: $d_{min} \geq 2t + 1$
Error-correcting capability: t bits.

The weight structure of BCH codes is still unknown in general, except for double-error and triple-error correcting, and for some low-rate BCH codes. The (127,112) BCH code is used in the *INTELSAT V* TDMA system.

Reed-Solomon codes are an important class of BCH codes with the following parameters.

Symbol length: m bits per symbol
Code length: $n = 2^m - 1$ symbols
Information block length: $k = n - 2t$ symbols
Minimum distance: $d_{min} = 2t + 1$ symbols
Error-correcting capability: t symbols.

Reed-Solomon codes provide correction for 2^m symbols, hence for burst errors. The weight structure of an (n,k) Reed-Solomon code is

$$A_0 = 1$$

$$A_j = 0 \qquad 1 \le j \le n - k$$

$$A_j = \binom{n}{j} \sum_{h=0}^{j-1-n+k} (-1)^h \binom{j}{h} [2^{m(j-h-n+k)} - 1] \qquad n - k + 1 \le j \le n$$

Maximal-length codes have the following parameters.

Code length: $n = 2^m - 1,\ m \ge 3$
Information block length: $k = m$
Minimum distance: $d_{min} = 2^{m-1}$.

A maximal-length code has $2^m - 1$ nonzero code words of the same weight 2^{m-1}.

The theory of encoding and decoding for linear block codes is well-developed. The reader can refer to the excellent texts [7–9].

9.10.3 Error Rate with Linear Block Coding

A measure of coding effectiveness is obtained by comparing the average probability of code word error to the average probability of block error without coding under similar constraints of power and information rate. Consider a BSC with a transition probability p for the uncoded system and a transition probability p' for the coded system where a t-error-correcting (n,k) code is used. For proper comparison we assume that the information rate R_b and transmitted carrier power C are identical for both systems. Let E_c be the energy per coded bit in the coded system and let E_b be the familiar energy per bit in the uncoded system. Then $E_b = C/R_b$ and $E_c = C/R_c$. From (9.149) and (9.150) we have

$$E_c = RE_b < E_b \qquad (9.165)$$

Therefore, the energy per coded bit is reduced with the use of coding, but a net gain in performance is achieved by error-correcting capabilities.

The average probability of block error without coding P_e is simply 1 minus the probability that all k information bits will be received correctly. This yields

$$P_e = 1 - (1 - p)^k \qquad (9.166)$$

The average probability of code word error P'_e is upper-bounded by

$$P'_e \leq \sum_{i=t+1}^{n} \binom{n}{i} p'^i (1 - p')^{n-i} \qquad (9.167)$$

where $\binom{n}{i}$ = number of all possible patterns of i error in an n-bit code word. Equality holds in (9.167) if the linear block code is a perfect code. Recall that a nonperfect code can correct a total of 2^{n-k} error patterns including those with t or fewer errors.

For $np' \ll 1$, the first term in the summation in (9.167) dominates all the other terms, and we can approximate the upper bound of P'_e by

$$P'_e \leq \binom{n}{t+1} p'^{t+1} (1 - p')^{n-t-1} \approx \binom{n}{t+1} p'^{t+1} \qquad (9.168)$$

In summary, (9.166) and (9.168) permit a comparison of the average probabilities of block error and code word error. The performances of coded and uncoded systems can also be compared using the average probability of bit error. For an uncoded system, the average probability of bit error P_b is simply the transition probability p:

$$P_b = p \qquad (9.169)$$

To find the upper bound for the average probability of bit error in a coded system P'_b assume that a pattern of $i > t$ errors will cause the decoded code word to differ from the correct code word in $i + t$ positions and thus a fraction $(i + t)/n$ of k information symbols to be decoded erroneously. Thus

$$P'_b \leq \sum_{i=t+1}^{n} \frac{i+t}{n} \binom{n}{i} p'^i (1 - p')^{n-i}$$

$$\leq \frac{2t+1}{n} \binom{n}{t+1} p'^{t+1} \qquad (9.170)$$

Example 9.2 Consider a coded PSK signal using the well-known three-error-correcting Golay code (23,12) over a BSC. The transition probability of the uncoded PSK signal is the average probability of bit error given in (9.35). At $E_b/N_0 = 9.12$ (or 9.6 dB) it is

$$P_b = p = Q\left(\sqrt{\frac{2E_b}{N_0}}\right) = Q(\sqrt{18.24})$$

$$= 10^{-5}$$

The average probability of block error without coding is

$$P_e = 1 - (1 - p)^k = 1 - (1 - 10^{-5})^{12}$$

$$= 1.2 \times 10^{-4}$$

With coding, the transition probability is

$$p' = Q\left(\sqrt{\frac{2E_c}{N_0}}\right) = Q\left(\sqrt{\frac{2RE_b}{N_0}}\right)$$

$$= Q(3.08)$$

$$\approx 10^{-3}$$

Consequently, the average probability of code word error is (Golay code is a perfect code)

$$P'_e \approx \binom{23}{4}(10^{-3})^4 \approx 9 \times 10^{-9}$$

The coded average probability of bit error can be approximated by (9.170):

$$P'_b \approx 3 \times 10^{-9}$$

It is seen that a remarkable improvement in the bit error rate can be achieved by coding. Now suppose we want to find the ratio of energy per coded bit to noise density E_c/N_0 required for the coded system to achieve $P'_b \approx 10^{-5}$, that is, about the same bit error rate as that of the uncoded system. Then (9.170) yields

$$\frac{7}{23}\binom{23}{4}p'^4 \approx 10^{-5}$$

The coded transition probability p' now becomes

$$p' = 8 \times 10^{-3}$$

The corresponding E_c/N_0 that yields p' is

$$\frac{E_c}{N_0} = 2.93$$

and the corresponding E_b/N_0 for the coded system is

$$\frac{E_b}{N_0} = \frac{E_c}{RN_0} = 5.61 \text{ or } 7.5 \text{ dB}$$

Recall that the uncoded system requires $E_b/N_0 = 9.6$ dB to achieve $P_b = 10^{-5}$. Thus the coding gain at 10^{-5} average probability of bit error is

$$\text{Coding gain} = 9.6 - 7.5 = 2.1 \text{ dB}$$

REFERENCES

1. J. M. Wozencraft and I. M. Jacobs, *Principles of Communication Engineering.* New York: Wiley, 1965.
2. W. C. Lindsey and M. K. Simon, *Telecommunication Systems Engineering.* Englewood Cliffs, N.J.: Prentice-Hall, 1973.
3. R. W. Lucky, J. Salz, and E. J. Weldon, Jr., *Principles of Data Communications,* New York: McGraw-Hill, 1968.
4. S. Haykin, *Communication Systems,* 2d ed. New York: Wiley, 1983.

5. K. Sam Shanmugam, *Digital and Analog Communication Systems*. New York: Wiley, 1979.
6. G. Robinson et al., "PSK Signal Power Spectrum Spread Produced by Memoryless Nonlinear TWTs," *Comsat Tech. Rev.*, Fall 1973, pp. 227–256.
7. G. C. Clark, Jr., and J. B. Cain, *Error-Correction Coding for Digital Communications*. New York: Plenum, 1981.
8. S. Lin and D. J. Costello, Jr., *Error Control Coding: Fundamentals and Applications*. Englewood Cliffs, N.J.: Prentice-Hall, 1983.
9. A. J. Viterbi and J. K. Omura, *Principles of Digital Communication and Coding*. New York: McGraw-Hill, 1979.
10. J. G. Proakis, *Digital Communications*. New York: McGraw-Hill, 1983.
11. L. M. Couch, II, *Digital and Analog Communication Systems*. New York: Macmillan, 1983.
12. B. P. Lathi, *Modern Digital and Analog Communication Systems*. New York: Holt, 1983.

APPENDIX 9A GRAM-SCHMIDT ORTHOGONALIZATION

To form the orthonormal basis functions $u_n(t)$, $n = 1, 2, \ldots, N$, we define the first basis function as

$$u_1(t) = \frac{s_1(t)}{\sqrt{E_1}} \tag{9A.1}$$

where E_1 = energy of signal $s_1(t)$ [see (9.3)]. Then it is obvious that

$$s_1(t) = \sqrt{E_1}\, u_1(t)$$
$$= s_{11} u_1(t) \tag{9A.2}$$

where $s_{11} = \sqrt{E_1}$ and $u_1(t)$ has unit energy as required. Next we define the coefficient s_{21} as

$$s_{21} = \int_0^{T_s} s_2(t) u_1(t)\ dt \tag{9A.3}$$

Now we define the second basis function:

$$u_2(t) = \frac{s_2(t) - s_{21} u_1(t)}{s_{22}}$$
$$= \frac{s_2(t) - s_{21} u_1(t)}{\sqrt{E_2 - s_{21}^2}} \tag{9A.4}$$

where E_2 is the energy of the signal $s_2(t)$ and $s_{22} = \sqrt{E_2 - s_{21}^2}$. It is clear that

$$\int_0^{T_s} u_1(t) u_2(t)\ dt = 0 \tag{9A.5}$$

$$\int_0^{T_s} u_2^2(t) \; dt = 1 \tag{9A.6}$$

Continuing in this fashion, we obtain

$$u_m(t) = \frac{s_m(t) - \sum_{n=1}^{m-1} s_{mn} u_n(t)}{s_{mm}} \tag{9A.7}$$

where

$$s_{mm} = \left(E_m - \sum_{n=1}^{m-1} s_{mn}^2 \right)^{1/2} \tag{9A.8}$$

APPENDIX 9B POWER SPECTRAL DENSITY

A bandpass signal $s(t)$ can be represented by its complex envelope $\hat{s}(t) = s_B(t) \exp(j\theta)$ as

$$s(t) = \text{Re}\,[\hat{s}(t)\,\exp(j\omega_c t)]$$
$$= \text{Re}[s_B(t)\,\exp(j\omega_c t + j\theta)] \tag{9B.1}$$

where ω_c = carrier frequency. The independent random phase θ is assumed to be uniformly distributed over $(0, 2\pi)$. To obtain the power spectral density of $s(t)$ we need to evaluate its autocorrelation function:

$$R(\tau) = E\{s(t)\,s(t+\tau)\}$$
$$= E\{\text{Re}[s_B(t)\,\exp(j\omega_c t + j\theta)]\;\text{Re}[s_B(t+\tau)\,\exp(j\omega_c t + j\omega_c \tau + j\theta)]\} \tag{9B.2}$$

Using the identity $\text{Re}(c_1)\,\text{Re}(c_2) = \frac{1}{2}\,\text{Re}(c_1 c_2) + \frac{1}{2}\,\text{Re}(c_1^* c_2)$ and recalling that θ is an independent random variable, we obtain

$$R(\tau) = \frac{1}{2}\,\text{Re}\{E[s_B(t)s_B(t+\tau)\,\exp(j2\omega_c t + j\omega_c \tau)]E[\exp(j2\theta)]\}$$
$$+ \frac{1}{2}\,\text{Re}\{E[s_B(t)s_B(t+\tau)]\,\exp(j\omega_c \tau)\} \tag{9B.3}$$

But $E\{\exp(j2\theta)\} = 0$ and $E\{s_B(t)s_B(t+\tau)\} = R_B(\tau)$; thus

$$R(\tau) = \frac{1}{2}\,\text{Re}\{R_B(\tau)\,\exp(j\omega_c \tau)\}$$
$$= \frac{1}{2}\,R_B(\tau)\,\cos\,\omega_c \tau \tag{9B.4}$$

Taking the Fourier transform of $R(\tau)$ yields the power spectral density

$S(f)$ of $s(t)$ in terms of the power spectral density $S_B(f)$ of the baseband signal $s_B(t)$:

$$S(f) = \tfrac{1}{4}[S_B(f-f_c) + S_B(f+f_c)] \qquad (9B.5)$$

There are many digital signaling formats that can be used for the baseband signal:

1. *Bipolar signaling.* Binary 1 and 0 are represented by equal positive and negative levels.
2. *Manchester signaling.* Binary 1 is represented by a positive level of duration $T_b/2$ followed by a negative level of duration $T_b/2$. Binary 0 is represented by a negative level of duration $T_b/2$ followed by a positive level of duration $T_b/2$.
3. *Unipolar signaling.* Binary 1 is represented by a positive level and binary 0 by a zero level.
4. *Return-to-zero signaling.* Binary 1 is represented by a positive level only over most of a bit period and then returns to zero. Binary 0 is represented by a zero level.
5. *Pseudoternary signaling.* Binary 1 is represented by alternative positive and negative levels over a duration $T_b/2$. Binary 0 is represented by a zero level.

These various signaling formats are shown in Fig. 9B.1. In satellite communications, bipolar signaling is the common format, and its power spectral density is the topic to be discussed next.

A bipolar baseband waveform can be modeled by

$$s_B(t) = \sum_{n=-\infty}^{\infty} a_n g(t - nT_b) \qquad (9B.6)$$

where $g(t)$ = signaling pulse shape
$g(t) = A, |t| < T_b/2$, and 0 elsewhere
T_b = bit duration

The set $\{a_n\} = \{\pm 1\}$ represents the independent binary digits and $\Pr\{a_n = 1\} = \Pr\{a_n = -1\} = \tfrac{1}{2}$. The power spectral density of the infinite series $s_B(t)$ can be obtained from its truncated version

$$\hat{s}_B(t) = \sum_{n=-N}^{N} a_n g(t - nT_b)$$

as follows:

$$S_B(f) = \lim_{T \to \infty} \frac{E\{|X_B(f)|^2\}}{T} \qquad (9B.7)$$

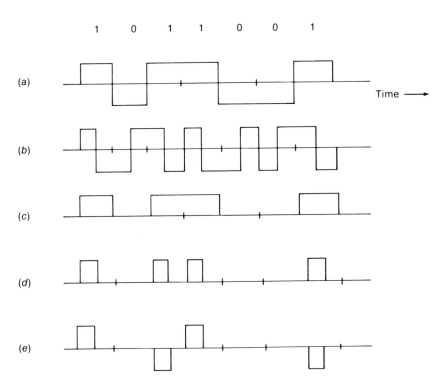

Figure 9B.1 Digital signaling formats. (*a*) Bipolar. (*b*) Manchester. (*c*) Unipolar. (*d*) Return-to-zero. (*e*) Pseudoternary.

$X_B(f)$ is the Fourier transform of $\hat{s}_B(t)$ defined as

$$X_B(f) = \mathscr{F}\,[\hat{s}_B(t)] = \sum_{n=-N}^{N} a_n \mathscr{F}[g(t - nT_b)]$$

$$= \sum_{n=-N}^{N} a_n G(f)\, \exp(-j2\pi fn T_b)$$

$$= G(f) \sum_{n=-N}^{N} a_n \exp(-j2\pi fn T_b) \qquad (9B.8)$$

where

$$G(f) = \mathscr{F}[g(t)]$$

and

$$T = (2N + 1)T_b$$

Substituting (9B.8) into (9B.7) yields

$$S_B(f) = \lim_{T \to \infty} \frac{1}{T} |G(f)|^2 E\left\{ \left| \sum_{n=-N}^{N} a_n \exp(-j2\pi f n T_b) \right|^2 \right\}$$

$$= |G(f)|^2 \lim_{T \to \infty} \frac{1}{T} \sum_{n=-N}^{N} \sum_{m=-N}^{N} E\{a_n a_m\} \exp[j(m-n)2\pi f T_b]$$

$$(9B.9)$$

Note that

$$E\{a_n a_m\} = \begin{cases} E\{a_n^2\} = (1^2)(\frac{1}{2}) + (-1)^2(\frac{1}{2}) = 1 & n = m \\ E\{a_n\}E\{a_m\} = [(1)(\frac{1}{2}) + (-1)(\frac{1}{2})]^2 = 0 & n \neq m \end{cases}$$

Using this result, (9B.9) becomes

$$S_B(f) = |G(f)|^2 \lim_{T \to \infty} \frac{1}{T} \sum_{n=-N}^{N} 1$$

$$= |G(f)|^2 \lim_{N \to \infty} \frac{2N+1}{(2N+1)T_b}$$

$$= \frac{|G(f)|^2}{T_b} \qquad (9B.10)$$

For the rectangular pulse shape $g(t) = A$ when $|t| < T_b/2$ and 0 elsewhere, the Fourier transform is given by

$$G(f) = AT_b\left(\frac{\sin \pi f T_b}{\pi f T_b}\right)$$

Hence

$$S_B(f) = A^2 T_b\left(\frac{\sin \pi f T_s}{\pi f T_s}\right)^2 \qquad (9B.11)$$

The power spectral density of the PSK signal is obtained by substituting $S_B(f)$ into (9B.5) and noting that $A^2 T_b/2 = E_b$.

The in-phase and quadrature components of the QPSK signal are statistically independent. Therefore, its baseband power spectral density equals the sum of the individual power spectral densities of the in-phase and quadrature components. For the QPSK signal in (9.40) with a bipolar baseband waveform, the pulse shape $g(t)$ is

$$g(t) = \begin{cases} \dfrac{A}{\sqrt{2}} & |t| < T_s/2 \\ 0 & \text{elsewhere} \end{cases}$$

Hence the baseband power spectral densities of the in-phase and quadrature components are

$$S_P(f) = S_Q(f) = \frac{A^2 T_s}{2}\left(\frac{\sin\pi f T_s}{\pi f T_s}\right)^2$$

The composite baseband power spectral density is therefore given by

$$S_B(f) = S_P(f) + S_Q(f) = A^2 T_s\left(\frac{\sin\pi f T_s}{\pi f T_s}\right)^2 \qquad (9B.12)$$

Substituting (9B.12) into (9B.5) and noting that $A^2 T_s/2 = E_s$ the QPSK power spectral density is obtained.

PROBLEMS

9.1 Show that the two coherent demodulators in Figs. 9.1 and 9.2 are equivalent.

9.2 Consider a coherent PSK demodulator where the recovered carrier phase is offset by ϕ from the initial phase θ of the received carrier. Derive the average probability of bit error in term of this static phase offset. Extend the result to a coherent QPSK signal.

9.3 Draw block diagrams of the phase ambiguity logic circuits for both PSK and QPSK using unique word detection.

9.4 Consider a coherent PSK system where the bit a priori probabilities are Pr(1 sent) = q and Pr(0 sent) = $1 - q$. Find the average probability of bit error in an AWGN channel.

9.5 Find the logic functions and implementations representing the differential encoding and decoding of QPSK in Tables 9.8 and 9.9.

9.6 Consider the following Gray encoding for QPSK symbols: $00 \rightarrow 0, 01 \rightarrow \pi/2, 11 \rightarrow \pi$, $10 \rightarrow 3\pi/2$.

(a) Find the four transmitted signals.

(b) Find the P- and Q-channel sequences for the following bipolar binary sequence 00101101110000110011011100.

(c) Is this symbol-phase mapping appropriate for differential encoding? Why? Give a simple example to justify your answer.

9.7 Derive the average probability of bit error for DPSK given in (9.37). Also, give a heuristic explanation why DPSK is 3 dB better in E_b/N_0 than noncoherent orthogonal FSK for the same P_b.

9.8 Find the power spectral density of FSK.

9.9 Show why the minimum frequency separation for coherent orthogonal FSK is $|\omega_1 - \omega_2| = \pi/T_b$ and for noncoherent orthogonal FSK is $|\omega_1 - \omega_2| = 2\pi/T_b$, where T_b is the bit duration.

9.10 Verify (9.86).

9.11 Show that the two noncoherent FSK demodulators in Figs. 9.25 and 9.26 are equivalent.

9.12 Find the optimum frequency separation such that the average probability of bit error for nonorthogonal coherent FSK is minimized for a given E_b/N_0.

9.13 Perform the differential encoding and decoding for OQPSK.

9.14 Consider the orthogonal M-ary FSK signal with signaling interval $T_s = T_b \log_2 M$, where T_b is the bit duration. Find the required bandwidth for its transmission in both coherent and noncoherent demodulation.

9.15 Consider a set of orthogonal waveforms $\{s_m(t)\}$, $m = 1,2,\ldots,M/2$, and its negative counterpart $\{-s_m(t)\}$ which is a set of M biorthogonal waveforms.

(a) What is the required channel bandwidth for transmitting a set of M biorthogonal waveforms relative to orthogonal waveforms?

(b) Find the average probability of symbol error in an AWGN channel.

9.16 The noise bandwidth of a bandpass signal is defined as the value of the bandwidth which satisfies the relation

$$B = \frac{C}{S(f_c)}$$

where B = *noise bandwidth*

$S(f_c)$ = value of power spectral density at carrier frequency f_c

C = Carrier power

Find the noise bandwidth of PSK, QPSK, and MSK.

9.17 Derive (9.63).

9.18 Consider an ideal satellite channel with the transfer function

$$H(f) = \begin{cases} H(0) & |f - f_c| \leq B \\ 0 & \text{elsewhere} \end{cases}$$

The channel filters are designed to have a raised cosine pulse spectrum to eliminate intersymbol interference and to minimize the average probability of bit error due to downlink Gaussian noise with zero mean and power spectral density $N_0/2$. The uplink noise effect is assumed to be much smaller than the downlink noise effect. Find the noise power at the output of the receiving filter.

9.19 Design the earth station discussed in Sec. 4.6.1 using BCH code (127,113) with the hard-decision decoding shown in Fig. P. 9.19.

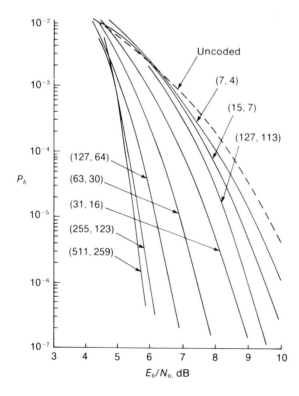

9.20 Design the earth station discussed in Sec. 4.6.2 using BCH code (127,113) with the hard-decision decoding shown in Fig. P. 9.19.

9.21 Repeat Prob. 4.21 using BCH code (127,113) with the hard-decision decoding shown in Fig. P.9.19.

9.22 Repeat Prob. 4.26 using BCH code (127,113) with the hard-decision decoding shown in Fig. P. 9.19

9.23 Find the link error probability in Example 4.2 assuming Golay code (23,12) with hard-decision decoding is used.

9.24 Repeat Prob. 9.23 assuming Hamming code (7,4) with hard-decision decoding is used.

9.25 Repeat Prob. 4.13 assuming (1) Golay code (23,12) and (2) Hamming code (7,4) are used (both with hard-decision decoding).

Chapter *10*

Carrier and Symbol Timing Synchronization

As detailed in Chap. 9, digital modulations that are efficient in power requirements employ coherent demodulation of the received signal. This requires a local carrier reference that closely matches the received carrier in frequency and phase. Also, the proper detection of data symbols requires a local clock that is accurately time-aligned with the received data pulses. In power-efficient modulation techniques, the transmitted power is devoted exclusively to data, and the discrete components of the carrier and symbol clock are suppressed completely. Therefore the demodulator must regenerate the carrier and clock from the received signal. In this chapter we will study techniques for synchronizing the carrier and symbol timing, especially for TDMA systems that employ M-ary PSK where the received signal is not continuous but occurs in short bursts.

10.1 CARRIER RECOVERY FOR M-ARY PSK

To remove the modulation, that is, to obtain the reference carrier for M-ary PSK signals, the frequency multiplication technique is used, as shown schematically in Fig. 10.1, where $M = 2$ for PSK, $M = 4$ for QPSK, $M = 8$ for 8-ary PSK, and so on. The bandpass filter BPF_i is the

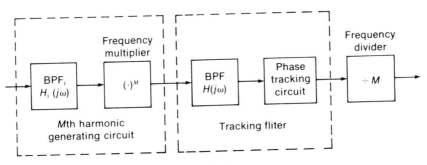

Figure 10.1 Concept of a carrier recovery circuit.

same filter as that shown in Fig. 9.7a for a PSK demodulator and in Fig. 9.15 for a QPSK demodulator. The noise bandwidth B of BPF_i (Hz) is given by

$$B = \frac{1}{2\pi} \int_0^\infty \frac{|H_i(j\omega)|^2}{H_i(j\omega_c)|^2} \, d\omega \qquad (10.1)$$

where $|H_i(j\omega)|^2 =$ gain response of BPF_i and $\omega_c =$ carrier frequency. Depending on the selection of BPF_i (e.g., to minimize carrier-to-noise ratio degradation due to the effects of the nonlinear satellite channel), the noise bandwidth B of BPF_i can be related to the M-ary symbol duration T_s:

$$T_s B = r \qquad (10.2)$$

where $r > 1$ is the symbol duration–bandwidth product. In practice r is normally chosen to be between 1 and 1.5. The bandpass filter BPF at the output of the time-M frequency multiplier is a very narrowband filter with the center frequency tuned to $M\omega_c$ to obtain a high signal-to-noise ratio at $M\omega_c$. Determination of the noise bandwidth B_n of BPF is extremely important in the design of the carrier recovery circuit for M-ary PSK signals, especially for TDMA systems which operate in the burst mode, and is the subject of discussion in subsequent sections. A phase tracking circuit is employed to adjust the carrier for phase coherence, that is, to keep the output phase coherent despite input frequency variations or mistuning of the center frequency of the narrowband bandpass filter. The combination of the bandpass filter BPF and the phase tracking circuit is called the *tracking filter.* In practice the tracking filter can be implemented by the following configuration which will be discussed in subsequent sections: (1) a phase-locked loop or (2) a narrowband bandpass filter with an automatic frequency control loop. The divide-by-M frequency divider is used at the output of the tracking filter to obtain the coherent reference carrier.

10.1.1 Analysis

The purpose of the Mth harmonic generating circuit is to generate an unmodulated Mth harmonic signal at frequency $M\omega_c$ from the modulated M-ary PSK carrier $s(t)$ at the input of the filter BPF$_i$. The modulated M-ary PSK carrier $s(t)$ can be represented by

$$s(t) = \text{Re}\,[Aa(t)\,\exp(j\omega_c t + j\theta)] \qquad (10.3)$$

where A = signal amplitude
 $a(t)$ = data signal; $|a(t)| = 1$
 θ = carrier phase

The output signal $s_M(t)$ of the time-M frequency multiplier is a combination process involving filtering and multiplication of $s(t)$ and can be characterized in terms of an equivalent operation on the envelope $a(t)$ of $s(t)$. Let $H_i(j\omega)$ denote the transfer function of the bandpass filter BPF$_i$ whose center frequency is the carrier frequency ω_c. The bandpass transfer function $H_i(j\omega)$ can be expressed in terms of its equivalent lowpass transfer function $\hat{H}_i(j\omega)$ as

$$H_i(j\omega) = \hat{H}_i(j\omega - j\omega_c) + \hat{H}_i{}^*(-j\omega - j\omega_c) \qquad (10.4)$$

If $H_i(j\omega)$ is symmetric about ω_c, then $\hat{H}_i(j\omega)$ is symmetric about $\omega = 0$ and the corresponding impulse response $\hat{h}_i(t)$ of $\hat{H}_i(j\omega)$ is real. If $H_i(j\omega)$ is assymmetric about ω_c, then $\hat{H}_i(j\omega)$ is also assymmetric about $\omega = 0$ and $\hat{h}_i(t)$ is complex. It can be shown that the output $y(t)$ of the bandpass filter BPF$_i$ expressed as

$$y(t) = \text{Re}\,[Am(t)\,\exp(j\omega_c t + j\theta)] \qquad (10.5)$$

where the envelope $m(t)$ of $y(t)$ is the convolution of $\hat{h}_i(t)$ and $a(t)$, that is,

$$m(t) = \hat{h}_i(t) \circledast a(t) \qquad (10.6)$$

Next consider the output $s_M(t) = y^M(t)$ of the time-M frequency multiplier:

$$s_M(t) = \{\text{Re}\,[Am(t)\,\exp(j\omega_c t + j\theta)]\}^M \qquad (10.7)$$

For the PSK signal ($M = 2$), we have

$$s_2(t) = \{\text{Re}\,[Am(t)\,\exp(j\omega_c t + j\theta)]\}^2$$
$$= \tfrac{1}{2}\,|Am(t)|^2 + \tfrac{1}{2}\,\text{Re}\,[A^2 m^2(t)\,\exp(j2\omega_c t + j2\theta)] \qquad (10.8)$$

As is commonly assumed for random data, $a(t)$ is a zero mean pulse amplitude modulating signal that takes the values ± 1 for $0 \leq t \leq T_b$. From (10.8) it is seen that the output of the bandpass filter BPF tuned to $2\omega_c$ is a sinusoidal signal with frequency $2\omega_c$, phase 2θ, and amplitude $\frac{1}{2}\overline{E\{A^2 m^2(t)\}}$. [We use $\overline{E\{A^2 m^2(t)\}}$ instead of $A^2 m^2(t)$ because band-limited filtering by the input filter BPF$_i$ causes slowly varying amplitude fluctuations.†] Thus the time-2 frequency multiplier (squarer) produces a periodic component of the received signal.

Consider next the QPSK signal $(M = 4)$, for which we have

$$s_4(t) = \{\text{Re}\,[Am(t)\,\exp(j\omega_c t + j\theta)]\}^4$$

$$= \frac{3}{8}\,|Am(t)|^4 + \frac{1}{2}\,\text{Re}[A^4|m(t)|^2 m^2(t)\,\exp(j2\omega_c t + j2\theta)]$$

$$+ \frac{1}{8}\,\text{Re}\,[A^4 m^4(t)\,\exp(j4\omega_c t + j4\theta)]$$

$$(10.9)$$

From (10.9) it is seen that the output of the bandpass filter tuned to $4\omega_c$ is a sinusoidal signal with frequency $4\omega_c$, phase 4θ, and amplitude $\frac{1}{8}\,\overline{E\{A^4 m^4(t)\}}$. The reader may ask why the signal $\text{Re}[A^4|m(t)|^2 m^2(t)\,\exp(j2\omega_c t + j2\theta)]$ is not a periodic component of the received signal. Consider a non-band-limited situation where the effect of the bandpass filter BPF$_i$ is negligible and

$$m(t) = a(t) = \frac{\pm 1 \pm j}{\sqrt{2}}$$

$$= \exp\left(j\,\frac{\pi}{4} + jk\,\frac{\pi}{2}\right) \qquad k = 0,1,2,3$$

as previously seen in (9.39). Then

$$m^4(t) = a^4(t) = -1$$

and the output of BPF is a pure sinusoid with no fluctuations in amplitude. Furthermore,

$$|m(t)|^2 m^2(t) = \exp\left(\pm j\,\frac{\pi}{2}\right)$$

implies that $\text{Re}[A^4|m(t)|^2 m^2(t)\,\exp(j2\omega_c t + j2\theta)]$ is a modulated signal. Thus we conclude that the squarer cannot produce a periodic component for a QPSK signal.

---

†In Sec. 10.1.2 we will show that $E\{m^2(t)\}$ is a periodic function with period T_b; therefore the time average represented by the overbar is needed to remove the time dependence. The component $\frac{1}{2} A^2 \text{Re}\{[m^2(t) - \overline{E\{m^2(t)\}}]\,\exp(j2\omega_c t + j2\theta)\}$ represents the self-noise from band-limited filtering of the modulated signal.

10.1.2 Performance in Noise

In this section we analyze the effect of noise at the output of the squarer (the time-2 frequency multiplier), since it is the most tractable mathematically and gives the results for the quadrupler (the time-4 frequency multiplier). To simplify the mathematical analysis we assume that $\hat{h}_i(t)$ in (10.6) is real; that is, the input bandpass filter BPF_i is symmetric so that $m(t)$ in (10.6) is real. In this case the input signal to the squarer can be written as

$$y(t) = Am(t) \cos(\omega_c t + \theta) + n(t) \qquad (10.10)$$

where $n(t) =$ bandpass AWGN and $m(t) =$ data message with zero mean. The bandpass AWGN $n(t)$ can be represented by its in-phase and quadrature components as

$$n(t) = n_c(t) \cos(\omega_c t + \theta) - n_s(t) \sin(\omega_c t + \theta)$$

Substituting $n(t)$ into (10.10) and calculating $y^2(t)$ we have

$$y^2(t) = y^2(t)_{\text{lowpass}} + y^2(t)_{\text{bandpass}} \qquad (10.11a)$$

where

$$y^2(t)_{\text{lowpass}} = \underbrace{\tfrac{1}{2} A^2 m^2(t)}_{s \times s} + \underbrace{An_c(t)m(t)}_{s \times n} + \underbrace{\tfrac{1}{2} n_c^2(t) + \tfrac{1}{2} n_s^2(t)}_{n \times n} \qquad (10.11b)$$

$$y^2(t)_{\text{bandpass}} = \underbrace{\tfrac{1}{2} A^2 m^2(t) \cos 2(\omega_c t + \theta)}_{s \times s}$$

$$\left.\begin{array}{l} + An_c(t)m(t) \cos 2(\omega_c t + \theta) \\ - An_s(t)m(t) \sin 2(\omega_c t + \theta) \end{array}\right\} s \times n$$

$$\left.\begin{array}{l} + [\tfrac{1}{2} n_c^2(t) - \tfrac{1}{2} n_s^2(t)] \cos 2(\omega_c t + \theta) \\ - n_c(t)n_s(t) \sin 2(\omega_c t + \theta) \end{array}\right\} n \times n \qquad (10.11c)$$

Note the various terms appearing in (10.11b) and (10.11c): The $s \times s$ term is the signal component mixed with itself, the $s \times n$ term corresponds to the signal mixed with noise, and the $n \times n$ term corresponds to the noise mixing with itself. Since the in-phase and quadrature Gaussian noise $n_c(t)$ and $n_s(t)$ are independent, $n_c^2(t)$ and $n_s^2(t)$ are also independent. Also, because $n_c(t)$ and $n_s(t)$ are Gaussian and have zero mean, their third moment is zero [1, p. 64]:

$$E\{n_c^3(t)\} = E\{n_s^3(t)\} = 0$$

Therefore the three terms $s \times s$, $s \times n$, and $n \times n$ in (10.11b) and (10.11c) are independent, and the autocorrelation function of $y^2(t)_{\text{bandpass}}$ [or

$y^2(t)_{\text{lowpass}}$] is the sum of the autocorrelation functions of each of the three terms.

Since the lowpass component $y^2(t)_{\text{lowpass}}$ will be rejected by the tracking filter, it is only necessary to look at the bandpass component $y^2(t)_{\text{bandpass}}$.

First consider the $s \times s$ term in $y^2(t)_{\text{bandpass}}$ in (10.11c). Since $m^2(t)$ is not a constant due to band-limited filtering, we can represent $m^2(t)$ by a constant term which is its average value and by another term which is time-varying. Let

$$m^2(t) = E\{m^2(t)\} + m_0^2(t)$$

where $\overline{E\{m_0^2(t)\}} = 0$. Then the $s \times s$ term in (10.11c) becomes

$$\tfrac{1}{2} A^2 m^2(t) \cos 2(\omega_c t + \theta) = \tfrac{1}{2} A^2 \overline{E\{m^2(t)\}} \cos 2(\omega_c t + \theta)$$
$$+ \tfrac{1}{2} A^2 m_0^2(t) \cos 2(\omega_c t + \theta)$$

$$(10.12a)$$

The term $\tfrac{1}{2} A^2 m_0^2(t) \cos 2(\omega_c t + \theta)$ in (10.12a) is the self-noise resulting from band-limited filtering of a modulated signal. The self-noise is small when the noise bandwidth of the input bandpass filter is greater than the symbol rate, $B_i > 1/T_s$, and can be neglected. This is our assumption in subsequent analyses. Let $n'(t)$ represent the noise component of $y^2(t)_{\text{bandpass}}$ in (10.11c):

$$n'(t) = A n_c(t) m(t) \cos 2(\omega_c t + \theta) - A n_s(t) m(t) \sin 2(\omega_c t + \theta)$$
$$+ [\tfrac{1}{2} n_c^2(t) - \tfrac{1}{2} n_s^2(t)] \cos 2(\omega_c t + \theta) - n_c(t) n_s(t) \sin 2(\omega_c t + \theta)$$

$$(10.12b)$$

Then, in order to determine this noise power within the bandwidth of the tracking filter, we need to derive the power spectral density $S_n'(f)$ of $n'(t)$ which is the Fourier transform of the autocorrelation of $n'(t)$. Taking the statistical average of $n'(t)n'(t + \tau)$, that is, $E\{n'(t)n'(t + \tau)\}$, yields

$$E\{n'(t)n'(t + \tau)\} = \tfrac{1}{2} A^2 E\{n_c(t)n_c(t + \tau)\} \; E\{m(t)m(t + \tau)\} \cos 2\omega_c \tau$$
$$+ \tfrac{1}{2} A^2 E\{n_s(t)n_s(t + \tau)\} \; E\{m(t)m(t + \tau)\} \cos 2\omega_c \tau$$
$$+ \tfrac{1}{8} E\{[n_c^2(t) - n_s^2(t)][n_c^2(t + \tau) - n_s^2(t + \tau)]\} \cos 2\omega_c \tau$$
$$+ \tfrac{1}{2} E\{n_c(t)n_c(t + \tau)\} E\{n_s(t)n_s(t + \tau)\} \cos 2\omega_c \tau$$

$$(10.13)$$

To evaluate $E\{n'(t)n'(t + \tau)\}$ in (10.13) we note that $n_c(t)$ and $n_s(t)$ are statistically alike, hence they possess the same autocorrelation function:

$$R_{n_c}(\tau) = R_{n_s}(\tau) = E\{n_c(t)n_c(t+\tau)\} = E\{n_s(t)n_s(t+\tau)\}$$

$$R_{n_c}(0) = R_{n_s}(0) = E\{n_c^2(t)\} = E\{n_s^2(t)\}$$

$$= E\{n_c^2(t+\tau)\} = E\{n_s^2(t+\tau)\}$$

Also, it can be shown that [1, p. 264]

$$E\{n_c^2(t)n_c^2(t+\tau)\} = E\{n_s^2(t)n_s^2(t+\tau)\} = R_{n_c}^2(0) + 2R_{n_c}^2(\tau)$$

By using the above results, $E\{n'(t)n'(t+\tau)\}$ in (10.13) can be expressed as

$$E\{n'(t)n'(t+\tau)\} = [R_{n_c}^2(\tau) + A^2 R_{n_c}(\tau)E\{m(t)m(t+\tau)\}]\cos 2\omega_c\tau$$

$$(10.14)$$

It remains to evaluate $E\{m(t)m(t+\tau)\}$ in (10.14). Since $m(t)$ is not stationary, its statistical average is time-dependent and so is $E\{n'(t)n'(t+\tau)\}$. Therefore to obtain the autocorrelation function of $n'(t)$ we need to take the time average of $E\{m(t)m(t+\tau)\}$. This is a characteristic of many binary data signals. They are not stationary but are cyclostationary [2]. A cyclostationary process has statistical moments that are periodic in time rather than constant as in the case of a stationary process. As we know, each data symbol is represented by a pulse, and all pulses have identical waveforms. Let $g(t)$ denote the data symbol pulse; then the message signal $m(t)$ given in (10.6) can be expressed as

$$m(t) = \sum_{k=-\infty}^{\infty} a_k p(t - kT_s) \qquad (10.15a)$$

$$p(t - kT_s) = \hat{h}_i(t)\circledast g(t - kT_s) \qquad (10.15b)$$

where T_s = symbol duration and $a_k = \pm 1$ for PSK. The data a_k is assumed to be random and to have zero mean, and each symbol is independent of the others; that is,

$$E\{a_k\} = 0$$

$$E\{a_k a_n\} = \begin{cases} 1 & n = k \\ 0 & n \neq k \end{cases}$$

By using (10.15) we can express $E\{m(t)m(t+\tau)\}$ as

$$E\{m(t)m(t+\tau)\} = E\left\{ \sum_{n=-\infty}^{\infty} a_n p(t - nT_s) \sum_{k=-\infty}^{\infty} a_k p(t + \tau - kT_s) \right\}$$

$$= \sum_{n=-\infty}^{\infty} \sum_{k=-\infty}^{\infty} p(t - nT_s)p(t + \tau - kT_s)E\{a_k a_n\}$$

$$= \sum_{k=-\infty}^{\infty} p(t - kT_s)p(t + \tau - kT_s) \qquad (10.16)$$

From (10.16) it is seen that $E\{m(t)m(t + \tau)\}$ is a periodic function in t with period T_s. Therefore the time average of $E\{m(t)m(t + \tau)\}$ needs only to be computed over one period T_s, and so we obtain the autocorrelation function of $m(t)$ as

$$R_m(\tau) = \overline{E\{m(t)m(t + \tau)\}}$$

$$= \frac{1}{T_s} \int_0^{T_s} E\{m(t)m(t + \tau)\} \, dt$$

$$= \frac{1}{T_s} \sum_{k=-\infty}^{\infty} \int_0^{T_s} p(t - kT_s)p(t + \tau - kT_s) \, dt$$

$$= \frac{1}{T_s} \int_{-\infty}^{\infty} p(t)p(t + \tau) \, dt$$

where the infinite sum of adjoining integrals is combined into a single infinite integral. The above expression is simply the autocorrelation of a random pulse train. In the case of non-band-limited filtering, $p(t)$ is simply a rectangular pulse train (or more precisely, an approximation of the rectangular pulse train), and its autocorrelation function is given by

$$R_m(\tau) = \begin{cases} 1 - \dfrac{|\tau|}{T_s} & |\tau| \le T_s \\ 0 & |\tau| > T_s \end{cases} \qquad (10.17a)$$

The corresponding Fourier transform of $R_m(\tau)$, which is the power spectral density of $m(t)$, is simply

$$S_m(f) = T_s \left(\frac{\sin \pi f T_s}{\pi f T_s}\right)^2 \qquad (10.17b)$$

Again with reference to (10.14), the autocorrelation function of the noise component $n'(t)$ of $y^2(t)_{bandpass}$ in (10.11c) can be obtained by taking the time average of $E\{n'(t)n'(t + \tau)\}$:

$$R_{n'}(\tau) = \overline{E\{n'(t)n'(t + \tau)\}}$$

$$= [R_{n_c}^2(\tau) + A^2 R_{n_c}(\tau)R_m(\tau)] \cos 2\omega_c \tau$$

$$= [R_1(\tau) + A^2 R_2(\tau)] \cos 2\omega_c \tau \qquad (10.18a)$$

where

$$R_1(\tau) = R_{n_c}^2(\tau) \qquad (10.18b)$$

$$R_2(\tau) = R_{n_c}(\tau)R_m(\tau) \qquad (10.18c)$$

Let $S_{n_c}(f)$ and $S_m(f)$ be the Fourier transforms of $R_{n_c}(\tau)$ and $R_m(\tau)$, respectively. In other words, $S_{n_c}(f)$ is the power spectral density of the lowpass noise $n_c(t)$ or $n_s(t)$, and $S_m(f)$ is the power spectral density of the message signal $m(t)$. Then, according to (10.18b) and (10.18c), the Fourier transforms $S_1(f)$ of $R_1(\tau)$ and $S_2(f)$ of $R_2(\tau)$ are given by

$$S_1(f) = S_{n_c}(f) \circledast S_{n_c}(f)$$

$$= \int_{-\infty}^{\infty} S_{n_c}(f - f') S_{n_c}(f') \, df' \qquad (10.19a)$$

$$S_2(f) = S_{n_c}(f) \circledast S_m(f)$$

$$= \int_{-\infty}^{\infty} S_{n_c}(f - f') S_m(f') \, df' \qquad (10.19b)$$

And consequently, the power spectral density of the noise component $n'(t)$, which is the Fourier transform of $R_{n'}(\tau)$ in (10.18a), is

$$S_{n'}(f) = \tfrac{1}{2} \left[S_1(f - 2f_c) + S_1(f + 2f_c) \right]$$

$$+ \tfrac{1}{2} A^2 [S_2(f - 2f_c) + S_2(f + 2f_c)] \qquad (10.20)$$

To simplify the evaluation of (10.20) we assume that the input bandpass filter BPF_i has a rectangular passband with bandwidth B hertz centered on $2f_c$ $(f > 0)$ and $-2f_c$ $(f < 0)$. Let the power spectral density of the bandpass noise $n(t)$ be $N_0/2$ (W/Hz) for $2f_c - B/2 \leq |f| \leq 2f_c + B/2$. Then the power spectral density of the lowpass noise $n_c(t)$ or $n_s(t)$ is given by

$$S_{n_c}(f) = \begin{cases} N_0 & -\dfrac{B}{2} \leq f \leq \dfrac{B}{2} \\ 0 & \text{elsewhere} \end{cases}$$

Now we make the following observation: The noise bandwidth B_n of the tracking filter is much smaller than that of the input bandpass filter in practical systems, that is,

$$B_n << B$$

Therefore the power spectral density of the noise component n' in (10.21) within the bandwidth B_n can be approximated by the value at $2f_c$ (for $f > 0$) and at $-2f_c$ (for $f < 0$); that is,

$$S_{n'}(f) \approx \tfrac{1}{2} \left[S_1(0) \Big|_{f>0} + S_1(0) \Big|_{f<0} \right]$$

$$+ \tfrac{1}{2} A^2 \left[S_2(0) \Big|_{f>0} + S_2(0) \Big|_{f<0} \right]$$

$$2f_c - \frac{B_n}{2} \leq |f| \leq 2f_c + \frac{B_n}{2} \qquad (10.21)$$

It remains to evaluate $S_1(0)$ and $S_2(0)$. From (10.19a) we have

$$S_1(0) = \int_{-B/2}^{B/2} S_{n_c}(-f')S_{n_c}(f')\, df'$$

$$= \int_{-B/2}^{B/2} N_0^2\, df'$$

$$= N_0^2 B \qquad (10.22a)$$

$$S_2(0) = \int_{-B/2}^{B/2} S_{n_c}(-f')S_m(f')\, df'$$

$$= \int_{-B/2}^{B/2} N_0 S_m(f')\, df' = N_0 \int_{-B/2}^{B/2} S_m(f')\, df'$$

$$= N_0 \overline{E\{m^2(t)\}} \qquad (10.22b)$$

where

$$\overline{E\{m^2(t)\}} = \int_{-B/2}^{B/2} S_m(f')\, df'$$

is the power of the message signal $m(t)$ within the noise bandwidth B of the input bandpass filter. Substituting (10.22a) and (10.22b) into (10.21) and integrating $S_{n'}(f)$ from $-\infty$ to ∞ yields the noise power of $n'(t)$ as

$$\sigma_{n'}^2 = \int_{-\infty}^{\infty} S_{n'}(f)\, df$$

$$\approx \int_{-2f_c - B_n/2}^{-2f_c + B_n/2} [\tfrac{1}{2} N_0^2 B + \tfrac{1}{2} A^2 N_0 \overline{E\{m^2(t)\}}]\, df$$

$$+ \int_{2f_c - B_n/2}^{2f_c + B_n/2} [\tfrac{1}{2} N_0^2 B + \tfrac{1}{2} A^2 N_0 \overline{E\{m^2(t)\}}]\, df$$

$$\approx N_0^2 B B_n + A^2 N_0 B_n \overline{E\{m^2(t)\}} \qquad (10.23)$$

From (10.12) we note that the power of the periodic component $\tfrac{1}{2} A^2 \overline{E\{m^2(t)\}} \cos 2(\omega_c t + \theta)$ is

$$S_0 = \tfrac{1}{8} A^4 \overline{E\{m^2(t)\}}^2 \qquad (10.24)$$

Therefore the output signal-to-noise ratio of the squarer is

$$\left(\frac{S}{N}\right)_0 = \frac{S_0}{\sigma_{n'}^2} = \frac{A^4 \overline{E\{m^2(t)\}}^2/8}{N_0^2 B B_n + A^2 N_0 B_n \overline{E\{m^2(t)\}}} \qquad (10.25)$$

To evaluate $(S/N)_0$ in terms of the carrier-to-noise ratio C/N at the input of the squarer we note that $N_0 B$ is the input noise power and $A^2 \overline{E\{m^2(t)\}}/2$ is the input carrier power; thus

$$\frac{C}{N} = \frac{A^2 \overline{E\{m^2(t)\}}}{2 N_0 B}$$

Consequently,

$$\left(\frac{S}{N}\right)_0 = \frac{1}{4}\left[\frac{A^2\overline{E\{m^2(t)\}}}{2N_0B_n}\right]\left[\frac{1}{1+\frac{1}{2}(C/N)^{-1}}\right]$$

$$= \frac{\frac{1}{4}(B/B_n)(C/N)}{1+\frac{1}{2}(C/N)^{-1}} \qquad (10.26)$$

Analysis of the time-4 frequency multiplier (quadrupler) is much more complicated. The result has been derived in [3]:

$$\left(\frac{S}{N}\right)_0 = \frac{\frac{1}{16}(B/B_n)(C/N)}{1+4.5(C/N)^{-1}+6(C/N)^{-2}+1.5(C/N)^{-3}} \qquad (10.27)$$

Now we consider the effect of the noise component $n'(t)$ on the phase angle 2θ of the periodic component at frequency $2\omega_c$. First we ignore the effect of self-noise. From (10.11c), (10.12a), and (10.12b) we have

$$s_2(t) = \left[\frac{1}{2}A^2\overline{E\{m^2(t)\}} + n'_c(t)\right]\cos 2(\omega_c t + \theta)$$
$$- n'_s(t)\sin 2(\omega_c t + \theta)$$

where $n'(t)$ in (11.12b) is expressed in terms of its in-phase and quadrature components, respectively, as

$$n'_c(t) = An_c(t)m(t) + \frac{1}{2}[n_c^2(t) - n_s^2(t)]$$
$$n'_s(t) = An_s(t)m(t) + n_c(t)n_s(t)$$

By using the trigonometric relation

$$A\cos x - B\sin x = \sqrt{A^2 + B^2}\cos\left[x + \tan^{-1}\left(\frac{B}{A}\right)\right]$$

we obtain

$$s_2(t) = \sqrt{\left[\frac{1}{2}A^2\overline{E\{m^2(t)\}} + n'_c(t)\right]^2 + n_s'^2(t)}\cos 2(\omega_c t + \theta + \theta_e)$$

where

$$2\theta_e = \tan^{-1}\left[\frac{n'_s(t)}{\frac{1}{2}A^2\overline{E\{m^2(t)\}} + n'_c(t)}\right]$$

The parameter $2\theta_e$ is the phase error (commonly called the phase jitter) at the output of the squarer caused by the noise term $n'(t)$. For a small phase error, that is, when $n'_c(t)$ and $n'_s(t)$ are much smaller than $\frac{1}{2}A^2\overline{E\{m^2(t)\}}$, we can approximate $2\theta_e$ by

$$2\theta_e \approx \frac{n'_s(t)}{\frac{1}{2}A^2\overline{E\{m^2(t)\}} + n'_c(t)}$$

$$\approx \frac{n'_s(t)}{\frac{1}{2}A^2\overline{E\{m^2(t)\}}}$$

Thus in this case the variance of the output phase jitter of the tracking filter is

$$E\{(2\theta_e)^2\} = \frac{E\{n_s'^2(t)\}}{\frac{1}{4}A^4\overline{E\{m^2(t)\}}^2}$$

$$= \frac{E\{n'^2(t)\}}{\frac{1}{4}A^4\overline{E\{m^2(t)\}}^2}$$

$$= \frac{\sigma_{n'}^2}{2S_0} = \frac{1}{2(S/N)_0}$$

$$= 2\frac{1 + \frac{1}{2}(C/N)^{-1}}{(B/B_n)(C/N)}$$

The variance of the phase jitter at the output of the divide-by-2 frequency divider (Fig. 10.1) is then

$$E\{\theta_e^2\} = \frac{1}{2}\frac{1 + \frac{1}{2}(C/N)^{-1}}{(B/B_n)(C/N)} \tag{10.28}$$

Similarly, the variance of the phase jitter at the output of the divide-by-4 frequency divider in the carrier recovery circuit using a quadrupler is

$$E\{\theta_e^2\} = \frac{1}{2}\frac{1 + 4.5(C/N)^{-1} + 6(C/N)^{-2} + 1.5(C/N)^{-3}}{(B/B_n)\ (C/N)} \tag{10.29}$$

The variance of the phase jitter for PSK signals in (10.28) and for QPSK signals in (10.29) has been derived without considering the effect of *self-noise*. These results are valid for a non-band-limited channel and for a band-limited channel when the received carrier-to-noise ratio C/N is low, that is, when the AWGN is dominant. At a high C/N self-noise becomes dominant, and the results in (10.28) and (10.29) no longer hold. Gardner [4] has investigated the effect of self-noise in a QPSK carrier recovery loop and has shown that the spectrum for the quadrature self-noise vanishes at $4f_c$ and peaks approximately $1/T_s$ hertz away from $4f_c$ for the raised-cosine data symbol pulse. [This pulse shape takes the form $p(t) = (1 + \cos \pi t/T_s)/2$ for $|t| \le T_s$, and 0 for $|t| > T_s$.] Thus if the noise bandwidth B_n of the narrowband bandpass filter BPF in Fig. 10.1 is much smaller than $1/T_s$, which is the case in a practical carrier recovery loop, the effect of the quadrature self-noise can be neglected. Gardner has also shown that if the steady-state phase error is zero (which can be achieved by the phase tracking circuit in Fig. 10.1), the effect of the in-phase self-noise, which has a spectrum centered at $4f_c$, can be nullified.

10.2 PHASE-LOCKED LOOP

In this section we discuss a negative feedback system called a *phase-locked loop* (PLL) that can be employed as a filter (Fig. 10.1) for tracking the phase of the received signal (for an M-ary PSK signal, this is the phase $M\theta$ of the recovered Mth harmonic component). The popularity and success of the phase-locked loop can be attributed to its ability to track accurately a signal immersed in AWGN even though the signal-to-noise ratio can be very low.

10.2.1 Principle of Operation

A phase-locked loop basically consists of three major components—a multiplier (phase detector), a loop filter, and a voltage-controlled oscillator (VCO)—arranged in the form of a feedback configuration as shown in Fig. 10.2. We assume that initially the frequency of the VCO is set to the frequency ω_0 of the input signal $x(t)$ and that it has a 90° phase shift with respect to $x(t)$. Suppose the input signal $x(t)$ is defined by

$$x(t) = A \cos[\omega_0 t + \phi(t)] + n(t)$$

where A = input signal amplitude and $\phi(t)$ = input signal phase. The term $n(t)$ is the AWGN with power spectral density $N_0/2$ and can be represented in terms of the in-phase and quadrature components as

$$n(t) = n_c(t) \cos \omega_0 t - n_s(t) \sin \omega_0(t)$$

Let the VCO output be defined by

$$y(t) = A_r \sin[\omega_0 t + \hat{\phi}(t)] \qquad (10.30)$$

where A_r = amplitude and $\hat{\phi}(t)$ = phase estimate of $\phi(t)$. With the control voltage $v(t)$ applied to the VCO input the phase $\hat{\phi}(t)$ can be expressed as

$$\hat{\phi}(t) = -K_r \int_0^t v(u) \, du \qquad (10.31)$$

where K_r = VCO sensitivity (rad/s-V).

The input signal $x(t)$ and the VCO output $y(t)$ are applied to the multiplier or phase detector, producing the product signal

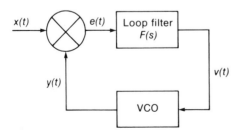

Figure 10.2 A phase-locked loop.

$$x(t)y(t) = \tfrac{1}{2} K_m A_v A \, \sin\left[\hat{\phi}(t) - \phi(t)\right]$$
$$+ \tfrac{1}{2} K_m A_v A \, \sin[2\omega_0 t + \phi(t) + \hat{\phi}(t)]$$
$$+ \tfrac{1}{2} K_m A_v [n_c(t) \, \sin \hat{\phi}(t) - n_s(t) \, \cos \hat{\phi}(t)]$$
$$+ \tfrac{1}{2} K_m A_v \{n_c(t) \, \sin[2\omega_0 t + \hat{\phi}(t)]$$
$$+ n_s(t) \, \cos[2\omega_0 t + \hat{\phi}(t)]\} \tag{10.32}$$

where $K_m A_v A = $ multiplier or phase detector sensitivity (V/rad).

The harmonic component at $2\omega_0$ is eliminated by the loop filter; therefore the error signal at the input of the loop filter is

$$e(t) = \tfrac{1}{2} K_m A_v A \, \sin \phi_e(t) + \tfrac{1}{2} K_m A_v A n'(t) \tag{10.33}$$

where $\phi_e(t)$ is the phase error defined by

$$\phi_e(t) = \hat{\phi}(t) - \phi(t) \tag{10.34}$$

and

$$n'(t) = \frac{n_c(t)}{A} \sin \hat{\phi}(t) - \frac{n_s(t)}{A} \cos \hat{\phi}(t) \tag{10.35}$$

The output of the loop filter is then given by

$$v(t) = \int_{-\infty}^{\infty} e(\tau) f(t - \tau) \, d\tau \tag{10.36}$$

where $f(t) = $ impulse response of loop filter. Substituting (10.33), (10.34), and (10.36) into (10.31) yields

$$\phi(t) = -\phi_e(t) - K_v \int_0^t \int_{-\infty}^{\infty} e(\tau) f(u - \tau) \, d\tau \, du$$
$$= -\phi_e(t) - \tfrac{1}{2} K_v K_m A_v A \int_0^t \int_{-\infty}^{\infty} [\sin \phi_e(\tau) + n'(\tau)] f(u - \tau) \, d\tau \, du$$

$$\tag{10.37}$$

or equivalently,

$$\frac{d\phi(t)}{dt} = -\frac{d\phi_e(t)}{dt} - \int_{-\infty}^{\infty} K[\sin \phi_e(\tau) + n'(\tau)] f(t - \tau) \, d\tau$$

$$\tag{10.38}$$

where

$$K = \tfrac{1}{2} K_v K_m A_v A$$

The constant K is called the *loop gain* and has dimensions of frequency (Hz).

When the phase error $\phi_e(t)$ is at its steady-state value, the loop is said to be in phase lock. When $\phi_e(t) \leq \pi/6$ radians for all time t, we can make the assumption

$$\sin \phi_e(t) \approx \phi_e(t) \qquad (10.39)$$

With this small error approximation, the integrodifferential equation (10.38) reduces to

$$\frac{d\phi(t)}{dt} = -\frac{d\phi_e(t)}{dt} - \int_{-\infty}^{\infty} K[\phi_e(\tau) + n'(\tau)]f(t-\tau)\, d\tau$$

$$(10.40)$$

Equation (10.40) represents a linearized model of a phase-locked loop. By setting $n'(t) = 0$ and using the Laplace transform we can determine the closed-loop transfer function:

$$s\Phi(s) = -s\Phi_e(s) - KF(s)\, \Phi_e(s) \qquad (10.41)$$

where $F(s) = s$-domain (Laplace) transfer function of loop filter [Laplace transform of $f(t)$]. Using $\Phi_e(s) = \hat{\Phi}(s) - \Phi(s)$ we obtain

$$s\Phi(s) = s\Phi(s) - s\hat{\Phi}(s) - KF(s)[\hat{\Phi}(s) - \Phi(s)]$$

or, equivalently,

$$\hat{\Phi}(s) = \frac{KF(s)}{KF(s) + s}\, \Phi(s) \qquad (10.42)$$

Thus the closed-loop transfer function of the phase-locked loop is

$$H(s) = \frac{\hat{\Phi}(s)}{\Phi(s)} = \frac{KF(s)}{KF(s) + s} \qquad (10.43)$$

Table 10.1 shows the calculation of $H(s)$ using the Laplace transform approach and the linearization assumption. When the transfer function $F(s)$ of the loop filter has n poles, the closed-loop transfer function $H(s)$ has $n + 1$ poles and the phase-locked loop is said to be an $(n + 1)$th-order loop.

A first-order loop is obtained if $F(s) = 1$, that is, if the loop filter is omitted. In this case the closed-loop transfer function is

$$H(s) = \frac{K}{K + s} \qquad (10.44)$$

It is seen that the loop gain K is the only parameter available to the designer. A first-order loop is not often used because good tracking requires a large K, while narrow noise bandwidth requires a small K. [The loop noise bandwidth will be defined in (10.61).] The most popular phase-locked loop is the second-order loop where the loop filter has one pole.

Table 10.1 Transfer function of a phase-locked loop

$$\hat{\Phi}(s) = -\frac{K_r V(s)}{s}$$

$$V(s) = F(s)E(s)$$

$$E(s) = \tfrac{1}{2} K_m A_r A \; \Phi_e(s)$$

$$K = \tfrac{1}{2} K_r k_m A_r A$$

$$\hat{\Phi}(s) = -\frac{K_r F(s)E(s)}{s} = \frac{-\tfrac{1}{2} K_r K_m A_r A F(s) \Phi_e(s)}{s}$$

$$= -\frac{KF(s) \; \Phi_e(s)}{s} = -\frac{KF(s)}{s} [\hat{\Phi}(s) - \Phi(s)]$$

$$\hat{\Phi}(s) = \frac{[KF(s)/s] \; \Phi(s)}{1 + [KF(s)/s]} = \frac{KF(s)}{KF(s) + s} \Phi(s)$$

$$H(s) = \frac{\hat{\Phi}(s)}{\Phi(s)} = \frac{KF(s)}{KF(s) + s}$$

For example,

$$F(s) = \frac{\tau_2 s + 1}{\tau_1 s + 1} \tag{10.45}$$

and can be realized by the passive RC filter shown in Fig. 10.3, where τ_1 and τ_2 are given by

$$\tau_1 = (R_1 + R_2)C$$

$$\tau_2 = R_2 C$$

Another form of $F(s)$ is

$$F(s) = \frac{\tau_2 s + 1}{\tau_1 s} \tag{10.46}$$

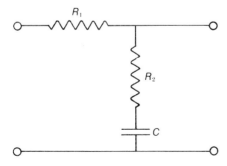

Figure 10.3 A passive loop filter.

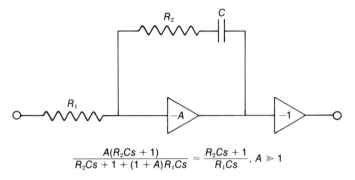

$$\frac{A(R_2Cs + 1)}{R_2Cs + 1 + (1 + A)R_1Cs} \approx \frac{R_2Cs + 1}{R_1Cs}, \, A \gg 1$$

Figure 10.4 An active loop filter.

and can be approximated by the active RC filter shown in Fig. 10.4, where τ_1 and τ_2 are given by

$$\tau_1 = R_1C$$

$$\tau_2 = R_2C$$

For the passive loop filter given in (10.44) the closed-loop transfer function is

$$H_p(s) = \frac{K(\tau_2 s + 1)/\tau_1}{s^2 + (1 + K\tau_2)s/\tau_1 + K/\tau_1}$$

$$= \frac{(2\zeta\omega_n - \omega_n^2/K)s + \omega_n^2}{s^2 + 2\zeta\omega_n s + \omega_n^2} \tag{10.47a}$$

where

$$\omega_n = \sqrt{\frac{K}{\tau_1}} \tag{10.47b}$$

$$\zeta = \frac{1}{2}\sqrt{\frac{K}{\tau_1}}\left(\tau_2 + \frac{1}{K}\right) \tag{10.47c}$$

The parameter ω_n is called the *natural frequency* of the loop, and ζ is the *damping factor*.

For the active loop filter given in (10.46), the closed-loop transfer function is approximated by

$$H_a(s) \approx \frac{K(\tau_2 s + 1)/\tau_1}{s^2 + K\tau_2 s/\tau_1 + K/\tau_1}$$

$$\approx \frac{2\zeta\omega_n s + \omega_n^2}{s^2 + 2\zeta\omega_n s + \omega_n^2} \tag{10.48a}$$

where

$$\omega_n = \sqrt{\frac{K}{\tau_1}} \qquad (10.48b)$$

$$\zeta = \frac{\tau_2}{2}\sqrt{\frac{K}{\tau_1}} \qquad (10.48c)$$

10.2.2 Steady-State Tracking Performance

To study the steady-state tracking performance we examine the phase error $\phi_e(t) = \hat{\phi}(t) - \phi(t)$ as $t \to \infty$ that results from a given input phase $\phi(t)$. By using the final value theorem of the Laplace transform we obtain

$$\lim_{t \to \infty} \phi_e(t) = \lim_{s \to 0} s\Phi_e(s)$$
$$\lim_{s \to 0} s[\hat{\Phi}(s) - \Phi(s)]$$
$$= -\lim_{s \to 0} s[1 - H(s)]\Phi(s)$$
$$= -\lim_{s \to 0} \frac{s^2\Phi(s)}{KF(s) + s} \qquad (10.49)$$

Consider first a step change in the input phase of magnitude $-\Delta\phi$; that is, $\Phi(s) = -\Delta\phi/s$. Then, for $F(0) > 0$,

$$\lim_{t \to \infty} \phi_e(t) = -\lim_{s \to 0} \frac{-s\,\Delta\phi}{KF(s) + s} = 0 \qquad (10.50)$$

Thus there is no steady-state phase error resulting from a step change in phase.

Now consider a step change in the frequency of magnitude $-\Delta\omega$; that is, $\phi(t) = -(\Delta\omega)t$ or $\Phi(s) = -\Delta\omega/s^2$. Then

$$\lim_{t \to \infty} \phi_e(t) = -\lim_{s \to 0} \frac{-\Delta\omega}{KF(s) + s} = \frac{\Delta\omega}{KF(0)} \qquad (10.51)$$

From (10.51) it is seen that, if the frequency of the input signal to the phase-locked loop is different from the VCO frequency (which is almost the case in practice), there is always a steady-state phase error. But it is not difficult to make $\Delta\omega/KF(0)$ small enough, say a few degrees, for a given maximum frequency offset $\Delta\omega_{max}$. This can be done by using the active loop filter shown in Fig. 10.4 with a large A, since $F(0) = A$.

10.2.3 Transient Response

To study transient response, we note that the Laplace transform of phase error $\phi_e(t)$ is

$$\Phi_e(s) = -[1 - H(s)]\Phi(s) \qquad (10.52)$$

For the second-order closed-loop transfer function $H(s)$ given in (10.48a) we have

$$\Phi_e(s) = \frac{-s^2}{s^2 + 2\zeta\omega_n s + \omega_n^2} \Phi(s) \tag{10.53}$$

Thus given $\Phi(s)$, we can evaluate the transient response of $\phi_e(t)$ by taking the inverse Laplace transform of (10.53):

1. Phase step input: $\Phi(s) = -\Delta\phi/s$
   $\zeta < 1$:

$$\phi_e(t) = \Delta\phi\left(\cos\sqrt{1-\zeta^2}\,\omega_n t \right.$$
$$\left. - \frac{\zeta}{\sqrt{1-\zeta^2}} \sin\sqrt{1-\zeta^2}\,\omega_n t\right) \exp(-\zeta\omega_n t) \tag{10.54a}$$

   $\zeta = 1$:

$$\phi_e(t) = \Delta\phi(1 - \omega_n t) \exp(-\omega_n t) \tag{10.54b}$$

   $\zeta > 1$:

$$\phi_e(t) = \Delta\phi\left(\cosh\sqrt{\zeta^2-1}\,\omega_n t \right.$$
$$\left. - \frac{\zeta}{\sqrt{\zeta^2-1}} \sinh\sqrt{\zeta^2-1}\,\omega_n t\right) \exp(-\zeta\omega_n t) \tag{10.54c}$$

2. Frequency step input: $\Phi(s) = -\Delta\omega/s^2$ [not including steady-state error $\Delta\omega/KF(0)$]
   $\zeta < 1$:

$$\phi_e(t) = \frac{\Delta\omega}{\omega_n}\left(\frac{1}{\sqrt{1-\zeta^2}} \sin\sqrt{1-\zeta^2}\,\omega_n t\right) \exp(-\zeta\omega_n t) \tag{10.55a}$$

   $\zeta = 1$:

$$\phi_e(t) = \frac{\Delta\omega}{\omega_n}(\omega_n t) \exp(-\omega_n t) \tag{10.55b}$$

   $\zeta > 1$:

$$\phi_e(t) = \frac{\Delta\omega}{\omega_n}\left(\frac{1}{\sqrt{\zeta^2-1}} \sinh\sqrt{\zeta^2-1}\,\omega_n t\right) \exp(-\zeta\omega_n t) \tag{10.55c}$$

10.2.4 Phase Jitter Due to Noise

In this section we study the phase jitter due to noise in the squaring loop shown in Fig. 10.5 where the phase-locked loop is employed as a tracking filter. The analysis is similar to the one in Sec. 10.1.2. We let the signal at

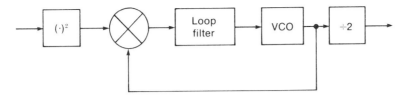

Figure 10.5 A squaring and phase-locked loop.

the input of the squarer be

$$y(t) = Am(t) \cos(\omega_c t + \theta) + n(t) \qquad (10.56)$$

where $n(t) = $ AWGN given in (10.11). The output of the squarer at the frequency $2\omega_c$ is given in (10.12) and (10.13).

Let the output of the VCO be represented as

$$v(t) = A_v \sin 2(\omega_c t + \hat{\theta}) \qquad (10.57)$$

After discarding the $2\omega_c$ components, the output of the phase detector with gain K_m is given by

$$e(t) = \tfrac{1}{4} K_m A_v A^2 \overline{E\{m^2(t)\}} \left[\sin 2(\hat{\theta} - \theta) + \hat{n}(t)\right] \qquad (10.58a)$$

where

$$\hat{n}(t) = \frac{1}{\tfrac{1}{4} A^2 \overline{E\{m^2(t)\}}} \{\frac{1}{4} A^2 m_0^2(t) \sin(\hat{\theta} - \theta)$$

$$+ \tfrac{1}{4} \left[n_c^2(t) - n_s^2(t)\right] \sin 2\hat{\theta}$$

$$+ \tfrac{1}{2} n_c(t) n_s(t) \cos 2\hat{\theta} + \tfrac{1}{2} A n_c(t) m(t) \sin(2\hat{\theta} - \theta)$$

$$- \tfrac{1}{2} A n_s(t) m(t) \cos(2\hat{\theta} - \theta)\} \qquad (10.58b)$$

$$m^2(t) = m_0^2(t) + \overline{E\{m^2(t)\}} \qquad (10.58c)$$

Assuming that the phase error $2(\hat{\theta} - \theta)$ is small, we can linearize (10.58a) to obtain

$$e(t) \approx \tfrac{1}{4} K_m A_v A^2 \overline{E\{m^2(t)\}} \left[(2\hat{\theta} - 2\theta) + \hat{n}(t)\right]$$

$$\approx \tfrac{1}{4} K_m A_v A^2 \overline{E\{m^2(t)\}} \{2\hat{\theta} - [2\theta - \hat{n}(t)]\} \qquad (10.59)$$

From (10.59) it is seen that the noise component $\hat{n}(t)$ is simply additive to the input phase 2θ. We know that $2\hat{\theta}$ is related to 2θ by the closed-loop transfer function $H(jf)$; therefore the same transfer function relates the power spectral density $\hat{S}(f)$ of the VCO phase jitter to the power spectral density $S_{\hat{n}}(f)$ of the noise $\hat{n}(t)$:

$$\hat{S}(f) = S_{\hat{n}}(f) |H(jf)|^2$$

The variance of the VCO phase jitter is †

$$\sigma_{2\hat{\theta}}^2 = \int_0^\infty 2S_{\hat{n}}(f) \, |H(jf)|^2 \, df$$

$$\approx 2S_{\hat{n}}(0) B_{\text{L}} \tag{10.60}$$

where the parameter B_{L} is the noise bandwidth of the phase-locked loop in hertz defined as

$$B_{\text{L}} = \frac{1}{2\pi} \int_0^\infty |H(j\omega)|^2 \, d\omega = \int_0^\infty |H(jf)|^2 \, df \tag{10.61}$$

The approximation in (10.60) is made because B_{L} is usually much smaller than the input noise bandwidth B in practical systems. The noise bandwidth of the phase-locked loop with transfer function $H_p(s)$ in (10.47) can be found to be

$$B_{\text{L}} = \frac{\omega_n}{8\zeta} \left(1 + 4\zeta^2 + \frac{4\zeta\omega_n}{K} + \frac{\omega_n^2}{K^2} \right) \tag{10.62}$$

and that of the transfer function $H_a(s)$ in (10.48a) is

$$B_{\text{L}} = \frac{\omega_n}{8\zeta} \, (1 + 4\zeta^2) \tag{10.63}$$

If we neglect the self-noise component $\frac{1}{2} A^2 m_0^2(t) \sin 2(\hat{\theta} - \theta)$ in (10.58b) and carry out an analysis similar to that in Sec. 10.1.2 the variance of the VCO phase jitter is given by

$$\sigma_{2\hat{\theta}}^2 = 4 \, \frac{1 + \frac{1}{2}(C/N)^{-1}}{(B/B_{\text{L}})(C/N)} \tag{10.64}$$

where $C/N = $ input carrier-to-noise ratio.

For a quadrupler loop, the variance of the VCO phase jitter is

$$\sigma_{4\hat{\theta}}^2 = 16 \, \frac{1 + 4.5(C/N)^{-1} + 6(C/N)^{-2} + 1.5(C/N)^{-3}}{(B/B_{\text{L}})(C/N)} \tag{10.65}$$

10.2.5 Hang-up

A phase-locked loop can be used as a tracking filter for M-ary PSK signals operating in a continuous mode but not in a TDMA burst mode because of the hang-up phenomenon [5]. Mathematically speaking, hang-

---

†This approximation is valid if the self-noise is neglected, as in the case of a non-band-limited signal or in the case of a band-limited signal operating at a low carrier-to-noise ratio where AWGN is dominant. At high carrier-to-noise ratios this approximation no longer holds.

up occurs when the phase error originates near 180° and remains in that vicinity for a long time before decaying toward equilibrium steady-state error. Therefore hang-up must be avoided in TDMA demodulators where carrier acquisitions must be completed within the carrier and clock recovery sequence at the beginning of the burst. Hang-up can cause error bunching, hence increase the unique word miss detection. In the next section we will study a tracking filter suitable for TDMA operation.

10.3 CARRIER RECOVERY CIRCUIT WITH NARROWBAND BANDPASS FILTER AND AUTOMATIC FREQUENCY CONTROL LOOP

To prevent the hang-up phenomenon that occurs in a phase-locked loop, the tracking filter in Fig. 10.1 can be used with the narrowband bandpass filter BPF with an automatic frequency control (AFC) loop, as illustrated in Fig. 10.6. The input signal to the mixer M1 is the periodic component at frequency $M\omega_c$ and is proportional to $\cos M(\omega_c t + \theta)$, where θ is the carrier phase. The other input signal to the mixer M1 is proportional to $\cos M(\omega_v t + \theta_v)$, where ω_v and θ_v are the VCO frequency and phase, respectively. These two signals are mixed in frequency by M1 which produces two output signals: One is proportional to cos

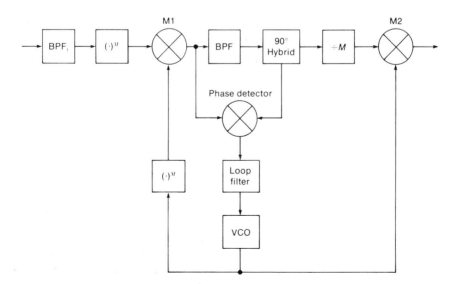

Figure 10.6 A carrier recovery circuit with narrowband bandpass filter and AFC loop.

$M[(\omega_c - \omega_v)t + \theta - \theta_v]$, and the other is proportional to cos $M[(\omega_c + \omega_v)t + \theta + \theta_v]$. The narrowband bandpass filter BPF with center frequency $M(\omega_c - \omega_v)$ passes only the former signal. A frequency divider is employed to obtain a signal proportional to $\cos[(\omega_c - \omega_v)t + \theta - \theta_v + \theta_e/M]$, where θ_e is the phase error introduced by the bandpass filter BPF which we will derive shortly. This signal is then mixed with the signal $\cos(\omega_v t + \theta_v)$ by the mixer M2 to produce two signals: One is proportional to $\cos(\omega_c t + \theta + \theta_e/M)$, which is the desired reference carrier, and the other is proportional to $\cos[(\omega_c - 2\omega_v)t + \theta - 2\theta_v + \theta_e/M]$, which can be rejected by a proper bandpass filter with center frequency ω_c.

The AFC loop is employed to correct the phase error θ_e introduced by the narrowband bandpass filter BPF. The inputs to the phase detector are two signals: One is proportional to $\cos M[(\omega_c - \omega_v)t + \theta - \theta_v]$, and the other is proportional to $\sin M[(\omega_c - \omega_v)t + \theta - \theta_v + \theta_e/M]$ due to 90° phase shift introduced by the 90° hybrid. The output of the phase detector is proportional to the error signal

$$e(t) = \sin \theta_e(t)$$

$$\approx \theta_e(t)$$

assuming that $\theta_e \leq \pi/6$ radians for all time t. This error signal is filtered by the loop filter, and the output signal is used to control the VCO frequency ω_v or, equivalently, the phase θ_v in such a way as to reduce the error θ_e to zero. Thus the AFC loop works on the same principle as the phase-locked loop discussed in Sec. 10.2, except that the phase error θ_e is always in the range $-90° \leq \theta_e \leq 90°$, hence the hang-up which occurs when the phase error originates near $\pm180°$ cannot exist.

To verify the above statement consider a symmetric bandpass filter with center frequency ω_0; then its real impulse response $h(t)$ can be expressed by

$$h(t) = \mathrm{Re}[2\hat{h}(t) \exp(j\omega_0 t)] \qquad (10.66)$$

where $2\hat{h}(t) = $ envelope of $h(t)$ and is real. [When the bandpass filter is assymmetric about its center frequency, then $h(t)$ is complex.] Let the input signal to the bandpass filter be defined as

$$x(t) = \mathrm{Re}\{A_x \exp[j(\omega_0 + \Delta\omega)t]\}$$

$$= \mathrm{Re}\{[A_x \exp(j\Delta\omega t)] \exp(j\omega_0 t)\} \qquad (10.67)$$

where $\Delta\omega = $ offset between input frequency $\omega_0 + \Delta\omega$ and center frequency ω_0 of bandpass filter. Note that $\Delta\omega$ can be positive or negative. The parameter A_x is a complex constant representing the amplitude $|A_x|$ and the phase $\theta = \tan^{-1}[\mathrm{Im}(A_x)/\mathrm{Re}(A_x)]$ of the input signal; that is,

$A_x = |A_x| \exp(j\theta)$. The signal $y(t)$ at the output of the bandpass filter is given by

$$y(t) = \text{Re}\{[\hat{h}(t) \circledast A_x \exp(j\Delta\omega t)] \exp(j\omega_0 t)\} \qquad (10.68)$$

With the assumption that A_x is a complex constant we have

$$\hat{h}(t) \circledast A_x \exp(j\Delta\omega t) = A_x \int_{-\infty}^{\infty} \hat{h}(\tau) \exp[j\Delta\omega(t-\tau)] \, d\tau$$

$$= A_x \exp(j\Delta\omega t) \int_{-\infty}^{\infty} \hat{h}(\tau) \exp(-j\Delta\omega\tau) \, d\tau$$

$$= A_x \hat{H}(j\Delta\omega) \exp(j\Delta\omega t)$$

$$(10.69)$$

where

$$\hat{H}(j\Delta\omega) = \int_{-\infty}^{\infty} \hat{h}(\tau) \exp(-j \, \Delta\omega\tau) \, d\tau$$

is simply the Fourier transform of $\hat{h}(t)$ evaluated at $\omega = \Delta\omega$; that is,

$$\hat{H}(j\Delta\omega) = \hat{H}(j\omega)|_{\omega=\Delta\omega}$$

The transfer function $\hat{H}(j\omega)$ is defined as the equivalent lowpass transfer function of the bandpass filter. The bandpass transfer function $H(j\omega)$, which is the Fourier transform of $h(t)$, is related to $\hat{H}(j\omega)$ by

$$H(j\omega) = \hat{H}[j(\omega - \omega_0)] + \hat{H}^* [-j(\omega + \omega_0)] \qquad (10.70)$$

Since $\hat{H}(j\Delta\omega)$ is a complex function we can represent it as

$$\hat{H}(j\Delta\omega) = |\hat{H}(j\Delta\omega)| \exp(j\theta_e) \qquad (10.71a)$$

where

$$\theta_e = \tan^{-1} \left\{ \frac{\text{Im}[\hat{H}(j\Delta\omega)]}{\text{Re}[\hat{H}(j\Delta\omega)]} \right\} \qquad (10.71b)$$

Substituting (10.69) and (10.71) into (10.68) yields

$$y(t) = \text{Re}\{A_x|\hat{H}(j\Delta\omega)| \exp(j[(\omega_0 + \Delta\omega)t + \theta_e])\} \qquad (10.72)$$

By comparing (10.72) to (10.67) we see that the bandpass filter has introduced a phase error θ_e given by (10.71b) when the input frequency is not coincident with the center frequency ω_0. Since $\tan^{-1}\{\text{Im}[\hat{H}(j\Delta\omega)]/\text{Re}[\hat{H}(j\Delta\omega)]\}$ belongs to the range $-90°$ to $90°$, the hang-up phenomenon does not exist.

The narrowband bandpass filter BPF in Fig. 10.8 is thus a very important component of the carrier recovery circuit. In practice, BPF can be a single- or double-tuned bandpass filter.

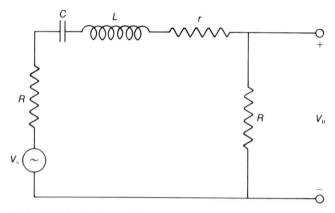

Figure 10.7 A single-tuned bandpass filter.

10.3.1 Single-Tuned Bandpass Filter

Consider a single-tuned LC bandpass filter with equal source and load resistance R as shown in Fig. 10.7 where the single-tuned resonator is represented by the series capacitor C, inductor L, and dissipation resistor r. The corresponding voltage transfer function $H(j\omega)$ is

$$H(j\omega) = \frac{V_0(j\omega)}{V_s(j\omega)} = \frac{R}{(2R + r) + j\omega L + 1/j\omega C} \tag{10.73}$$

Defining

$$\omega_0 = \frac{1}{\sqrt{LC}} \tag{10.74}$$

$$Q = \frac{1}{2R + r} \sqrt{\frac{L}{C}} \tag{10.75}$$

$$K = \frac{R}{2R + r}$$

we may write (10.73) as

$$H(j\omega) = \frac{K}{1 + jQ(\omega/\omega_0 - \omega_0/\omega)} \tag{10.76}$$

From (10.76) it is seen that the maximum power gain $|H(j\omega)|^2 = K^2$ occurs at the center frequency ω_0, and that the 3-dB points where the power gain $|H(j\omega)|^2 = K^2/2$ occur at

$$Q\left[\frac{\omega}{\omega_0} - \frac{\omega_0}{\omega}\right] = \pm 1 \tag{10.77}$$

Solving (10.77) and taking only the positive frequencies we have

$$\omega_{1,2} = \frac{\pm\omega_0/Q + \sqrt{4\omega_0^2 + \omega_0^2/Q}}{2} \tag{10.78}$$

From (10.78) the 3-dB bandwidth B_3 (rad/s) of the single-tuned bandpass filter is given as

$$B_3 = \omega_2 - \omega_1 = \frac{\omega_0}{Q} \text{ (rad/s)}$$

Thus the 3-dB bandwidth is the ratio of the center frequency to the Q factor. It is seen that, the higher the Q factor, the narrower the 3-dB bandwidth.

To calculate the noise bandwidth B_n (rad/s) of the single-tuned bandpass filter we can substitute $H(j\omega)$ in (10.76) into the following definition of noise bandwidth and assume that $\omega_0 \gg B_3$:

$$B_n = \int_0^\infty \frac{|H(j\omega)|^2}{|H(j\omega_0)|^2} \, d\omega$$

$$= \frac{\pi}{2} B_3$$

$$= \frac{\pi}{2} \left(\frac{\omega_0}{Q}\right) \text{(rad/s)} \tag{10.79}$$

If the Q factor of the bandpass filter is high compared with unity, that is, $Q \gg 1$, then we may approximate (10.76) by

$$H(j\omega) = \frac{K}{1 + j2Q(\omega - \omega_0)/\omega_0} + \frac{K}{1 + j2Q(\omega + \omega_0)/\omega_0} \tag{10.80}$$

Comparing (10.80) with (10.70) we conclude that the equivalent lowpass transfer function of the bandpass filter is

$$\hat{H}(j\omega) = \frac{K}{1 + j2Q\omega/\omega_0} = \frac{K}{1 + j2\omega/B_3} \tag{10.81}$$

If we replace the frequency $j\omega$ by the complex frequency $s = \sigma + j\omega$, then the s-domain equivalent lowpass transfer function is

$$\hat{H}(s) = \frac{K}{1 + (2/B_3)s} = \frac{KB_3/2}{s + B_3/2} \tag{10.82}$$

10.3.2 Double-Tuned Bandpass Filter

In the above analysis it was shown that the noise bandwidth of a single-tuned bandpass filter is equal to $\pi/2$ times the 3-dB bandwidth. To reduce the noise bandwidth so that the signal-to-noise ratio at the output of BPF2 can be improved without narrowing the 3-dB bandwidth, a double-tuned filter can be employed. If the double-tuned bandpass filter BPF2 is of the Butterworth type [6], the s-domain equivalent lowpass transfer function $\hat{H}(s)$ can be approximated by a second-order lowpass Butterworth filter with a 3-dB bandwidth B_3 as

$$
\begin{aligned}
\hat{H}(s) &= \frac{K}{(2/B_3)^2 s^2 + (2\sqrt{2}/B_3)s + 1} \\
&= \frac{K(B_3/2)^2}{(s + B_3/2\sqrt{2})^2 + (B_3/2\sqrt{2})^2}
\end{aligned}
\tag{10.83}
$$

where $K = $ dc gain of $\hat{H}(s)$. The corresponding frequency responses $\hat{H}(j\omega)$ and $|\hat{H}(j\omega)|^2$ are given by, respectively,

$$
\hat{H}(j\omega) = \frac{K(B_3/2)^2}{1 - (2/B_3)^2\omega^2 + j(2\sqrt{2}/B_3)\omega}
\tag{10.84a}
$$

and

$$
|\hat{H}(j\omega)|^2 = \frac{K^2(B_3/2)^4}{\omega^4 + (B_3/2)^4}
\tag{10.84b}
$$

Therefore on using (10.84b) the noise bandwidth B (rad/s) of the double-tuned bandpass filter can be computed as

$$
\begin{aligned}
B_n &= \int_{-\infty}^{\infty} \frac{|\hat{H}(j\omega)|^2}{|\hat{H}(0)|^2}\, d\omega \\
&= \left(\frac{B_3}{2}\right)^4 \int_{-\infty}^{\infty} \frac{d\omega}{\omega^4 + (B_3/2)^4} \\
&= \frac{\pi}{2\sqrt{2}} B_3
\end{aligned}
\tag{10.85}
$$

By comparing (10.85) to (10.79) we note that the noise bandwidth of a double-tuned bandpass filter is reduced by a factor of $\sqrt{2}$ over the noise bandwidth of a single-tuned bandpass filter of the same 3-dB bandwidth.

10.3.3 Cycle Slipping

In any carrier recovery loop the phase of the carrier occasionally slips one or more cycles as the result of a large noise event, and tracking returns to equilibrium n cycles away from the original condition. This phenomenon is called *cycle slipping* and is particularly destructive in TDMA operations because it can increase the probability of unique word miss detection considerably. The cycle slipping duration can be very long when the signal-to-noise ratio at the output of the narrowband bandpass filter BPF

(Fig. 10.6) is low. The occurrence of cycle slipping during the unique word detection period normally causes a miss detection and the entire burst is lost.

For a qualitative discussion of the cycle slipping phenomenon, consider the carrier recovery loop shown in Fig. 10.6 where the output to the narrowband bandpass filter BPF can be expressed as the sum of the periodic component $A_y \cos \omega_0 t$ and the narrowband noise $n(t) = n_c(t) \cos \omega_0 t - n_s(t) \sin \omega_0 t$. Without loss of generality we let the phase of the periodic component $\tan^{-1}[\text{Im}(A_y)/\text{Re}(A_y)] = 0$, since it is immaterial to our discussion of cycle slipping.

This output signal is

$$y(t) = [A_y + n_c(t)] \cos \omega_0 t - n_s(t) \sin \omega_0 t$$

$$= a(t) \cos[\omega_0 t + \psi(t)] \tag{10.86}$$

and its phasor diagram is sketched in Fig. 10.8. As the amplitudes and phases of $n_c(t)$ and $n_s(t)$ change with time in a random manner, the phase $\psi(t)$ occasionally contains steps of $\pm 2n\pi$ radians, $n = 1, 2, \ldots$. At the time when $\psi(t)$ contains steps of $\pm 2n\pi$ radians, the derivative $d\psi(t)/dt$ consists of impulselike components with areas nearly equal to $\pm 2\pi$ radians. These are shown in Fig. 10.9. When the signal $d\psi(t)/dt$ is passed through a lowpass filter, these impulselike components produce clicks. This is a phenomenon that occurs in FM receivers where the signal $d\psi(t)/dt$ which contains the message signal is obtained by passing $\psi(t)$ through a limiter-discriminator [7]. As the signal-to-noise ratio at the output of BPF decreases, the average number of clicks per unit time increases. Rice [8] has investigated noise in FM receivers and has shown that the average number of positive-going clicks (corresponding to a cycle slipping of $2n\pi$ radians) per second N_+ is the same as the average number of negative-going clicks (corresponding to a cycle slipping of $-2n\pi$ radians) per second N_- and is given by

$$N_+ = N_- = \frac{1}{2} B_{\text{rms}} \, \text{erfc}\left[\sqrt{\left(\frac{S}{N}\right)_0}\right] \tag{10.87}$$

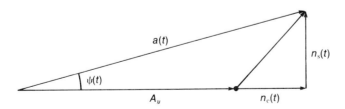

Figure 10.8 Signal and noise phasor diagram.

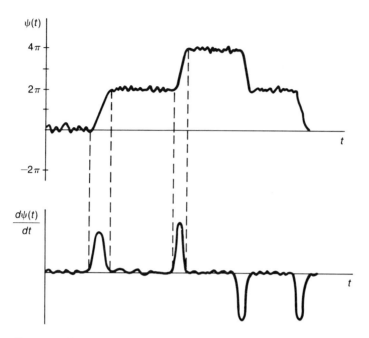

Figure 10.9 Impulse clicks due to cycle slipping.

where $(S/N)_0 =$ signal-to-noise ratio at output of narrowband bandpass filter BPF and $B_{rms} =$ rms bandwidth of BPF defined by

$$
B_{rms} = \frac{1}{2\pi} \left[\frac{\int_0^\infty (\omega - \omega_0)^2 \, |H(j\omega)|^2 \, d\omega}{\int_0^\infty |H(j\omega)|^2 \, d\omega} \right]^{1/2}
$$

$$
= \frac{1}{2\pi} \left[\frac{\int_{-\infty}^\infty \omega^2 |\hat{H}(j\omega)|^2 \, d\omega}{\int_{-\infty}^\infty |\hat{H}(j\omega)|^2 \, d\omega} \right]^{1/2} \quad \text{(Hz)} \qquad (10.88)
$$

where $H(j\omega)$ and $\hat{H}(j\omega)$ are the transfer function and the equivalent lowpass transfer function of BPF, respectively, and ω_0 is the center frequency of BPF. The average number of cycle slipping events per second in the carrier recovery loop is the sum N_+ and N_-:

$$
N = N_+ + N_-
$$

$$
= B_{rms} \, \text{erfc} \left[\sqrt{\left(\frac{S}{N}\right)_0} \right] \qquad (10.89)
$$

Consequently the probability of cycle slipping occurrence during a UW detection period is given by

$$P_{cs} = T_s LN$$

$$= T_s LB_{rms} \text{ erfc}\left[\sqrt{\left(\frac{S}{N}\right)_0}\right] \tag{10.90}$$

where $T_s = M$-ary symbol duration (s) and $L =$ unique word length (symbols).

The output signal-to-noise ratio $(S/N)_0$ has been derived in (10.26) for PSK and in (10.27) for QPSK where the effect of self-noise has been ignored for simplicity. Gardner [8] has investigated the effect of self-noise in the QPSK carrier recovery loop and has shown that the spectrum for the quadrature self-noise vanishes at $4f_c$ and peaks approximately $1/T_s$ hertz away from $4f_c$ for the raised-cosine data symbol pulse [this pulse shape takes the form $p(t) = (1 + \cos \pi t/T_s)/2$ for $|t| \leq T_s$ and 0 for $|t| > T_s$]. Thus, if the noise bandwidth B_n of the narrowband bandpass filter BPF is much smaller than $1/T_s$, which is the case in a practical carrier recovery loop, the effect of the quadrature self-noise can be neglected. Gardner also shows that, if the steady-state phase error is zero (this is achieved by the AFC loop in Fig. 10.6), the coupling of the in-phase self-noise into the loop (that in turn induces phase noise into the VCO and causes phase fluctuations) can be avoided.

When the output signal-to-noise ratio $(S/N)_0 \geq 2.5$, the complementary error function erfc$\sqrt{(S/N)_0}$ can be approximated by

$$\text{erfc}\left[\sqrt{\left(\frac{S}{N}\right)_0}\right] \approx \frac{\exp[-(S/N)_0]}{\sqrt{\pi(S/N)_0}} \tag{10.91}$$

Thus (10.89) and (10.90) become

$$N \approx \frac{B_{rms} \exp[-(S/N)_0]}{\sqrt{\pi(S/N)_0}} \tag{10.92}$$

$$P_{cs} \approx \frac{T_s LB_{rms} \exp[-(S/N)_0]}{\sqrt{\pi(S/N)_0}} \tag{10.93}$$

We leave it to the reader to show that P_{cs} decreases as the noise bandwidth B_n of the bandpass filter BPF is made smaller. In practice, the cycle slipping rate $N = P_{cs}/T_s L$ in (10.89) is selected to be smaller than the channel bit error rate at the same ratio of energy per bit to noise density E_b/N_0 [noting that $E_b/N_0 = T_b B(C/N)$, where T_b is the bit duration]. The typical carrier cycle slipping requirement for a QPSK demodulator operating in a TDMA mode is 10^{-4}/s (or 1 in 10^4 s) at $E_b/N_0 = 7$ dB.

In order to calculate N in (10.92) or P_{cs} in (10.93), the rms bandwidth B_{rms} must be known. Consider a single-tuned bandpass filter whose equivalent lowpass transfer function is given in (10.81). Then

$$|\hat{H}(j\omega)|^2 = \frac{K^2 B_3^2/4}{\omega^2 + B_3^2/4} \tag{10.94}$$

If $|\hat{H}(j\omega)|^2$ in (10.94) is substituted in (10.88), B_{rms} will diverge when the integration limits are taken from $-\infty$ to ∞. But since the carrier recovery loop is preceded by the input bandpass filter BPF_i whose noise bandwidth $B \gg B_3$ determines the integration limits, we have

$$B_{rms} = \frac{1}{2\pi} \left[\frac{\displaystyle\int_{-B/2}^{B/2} \frac{\omega^2}{\omega^2 + B_3^2/4}\, d\omega}{\displaystyle\int_{-B/2}^{B/2} \frac{1}{\omega^2 + B_3^2/4}\, d\omega} \right]^{1/2}$$

$$\approx \frac{1}{2\pi} \sqrt{\frac{BB_3}{2\pi}} \text{ (Hz)} \tag{10.95}$$

where B and B_3 are both in radians per second.

Example 10.1 Cycle Slipping Rate for a Single-Tuned Bandpass Filter
As an example, consider a TDMA-QPSK carrier recovery loop working at a bit rate of $R_b = 62$ Mbp. The input bandpass filter BPF_i in Fig. 10.6 is selected to satisfy the symbol duration–bandwidth product $T_s B = 1.16$, where $T_s = 2T_b = 2/R_b$ is the symbol duration and T_b is the bit duration. At the level of incoming $E_b/N_0 = 6$ dB, the corresponding input carrier-to-noise ratio is given by

$$\frac{C}{N} = \frac{E_b/N_0}{T_b B} = 6.86 \tag{10.96}$$

or, equivalently,

$$\frac{C}{N} = 8.36 \text{ dB}$$

Let BPF in Fig. 10.6 be a single-tuned bandpass filter with a noise bandwidth of $B_n = 360$ kHz. Consequently, its 3-dB bandwidth is given by (10.79) as

$$B_3 = \frac{2}{\pi} B_n = 229.2 \text{ kHz} \tag{10.97}$$

Also, the noise bandwidth B of the input bandpass filter BPF_i is

$$B = \frac{1.16}{T_s} = \frac{1.16}{2/R_b} = 36 \text{ MHz}$$

Substituting $C/N = 6.86$, $B_n = 360$ kHz, and $B = 36$ MHz into (10.27) yields the signal-to-noise ratio at the output of BPF as

$$\left(\frac{S}{N}\right)_0 = 24 \tag{10.98}$$

Also from (10.95) we get $B_{rms} = 1.15$ MHz. (Note that B and B_3 must be in radians per second in the evaluation of B_{rms}.) Therefore the cycle slipping rate N in (10.92) is given as

$$N = 5 \times 10^{-6} \qquad (10.99)$$

or, equivalently, the mean time between carrier cycle slips is $1/N = 2 \times 10^5$ s. If the unique word sequence has a length of $L = 128$ symbols, then the probability of cycle slipping occurrence during its detection time is

$$P_{cs} = T_s L N = 2.13 \times 10^{-11} \qquad (10.100)$$

which is extremely good. If a TDMA frame consists of 100 bursts, then it will take $1/2.13 \times 10^{-9} = 4.7 \times 10^8$ frames to lose one burst as a result of cycle slipping. Figure 10.10 shows the cycle slipping rate of QPSK signaling as a function of B_n (and B_3) for $E_b/N_0 = 6$ dB, $T_b B = 0.58$, and $R_b = 62$ Mbps.

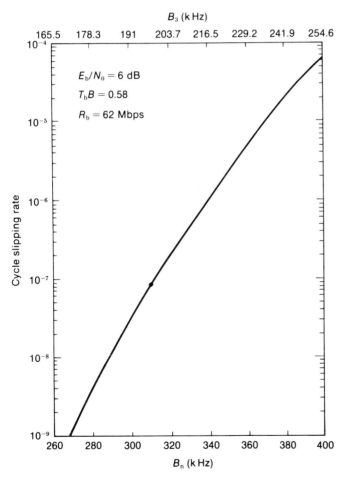

Figure 10.10 Cycle slipping rate of a QPSK signal as a function of the bandwidth of the single-tuned bandpass filter.

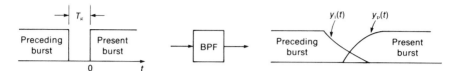

Figure 10.11 Concept of interburst interference.

10.3.4 Interburst Interference

The use of a narrowband bandpass filter BPF in the carrier recovery loop in Fig. 10.6 creates another problem in tracking the carriers of TDMA bursts. As is well-known, a narrowband filter induces a longer-lasting transient effect than a wideband filter. Since the 3-dB bandwidth of BPF is much smaller than that of the input filter BPF$_i$, the transient effect is dominated by BPF. When the recovered periodic components of two adjacent bursts pass through BPF, the transient response of the decaying periodic component of the preceding burst interacts with the rising periodic component of the present burst. This phenomenon is called *interburst interference* and is illustrated in Fig. 10.11. Interburst interference introduces degradation in the signal-to-noise ratio $(S/N)_0$ at the output of BPF and a phase error at the start of the unique word sequence. The intensity of the interference depends on the initial phase difference between the periodic components of the two adjacent bursts. In any case it is desirable to have a small output signal-to-noise degradation so that the cycle slipping rate does not increase considerably, and it is also desirable to have a small phase error so that the unique word miss detection rate which depends on the channel bit rate can be kept at a desired level.

To simplify the mathematical analysis of interburst interference, we use the equivalent lowpass (s domain) transfer function of the bandpass filter BPF. Thus we deal with the envelope of the periodic components of two adjacent bursts instead of the periodic components themselves. The equivalent lowpass (s domain) transfer function $\hat{H}(s)$ of the single-tuned BPF was derived in (10.82):

$$\hat{H}(s) = \frac{KB_3/2}{s + B_3/2} \tag{10.101}$$

For simplification, we assume that the frequency of the periodic components of the bursts are the same as the center frequency of BPF. [The phase error introduced by an offset frequency is given in (10.71b) and is another problem.] At the input of BPF the envelope $x_p(t)$ of the present burst and the envelope $x_i(t)$ of the preceding interfering burst can be modeled as

$$x_p(t) = \hat{A}_p u(t) \tag{10.102}$$

and

$$x_i(t) = \hat{A}_i \exp(j\phi)[1 - u(t)] \tag{10.103}$$

where $\hat{A}_p, \hat{A}_i$ = corresponding amplitudes

ϕ = initial phase difference between periodic components of preceding burst and present burst

$u(t)$ = unit step function defined as

$$u(t) = \begin{cases} 1 & t > 0 \\ 0 & t < 0 \end{cases}$$

Also, we have assumed the guard time $T_g = 0$ as the worst case. At the output of BPF the envelope $y_p(t)$ of the present burst is given by the inverse Laplace transform

$$y_p(t) = \mathscr{L}^{-1}\{\hat{H}(s)X_p(s)\} \tag{10.104}$$

where $X_p(s) = \hat{A}'_p/s$ is the Laplace transform of $x_p(t)$. By using $\hat{H}(s)$ in (10.101) we have

$$y_p(t) = A_p\left[1 - \exp\left(\frac{-B_3 t}{2}\right)\right] \tag{10.105}$$

where $A_p = K\hat{A}_p$ = amplitude of $y_p(t)$. Similarly, it can be shown that the envelope of the preceding interfering burst at the output of BPF is given by

$$y_i(t) = A_i \exp(j\phi) \exp\left(\frac{-B_3 t}{2}\right) \tag{10.106}$$

where $A_i = K\hat{A}_i$ = amplitude of $y_i(t)$.

From (10.105) and (10.106) the signal-to-interference ratio at the output of BPF is given as

$$\left(\frac{S}{I}\right)_0 = \frac{|y_p(t)|^2}{|y_i(t)|^2} = \left[\frac{\exp(B_3 t/2) - 1}{A_i/A_p}\right]^2 \tag{10.107}$$

where the ratio A_i/A_p represents the burst-to-burst amplitude variation. It is obvious that $(S/I)_0$ is high when $B_3 t$ is large enough. Therefore it is necessary that the bandpass filter BPF have a large 3-dB bandwidth. But this in turn implies a large noise bandwidth B_n and increases the cycle slipping rate, as we saw in Sec. 10.3.3. Thus cycle slipping and interburst interference requirements are two conflicting requirements, and consequently selection of the 3-dB bandwidth of the bandpass filter BPF must be a compromise. In practical TDMA systems, the burst-to-burst amplitude variation A_i/A_p can be as high as 2 or 20 $\log(A_i/A_p) = 6$ dB, and it is normally desired to keep the output signal-to-interference ratio $(S/I)_0 \geq 27$ dB. To evaluate the degradation in the output signal-to-noise ratio $(S/N)_0$ due to $(S/I)_0$ we note that the combined envelope $y(t)$ at the output of

BPF is simply the sum of $y_p(t)$ and $y_i(t)$ (the superposition principle applies here because BPF is a linear two-port network). Thus

$$y(t) = y_p(t) + y_i(t)$$

$$= A_p \left[1 - \exp\left(\frac{-B_3 t}{2}\right) \right]$$

$$+ A_i [\cos\phi + j \sin\phi] \exp\left(\frac{-B_3 t}{2}\right) \qquad (10.108)$$

The worst-case amplitude reduction occurs when the phase $\phi = \pm 180°$, that is, $y_i(t) = -A_i \exp(-B_3 t/2)$. This yields

$$y(t) = A_p \left[1 - \exp\left(\frac{-B_3 t}{2}\right) \right] - A_i \exp\left(\frac{-B_3 t}{2}\right) \qquad (10.109)$$

Recall that $y_p(t) = A_p [1 - \exp(-B_3 t/2)]$ is the output envelope where there is no interburst interference. Therefore the maximum signal power reduction in decibels is simply

$$\Delta = \frac{|y(t)|^2}{|y_p(t)|^2} = \left[1 - \frac{1}{\sqrt{(S/I)_0}} \right]^2 \qquad \left(\frac{S}{I}\right)_0 \geq 1 \qquad (10.110)$$

Note that the expression for Δ in (10.110) is meaningful only when $(S/I)_0 \geq 1$; that is, we consider the reduction only when the envelope of the present burst exceeds the envelope of the preceding interfering burst. With $(S/I)_0 \geq 1$ we see that $\Delta \to 0$ dB when $B_3 t$ is large enough. Consider the case when $(S/I)_0 = 27$ dB; then $\Delta = 0.913$, or $\Delta = -0.4$ dB. This results in 0.4-dB degradation in the output $(S/N)_0$. Thus when interburst interference is taken into account, the cycle slipping rate will increase. For example, when $\Delta = 0.913$, $(S/N)_0$ in (10.98) becomes $\Delta(S/N)_0 = 21.9$ and the corresponding cycle slipping rate in (10.99) becomes $N = 4.3 \times 10^{-5}$, almost a 10-fold increase.

The time t_Δ required to achieve a given Δ at a specified A_i/A_p is

$$t_\Delta = \frac{2}{B_3} \ln\left(1 + \frac{A_i/A_p}{1 - \sqrt{\Delta}} \right) \qquad (10.111)$$

where B_3 is in radians per second. For $\Delta = 0.913$ (or -0.4 dB), $A_i/A_p = 2$, and $B_3 = 229.2$ kHz (10.97), we have

$$t_\Delta = 5.3 \times 10^{-6} \text{ s} \qquad (10.112)$$

Therefore, in order to have 0.4-dB degradation in $(S/N)_0$ at the start of the unique word sequence, the length of the carrier and clock recovery sequence must have a minimum length in QPSK symbols:

$$L_{CCR} \geq \frac{t_\Delta}{T_s} = t_\Delta R_s \qquad (10.113)$$

where T_s = symbol duration and R_s = symbol rate. For $t_\Delta = 5.3$ μs, as in (10.112), and $R_s = 31$ Msps we have

$$L_{CCR} \geq 164 \text{ symbols} \tag{10.114}$$

The carrier and clock recovery sequence calculated in (10.114) is very pessimistic, since it deals with the worst case where

The periodic component of the preceding interfering burst is assumed to be exactly 180° out-of-phase, and this happens only on a fraction of bursts.
The amplitude of the preceding interfering burst is 6 dB higher.
The guard time is zero.

Besides causing signal-to-noise degradation, the interburst interference also introduces a transient phase error at the output of the bandpass filter BPF when the initial phase error ϕ between the periodic components of adjacent bursts is not $\pm 180°$. From (10.108) this transient phase error can be calculated as

$$\theta_t(\phi) = \tan^{-1}\left[\frac{(A_i/A_p) \sin\phi \exp(-B_3 t/2)}{1 - \exp(-B_3 t/2) + (A_i/A_p) \cos\phi \exp(-B_3 t/2)}\right]$$
$$\tag{10.115}$$

From (10.115) it can be seen that θ_t is maximum when $\phi = 90°$, which yields the time t_θ required to achieve a given $\theta_{t,\max}$ at a given A_i/A_p:

$$t_\theta = -\frac{2}{B_3} \ln\left(\frac{\tan\theta_{t,\max}}{\tan\theta_{t,\max} + A_i/A_p}\right) \tag{10.116}$$

where B_3 is in radians per second. The transient phase error of course will cause an increase in the detection bit error rate. For a QPSK demodulator the average probability of bit error due to a phase error α [9] is given by

$$P_b = \frac{1}{4}\left\{\text{erfc}\left[\sqrt{\frac{2E_b}{N_0}} \cos\left(\frac{\pi}{4} + \alpha\right)\right] + \text{erfc}\left[\sqrt{\frac{2E_b}{N_0}} \sin\left(\frac{\pi}{4} + \alpha\right)\right]\right\}$$
$$\tag{10.117}$$

and is plotted in Fig. 10.12. It is seen that a phase error $\alpha = 5°$ can cause an increase of about 0.4 dB in E_b/N_0 at $E_b/N_0 = 6$ dB, and an increase of about 0.6 dB in E_b/N_0 at $E_b/N_0 = 11$ dB. Given $A_i/A_p = 2$ and $B_3 = 229.2$ kHz (10.97), the time t_θ in (10.116) required to achieve $\theta_{t,\max} = 5°$ is

$$t_\theta = 4.4 \times 10^{-6} \text{ s} \tag{10.118}$$

For a bit rate of $R_b = 62$ Mbps, this translates to 136 QPSK symbols. Therefore, to achieve a resonable bit error rate at the beginning of the

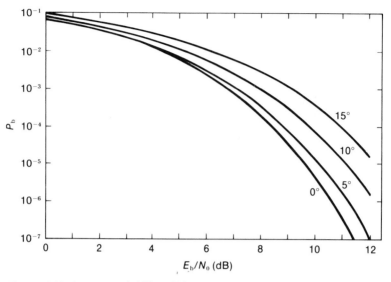

Figure 10.12 Average probability of bit error of a QPSK signal due to a phase error.

unique word, the carrier and clock recovery sequence must be at least 136 QPSK symbols long

$$L_{\text{CCR}} \geq \frac{t_\theta}{T_s} = t_\theta R_s = 136 \text{ symbols} \tag{10.119}$$

By comparing (10.114) and 10.119), we see that $L_{\text{CCR}} = 164$ symbols can be chosen as a conservative length for the carrier and clock recovery sequence to achieve both objectives—cycle slipping performance and bit error rate performance. Table 10.2 summarizes these two objectives for the single-tuned bandpass filter.

10.3.5 Burst-to-Burst Frequency Variations

As discussed at the beginning of Sec. 10.3, a phase error θ_e is introduced at the output of the narrowband bandpass filter BPF when the frequency of its input signal is not the same as its center frequency. The error θ_e can be evaluated according to (10.71b). This situation occurs in a TDMA operation because the carrier recovery loop sees bursts from different stations. Although all the bursts have the same nominal carrier frequency ω_c established by the reference station, satellite movement introduces a slight frequency offset between these carriers. Furthermore, the oscillator frequency in various satellite transponders may drift and further worsen the problem when transponder hopping is involved. In practical systems, the burst-to-burst carrier frequency variation can be as high as $\Delta f_c = \Delta\omega_c/2\pi = 6$ kHz, and the nominal carrier frequency of the bursts may drift as much as 60 kHz over the long term. Thus the AFC loop must be able to track a periodic component whose frequency can be offset as much as 60 kHz from the center frequency of the bandpass filter BPF. If BPF is a single-tuned bandpass filter, the phase error θ_e at its output due to a frequency offset $\Delta\omega$ at its input is given by (10.71b) and (10.81) as

$$\theta_e = -\tan^{-1}\left[\frac{2(\Delta\omega)}{B_3}\right] = -\tan^{-1}\left[\frac{2M(\Delta\omega_c)}{B_3}\right] \qquad (10.120a)$$

where

$$\Delta\omega = M(\Delta\omega_c) \qquad (10.120b)$$

and where $\Delta\omega_c =$ burst-to-burst carrier frequency variation and $B_3 = 3$-dB bandwidth of BPF (rad/s).

Note that θ_e can be positive or negative depending on whether $\Delta\omega$ is negative (the input frequency is smaller than the center frequency of BPF) or positive (the input frequency is larger than the center frequency of BPF). Because of the divide-by-M frequency divider (Fig. 10.6) the phase error of the recovered reference periodic component is

$$\alpha = \frac{\theta_e}{M} = -\frac{1}{M}\tan^{-1}\left[\frac{2M(\Delta\omega_c)}{B_3}\right] \qquad (10.121a)$$

$$\alpha \approx -\frac{2(\Delta\omega_c)}{B_3} \qquad \frac{2M(\Delta\omega_c)}{B_3} \leq \tan\left(\frac{\pi}{6}\right) \qquad (10.121b)$$

By taking $B_3 = 229.2$ kHz (Table 10.2), the phase error α in a QPSK mod-

Table 10.2 Performance of carrier recovery circuit using single-tuned bandpass filter

Modulation: QPSK
Ratio of energy per bit to noise density: $E_b/N_0 = 6$ dB
Bit duration–bandwidth product: $T_bB = 0.58$
Bit rate: $R_b = 62$ Mbps
Noise bandwidth of single-tuned BPF: $B_n = 360$ kHz
3-dB bandwidth of single-tuned BPF: $B_3 = 229.2$ kHz
Signal-to-noise ratio at output of BPF: $(S/N)_0 = 24$ or 13.8 dB
Cycle slipping rate: $N = 5 \times 10^{-6}$
Signal-to-noise degradation at the start of a UW sequence: $\Delta = -0.4$ dB
Transient phase error at the start of a UW sequence: $\theta_t = 5°$
Worst-case length of carrier and clock recovery sequence: $L_{CCR} = 164$ symbols
Steady-state phase error: $\lim t \to 0 \; \alpha(t) \approx 0°$ (Example 10.3)
Root-mean-square phase error $\sigma_\alpha = 3.66°$ (Example 10.2)

ulation scheme for $\Delta f_c = \Delta \omega_c / 2\pi = 6$ kHz is

$$\alpha = -\tfrac{1}{4} \tan^{-1} \left[\frac{2 \times 4 \times 2\pi \times 6 \text{ kHz}}{2\pi \times 229.2 \text{ kHz}} \right] = -3° \qquad (10.122)$$

As far as the unique word detection is concerned, this phase error induces a very small degradation in E_b/N_0 (0.2 dB at $E_b/N_0 = 6$ dB, and 0.4 dB at $E_b/N_0 \geq 11$ dB as seen from Fig. 10.12).

So far we have seen that the effect of the burst-to-burst carrier frequency variations can be made small while maintaining a low cycle slipping rate and a small interburst interference error (Table 10.2). This is not the case when the long-term drift of the carrier frequency is taken into account. Consider the case when $\Delta f_c = \Delta \omega_c / 2\pi = 60$ kHz; then according to (10.121a) we have

$$\alpha = -16.1° \qquad (10.123)$$

The above phase error would induce an unacceptably large degradation in E_b/N_0 (2.5 dB at $E_b/N_0 = 7$ dB, and 3 dB at $E_b/N_0 = 10$ dB) if left uncorrected. The purpose of the AFC loop in Fig. 10.6 is to correct such a large phase error. To analyze the tracking performance of the AFC loop, we need to derive the closed-loop transfer function of the entire carrier recovery loop shown in Fig. 10.6. As in the analysis of the phase-locked loop, we assume that the phase error θ_e at the output of the phase detector is small (less than $\pi/6$ radians or 30°) so that a linear model can be used. Note that the error $\theta_e = 4\alpha \approx -64°$, where α as given in (10.123) is the cumulative phase error if left uncorrected. In practice, the AFC loop will correct a phase error as large as this, although it may take a much longer time than the length of the carrier and clock recovery sequence. Using the Laplace transform, defining

$\theta_i(s) = $ input phase
$\theta_0(s) = $ output phase
$\theta_1(s) = $ input phase of BPF
$\theta_2(s) = $ output phase of BPF
$V(s) = $ output of loop filter
$\theta_v(s) = $ output phase of VCO
$F(s) = $ transfer function of loop filter
$\hat{H}(s) = $ equivalent lowpass transfer function of BPF

as shown in Fig. 10.6, and following the derivation in Table 10.1 we have

$$\theta_v(s) = \frac{KF(s)}{s} \left[\theta_2(s) - \theta_1(s) \right] \qquad (10.124)$$

$$\theta_2(s) = \frac{\hat{H}(s)}{\hat{H}(0)} \theta_1(s) \qquad (10.25)$$

$$\boldsymbol{\theta}_1(s) = M[\boldsymbol{\theta}_i(s) - \boldsymbol{\theta}_v(s)] \tag{10.126}$$

$$\boldsymbol{\theta}_0(s) = \frac{\boldsymbol{\theta}_2(s)}{M} + \boldsymbol{\theta}_v(s) \tag{10.127}$$

where $K = K_d K_v$

K_d = phase detector gain (V/rad)

K_v = VCO sensitivity (rad/s-V)

The plus sign in the expression for $\boldsymbol{\theta}_v(s)$ is used since the output of the VCO is $\cos(\omega_v t + \theta_v)$ instead of $\sin(\omega_v t + \theta_v)$. Also, the normalized constant $\hat{H}(0)$ is employed in the expression for $\boldsymbol{\theta}_2(s)$, since only the phase is considered and not the amplitude. From (10.124) to (10.127) we have

$$\begin{aligned}
\boldsymbol{\theta}_0(s) &= \frac{\hat{H}(s)}{M\hat{H}(0)} \boldsymbol{\theta}_1(s) + \boldsymbol{\theta}_v(s) \\[2mm]
&= \frac{\hat{H}(s)}{\hat{H}(0)} [\boldsymbol{\theta}_i(s) - \boldsymbol{\theta}_v(s)] + \boldsymbol{\theta}_v(s) \\[2mm]
&= \frac{\hat{H}(s)}{\hat{H}(0)} \boldsymbol{\theta}_i(s) + \left[1 - \frac{\hat{H}(s)}{\hat{H}(0)}\right] \boldsymbol{\theta}_v(s) \\[2mm]
&= \frac{\hat{H}(s)}{\hat{H}(0)} \boldsymbol{\theta}_i(s) + \frac{KF(s)}{s} \left[1 - \frac{\hat{H}(s)}{\hat{H}(0)}\right] [\boldsymbol{\theta}_2(s) - \boldsymbol{\theta}_1(s)] \\[2mm]
&= \frac{\hat{H}(s)}{\hat{H}(0)} \boldsymbol{\theta}_i(s) + \frac{KF(s)}{s} \left[1 - \frac{\hat{H}(s)}{\hat{H}(0)}\right] \\[2mm]
&\quad [M\boldsymbol{\theta}_0(s) - M\boldsymbol{\theta}_v(s) - M\boldsymbol{\theta}_i(s) + M\boldsymbol{\theta}_v(s)]
\end{aligned} \tag{10.128}$$

Rearranging the terms on both sides of (10.128) yields

$$\begin{aligned}
\left\{1 - \frac{MKF(s)}{s} \left[1 - \frac{\hat{H}(s)}{\hat{H}(0)}\right]\right\} \boldsymbol{\theta}_0(s) = \\[2mm]
\left\{\frac{\hat{H}(s)}{\hat{H}(0)} - \frac{MKF(s)}{s} \left[1 - \frac{\hat{H}(s)}{\hat{H}(0)}\right]\right\} \boldsymbol{\theta}_i(s)
\end{aligned} \tag{10.129}$$

Therefore the closed-loop transfer function of the carrier recovery loop in Fig. 10.6 is

$$G(s) = \frac{\boldsymbol{\theta}_0(s)}{\boldsymbol{\theta}_i(s)} = \frac{\hat{H}(s)/\hat{H}(0) - MKF(s)[1 - \hat{H}(s)/\hat{H}(0)]/s}{1 - MKF(s)[1 - \hat{H}(s)/\hat{H}(0)]/s} \tag{10.130}$$

Consider the case when BPF is a single-tuned filter with an equivalent lowpass transfer function $\hat{H}(s)$ in (10.82); then,

$$\frac{\hat{H}(s)}{\hat{H}(0)} = \frac{B_3/2}{s + B_3/2} \tag{10.131}$$

where $B_3 = 3$-dB bandwidth (rad/s). Let the loop filter be the active lowpass filter shown in Fig. 10.4 without the inverter at its output; then,

$$F(s) = -\frac{\tau_2 s + 1}{\tau_1 s} \tag{10.132a}$$

where

$$\tau_1 = R_1 C \tag{10.132b}$$
$$\tau_2 = R_2 C \tag{10.132c}$$

Substituting (10.131) and (10.132) into (10.130) yields

$$G(s) = \frac{2\zeta\omega_n s + \omega_n^2}{s^2 + 2\zeta\omega_n s + \omega_n^2} \tag{10.133a}$$

where

$$\omega_n = \sqrt{\frac{MK}{\tau_1}} \tag{10.133b}$$

$$2\zeta\omega_n = \frac{B_3}{2} + \frac{MK\tau_2}{\tau_1} \tag{10.133c}$$

and B_3 is in radians per second. Note the similarity between the closed-loop transfer function $G(s)$ of the AFC and bandpass filter loop and the closed-loop transfer function $H(s)$ of a second-order phase-locked loop in (10.48). The corresponding noise bandwidth of the carrier recovery loop in Fig. 10.6 is then given by (10.63) with ω_n and ζ given in (10.133b) and (10.133c):

$$B_L = \frac{\omega_n}{8\zeta}(1 + 4\zeta^2) \tag{10.134}$$

where B_L is in radians per second.

The phase error function of the carrier recovery loop is of interest since it determines both steady-state and transient phase errors due to a frequency offset at the input. It is given by

$$\alpha(s) = \theta_0(s) - \theta_i(s)$$
$$= -[1 - G(s)]\theta_i(s) \tag{10.135}$$

The steady-state phase error is obtained from the final value theorem of the Laplace transform as

$$\lim_{t \to \infty} \alpha(t) = \lim_{s \to 0} s\boldsymbol{\alpha}(s) \tag{10.136}$$

If the exact transfer function $F(s)$ is employed instead of the approximate one given in (10.132), that is, when

$$F(s) = -\frac{A(\tau_2 s + 1)}{[(1 + A)\,\tau_1 + \tau_2]s + 1} \tag{10.137}$$

as seen in Fig. 10.4, then for a burst-to-burst carrier frequency step $-\Delta\omega_c$, that is, $\theta_i = -(\Delta\omega_c)t$ or $\boldsymbol{\theta}_i(s) = -\Delta\omega_c/s^2$, the steady-state phase error given in (10.136) is found to be

$$\lim_{t \to \infty} \alpha(t) = \frac{\Delta\omega_c}{B_3/2 + MKA} \tag{10.138}$$

Thus if $A = F(0)$ is selected to be large enough, the steady-state phase error can be made negligible (note that $K > 0$). The reader may compare (10.138) to (10.121b) where the phase error is left uncorrected (the sign of the phase error being immaterial in this case).

The phase error transient response of the second-order loop in (10.133a) was derived in (10.55) for a frequency step input $\theta_i(s) = -\Delta\omega_c/s^2$. It is seen that the phase error is always bounded by (excluding the steady-state error)

$$\alpha(t) \leq \frac{\Delta\omega_c}{\omega_n} \tag{10.139}$$

Substituting (10.133b) into (10.139) we obtain

$$\alpha(t) \leq \frac{\Delta\omega_c}{\sqrt{MK/\tau_1}} = \frac{\Delta\omega_c\,(2\zeta)}{(B_3/2 + MK\tau_2/\tau_1)} \tag{10.140}$$

Consider the case where $\zeta = 0.5$ and note that $K > 0$; then (10.140) becomes

$$\alpha(t) \leq \frac{\Delta\omega_c}{B_3/2 + MK\tau_2/\tau_1} \tag{10.141}$$

hence

$$\alpha(t) \leq \frac{\Delta\omega_c}{B_3/2} \tag{10.142}$$

which is upper-bounded by the magnitude of the phase error if left uncorrected [see (10.121b)].

Example 10.2 Design of AFC Loop with a Single-Tuned Bandpass Filter. In practice, $K_d = 1.6$ V/rad, and $K_v = 10^6$ rad/s-V can be readily obtained for the phase detector gain and the VCO sensitivity. Thus $K = K_d K_v = 1.6 \times 10^6$ Hz is the AFC loop gain. Let $M = 4$, $B_3 = 229.2$ kHz, $\zeta = 0.5$, and $\tau_2/\tau_1 = 1$ and consider $\Delta f_c = \Delta\omega_c/2\pi = 16$ kHz. [This is the maximum frequency offset allowable for the linear model so that the phase error θ_e given by (10.120a) at the output of the phase detector will be less than $\pi/6$ radians (30°) and $\sin \theta_e$ can be approximated by θ_e.] Then according to (10.141) the phase error will be upper- bounded by

$$\alpha(t) \leq 0.8°$$

The natural frequency ω_n of the loop is given by (10.133c) as

$$\omega_n = 7.12 \times 10^6 \text{ rad/s}$$

Since $\tau_2/\tau_1 = 1$, the time constants $\tau_1 = R_1 C$ and $\tau_2 = R_2 C$ (10.132) can be calculated from (10.133b):

$$\tau_1 = \tau_2 = 0.126 \times 10^{-6}$$

Let $R_1 = R_2 = 10 \text{ k}\Omega$; then $C = 12.6 \text{ pF}$.

Now consider the case where the carrier frequency might drift as much as 60 kHz from its nominal value in the long term. We would like to know whether the AFC loop will lock to such a frequency offset. Note that the linear model does not apply in this case because the error at the output of the phase detector is about $-64°$ by virtue of (10.120a). To find the transient response we are required to solve an integrodifferential equation similar to (10.38), excluding the noise effect. To study the effect of a large frequency offset we note that the ability of the AFC loop to track a frequency variation depends on its loop gain $K = K_d K_v$, where K_d is the sensitivity of the phase detector and K_v is the sensitivity of the VCO. The parameter K_d depends on the amplitudes of the two input signals, namely, the signal at the input of BPF and that at its output (Fig. 10.6). Based on (10.67) to (10.72) we can write K_d as

$$K_d \propto |\hat{H}(j\Delta\omega)| \qquad (10.143)$$

where $\hat{H}(j\omega) = $ equivalent low-pass transfer function of BPF
$\quad \Delta\omega = M(\Delta\omega_c)$
$\quad \Delta\omega_c = $ carrier frequency offset

At the beginning of this example we selected $K_d = 1.6$ V/rad for $\Delta f_c = \Delta\omega_c/2\pi = 16$ kHz. Thus as $\Delta f_c = \Delta\omega_c/2\pi = 60$ kHz, K_d will be reduced by an amount equal to $|\hat{H}(j2\pi \times 64 \text{ kHz})| / |\hat{H}(j2\pi \times 240 \text{ kHz})| = 2$. Consequently, K is reduced from 1.6×10^6 to 0.8×10^6 Hz. The reduction of the loop gain by a factor of 2 should not prevent the AFC loop from tracking a carrier frequency offset as large as 60 kHz, although it may take a much longer time to reach the steady-state error at $\approx 0°$.

The loop noise bandwidth can be evaluated based on (10,133b) to (10.134) as

$$B_L = 566.6 \text{ kHz}$$

The variance of the phase error α at the output of the loop is $\frac{1}{16}$ times the variance given by (10.65) where B_L is given as above. Using the data in Table 10.2, with $C/N = 6.86$ ($E_b/N_0 = 6$ dB) and $B = 36$ MHz, we have

$$\sigma_\alpha^2 = \frac{1 + 4.5(C/N)^{-1} + 6(C/N)^{-2} + 1.5(C/N)^{-3}}{(B/B_L)(C/N)}$$

$$= 4.1 \times 10^{-3} \text{ rad}^2$$

or

$$\sigma_\alpha = 0.064 \text{ rad} = 3.66°$$

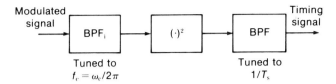

Figure 10.13 Symbol timing recovery circuit.

10.4 SYMBOL TIMING RECOVERY CIRCUIT

Symbol timing is required at the decision circuits in the demodulator to re-generate data streams. The symbol timing clock is obtained by the symbol timing recovery circuit from the transitions of the bipolar data. A nonlinear operation on the received signal is necessary to obtain the reference clock. For an M-ary PSK signal, a second-order nonlinearity is sufficient to extract the symbol timing from the modulated signal. A block diagram of the symbol timing recovery circuit is shown in Fig. 10.13; it employs a squarer followed by a bandpass filter whose center frequency is $f_0 = \omega_0/2\pi = R_s = 1/T_s$, where R_s is the symbol rate and T_s is the symbol duration. The bandpass filter BPF_i is the input filter mentioned often in analysis of the carrier recovery circuit (Figs. 10.1 and 10.6). It determines the carrier-to-noise ratio C/N at the input of the squarer.

To analyze the symbol recovery circuit consider the M-ary PSK signal given previously in (10.3):

$$s(t) = \text{Re}\,[Aa(t)\,\exp\,(j\omega_c t + \theta)] \qquad (10.144)$$

This signal is filtered by the input bandpass filter BPF_i to yield an input signal $y(t)$ at the input of the squarer as indicated by (10.5) and (10.6):

$$y(t) = \text{Re}\,[Am(t)\,\exp\,(j\omega_c t + \theta)] \qquad (10.145)$$

$$m(t) = \hat{h}_i(t) \circledast a(t) \qquad (10.146)$$

Consequently, the output of the squarer is

$$y^2(t) = \frac{A^2}{2}\,|m(t)|^2 + \frac{1}{2}\,\text{Re}\,[A^2m^2(t)\,\exp\,(j\,2\omega_c t + 2\theta)] \qquad (10.147)$$

With the assumption that $f_c = \omega_c/2\pi > R_s$, as is the case in practice, the component of $y^2(t)$ at $2\omega_c$ is rejected by the bandpass filter BPF (Fig. 10.13), leaving the signal $A^2|m(t)|^2/2$ that contains the clock component, as we will show. As a result of filtering by BPF, the output signal of the symbol timing recovery circuit is

$$z(t) = \frac{A^2}{2} \, \mathrm{Re} \, [h(t) \circledast |m(t)|^2] \tag{10.148}$$

where $h(t)$ = impulse response of BPF.

As discussed in Sec. 10.1.2, the data signal $a(t)$ is a cyclostationary process and can be represented by

$$a(t) = \sum_{k=-\infty}^{\infty} a_k g(t - kT_s) \tag{10.149}$$

where $\{a_k\}$ = zero mean independent sequence and $g(t)$ = pulse shape. Substituting (10.149) into (10.146) yields

$$m(t) = \sum_{k=-\infty}^{\infty} a_k p(t - kT_s) \tag{10.150}$$

where

$$p(t - kT_s) = \hat{h}_i(t) \circledast g(t - kT_s) \tag{10.150b}$$

is assumed to be real (i.e., we assume that BPF$_i$ is symmetric). Also, we note that

$$|m(t)|^2 = m(t) m^*(t)$$

$$= \sum_{n=-\infty}^{\infty} \sum_{k=-\infty}^{\infty} a_k a_n^* \, p(t - kT_s) p^*(t - nT_s)$$

$$= \sum_{n=-\infty}^{\infty} \sum_{k=-\infty}^{\infty} a_k a_n^* \, p(t - kT_s) p(t - nT_s) \tag{10.151}$$

Therefore, on substituting (10.151) into (10.148) we have

$$z(t) = \frac{A^2}{2} \, \mathrm{Re} \left\{ \sum_{n=-\infty}^{\infty} \sum_{k=-\infty}^{\infty} a_k a_n^* \, [h(t) \circledast p(t - kT_s) \, p(t - nT_s)] \right\}$$

$$\tag{10.152}$$

Since the data sequence $\{a_k\}$ is statistically independent, we have

$$E\{a_k a_n^*\} = \begin{cases} 1 & k = n \\ 0 & k \neq n \end{cases} \tag{10.153}$$

Consequently, the mean value of $z(t)$ is given by

$$E\{z(t)\} = \frac{A^2}{2} \, \mathrm{Re} \left\{ \sum_{k=-\infty}^{\infty} h(t) \circledast p^2(t - kT_s) \right\} \tag{10.154}$$

It is convenient to express the mean of $z(t)$ in (10.154) in terms of its Fourier series, since $E\{z(t)\}$ is a periodic function. Using the Poisson sum formula which states that

$$\sum_{k=-\infty}^{\infty} u(t - kT_s) = \frac{1}{T_s} \sum_{l=-\infty}^{\infty} U\left(\frac{l}{T_s}\right) \exp\left(\frac{j2\pi lt}{T_s}\right) \qquad (10.155)$$

where U is the Fourier transform of u, we can write (10.154) as

$$E\{z(t)\} = \frac{A^2}{2T_s} \operatorname{Re}\left\{ \sum_{l=-\infty}^{\infty} \left[H\left(\frac{jl}{T_s}\right) \int_{-\infty}^{\infty} P(f) P\left(\frac{l}{T_s} - f\right) df \right] \exp\left(\frac{j2\pi lt}{T_s}\right) \right\}$$

$$(10.156)$$

where $P(f)$ = Fourier transform of $p(t)$ in (10.150b). When the bandwidth of the bandpass filter BPF at the output of the squarer in Fig. 10.13 is much smaller than the data rate (which is the case in practical systems), then

$$H\left(\frac{jl}{T_s}\right) = \begin{cases} H(\pm j/T_s) & l = \pm 1 \\ 0 & l \neq \pm 1 \end{cases} \qquad (10.157)$$

and consequently we see that only the $l = \pm 1$ terms are retained in (10.156) and the mean timing signal at the output of the clock filter BPF is a purely sinusoidal signal $w(t)$ with frequency $1/T_s$ given by (with the assumption that BPF is symmetric)

$$\frac{w(t)}{|H(j/T_s)|} = \frac{A^2}{T_s} |\mu| \cos\left\{\frac{2\pi t}{T_s} + \tan^{-1}\left[\frac{\operatorname{Im}(\mu)}{\operatorname{Re}(\mu)}\right]\right\} \qquad (10.158a)$$

where

$$\mu = \frac{H(j/T_s)}{|H(j/T_s)|} \int_{-\infty}^{\infty} P(f) P\left(\frac{1}{T_s} - f\right) df$$

$$= \frac{H(j/T_s)}{|H(j/T_s)|} \int_{-\infty}^{\infty} G(f) G\left(\frac{1}{T_s} - f\right) \hat{H}_i(jf) \hat{H}_i\left(\frac{j}{T_s} - jf\right) df$$

$$(10.158b)$$

and where $G(f)$ = Fourier transform of pulse shape $g(t)$. When $g(t)$ is rectangular with width T_s and amplitude 1, then

$$G(f) = T_s \frac{\sin(\pi f T_s)}{\pi f T_s} \qquad (10.159)$$

To evaluate the performance in noise of the clock signal $w(t)$ in (10.158a) we consider the noise component given in (10.11b) which consists of the $s \times n$ and $n \times n$ terms†

$$\hat{n}(t) = An_c(t) m(t) + \tfrac{1}{2} n_c^2(t) + \tfrac{1}{2} n_s^2(t) \qquad (10.160)$$

Note that all three terms on the right side of (10.160) are mutually un-

---

†Here we ignore the self-noise effect. This assumption is valid for non-band-limited signals and for band-limited signals operating at a low carrier-to-noise ratio.

correlated. In order to determine the noise power within the bandwidth of the clock filter BPF we need to derive the power spectral density of $\hat{n}(t)$ which is the Fourier transform of the autocorrelation function of $\hat{n}(t)$. First we take the statistical average of $n(t)\hat{n}(t + \tau)$ and then the time average [because $m(t)$ is cyclostationary and not stationary] to obtain the autocorrelation function of $\hat{n}(t)$:

$$
\begin{aligned}
R_{\hat{n}}(\tau) &= \overline{E\{\hat{n}(t)\hat{n}(t + \tau)\}} \\
&= \tfrac{1}{4} \left[E\{n_c^2(t)n_c^2(t + \tau)\} + E\{n_s^2(t)n_s^2(t + \tau)\} \right. \\
&\quad + E\{n_c^2(t)\}\, E\{n_s^2(t + \tau)\} + E\{n_s^2(t)\}\, E\{n_c^2(t + \tau)\}\Big] \\
&\quad + A^2 E\{n_c(t)n_c(t + \tau)\}\, \overline{E\{m(t)m(t + \tau)\}} \\
&= R_{n_c}^2(0) + R_{n_c}^2(\tau) + A^2 R_{n_c}(\tau)R_m(\tau)
\end{aligned}
$$

(10.161)

where $R_{n_c}(\tau)$ = autocorrelation of lowpass noise $n_c(t)$ or $n_s(t)$ and $R_m(\tau)$ = autocorrelation of message signal $m(t)$. Consequently, the power spectral density of $\hat{n}(t)$ is

$$
S_{\hat{n}}(f) = R_{n_c}^2(0)\,\delta(f) + S_{n_c}(f) \circledast S_{n_c}(f) + A^2 S_{n_c}(f) \circledast S_m(f)
$$

(10.162)

Since the clock filter BPF is a narrowband bandpass filter tuned to $1/T_s$ whose noise bandwidth B_n is much smaller than the bandwidth of the input bandpass filter or the symbol rate $1/T_s$, the dc term $R_{n_c}^2(0)\,\delta(f)$ in (10.162) will be rejected by BPF. Furthermore, we only need to evaluate $S_{\hat{n}}(f) - R_{n_c}^2(0)\,\delta(f)$ at the frequency $f = 1/T_s$. Ignoring the dc term in $S_{\hat{n}}(f)$ we have

$$
S_{\hat{n}}(f) = \int_{-B/2}^{B/2} S_{n_c}(f - f') S_{n_c}(f')\, df' + A^2 \int_{-B/2}^{B/2} S_{n_c}(f - f') S_m(f')\, df'
$$

$$
1/T_s - B_n/2 \leq |f| \leq 1/T_s + B_n/2
$$

$$
\begin{aligned}
S_{\hat{n}}\!\left(\frac{1}{T_s}\right) &\approx \int_{-B/2}^{B/2} S_{n_c}\!\left(\frac{1}{T_s} - f'\right) S_{n_c}(f')\, df' \\
&\quad + A^2 \int_{-B/2}^{B/2} S_{n_c}\!\left(\frac{1}{T_s} - f'\right) S_m(f')\, df' \\
&\approx N_0^2 B\!\left(1 - \frac{|f|}{B}\right)\Bigg|_{f=1/T_s} + A^2 \int_{-B/2}^{-B/2+(B-1/T_s)} N_0 S_m(f')\, df' \\
&\approx N_0^2\!\left(B - \frac{1}{T_s}\right) + N_0 A^2 \int_{-B/2}^{-B/2+(B-1/T_s)} S_m(f')\, df'
\end{aligned}
$$

(10.163)

The noise power within the bandwidth of the clock filter BPF is given by

$$\sigma_n{}^2 \approx 2B_n\, S_{\hat{n}}\!\left(\frac{1}{T_s}\right)$$

$$\approx 2N_0^2\, B_n\!\left(B - \frac{1}{T_s}\right) + 2N_0 B_n A^2 \int_{-B/2}^{-B/2+(B-1/T_s)} S_m(f')\; df' \tag{10.164}$$

To evaluate the output signal-to-noise ratio, recall that the mean timing signal is given by (10.158a). [The normalization constant $|H(j/T_s)|$ in (10.158a) is used because the noise bandwidth B_n is defined with the same normalization constant, $B_n = \int_0^\infty |H(jf)|^2/|H(j/T_s)|^2\, df$.] Hence the signal power is

$$S_0 = \frac{A^4}{2T_s^2}\, |\mu|^2 \tag{10.165}$$

where μ is given in (10.158b). Therefore the output signal-to-noise ratio is

$$\left(\frac{S}{N}\right)_0 = \frac{S_0}{\sigma_n^2}$$

$$= \frac{A^4 |\mu|^2}{2T_s^2 \left[2N_0^2 B_n (B - 1/T_s) + 2N_0 B_n A^2 \displaystyle\int_{-B/2}^{-B/2+(B-1/T_s)} S_m(f')\; df'\right]}$$

$$= \frac{|\mu|^2}{E\{m^2(t)\}^2}\, \frac{(B/B_n)\,(C/N)}{2T_s^2\left[\frac{1}{2}\!\left(1 - \dfrac{1}{T_s B}\right)\!\left(\dfrac{C}{N}\right)^{-1}\right.}$$

$$\left. + \frac{1}{E\{m^2(t)\}} \int_{-B/2}^{-B/2+(B-1/T_s)} S_m(f')\; df' \right] \tag{10.166}$$

where C/N is the input carrier-to-noise ratio given by

$$\frac{C}{N} = \frac{A^2 \overline{E\{m^2(t)\}}}{2N_0 B} \tag{10.167}$$

and

$$\overline{E\{m^2(t)\}} = \int_{-B/2}^{B/2} S_m(f)\; df \tag{10.168}$$

Knowing the output signal-to-noise ratio, the variance of the timing jitter can be calculated as in (10.28)†:

$$\sigma_\tau^2 = \frac{1}{2(S/N)_0}\ (\text{radian})^2 = \left(\frac{T_s}{2\pi}\right)^2 \frac{1}{2(S/N)_0}\ (\text{second})^2 \tag{10.169}$$

†This result is valid for non-band-limited signals and for band-limited signals at a low carrier-to-noise ratio where the effect of self-noise can be neglected.

The evaluation of $(S/N)_0$ in (10.166) or σ_τ^2 in (10.169) must be carried out by numerical integration using the value of $G(f)$ in (10.15) and $S_m(f)$ in (10.17b) for a rectangular pulse shape and a random independent data pulse train. The equivalent lowpass transfer function $\hat{H}_i(jf)$ in (10.158b) depends on the choice of the input bandpass filter (a Butterworth filter is commonly used in practical systems). Also, from (10.166) we note that the signal-to-noise ratio at the output of the clock filter BPF depends on the amount of excess bandwidth of the pulse $p(t)$ which is indicated by the quantity $1 - 1/T_s B$. The smaller the excess bandwidth, the smaller the output signal-to-noise ratio, (note that μ also depends on the excess bandwidth, since the integrand in (10.158b) is a product of $G(f)\hat{H}_i(jf)$ and $G[1/T_s - f)\hat{H}_i(j/T_s - jf)]$ and a poor timing recovery performance can be expected.

Once the output signal-to-noise ratio $(S/N)_0$ in (10.166) is known, and the clock filter BPF tuned to $1/T_s$ has been selected (single- or double-tuned), the cycle slipping rate and the interburst interference effect can be evaluated in the same manner used for the carrier recovery circuit.

REFERENCES

1. J. B. Thomas, *An Introduction to Statistical Communication Theory*. New York: Wiley, 1969.
2. L. E. Franks, *Signal Theory*. Engelwood Cliffs, N.J.: Prentice-Hall, 1969.
3. S. A. Butman and J. R. Lesh, "The Effects of Bandpass Limiters on *n*-Phase Tracking Systems," *IEEE Trans. Commun.*, Vol. COM-25, No. 6, June 1977, pp. 569–576.
4. F. M. Gardner, "Self-Noise in Synchronizers," *IEEE Trans. Commun.*, Vol. COM-28, No. 8, Aug. 1980, pp. 1159–1163.
5. F. M. Gardner, "Hangup in Phase-Locked Loop," *IEEE Trans. Commun.*, Vol. COM-25, No. 10, Oct. 1977, pp. 1210–1214.
6. G. C. Temes and J. W. Lapatra, *Introduction to Circuit Synthesis and Design*. New York: McGraw-Hill, 1977.
7. S. Haykin, *Communication Systems*, 2d ed. New York, Wiley, 1983.
8. S. O. Rice, "Noise in FM Receivers," in M. Rosenblatt (ed.), *Symposium Proceedings of Time Series Analysis*. New York: Wiley, 1963, pp. 395–422.
9. R. J. Sherman, "Quadriphase Shift-Keyed Detection with a Noisy Reference Signal," IEEE Eascon Convention Record, 1969, pp. 45–52.

PROBLEMS

10.1 Consider a PSK carrier recovery circuit using a squarer and a tracking filter with noise bandwidth B_n. Assume that the bit rate $R_b = 64$ kbps and the channel noise bandwidth $B = 77$ kHz. Find B_n so that the output signal-to-noise ratio is 14 dB for an input E_b/N_0 of 7 dB. What is the rms phase jitter of the recovered carrier?

10.2 Consider a QPSK carrier recovery circuit using a quadrupler and a tracking filter. The noise bandwidth of the input bandpass filter is 36 MHz, and the signal bit rate is 62 Mbps. It is required that the rms phase jitter of the recovered carrier be at most 5° at an input $E_b/N_0 = 6$ dB. Find the noise bandwidth of the tracking filter.

10.3 A satellite PSK carrier is received at an earth station with a power of $C = -130$ dBm. The carrier is recovered by a second-order phase-locked loop using an active loop filter. The damping factor is 0.5, the loop noise bandwidth is 550 kHz, and the loop gain is 1.6×10^6 Hz. Find the time constants of the loop filter. It is required that the rms phase jitter of the recovered carrier be at most 2° at $C/N = 11$ dB. What is the carrier noise bandwidth? What is the system's noise temperature?

10.4 Repeat Prob. 10.3 for a QPSK signal at $C/N = 12$ dB.

10.5 Find the transient response of the second-order phase-locked loop with the transfer function given in (10.47) for a step frequency input.

10.6 A single-tuned bandpass filter is designed with a center frequency of 64 MHz and an equivalent noise bandwidth of 400 kHz. What is the required Q factor? Repeat the question for a double-tuned bandpass filter.

10.7 Find the rms bandwidth of an ideal bandpass filter of bandwidth W.

10.8 Consider a TDMA-QPSK carrier recovery loop working at a bit rate of $R_b = 120$ Mbps and a symbol duration–bandwidth product of $T_sB = 1.2$. Design a single-tuned bandpass filter to achieve a mean time between cycle slips of at least 10^4 s at an effective E_b/N_0 of 7 dB.

10.9 Consider a TDMA-QPSK carrier recovery circuit working at a bit rate of 43 Mbps and an intermediate frequency of 70 MHz. The channel noise bandwidth is 36 MHz, and the VCO frequency in the AFC loop is 54 MHz.

  (a) Find the center frequency of the single-tuned bandpass filter.

  (b) What Q factor is required for this filter to achieve a mean time between cycle slips of at least 10^4 s at an E_b/N_0 of 5.5 dB and an interburst interference of 6 dB?

  (c) Find the maximum transient phase error.

  (d) Find the carrier and clock recovery length.

10.10 Design a TDMA-QPSK carrier recovery circuit with input and output frequencies of 140 MHz using a single-tuned bandpass filter with a center frequency of 64 MHz to achieve a mean time between cycle slips of at least 10^4 s at an effective $E_b/N_0 = 6$ dB assuming an interburst interference of 6 dB. Find the maximum transient phase error and the required length of the carrier and clock recovery sequence. If the active loop filter is chosen such that $\tau_2/\tau_1 = 1$, $\zeta = 0.707$ and $K = 2 \times 10^6$ Hz, find the phase error due to a step frequency input of 16 kHz. Also, find the rms phase jitter due to noise at $E_b/N_0 = 6$ dB.

10.11 Find the rms bandwidth B_{rms} of the double-tuned bandpass filter.

10.12 Consider the TDMA-QPSK carrier recovery loop in Fig. 10.6 working at a bit rate of $R_b = 62$ Mbps. The input bandpass filter BPF_i is selected with $T_bB = 0.58$. Let BPF be a double-tuned bandpass filter with noise bandwidth $B_n = 360$ kHz. Find the cycle slipping rate at $E_b/N_0 = 6$ dB. Find the 3-dB bandwidth of BPF to achieve a cycle slipping rate of 5×10^{-6} at the same E_b/N_0. Compare the result with the 3-dB bandwidth of BPF in Example 10.1.

10.13 Consider the TDMA-QPSK carrier recovery loop in Fig. 10.6 using a double-tuned bandpass filter BPF. Find

  (a) Envelopes of the present and interfering bursts.

  (b) Signal-to-interference ratio at the output of BPF

  (c) $t_{\theta,max}$ to achieve a maximum transient phase error $\theta_{t,max} = 5°$.

  (d) $t_{\Delta,max}$ required to achieve a signal-to-interference degradation $\Delta = -0.4$ dB for $B_3 = 351$ kHz, $A_1/A_p = 2$.

10.14 For the TDMA-QPSK carrier recovery loop in Fig. 10.6 using a double-tuned bandpass filter BPF. Find

(a) The phase error θ_e at the output of BPF due to a burst-to-burst carrier frequency variation $\Delta f_c = 6$ kHz and $\Delta f_c = 60$ kHz given by $B_3 = 351$ kHz.

(b) Find the corresponding degradations in E_b/N_0 at $E_b/N_0 = 6$ dB.

10.15 Find the closed-loop transfer function of the carrier recovery loop in Fig. 10.6 if BPF is a double-tuned bandpass filter, and the loop filter is selected to be $F(s) = -1$. What is the noise bandwidth of the loop, the steady-state phase error, and the transient phase error?

10.16 Design the AFC loop with a double-tuned bandpass filter for the TDMA-QPSK carrier recovery circuit in Fig. 10.6 with the following parameters: $K = 1.6 \times 10^6$ Hz, $B_3 = 351$ kHz, $\Delta f_c = 16$ kHz, $E_b/N_0 = 6$ dB.

(a) Find ω_n and ζ.

(b) Find $\alpha(t)$ as $t \to \infty$.

(c) Find the loop noise bandwidth B_L.

(d) Find the variance of the phase error α.

10.17 Design a TDMA-QPSK clock recovery circuit to achieve a mean time between cycle slips of at least 10^5 s at an E_b/N_0 of 5.5 dB, assuming an interburst interference of 6 dB. The bit rate is 60 Mbps, and the symbol timing-bandwidth product is 1.2.

Chapter *11*

Satellite Spread Spectrum Communications

So far we have studied satellite systems for commercial usage where a link signal is contaminated by AWGN and other unintentional interferences, such as intermodulation interference, adjacent channel interference, cochannel interference, and multipath interference, all of which are treated as AWGN. In military satellite systems, a link signal may encounter intentional interference or *jamming* whose effect is much more devastating than noise itself. To provide protection for such a link, spread spectrum techniques are employed to greatly expand or spread the carrier spectrum relative to the information rate. Recall that the bit duration–bandwidth product T_bB of a satellite PSK signal is about 1.2 to 1.4. For a spread spectrum PSK, this is on the order of a thousand or more. The *spread spectrum* signal attains an antijamming capability by forcing the jammer to deploy the transmitted jamming power over a much wider bandwidth than would be necessary for a conventional system. In other words, for a given jammer power, the jamming power spectral density is reduced in proportion to the ratio of the spread bandwidth B_s to the unspread bandwidth B.

Besides antijamming protection, another important feature of a spread spectrum signal is the pseudorandomness which makes the signal appear similar to random noise and makes interception by the enemy difficult. There are many ways to spread the carrier spectrum relative to the information rate. They include *direct sequence* (DS) *spreading, frequency hopping* (FH), and *hybrid* techniques. In this chapter we will discuss DS and FH spread spectrum systems.

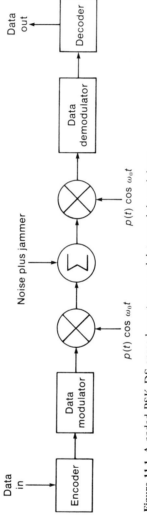

Figure 11.1 A coded PSK-DS spread spectrum modulator and demodulator

11.1 DIRECT SEQUENCE SPREAD SPECTRUM SYSTEMS

A block diagram of a coded DS *spread spectrum system* using coherent PSK as carrier modulation is shown in Fig. 11.1. Unlike the standard unspread modem discussed in Chap. 9, here the resulting PSK carrier at the modulator output is spread by multiplying it by another carrier that has been modulated by a *pseudorandom* or *pseudonoise* (PN) sequence represented by the bipolar waveform $p(t)$ with a chip rate R_p that is much larger than the information rate R_b. At the receiver, the information is recovered by multiplying the channel waveform by a synchronized replica of the PN sequence. This operation therefore spreads the jamming signal over the spread bandwidth determined by the PN sequence. It is important that the jammer have no knowledge of the PN sequence.

11.1.1 PN Sequence

By far the most widely known binary PN sequences are the maximal-length linear feedback shift register sequences, or m sequences, that can be generated with m-stage shift registers. The m sequence $\{m_j\}$ has a period of $2^m - 1$ and can be generated by a primitive polynomial $h(x)$:

$$h(x) = x^n + h_{n-1}x^{n-1} + \cdots + h_1 x + 1 \tag{11.1}$$

where h_i takes on the binary value 0 or 1 according to the relationship

$$m_j = h_1 m_{j-1} \oplus h_2 m_{j-2} \oplus \cdots \oplus h_{n-1} m_{j-n+1} \oplus m_{j-n} \tag{11.2}$$

Figure 11.2 shows the m-sequence generation. The feedback taps for a given $h(x)$ can be found based on [1,2]. Each m sequence generated by $h(x)$ has 2^{m-1} ones and $2^{m-1} - 1$ zeros. An m sequence possess a two-valued periodic autocorrelation function given by

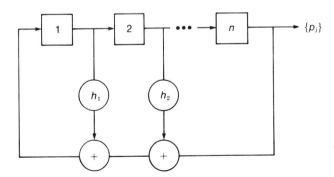

Figure 11.2 The generation of an m-sequence.

$$R_m(j) = 2^m - 1 - 2w\{m_i \oplus m_{i+j}\} = \begin{cases} 2^m - 1 & j = 0 \ (\text{mod } 2^m - 1) \\ -1 & j \neq 0 \ (\text{mod } 2^m - 1) \end{cases}$$

$$(11.3)$$

where $\oplus$ means addition modulo-2 and $w\{m_i \oplus m_{i+j}\}$ is the weight of the sequence $\{m_i \oplus m_{i+j}\}$ (i.e., the number of 1s in $\{m_i \oplus m_{i+j}\}$).

In practice the $\{0,1\}$-valued sequence $\{m_j\}$ is normally transmitted by a bipolar waveform $p(t)$ of positive and negative amplitudes according to the relationship

$$p(t) = \sum_{j=-\infty}^{\infty} (2m_j - 1)g(t - jT_c) = \sum_{j=-\infty}^{\infty} p_j g(t - jT_c) \quad (11.4)$$

where $g(t)$ = rectangular pulse of chip duration T_c and unit amplitude. Note that the sequence $\{p_j\}$ is a $\{-1,1\}$-valued sequence with an autocorrelation identical to that of $\{m_j\}$:

$$R_p(j) = \sum_{i=1}^{2^m - 1} p_i p_{i+j} = \begin{cases} 2^m - 1 & j = 0 \\ -1 & 0 < j < 2^m - 1 \end{cases}$$

since modulo-2 addition on $\{m_j\}$ has become multiplication on $\{p_j\}$. The normalized autocorrelation function of the periodic bipolar waveform $p(t)$ representing an m sequence is

$$\rho(\tau) = \frac{1}{(2^m - 1)T_c} \int_0^{(2^m - 1)T_c} p(t)p(t + \tau) \, dt$$

$$= -\frac{1}{2^m - 1} + \frac{2^m}{2^m - 1} q(\tau) \circledast \sum_{j=-\infty}^{\infty} \delta[\tau - j(2^m - 1) \ T_c]$$

$$(11.5a)$$

where

$$q(\tau) = \begin{cases} 1 - |\tau|/T_c & |\tau| \leq T_c \\ 0 & |\tau| > T_c \end{cases} \quad (11.5b)$$

The plot of $\rho(\tau)$ is shown in Fig. 11.3. The power spectral density of the m-sequence waveform $p(t)$ is the Fourier transform of $\rho(\tau)$ and is given by

$$S(\omega) = \frac{1}{(2^m - 1)^2} \delta(\omega) + \frac{2^m}{(2^m - 1)^2} \left[\frac{\sin(\omega T_c/2)}{\omega T_c/2} \right]^2$$

$$\sum_{\substack{j=-\infty \\ j \neq 0}}^{\infty} \delta\left(\omega - \frac{2\pi j}{(2^m - 1)T_c}\right)$$

$$(11.6)$$

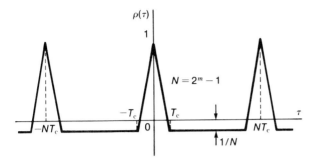

Figure 11.3 Normalized autocorrelation function of an m-sequence waveform.

$S(\omega)$ is plotted in Fig. 11.4. It is noted that the spectral lines get closer as the m-sequence period gets longer.

Although the properties of m sequences are well-understood, the m sequence can be copied by a jammer who can observe $2m$ chips of $p(t)$ and can compute the feedback connections of the linear feedback shift register using the *Berlekamp algorithm* [3]. To enhance the antijamming capability, the output sequence from the linear feedback shift register is not used directly. Instead, the outputs from many stages are combined by a nonlinear logic network to produce the output PN sequence, making it computationally impossible to determine the sequence generator by ob-

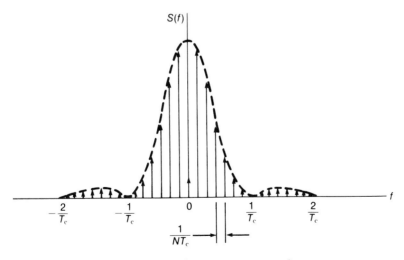

Figure 11.4 Power spectral density of an m-sequence waveform.

Table 11.1 Properties of sequence sets with a period of $2^m - 1$

Family	m	Set size	$|R_{uv,\max}|$
Gold	Odd	$2^m + 1$	$1 + 2^{(m+1)/2}$
Gold	2 (mod 4)	$2^m + 1$	$1 + 2^{(m+2)/2}$
Kasami (small set)	Even	$2^{m/2}$	$1 + 2^{m/2}$
Kasami (large set)	Even	$2^{m/2}(2^m + 1)$	$1 + 2^{(m+2)/2}$
Bent	0 (mod 4)	$2^{m/2}$	$1 + 2^{m/2}$

serving a portion of the sequence [4]. The PN sequence is changed frequently to enhance security.

In applications such as direct sequence code division multiple access (DS-CDMA), to be discussed in subsequent sections, the cross-correlation properties of PN sequences are as important as their autocorrelation properties. For m sequences the ratio of the peak magnitude of the cross-correlation function between two m sequences $\{u_j\}$ and $\{v_j\}$, namely $R_{uv,\max}$, to the peak magnitude of the autocorrelation function $R(0) = 2^m - 1$ is a large number ranging from 0.71 for m = 3 to 0.14 for $m = 11$. Although it is possible to select a small subset of m sequences that exhibit smaller cross-correlation peak values, the number of m sequences in the subset is too small for CDMA applications. Gold, Kasami, and Bent sequences [5], all with a period of $2^m - 1$ exhibit much smaller peak cross-correlation values that are appropriate for DS-CDMA. Table 11.1 lists the peak values of the peak cross-correlation for these sequences.

11.1.2 Error Rate Performance in Uniform Jamming

The performance of a DS spread spectrum system in the presence of jamming may be estimated from the average probability of bit error. Let C be the average carrier power and J be the average jammer power at the receiver input. The despreading process of the receiver will collapse the spread signal back to its original bandwidth B. At the same time, the jamming signal will be multiplied by the PN waveform $p(t)$, resulting in a signal whose power spectral density is the convolution of the power spectral densities of the above two signals. Depending on the bandwidth of the jamming signal, this resulting signal will have a bandwidth at least equal to the spread bandwidth B_s. When PSK is used, the spread bandwidth B_s is commonly taken to be the null-to-null bandwidth, hence $B_s = 2R_p$, where R_p is the chip rate. Let $J_0/2$ be the effective jamming power spectral density. For a tone jammer, J_0 is given by

$$J_0 = \frac{J}{R_p} \qquad (11.8)$$

where $R_p = B_s/2 =$ noise bandwidth of PSK spread signal [6].

If the information bit rate is R_b, the received energy per information bit is

$$E_b = \frac{C}{R_b} \qquad (11.9)$$

Therefore the ratio of the energy per information bit to the jamming density after despreading is given by

$$\frac{E_b}{J_0} = \frac{C}{J} \frac{R_p}{R_b} \qquad (11.10)$$

hence the jammer-to-carrier ratio is

$$\frac{J}{C} = \frac{R_p/R_b}{E_b/J_0} \qquad (11.11)$$

The spreading ratio R_p/R_b is called the processing gain since it makes the ratio of the energy per bit to the jammer density E_b/J_0 larger than the actual carrier-to-jammer ratio C/J by a factor of R_p/R_b:

$$\text{Processing gain} = \frac{R_p}{R_b} \qquad (11.12)$$

The jammer-to-carrier ratio J/C defines the jamming margin of the receiver:

$$(\text{Jamming margin})_{dB} = (\text{processing gain})_{dB} - \left(\frac{E_b}{J_0}\right)_{dB} \qquad (11.13)$$

Taking into account the AWGN of the channel, the required $E_b/\mathcal{N}_0$ after despreading is

$$\frac{E_b}{\mathcal{N}_0} = \frac{E_b}{N_0 + J_0} = \left[\left(\frac{E_b}{N_0}\right)^{-1} + \left(\frac{E_b}{J_0}\right)^{-1}\right]^{-1}$$

$$= \left[\left(\frac{E_b}{N_0}\right)^{-1} + \frac{R_b}{R_p} \frac{J}{C}\right]^{-1} \qquad (11.14a)$$

where $N_0/2 =$ power spectral density of AWGN. In terms of the received carrier-to-noise plus interference ratio $C/\mathcal{N} = C/(N+J)$, we can express $E_b/\mathcal{N}_0$ as

$$\frac{E_b}{\mathcal{N}_0} = \frac{R_p}{R_b} \frac{C}{\mathcal{N}} = \frac{R_p}{R_b} \frac{C}{N+J} = \frac{R_p}{R_b} \left[\left(\frac{C}{N}\right)^{-1} + \left(\frac{C}{J}\right)^{-1}\right]^{-1}$$

$$(11.14b)$$

Note that (11.14b) reduces to $E_b/\mathcal{N}_0 = (T_b B) C/\mathcal{N}$, which is the relationship in a nonspread spectrum system where $T_b B$ is the familiar bit duration–bandwidth product ($1 \le T_b B \le 2$). Equation (11.14a) indicates an interference rejection capability in a DS spread spectrum system. Although the jammer at the output of the demodulator is not Gaussian, it can be approximated as Gaussian noise if the processing gain is large. (Based on the central limit theorem, when a broadband signal with any statistics is passed through the integrate-and-dump filter of the PSK demodulator, the output is approximately Gaussian.) Therefore the average probability of bit error can be evaluated from the relationship

$$P_b = Q\left(\sqrt{\frac{2E_b}{\mathcal{N}_0}}\right) \tag{11.15}$$

Example 11.1 Consider an uncoded PSK-DS spread spectrum signal. At $P_b = 10^{-5}$, the required $E_b/\mathcal{N}_0$ is given by

$$\frac{E_b}{\mathcal{N}_0} = 9.6 \text{ dB}$$

Suppose that $E_b/N_0 = 20$ (or 13 dB) and the processing gain $R_p/R_b = 10^4$ (or 40 dB). Then the jamming margin is given by (11.14) as

$$\frac{J}{C} = 27.76 \text{ dB}$$

When $E_b/J_0 << E_b/N_0$, the jamming margin approaches the limit

$$\lim_{E_b/N_0 \to \infty} \frac{J}{C} = 30.4 \text{ dB}$$

11.1.3 Error Rate Performance in Pulsed Jamming

So far we have studied the performance of a DS spread spectrum system in the presence of *uniform jamming* (a continuous-wave or a tone jammer, and a noise jammer). When the jammer instead concentrates the power into *pulses* many bits long with a duty cycle $0 < \delta < 1$, the jamming peak power increases from J to J/δ and the result can be dramatic. Assume that the pulsed jammer is much stronger than the channel noise; that is, its jamming density $J_0/\delta = J/\delta R_p >> N_0$. Since only a fraction δ of the information bits are jammed, the average probability of bit error is

$$P_b = \delta Q\left(\sqrt{\frac{2\delta E_b}{J_0}}\right) \tag{11.16}$$

Clearly, the jammer would choose a duty cycle δ to maximize P_b. This occurs at $dP_b/d\delta = 0$, when

$$\delta = 0.71 \left(\frac{E_b}{J_0}\right)^{-1} \tag{11.17}$$

at which value

$$\max_{0<\delta<1} P_{\rm b} = 0.083 \left(\frac{E_{\rm b}}{J_0}\right)^{-1} = 0.083 \frac{R_{\rm b}}{R_{\rm p}} \frac{J}{C} \qquad (11.18)$$

Thus pulsed jamming, even with spread spectrum, changes an exponential relation into an inverse one.

Example 11.2 Consider the uncoded PSK-DS spread spectrum signal in Example 11.1. In the presence of the worst-case pulse jammer, the jamming margin at $P_{\rm b} = 10^{-5}$ is reduced from $J/C = 30.4$ dB to

$$\frac{J}{C} = \frac{10^{-5}}{0.083 \, R_{\rm b}/R_{\rm p}} = 0.81 \text{ dB}$$

It is seen that the jammer can inflict a devastating effect on an uncoded system. Surprisingly, coding can fully correct this deplorable situation if it is used appropriately. To avoid the case where the jammer uses pulses on the order of n coded bits long to destroy the advantage of coding, a pseudorandom interleaver (scrambler) can be used after the encoder. A deinterleaver (descrambler) before the decoder restores the order of the coded bits and at the same time scrambles the jamming pulses into random patterns, making the random-error-correcting capability of the code effective (restoring the memoryless property of the channel).

Numerical evaluation of the pulsed jamming margin can be achieved for any code whose standard bit error probability curve is available. The basic modulation discussed here is PSK or QPSK, but the procedure applies to any other modulation.

1. From the $P_{\rm b}$-versus-$E_{\rm b}/N_0$ curve of a coded PSK signal, find $E_{\rm b}/N_0$ for a desired $P_{\rm b}$.
2. Evaluate $E_{\rm c}/N_0 = E_{\rm b}/N_0 + 10 \log R$, where R is the code rate.
3. Evaluate the transition probability p' of the coded channel from $p' = Q(\sqrt{2E_{\rm c}/N_0})$.
4. For a given jamming duty cycle δ, find the required pulsed jamming $E_{\rm c}/J_0$ from $p' = \delta Q(\sqrt{2\delta E_{\rm c}/J_0})$.
5. Find the required pulse jamming $E_{\rm b}/J_0 = (E_{\rm c}/J_0)/R$.
6. Find $J/C = (R_{\rm p}/R_{\rm b})(E_{\rm b}/J_0)^{-1}$.

Example 11.3 Consider the PSK-DS spread spectrum signal in Example 11.1 using a rate $R = \frac{1}{2}$ convolutional code of constraint length $K = 7$ with Viterbi soft-decision decoding as shown in Fig. 11.5. (See [3] for a description of Viterbi soft-decision decoding.) At $P_{\rm b} = 10^{-5}$, the required $E_{\rm b}/N_0 = 4.5$ dB, which yields $E_{\rm c}/N_0 = 1.5$ dB. The corresponding transition probability $p' \approx 0.047$. In the worst-case pulsed jamming the required pulsed jamming $E_{\rm c}/J_0$ [see (11.18)] is

$$\frac{E_{\rm c}}{J_0} = 2.5 \text{ dB}$$

This yields, for a rate of $R = \frac{1}{2}$ code,

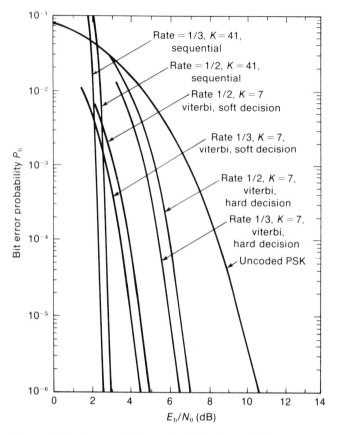

Figure 11.5 Bit error probability of coded coherent PSK.

$$\frac{E_b}{J_0} = 5.5 \text{ dB}$$

With the processing gain of 40 dB, the maximum tolerable jamming margin is

$$\frac{J}{C} = 34.5 \text{ dB}$$

Without coding, the jamming margin is 0.81 dB (Example 11.2). Thus a coding gain of 33.7 dB is achieved.

11.2 DIRECT SEQUENCE CODE DIVISION MULTIPLE ACCESS

One of the promising applications of DS spread spectrum to *commercial* satellite communications is DS code division multiple access. From our earlier discussion, although DS systems provide protection against inter-ference, the bandwidth is greatly wasted. To improve the bits-per-hertz throughput, one can take advantage of the low peak cross-correlation value in a set of DS addressed codes such as Gold sequences. In a DS-CDMA satellite system, each uplink earth station has its own addressed PN sequence and, unlike the situation in TDMA or FDMA where the carriers are separated by time or frequency, all active stations transmit their carriers on the same allocated bandwidth and overlap in time. Carrier separation is achieved at the earth station by correlation with the proper addressed PN sequence. Therefore, in DS-CDMA each carrier in the group represents a low interference signal to the others. The level of interference is directly determined by the ratio of the peak cross-correla-tion value to the peak autocorrelation value of the PN sequences. DS-CDMA consists of two types: (1) sequence-synchronous and (2) sequence-asynchronous.

11.2.1 Sequence-Synchronous DS-CDMA

In a *sequence-synchronous* system, the information bit duration $T_b = 1/R_b$ is chosen to be the period NT_c of the addressed PN sequences, where $T_c = 1/R_p$ is the chip duration. In other words the number of chips in a period N is equal to the processing gain of the system:

$$N = \frac{T_b}{T_c} = \frac{R_p}{R_b}$$

Furthermore, the entire system must be synchronized such that the PN sequence period or bit duration of each carrier in the system is time-aligned at the satellite. That is, a synchronous DS-CDMA system requires the type of *network synchronization* seen in a TDMA system, al-though at a much simpler level. Such a synchronization must be on the order of one-fifth of a chip duration T_c. Thus for a given satellite Doppler effect, this determines the maximum chip rate of the system so that a sta-tion can make a correction once per satellite roundtrip delay. For ex-ample, assume that the satellite Doppler effect is 40 ns/s; then the error in the satellite roundtrip delay of 0.25 s is 10 n. Since the required error can-not exceed $0.2 T_c = 0.2/R_p$, the maximum chip rate is $R_{p,max} = 20$ Mbps. To increase $R_{p,max}$ better satellite stationkeeping must be considered.

To study the performance of a synchronous DS-CDMA system in terms of the average probability of bit error, we consider PSK modulation and assume that the system allows $k + 1$ simultaneous users, each trans-mitting a carrier:

$$c_m(t) = Ab_m(t)\, p_m(t) \cos(\omega_c t + \theta_m) \qquad m = 1, 2, \ldots, k + 1$$

Here $b_m(t) = \pm 1$, $0 \leq t \leq T_b$ is the independent and identically distributed binary information, and $p_m(t) = \pm 1, 0 \le t \le T_c$ is the addressed PN waveform. Let us consider a receiver that tunes to the signal $c_1(t)$. It will receive $c_1(t)$ contaminated by noise and other k interferers,

$$\sum_{m=2}^{k+1} c_m(t)$$

Assume that the interference due to these k signals dominates the additive white Gaussian noise. This is the practical case, because one would like the system to be user interference–limited rather than noise-limited to maximize the throughput. Assume also that the carrier phases θ_m, $m = 1,2,\ldots,k+1$, are random variables uniformly distributed over $(0,2\pi)$. Without a loss of generality we let $\theta_1 = 0$. After PN sequence correlation and coherent demodulation of the received signal $c_1(t)$, the receiver bases its decision on the data $b_1(t)$ on the sign of

$$\hat{b}_1(t) = \frac{1}{T_b} \int_0^{T_b} b_1(t) p_1^2(t) \ dt + \frac{1}{T_b} \sum_{m=2}^{k+1} \int_0^{T_b} b_m(t) p_1(t) p_m(t) \cos \theta_m \ dt$$

$$= \pm 1 + \sum_{m=2}^{k+1} (\pm 1) \rho_{1m} \cos \theta_m \qquad 0 \le t \le T_b$$

$$(11.19a)$$

where

$$\rho_{1m} = \frac{1}{T_b} \int_0^{T_b} p_1(t) p_m(t) \ dt \qquad (11.19b)$$

is the normalized cross-correlation value of the PN waveforms $p_1(t)$ and $p_m(t)$, $m = 2,3,\ldots,k+1$. Let

$$p_m(t) = \sum_{j=1}^{N} p_j^{(m)} g(t - jT_c) \qquad 0 < t < T_b \qquad (11.20)$$

where $\{p_j^{(m)}\} = \{+1,-1\}$-valued sequence and $g(t) = $ rectangular pulse of chip duration T_c and unit amplitude. Equation (11.20) represents one period $(NT_c = T_b)$ of the PN waveform $p_m(t)$. Substituting (11.20) into (11.19b) yields

$$\rho_{1m} = \frac{1}{T_b} \sum_{j=1}^{N} p_j^{(1)} p_j^{(m)} \int_{jT_c}^{(j+1)T_c} g(t - jT_c) \ dt = \frac{T_c}{T_b} \sum_{j=1}^{N} p_j^{(1)} p_j^{(m)}$$

$$= \frac{R_{1m}}{N}$$

where

$$R_{1m} = \sum_{j=1}^{N} p_j^{(1)} p_j^{(m)}$$

is the cross-correlation value of sequences $\{p_j^{(1)}\}$ and $\{p_j^{(m)}\}$. It is seen that ρ_{1m} is indeed the normalized cross-correlation value of the PN sequences $\{p_j^{(1)}\}$ and $\{p_j^{(m)}\}$.

The probability of bit error for carrier $c_1(t)$ is

$$P_{\mathrm{b}}(c_1) = \Pr\left\{1 + \sum_{m=2}^{k+1} (\pm 1)\rho_{1m} \cos \theta_m < 0 \mid 1\right\}$$

or

$$P_{\mathrm{b}}(c_1) = \Pr\left\{-1 + \sum_{m=2}^{k+1} (\pm 1)\rho_{1m} \cos \theta_m > 0 \mid -1\right\}$$

By using the Chernoff bound on the random variable (Appendix 11A),

$$z = -1 + \sum_{m=2}^{k+1} (\pm 1)\rho_{1m} \cos \theta_m$$

the probabiiity of bit error is shown to be upper-bounded by

$$P_{\mathrm{b}}(c_1) \le P_{\mathrm{c}} = \min_{\lambda \ge 0} \exp(-\lambda) \prod_{m=2}^{k+1} \frac{1}{4\pi} \int_0^{2\pi} [\exp(\lambda\rho_{1m} \cos \theta_m$$

$$+ \exp(-\lambda\rho_{1m} \cos \theta_m)] \, d\theta_m$$

$$(11.21)$$

This Chernoff bound is a tight upper bound and for all practical purposes closely approximates the probability of bit error when z is considered a Gaussian random variable (i.e., the interuser interference is considered additive white Gaussian noise). In such a case the mean and variance of z are given by

$$\mu = E\{z\} = -1$$

$$\sigma^2 = E\{(z - \mu)^2\} = \tfrac{1}{2} \sum_{m=2}^{k+1} \rho_{1m}^2$$

and consequently the detection signal-to-interference ratio and the probability of bit error with a Gaussian assumption are

$$\frac{S}{I} = \frac{\mu^2}{\sigma^2} = \frac{2}{\displaystyle\sum_{m=2}^{k+1} \rho_{1m}^2}$$

$$P_{\mathrm{b}}(c_1) = Q\left(\sqrt{\frac{S}{I}}\right) = Q\left[\left(\frac{2}{\displaystyle\sum_{m=2}^{k+1} \rho_{1m}^2}\right)^{1/2}\right] \qquad (11.22)$$

where

$$Q(y) = \frac{1}{\sqrt{2\pi}} \int_y^\infty e^{-x^2/2} \, dx$$

is the standard Gaussian integral.

Suppose we select a set of addressed PN sequences such that the cross-correlation ρ_{nm} is bounded by ρ_{max}:

$$\max_{m \neq n} |\rho_{nm}| = |\rho_{max}|$$

then the probability of bit error for the entire population of users is upper-bounded by

$$P_b \leq P_c = \min_{\lambda \geq 0} \frac{1}{(4\pi)^k} \exp(-\lambda) \left[\int_0^{2\pi} [\exp(\lambda |\rho_{max}| \cos \theta) \right.$$

$$\left. + \exp(-\lambda |\rho_{max}| \cos \theta)] \, d\theta \right]^k \tag{11.23}$$

For a Gaussian assumption, P_b is given by

$$P_b = Q\left(\sqrt{\frac{2}{k\rho_{max}^2}} \right) \tag{11.24}$$

When additive white Gaussian noise, and other sources of interference (besides interuser interference) are taken into account, the Chernoff bound in (11.23) can be generalized to include them (Appendix 11A):

$$P_b \leq P_c = \min_{\lambda \geq 0} \frac{1}{(4\pi)^k} \exp(-\lambda) \exp\left(\frac{\mathcal{N}_0}{4E_b} \lambda^2 \right)$$

$$\left[\int_0^{2\pi} [\exp(\lambda |\rho_{max}| \cos \theta) + \exp(-\lambda |\rho_{max}| \cos \theta)] \, d\theta \right]^k \tag{11.25a}$$

where

$$\sigma_n^2 = \frac{\mathcal{N}_0}{2E_b} \tag{11.25b}$$

is the variance of the above additional noise. Note that $E_b/\mathcal{N}_0$ is the overall ratio of energy per bit to noise density. In terms of the processing gain R_p/R_b and the carrier-to-noise ratio, $E_b/\mathcal{N}_0$ is given by

$$\frac{E_b}{\mathcal{N}_0} = \frac{R_p}{R_b} \left(\frac{C}{\mathcal{N}} \right) \tag{11.26}$$

Here $C/\mathcal{N}$ is the overall carrier-to-noise ratio. If a Gaussian assumption is invoked, then P_b is given by

$$P_b = Q\left[\left(\frac{2}{k\rho_{max}^2 + (E_b/\mathcal{N}_0)^{-1}} \right)^{1/2} \right] \tag{11.27}$$

Because of the despreading process DS-CDMA has the capability to reject narrow band interference from adjacent satellite and terrestrial radio systems. Problems involving the broad beamwidth of a small antenna that receives signals from many geostationary satellites and terrestrial systems are solved by spreading the spectrum. Also, the problem caused by the sun appearing behind the satellite twice a year and greatly increasing the earth station noise temperature is reduced in proportion to a decrease in antenna gain. For an antenna less than 1 m in diameter, the sun effect adds about 1 dB to the system noise temperature. Thus DS-CDMA can avoid the periodic sun outages encountered in large earth stations.

Example 11.4 Consider a synchronous DS-CDMA system using Kasami sequences (small set) of period $2^8 - 1$ (the processing gain is $N = R_p/R_b = 255$ or 24 dB). The set size is 16, and the normalized maximum cross-correlation between any pair of sequences is $|\rho_{max}| = 17/255 = 0.0667$. Assume that the satellite link $E_b/\mathcal{N}_0 = 17$ dB; then the Chernoff bound of (11.25) yields a guaranteed probability of bit error $P_b \leq 4 \times 10^{-6}$ for k = 15. If the interuser interference is considered to be Gaussian, $P_b = 10^{-6}$.

Example 11.5 Consider a synchronous DS-CDMA system using Gold sequences of period $2^{11} - 1 = 2047$ (the processing gain is 2047 or 33 dB). The set size is 2049, and the normalized maximum cross-correlation between any pair of sequences is $|\rho_{max}| = 65/2047 = 0.0318$. Assume that the satellite link $E_b/\mathcal{N}_0 = 17$ dB; then the Chernoff bound of (11.25) yields a guaranteed probability of bit error $P_b \leq 10^{-5}$ for k = 70. When Golay code (23,12) with hard-decision decoding is used (t = 3, n = 23), the channel transition probability p for $P_{b,min} = 10^{-5}$ is $p = 8 \times 10^{-3}$ which by (11.25) (where $E_b/\mathcal{N}_0 = 17$ dB is replaced by the ratio of the energy per coded bit to the noise density $E_c/\mathcal{N}_0 = 14.2$ dB) yields k = 170. Thus the use of error-correction coding can increase the number of simultaneous users.

The above two examples illustrate DS-CDMA systems using nonorthogonal PN sequence sets. Of course, if orthogonal PN sequences are used, the number of simultaneous users would be equal to the sequence period N, but the user population would also be limited to N. On the other hand, a nonorthogonal PN sequence set with period $N = 2^m - 1$ such as the large Kasami set can provide a larger user population of $2^{m/2}(2^m + 1)$ at the expense of a sharp drop in the number of simultaneous users. Furthermore, if a few of the N simultaneous users lose network synchronization in an orthogonal synchronous DS-CDMA, the system failure could be catastrophic, because a pair of orthogonal sequences may have a high partial cross-correlation.

When the user population U is large ($U >> N$), the designer may wish to settle for an ensemble average performance instead of a guaranteed performance, as discussed above, knowing that some combinations of simultaneous users may have a worse performance than the average performance, while the performance of some combination of simultaneous users may exceed the average performance. This is the case in which the mean-squared value of ρ_{nm}^2, $m \neq n$ is considered instead of ρ_{max}^2. Since the PN sequence period is assumed to be N, we have

$$E\{\rho_{nm}^2\} = \frac{1}{N}$$

Thus the ensemble average probability of bit error of a synchronous DS-CDMA system with sequence period N is given (assuming interference is Gaussian) by

$$\bar{P}_b = Q\left(\sqrt{\frac{2}{k/N + (E_b/\mathcal{N}_0)^{-1}}}\right)$$

This ensemble average performance can also be derived as follows. Let us again assume equal carrier power C for each user. The interuser interference power spectral density is $kC/R_p = kE_b R_b/R_p = kE_b/N$. Therefore; the total noise power spectral density is $kE_b/N + \mathcal{N}_0$. Consequently, the energy per bit-to-noise density ratio is

$$\frac{E_b}{kE_b/N + \mathcal{N}_0} = \frac{1}{k/N + (E_b/\mathcal{N}_0)^{-1}}$$

and hence

$$\bar{P}_b = Q\left(\sqrt{2/[k/N + (E_b/\mathcal{N}_0)^{-1}]}\right)$$

11.2.2 Sequence-Asynchronous DS-CDMA

In an *asynchronous* DS-CDMA satellite system the information bit duration T_b is again chosen to be the period of the addressed PN sequences NT_c. No attempt is made to align the sequence period at the satellite. Therefore the overall system is much less complex than a synchronous DS-CDMA system. The only difference between them is the time shifts $0 \leq \tau_m \leq T_b$ that exist between carriers $c_m(t)$, $m = 1,2,\ldots,k+1$:

$$c_m(t) = Ab_m(t - \tau_m)p_m(t - \tau_m)\cos[\omega_c(t - \tau_m) + \theta_m]$$

For simplicity we consider $\tau_1 = 0$ and $\theta_1 = 0$. It is seen that, during the detection interval T_b of the carrier $c_1(t)$, the carrier $c_m(t)$, $m \neq 1$, arrives with a relative delay τ_m that involves the previous and present bits. Therefore the detection bit $b_1(t)$, $\theta \leq t \leq T_b$, in (11.19a) becomes

$$\hat{b}_1(t) = \frac{1}{T_b}\int_0^{T_b} b_1(t)p_1^2(t)\,dt$$

$$+ \sum_{m=2}^{k+1} \frac{1}{T_b}\left[\int_0^{\tau_m} (\pm 1)p_1(t)p_m(t - \tau_m)\cos\theta_m\,dt\right.$$

$$\left. + \int_{\tau_m}^{T_b} (\pm 1) p_1(t) p_m(t - \tau_m) \cos \theta_m \, dt \right]$$

$$= (\pm 1) + \sum_{m=2}^{k+1} \left[(\pm 1) \; \gamma_{1m}(\tau_m) + (\pm 1) \; \hat{\gamma}_{1m}(\tau_m) \cos \theta_m \right]$$

$$(11.28a)$$

where $\gamma_{1m}(\tau_m)$ and $\hat{\gamma}_{1m}(\tau_m)$ are partial cross-correlation functions of the addressed PN sequences $p_1(t)$ and $p_m(t)$ defined as

$$\gamma_{1m}(\tau_m) = \frac{1}{T_b} \int_0^{\tau_m} p_1(t) p_m(t - \tau_m) \, dt \qquad (11.28b)$$

$$\hat{\gamma}_{1m}(\tau_m) = \frac{1}{T_b} \int_{\tau_m}^{T_b} p_1(t) p_m(t - \tau_m) \, dt \qquad (11.28c)$$

Now assume that the random variables τ_m, $m = 2, 3, \ldots, k+1$, are uniformly distributed over $(0, T_b)$. Using the Chernoff bound it can be shown that the probability of bit error is upper-bounded by

$$P_b(c_1) \leq P_c = \min_{\lambda \geq 0} \exp(-\lambda) \, \exp\left(\frac{N_0}{4E_b} \lambda^2\right) \prod_{m=2}^{k+1} \frac{1}{8\pi T_b}$$

$$\int_0^{T_b} \int_0^{2\pi} \left[\exp\{\lambda[\gamma_{1m}(\tau_m) + \hat{\gamma}_{1m}(\tau_m)] \cos \theta_m\} \right.$$

$$+ \exp\{-\lambda[\gamma_{1m}(\tau_m) + \hat{\gamma}_{1m}(\tau_m)] \cos \theta_m\}$$

$$+ \exp\{\lambda[\gamma_{1m}(\tau_m) - \hat{\gamma}_{1m}(\tau_m)] \cos \theta_m\}$$

$$+ \exp\{-\lambda[\gamma_{1m}(\tau_m) - \hat{\gamma}_{1m}(\tau_m)] \cos \theta_m\} \, d\theta_m \, d\tau_m \right]$$

$$(11.29)$$

If we select a set of addressed PN sequences such that

$$\max_{m \neq n} |\gamma_{nm}(\tau_m - \tau_n) + \hat{\gamma}_{nm}(\tau_m - \tau_n)| = \max_{m \neq n} |\rho_{nm}(\tau)| = |\rho_{max}|$$

$$\max_{m \neq n} |\gamma_{nm}(\tau_m - \tau_n) - \hat{\gamma}_{nm}(\tau_m - \tau_n)| = \max_{m \neq n} |\hat{\rho}_{nm}(\tau)| = |\hat{\rho}_{max}|$$

where $\hat{\rho}_{max} = $ odd cross-correlation peak value [5]. Then the probability of bit error of the entire population of users is upper-bounded by

$$P_b \leq P_c = \min_{\lambda \geq 0} \frac{1}{(8\pi)^k} \exp(-\lambda) \, \exp\left(\frac{N_0}{4E_b} \lambda^2\right) \left[\int_0^{2\pi} \left[\exp(\lambda|\rho_{max}|\cos \theta) \right. \right.$$

$$+ \exp(-\lambda|\rho_{max}|\cos \theta) + \exp(\lambda|\hat{\rho}_{max}|\cos \theta)$$

$$\left. \left. + \exp(-\lambda|\hat{\rho}_{max}|\cos \theta) \right] \, d\theta \right]^k$$

$$(11.30)$$

Note that (11.30) reduces to (11.23) when $\tau_m = 0$, $m = 2,3,\ldots,k+1$. The evaluation of $|\hat{\rho}_{max}|$ can be found in [7]. Unfortunately, $|\hat{\rho}_{max}|$ is not available for Gold or Kasami sequences as ρ_{max}. Therefore analysis of asynchronous DS-CDMA is not as easily achieved as that of synchronous DS-CDMA. The latter can be viewed as an upper bound for the former. If a Gaussian assumption is invoked, then the probability of bit error is given in terms of the detection signal-to-noise ratio as

$$P_b = Q\left[\left(\frac{1}{(S/I)^{-1} + (2E_b/N_0)^{-1}}\right)^{1/2}\right] \tag{11.31a}$$

where S/I is the signal-to-user interference ratio given by

$$\frac{S}{I} = \frac{2}{k\beta_{max}^2} \tag{11.31b}$$

$$\beta_{max}^2 = \max_{m \neq n} \frac{1}{T_b} \int_0^{T_b} [\gamma_{nm}^2(\tau) + \hat{\gamma}_{nm}^2(\tau)] \, d\tau \tag{11.31c}$$

When the user population U is large, $(U >> N)$, the ensemble average performance may be considered. The ensemble average probability of bit error of an asynchronous DS-CDMA system with sequence period N is the same as that of a synchronous DS-CDMA system, namely,

$$\bar{P}_b = Q(\sqrt{2/[k/N + (E_b/N_0)^{-1}]})$$

In other words

$$E\{\beta^2\} = \frac{1}{T_b} \int_0^{T_b} [\gamma_{nm}^2(\tau) + \hat{\gamma}_{nm}^2(\tau)] \, d\tau = \frac{1}{N}$$

Therefore, synchronous DS-CDMA offers no advantage over asynchronous DS-CDMA when the user population is much larger than N.

The use of DS-CDMA offers an inexpensive alternative to FDMA and TDMA for low-bit-rate applications. With the new orbital spacing of $2°$ and the increase in bandwidth allocation by the WARC-79, DS-CDMA represents a new breed of satellite service that presents an extremely low interference to adjacent satellites and terrestrial systems. The commercial success of DS-CDMA [8] is a proof that DS-CDMA may be of use to low-data-rate applications. Although DS-CDMA cannot compete with FDMA or TDMA in medium- to high-data-rate applications, it can surely make satellite services affordable for many low-data-rate users.

11.2.3 DS-CDMA Link Analysis

In this section the link analysis of a synchronous DS-CDMA system is presented to illustrate that DS-CDMA is of use to small earth terminals (SET). The system consists of 448 small earth terminals communicating via a satellite transponder of 36 MHz. The following system parameters are used:

Antenna diameter = 1.2 m
Antenna noise temperature = 70 K
Low-noise-amplifier temperature = 90 K
High-power amplifier = 5 W or 7 dBW
EIRP to saturate transponder = 80 dBW
Transponder saturation EIRP = 36 dBW
Satellite G/T = −3 dB/K
Modulation: PSK
Information bit rate = 2400 bps
PN sequence period = $2^8 - 1 = 255$ chips
DS-CDMA chip rate = 612 kbps
Filtering: square root 100% cosine roll-off
Spread bandwidth = 1.224 MHz
Number of channels = 28
Number of simultaneous access per channel = 16
Total number of simultaneous access = $28 \times 16 = 448$

The link budget is given as follows:

Uplink (6 GHz)
Earth station EIRP/carrier = 42 dBW
Free space loss = 199.7 dB
Atmospheric attenuation and antenna pointing loss = 0.5 dB
Satellite G/T = −3 dB/K
Boltzmann's constant = −228.6 dB/K-Hz
$(C/N_0)_u = 67.4$ dB-Hz
$(C/N)_u = 6.5$ dB
Downlink (4 GHz)
Satellite EIRP/carrier = 6 dBW
Free space loss = 196 dB
Atmospheric attenuation and antenna pointing loss = 0.3 dB
Earth station G/T = 9.3 dB/K
Boltzmann's constant = −228.6 dB/K-Hz
$(C/N_0)_d = 47.6$ dB-Hz
$(C/N)_d = -13.3$ dB
Link $C/N = -13.3$ dB

Now assume that there are two adjacent satellites locating 4° east and west of the above satellite. Each interfering satellite has an EIRP of 36 dBW per 36 MHz. Thus the total interfering EIRP into the small earth terminal is 39 dBW per 36 MHz, or 24.3 dBW per 1.224 MHz (DS-

CDMA spread bandwidth). Assuming the SET antenna has uniform aperture distribution (to reduce the beamwidth). The normalized gain pattern is $G(\theta) = 4|J_1(\pi D \sin \theta/\lambda)/(\pi D \sin \theta/\lambda)|^2$ where $D = 1.2$ m is the antenna diameter, λ is the downlink wavelength, and J_1 is the Bessel function of first order. At $\theta = 4°$, the gain is 22 dB lower than the on-axis gain. Thus the carrier-to-adjacent satellite interference ratio is

$$\frac{C}{I} = 6 - 24.3 + 22 = 3.7 \text{ dB}$$

Combining with the link carrier-to-thermal noise ratio $C/N = -13.3$ dB yields the link carrier-to-thermal noise plus adjacent satellite interference ratio $C/N = -13.4$ dB. This corresponds to $E_b/N_0 = 13.7$ dB.

The other interference is the DS-CDMA cochannel interference. If Gaussian assumption is used, the energy per bit-to-cochannel interference density is $1/k\rho_{max}^2 = 1/15 \times 0.0667^2 = 11.8$ dB. Therefore the overall $(E_b/N_0)_t = 9.6$ dB. By (11.27) the probability of bit error is $P_b = 10^{-5}$.

The error rate can be improved by the use of error-correction coding. For example, rate 7/8 convolutional code with Viterbi decoding can offer a coding gain of 4 dB at $P_b < 10^{-5}$. The channel efficiency can be improved by using square root 87% cosine roll-off filter. The information bit rate may be fixed at 2401 bps to yield the coded bit rate of 2744 bps.

The DS-CDMA system discussed here is capable of rejecting narrowband interference from terrestrial radio systems. The interference reduction is $10 \log 255 = 24$ dB. Also, DS-CDMA systems avoid the difficulty for phase noise performance on the satellite that arises in single-channel-per-carrier (SCPC) systems, especially at low bit rate. In addition, an SCPC system must use very stable frequency sources and a frequency synthesizer that can accommodate a large number of carriers. The DS-CDMA system, on the other hand, requires no frequency synthesizer for SET that communicates only with those in the same channel and does not suffer the above disadvantages of SCPC systems. The price paid for is a low efficiency (0.03 bits/Hz in this case). The number of simultaneous access per channel can be increased by using PN sequences with a longer period than 255 chips.

11.3 FREQUENCY HOP SPREAD SPECTRUM SYSTEMS

An alternative to a DS spread spectrum is a *frequency hop spread spectrum*. A block diagram of a coded FH spread spectrum system is shown in Fig. 11.6. With the use of a frequency synthesizer controlled by a PN sequence generator, the encoded carrier is sequentially hopped into a

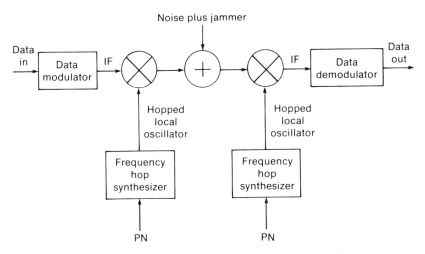

Figure 11.6 A frequency hop spread spectrum modulator and demodulator.

series of frequency slots into which the satellite bandwidth is partitioned. In order to recover the data, an identical frequency synthesizer controlled by an identical synchronized PN sequence generator must be used at the receiver. As in DS systems, the random hopping pattern determined by the PN sequence must be selected in such a way that it cannot be predicted. To an observer, the carrier appears to use the entire satellite bandwidth over a PN sequence period, although it occupies only a frequency slot at any instant in time (hence spread spectrum). The hopping rate R_h of the carrier may occur at the coded symbol rate R_c (slow hop) or several times the coded symbol rate (fast hop). The commonly used modulation schemes in FH systems are noncoherent M-ary FSK and DQPSK. Noncoherent demodulation is used because it is very difficult to maintain phase coherence between hops.

Against a FH spread spectrum link, a jammer can inflict severe damage on an uncoded system and wipe out the advantage of random hopping by using a *partial band jamming* strategy. Let B_s be the total satellite spread bandwidth and assume that the jammer concentrates his Gaussian interference with average power J into a fraction $0 < \delta < 1$ of the spread bandwidth B_s to yield an effective jamming density $J_0/2 >> N_0/2$, where

$$J_0 = \frac{J}{B_s} \qquad (11.32)$$

Here $N_0/2$ is the channel AWGN density. Now consider the slow-hop noncoherent FSK signal. The average probability of bit error for such a signal is given by (9.91) as

$$P_b = \frac{\delta}{2} \exp\left(-\frac{\delta E_b}{2J_0}\right) + \frac{1-\delta}{2} \exp\left(-\frac{E_b}{2N_0}\right)$$

$$\approx \frac{\delta}{2} \exp\left(-\frac{\delta E_b}{2J_0}\right) \approx \frac{\delta}{2} \exp\left(-\frac{\delta}{2} \frac{B_s}{R_b} \frac{C}{J}\right) \tag{11.33}$$

The worst-case bit error probability $P_{b,max}$ occurs at

$$\delta = \min\left(2 \frac{R_b}{B_s} \frac{J}{C}, 1\right) \tag{11.34}$$

at which value

$$P_{b,max} = 0.368 \frac{R_b}{B_s} \frac{J}{C} = 0.368 \left(\frac{E_b}{J_0}\right)^{-1} \tag{11.35}$$

for $(B_s/R_b)C/J > 2$. Note that the worst-case FH partial band jamming is similar to the worst-case DS pulsed jamming in (11.18). The difference is that in FH jamming the jammer does not pay the price of high peak power. By comparing (11.35) and (11.18) it is seen that, for the same jamming margin J/C and the same processing gain (B_s/R_b for FH and R_p/R_b for DS), noncoherent FSK-FH systems are about 6.5 dB worse than coherent PSK-DS systems. This is tantamount to quadrupling the processing gain in FH. But this can be misleading, since FH technology can achieve band spreading many orders of magnitude greater than that achieved by DS technology and, furthermore, FH systems can employ higher signaling alphabets such as M-ary FSK. In contrast, PSK is considered the most effective antijam modulation for DS sytems [6].

We have investigated the worst-case partial band jamming of slow-hop noncoherent FSK. In antijamming applications when it is necessary to prevent a type of jammer called a repeat-back jammer, to have sufficient time to intercept the frequency and retransmit it along with adjacent frequencies so as to create interfering signal components, fast FH can be employed where $L > 1$ hops per symbol are used. Fast-hop noncoherent FSK performs poorer in worst-case partial band jamming than its slow-hop counterpart because of the penalty incurred in a noncoherent combining of L-hopped elements of the signal in a signaling interval.

To combat partial band jamming, a FH system must employ coding like that used in pulsed jamming of DS systems. Such a procedure for finding the jamming margin J/C in a coded FH system is similar to that used with a coded DS system in Example 11.3.

Example 11.6 Consider a slow-hop noncoherent FSK spread spectrum system using a rate $R = \frac{1}{2}$ convolutional code of constraint length $K = 7$ with Viterbi soft-decision decoding as shown in Fig. 11.7. At $P_b = 10^{-5}$, the required $E_b/N_0 = 9.4$ dB, which yields $E_c/N_0 = 6.4$ dB. The corresponding transition probability for noncoherent FSK is given by (9.91) as $p' = \frac{1}{2} \exp(-E_c/2N_0) = 0.0564$. For the worst-case partial band

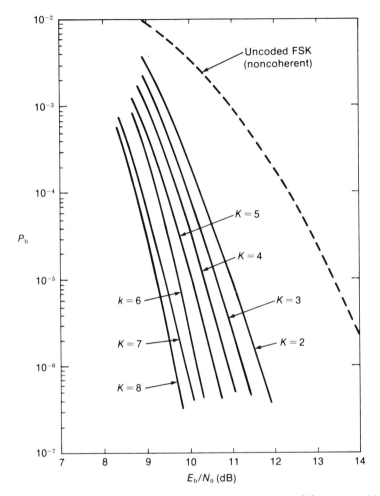

Figure 11.7 Bit error probability of coded noncoherent FSK [3] *(Reprinted by permission from Plenum)*

jamming, the required partial band jamming E_c/J_0 is [see (11.35)]:

$$\frac{E_c}{J_0} = 8.15 \text{ dB}$$

This yields, for rate $R = \frac{1}{2}$ code,

$$\frac{E_b}{J_0} = 11.15 \text{ dB}$$

Assume a processing gain $B_s/R_b = 40$ dB; then the jamming margin is $J/C = (B_s/R_b)(E_b/J_0)^{-1}$:

$$\frac{J}{C} = 28.85 \text{ dB}$$

Without coding, the jamming margin is given by (11.35) as

$$\left(\frac{J}{C}\right)_{\text{uncoded}} = 10 \log \frac{10^{-5}}{0.368} \times 10^4 = -5.65 \text{ dB}$$

Hence the coding gain is 34.5 dB.

11.4 FREQUENCY HOP CODE DIVISION MULTIPLE ACCESS

Like DS-CDMA, frequency hop can be used in a multiple access system where each earth station or user is assigned a distinct hop pattern. When a slow-hop noncoherent M-ary FSK system is used, each user's carrier occupies a minimum instantaneous bandwidth equal to $M/T_s = MR_b/\log_2 M$. When guard bands are taken into account, a carrier bandwidth of $2MR_b/\log_2 M$ is more likely. For a given spreading bandwidth B_s, the number of frequency slots is

$$n = \frac{B_s}{MR_b} \log_2 M \qquad (11.36)$$

The average probability of bit error in a slow-hop FH-CDMA link is (assuming random hopping for all users)

$$P_b = \frac{M/2}{M-1} \left[\frac{M-1}{M} P_h + (1 - P_h)P_s \right] \qquad (11.37)$$

where P_s = probability of symbol error for noncoherent M-ary FSK given in (9.115) and P_h = probability of a hit (another carrier hop into the same frequency slot). If there are $k + 1$ active users in n frequency slots, the probability of a hit for all possible arrangements is

$$P_h = 1 - \frac{(n-1)^k}{n^k} = 1 - \left(1 - \frac{1}{n}\right)^k \approx \frac{k}{n} \qquad k << n \qquad (11.38)$$

For $P_s << P_h << 1$, the bit error probability P_b is almost the same as $P_h/2$. To have a small P_h, one is required to increase the number of frequency slots n and reduce the number of users k. Unlike the situation in DS-CDMA, where the bit error probability is determined by the number of users and the full or partial cross-correlations of the PN sequences, in FH-CDMA the bit error probability is determined by the number of users, the spread bandwidth, and the bit rate. Since it is easier to increase the spread bandwidth with present technology, FH-CDMA can accommodate many users with a low bit rate and may be favored over DS-CDMA in many military applications. On the other hand, DS-CDMA may be more suitable for commercial users.

Example 11.7 Consider a slow-hop noncoherent FSK-FH-CDMA satellite system operating at a user bit rate of $R_b = 750$ bps. The spread bandwidth is $B_s = 15$ MHz. We wish to design a link with a bit error probability of $P_b = 10^{-5}$ to accommodate $k + 1 = 201$ users.

Without coding, the number of frequency slots in B_s is given by (11.36), $n = 10^4$. The link bit error probability is given by (11.37) and (11.38) as

$$P_b = \frac{1}{2} \frac{200}{10^4} + \left(1 - \frac{200}{10^4}\right) p = 0.01 + 0.98p$$

where $p = $ bit error probability due to noise and intermodulation. If $p << 0.01$, the link P_b is 0.01, which is not acceptable for communications.

To alleviate this situation, error-correction coding must be used. Consider the use of rate $R = \frac{1}{2}$ convolutional code with Viterbi decoding as shown in Fig. 11.7. The coded bit rate becomes $R_c = 2R_b = 1500$ bps. Therefore the number of frequency slots in B_s becomes $n = 5 \times 10^3$. The new link transition probability $\hat{p}$ due to all interferences is

$$\hat{p} = \frac{1}{2} \frac{200}{5 \times 10^3} + \left(1 - \frac{200}{5 \times 10^3}\right) p' = 0.02 + 0.96p'$$

where $p' = $ link transition probability due to noise and intermodulation. Now if we wish to make $\hat{p} = 0.04$, then the minimum tolerable p' must be $p' = 0.0208$. Since $\hat{p} = \frac{1}{2} \exp(-E_c/2\hat{\mathcal{N}}_0)$ we obtain a ratio of energy per coded bit to noise plus interference of $E_c/\hat{\mathcal{N}}_0 = 7$ dB. The corresponding ratio of energy per bit to noise plus interference for rate $\frac{1}{2}$ code is $E_b/\hat{\mathcal{N}}_0 = 10$ dB. Thus from Fig. 11.7, all convolutional codes with constraint length $K \geq 6$ yield $P_b < 10^{-5}$.

11.5 DS ACQUISITION AND SYNCHRONIZATION

To demodulate the data contained in a DS spread spectrum signal the receiver must correlate the received signal with a generated replica of the PN sequence such that the phases of the received PN sequence and its replica are synchronized in order to despread the spread signal. The process of initial searching through all possible phases until the received sequence phase is essentially the same as the generated sequence phase is called *acquisition*. A commonly used acquisition technique is the *stepped serial search* scheme shown in Fig. 11.8. Let the received signal be $c(t) = Ab(t)p(t) \cos(\omega_c t + \theta)$, where $b(t)$ is the bipolar data waveform and $p(t)$ is the PN sequence waveform. The bandpass filter is centered at the carrier frequency ω_c, and its bandwidth is equal to the data bandwidth. The output of the local PN sequence generator is $p(t + \Delta)$. Therefore the output of the bandpass filter is proportional to $p(t)p(t + \Delta)b(t) \cos(\omega_c t + \theta)$. The envelope detector outputs the enve-

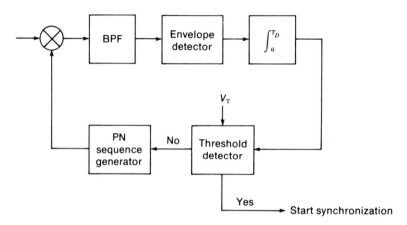

Figure 11.8 Stepped serial-search acquisition for a DS spread spectrum system.

lope $p(t)p(t + \Delta)$, hence after integration the detection voltage v is proportional to

$$v \propto \left| \int_0^{T_D} p(t)p(t + \Delta) \, dt \right|$$

Note that, when T_D is a multiple of the PN sequence period, v is proportional to the autocorrelation $R(\Delta)$ of $p(t)$. It is obvious that, when $\Delta = 0$, $p(t)$ and $p(t + \Delta)$ are aligned and v is maximized. In this system, the timing error, $\Delta = k(T_c/q)$, where q is the number of cells per chip, is initially set at $k = 0$ and a correlation is performed by examining $L = T_D/T_c$ chips, where T_D is the integration time. If the integrator output falls below the threshold V_T set to indicate that the phases of the two sequences are essentially the same, k is set to $k = 1$ and the search continues until the threshold V_T is exceeded. It is clear that in the worse case k might have been set at $k = 0, 1, 2, \ldots, qN - 1$, where N is the number of chips per period, before a correct value is found. Therefore, when the false alarm and detection probabilities are neglected, the worst-case acquisition time is simply $qNT_D = qNLT_c$. Note that an envelope detector is employed to obtain the envelope of the correlated signal. This eliminates the need for a coherent carrier to demodulate the spread signal, hence is a noncoherent acquisition.

It is intuitively clear that the acquisition time depends on the detection probability P_D and the false alarm probability P_F (probability of $v \geq V_T$ when the spread signal is not present). Assume that the received PN sequence and the generated replica PN sequence are initially offset by a random number of $0 \leq K \leq N - 1$ chips, where N is the number of chips per period. The acquisition time consists, in the mean, of two additive terms T_{a1} and T_{a2}. The term T_{a1} takes into account the mean time needed to search K chips, and the term T_{a2} represents the mean time required to search N addition chips if the correct phase is not obtained after searching K chips. Since the search is done by generating a $1/q$ chip delay and integrating an addition L chips, the total elapsed time is $qK(L + 1/q)T_c$, assuming that $v \geq V_T$ at the end of the search interval. However, if a false alarm occurs, no additional $1/q$ chip delay is generated until L additional chips are searched. Thus the mean time needed to search a random number of K chips is

$$T_{a1} = \frac{1}{N} \sum_{k=0}^{N-1} qK\left(L + \frac{1}{q}\right)T_c + LT_c P_F + 2LT_c P_F^2 + \cdots$$

$$= \frac{q(N-1)}{2} \left[\left(L + \frac{1}{q}\right)T_c + \frac{LT_c P_F}{(1 - P_F)^2}\right] \qquad (11.39)$$

If the correct phase is not obtained after searching the initial K chips, the search will continue with the next N chips with the mean time

$$T_{a2} = qN \left[\left(L + \frac{1}{q}\right)T_c + \frac{LT_c P_F}{(1 - P_F)^2}\right]$$

As the probability that the correct phase will be obtained during the jth search is $(1 - P_D)^j P_D$, the mean acquisition time is

$$T_{acq} = T_{a1} + T_{a2}(1 - P_D)P_D + 2T_{a2} (1 - P_D)^2 P_D + \cdots$$

$$= T_{a1} + \frac{1 - P_D}{P_D} T_{a2} \qquad (11.40)$$

In the serial search method it is important to set the threshold voltage V_T to obtain a high *detection probability* P_D and a low *false alarm probability* P_F. Both can be determined from the probability density function of the signal-plus-noise envelope given by (9.87) when the signal is present, and by (9.88) for the noise and jammer alone (again we assume the jammer is Gaussian). For convenience we write (9.87) and (9.88) in normalized form:

$$p(\text{signal} + \text{noise}) = r \exp\left(-\frac{r^2 + 2E/\mathcal{N}_0}{2}\right) I_0 \left(\sqrt{\frac{2E}{\mathcal{N}_0}} r\right)$$

$$p(\text{noise}) = r \exp\left(-\frac{r^2}{2}\right) \qquad (11.41)$$

where $E = $ carrier energy in $T_D = LT_c = nT_b$ seconds and $\mathcal{N}_0/2 = $ jammer power spectral density. We have

$$\frac{E}{\mathcal{N}_0} = n\left(\frac{E_b}{\mathcal{N}_0}\right) \qquad n = \frac{T_D}{T_b} \qquad (11.42)$$

Plots of (11.41) and (11.42) are shown in Fig. 11.9.

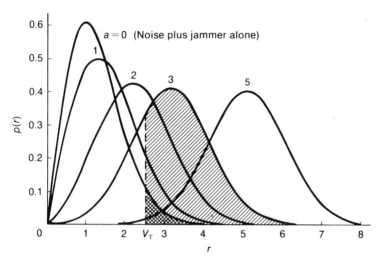

Figure 11.9 Probability density function of the envelope of signal plus noise and jammer.

The detection probability is the same as the probability that the envelope r will exceed the predetermined threshold V_T when the signal is present. The false alarm probability is the probability that the envelope r will exceed V_T for the noise and jammer alone:

$$P_D = \int_{V_T}^{\infty} p(\text{signal} + \text{noise})\ dr$$

$$= Q\left(\sqrt{2\frac{nE_b}{\mathcal{N}_0}},\ V_T\right) \tag{11.43}$$

$$P_F = \int_{V_T}^{\infty} p(\text{noise})\ dr$$

$$= Q(0, V_T) \tag{11.44}$$

where $Q(a,b)$ is the Marcum Q function defined as

$$Q(a,b) = \int_{b}^{\infty} r \exp\left(-\frac{r^2 + a^2}{2}\right) I_0(ar)\ dr \tag{11.45}$$

P_D is plotted in Fig. 11.10 as a function of $nE_b/\mathcal{N}_0$ with P_F as a parameter. For example, a 0.9 probability of detection and a 10^{-3} probability of a false alarm require $nE_b/\mathcal{N}_0 = 11$ dB. If the DS system operates at $E_b/\mathcal{N}_0 = 6$ dB (with coding), the required integration time must be about 3.2 times the bit duration T_b or sequence period NT_c if $NT_c = T_b$.

Once acquisition has been accomplished, the receiver must be able to maintain the phase alignment of the received PN sequence and the generated replica, normally within a fraction of the chip interval. This process of tracking the phase of the received PN sequence is called *syn-*

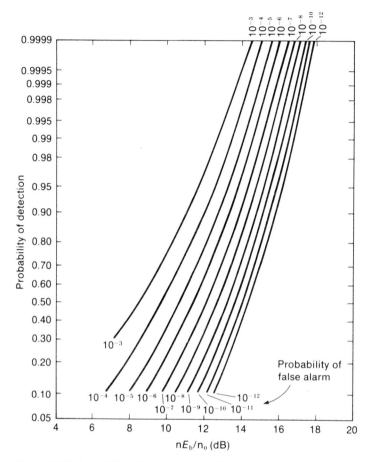

Figure 11.10 Probability of detection as a function of the probability of false alarm and energy per bit-to-noise density ratio *(M. I. Skolnik: Introduction to Radar Systems. Reprinted by permission from McGraw-Hill)*

chronization and can be achieved by a *delay-lock loop* as shown in Fig. 11.11. The received signal is multiplied by the early (half-chip interval) and the late (half-chip interval) replicas. The early and late product signals are then passed through bandpass filters to form the early and late correlation signals whose envelopes at the output of the envelope detectors are subtracted to yield a correction voltage which is then lowpass-filtered by the loop filter to remove the harmonic components and then used to drive a clock voltage-controlled oscillator. This clock in turn drives the PN sequence generator in such a way that, if the clock is lagging in phase, the correction signal will drive the clock faster and the replica PN sequence will speed up and run in coincidence with the received PN sequence, and vice versa. Thus the replica PN sequence tracks the received PN

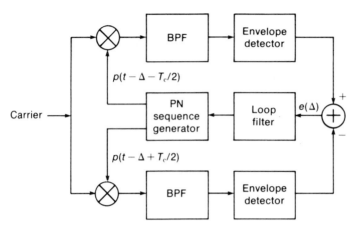

Figure 11.11 Delay-locked loop.

sequence provided that the initial phase error is small (the relative delay between the received PN sequence and the generated replica PN sequence is within a half-chip interval). The correction voltage $e(\Delta)$ is proportional to

$$e(\Delta) \propto \left| R\left(\Delta + \frac{T_c}{2}\right) \right| - \left| R\left(\Delta - \frac{T_c}{2}\right) \right|$$

and is plotted in Fig. 11.12. Note that $e(\Delta)$ is generated only if $|\Delta| < T_c/2$, that is, within a half-chip time of the clock. This is why acquisition is necessary to align the received PN sequence with its replica within at least half a chip.

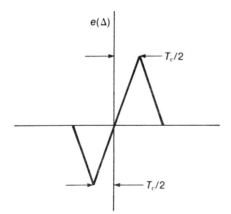

Figure 11.12 Correction voltage in the delay-locked loop.

11.6 FH ACQUISITION AND SYNCHRONIZATION

Acquisition for a FSK-FH system is similar to the serial search scheme shown in Fig. 11.8 for a DS system except that the PN sequence generator controls the frequency synthesizer that produces the hopping frequencies. The bandpass filter has a bandwidth $B = 1/T_h$, where T_h is the hopping interval and the integration time T_D is a multiple of T_h; that is, $T_D = nT_h$. Again, the timing period of the generated replica PN sequence is set, and the locally generated signal of the frequency synthesizer is correlated with the incoming signal. If at the end of the integration interval the threshold V_T is not exceeded, the search will continue with the next phase of the code. The acquisition performance of a FH system is similar to that of a DS system except that E_b in (11.43) is the carrier energy per hop E_h. Also, the output signal is the sum of n envelopes, each of which is a Rician random variable with a density function given by (11.41) and (11.42). The detection probability of the output signal can be expressed by the generalized Marcum Q function (Prob. 11.17) as

$$P_D = Q_n\left(\sqrt{2\frac{nE_h}{\mathcal{N}_0}}, V_T\right) \tag{11.46}$$

where

$$Q_n(a,b) = \int_b^\infty x\left(\frac{x}{a}\right)^{n-1} \exp\left(-\frac{x^2 + a^2}{2}\right) I_{n-1}(ax)\, dx \tag{11.47}$$

Here I_{n-1} is the modified Bessel function of the first kind of order $n - 1$.

The false alarm probability (Prob. 11.18) is simply

$$P_F = \int_{V_T}^\infty \frac{1}{(n-1)!} x\left(\frac{x^2}{2}\right)^{n-1} \exp\left(\frac{-x^2}{2}\right) dx \tag{11.48}$$

Once acquisition has been achieved, the synchronization system is activated to track the incoming hopping signals within a small error. The synchronization circuit for a FH system is shown in Fig. 11.13. The bandpass filter is wide enough to pass the product signal $V_1(t)V_2(t)$ when $V_1(t)$ and $V_2(t)$ are at the same frequency, but narrow enough to reject it when they are at different frequencies. Thus the output voltage $V_d(t)$ of the envelope detector is a constant when $V_1(t)$ and $V_2(t)$ are at the same frequency, and zero when they are at different frequencies, as shown in Fig. 11.14. The output $V_2(t)$ of the second mixer is a three-level signal. The lowpass filter voltage $V_f(t)$ will be negative if the hopping interval T_h of $V_2(t)$ occurs before the corresponding interval for $V_1(t)$, forcing the VCO to adjust its frequency to slow down the start of the hopping time. Similarly, when the hopping interval T_h of $V_2(t)$ starts after the corresponding interval for $V_1(t)$, the voltage $V_f(t)$ will be positive, forcing the VCO to adjust its frequency to speed up the start of the hopping time.

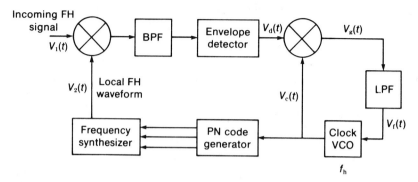

Figure 11.13 Synchronization for a FH spread spectrum system.

11.7 SATELLITE ON-BOARD PROCESSING

Throughout this book we have studied satellite links that involve a classical repeater where the error rate performance depends on the carrier-to-noise ratios of both the uplink and the downlink. The classical repeater is

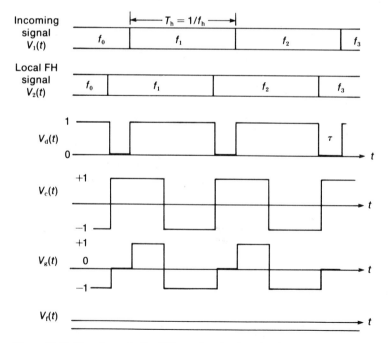

Figure 11.14 Waveforms in the FH synchronization circuit.

reliably field-proven and is used in all commercial applications today. For military applications, it may have certain drawbacks in regard to antijam capability. In military TWTA repeaters, the back-off is usually set by a hard limiter employed to prevent a strong jamming signal from driving the TWTA beyond saturation and seriously degrading the link performance. But the hard limiter itself produces a degradation on the uplink between 1 dB (strong Gaussian noise and a weak signal) and 6 dB (a strong signal and a weak signal). Furthermore, when the link employs a DS spread spectrum, the hard limiter also produces narrowband intermodulation products that are independent of the processing gain, and in many cases involving a small carrier-to-noise ratio, that can destroy the antijam capability.

To avoid the drawback of classical repeaters, many military satellites use *on-board processing* repeaters where waveform demodulation or remodulation, decoding or recoding, transponder or beam switching, interference reduction, and so on, are carried out at the satellite instead of on the ground. An on-board processing satellite essentially separates the uplink and downlink, hence reduces the link error rate P_b:

$$P_b = P_{b,u} + P_{b,d} - 2P_{b,u}P_{b,d} \approx P_{b,u} + P_{b,d} \qquad (11.49)$$

where $P_{b,u}$ and $P_{b,d} =$ uplink and downlink error rates. With on-board processing, the downlink does not have to be spread spectrum and can use any type of modulation or multiple access. For example, FDMA uplinks can be combined into a single TDMA downlink to achieve more efficient use of downlink power, especially for stations with mixed EIRP, and to avoid intermodulation. This results in a greater jamming margin for small stations, which cannot be achieved with a classical repeater.

As future military satellite communication goes to a higher frequency (EHF), the available bandwidth is increased to 2 GHz which is needed for a large J/C at higher data rates on the order of 100 kbps. Such large band spreading favors FH systems over DS systems because of the extremely high chip rate of PN signaling. Coupled with on-board processing, FH systems are attractive for moving platforms, while DS systems are desirable for fixed stations handling large amounts of traffic.

REFERENCES

1. R. C. Dixon, *Spread Spectrum Systems*, 2d. ed. New York: Wiley, 1984.
2. W. W. Peterson, *Error-Correcting Codes*. New York: Wiley, 1961.
3. G. C. Clark, Jr., and J. B. Cain, *Error-Correction Coding for Digital Communications*. New York: Plenum, 1981.
4. S. W. Golomb, *Shift Register Sequences*. San Francisco: Holden-Day, 1967.
5. D. V. Sarwate and M. B. Pursley, "Cross-Correlation Properties of Pseudorandom and Related Sequences," *Proc. IEEE*, Vol. 68, No. 5, May 1980, pp. 593–618.

6. F. Amoroso, "The Bandwidth of Digital Data Signals," *IEEE Commun. Mag.*, Nov. 1980, pp. 13–24.

7. M. B. Pursley and D. V. Sarwate, "Performance Evaluation for Phase-Coded Spread Spectrum Multiple Access communication. II. Code Sequence Analysis," *IEEE Trans. Commun.*, Vol. COM-25, No. 8, Aug. 1977, pp. 800–803.

8. E. B. Parker, "Micro Earth Stations as Personal Computer Accessories," *Proc. IEEE*, Vol. 72, No. 11, Nov. 1984, pp. 1526–1531.

9. J. G. Proakis, *Digital Communications*. New York: McGraw-Hill, 1983.

10. A. D. Whalen, *Detection of Signals in Noise*. New York: Academic Press, 1971.

11. C. W. Helstrom, *Statistical Theory of Signal Detection*. London: Pergamon, 1968.

12. R. Gagliardi, *Satellite Communications*. Lifetime Learning Publication, 1984.

APPENDIX 11A CHERNOFF BOUND

Given a real number $\lambda \geq 0$, the Chernoff bound states that

$$\Pr\{z \geq 0\} \leq E\{\exp(\lambda z)\} \qquad (11\text{A}.1)$$

Now let z be the random variable

$$z = -1 + \sum_{m=2}^{k+1} (\pm 1)\, \rho_{1m} \cos \theta_m + n \qquad (11\text{A}.2)$$

where n = independent Gaussian random variable with zero mean and variance σ_n^2. Substituting (11A.2) into (11A.1) yields

$$\Pr\{z \geq 0\} \leq \exp(-\lambda)\, E\{\exp(\lambda n)\}\, E\left\{\exp\left[\sum_{m=2}^{k+1} (\pm 1)\lambda \rho_{1m} \cos \theta_m\right]\right\}$$

$$\leq \exp(-\lambda) E\{\exp(\lambda n)\} \prod_{m=2}^{k+1} E\{\exp[(\pm 1)\, \lambda \rho_{1m} \cos \theta_m]\}$$

$$(11\text{A}.3)$$

Since

$$E\{\exp(\lambda n)\} = \frac{1}{\sigma_n \sqrt{2\pi}} \int_{-\infty}^{\infty} \exp(\lambda n)\, \exp\left(\frac{-n^2}{2\sigma_n^2}\right) dn$$

$$= \exp\left(\tfrac{1}{2}\, \sigma_n^2 \lambda^2\right) \qquad (11\text{A}.4)$$

$$E\{\exp[(\pm 1)\lambda \rho_{1m} \cos \theta_m]\} = \frac{1}{2\pi} \int_0^{2\pi} \left[\frac{1}{2} \exp(\lambda \rho_{1m} \cos \theta_m)\right.$$

$$\left. + \frac{1}{2} \exp(-\lambda \rho_{1m} \cos \theta_m)\right] d\theta_m$$

$$(11\text{A}.5)$$

Therefore

$$\Pr\{z \geq 0\} \leq \exp(-\lambda) \exp\left(\frac{\sigma_n^2 \lambda^2}{2}\right) \prod_{m=2}^{k+1} \frac{1}{4\pi}$$

$$\int_0^{2\pi} \left[\exp(\lambda \rho_{1m} \cos \theta_m) + \exp(-\lambda \rho_{1m} \cos \theta_m)\right] d\theta_m$$

$$(11A.6)$$

The tight upper bound is achieved by selecting the value of λ that minimizes the above bound.

PROBLEMS

11.1 Consider an uncoded PSK-DS spread spectrum signal and assume that the jammer at the output of the demodulator is not Gaussian. Using a Chernoff bound show that
   (a) $P_b \leq \exp(-E_b/J_0)$ if the jammer is continous.
   (b) $P_b \leq \delta \exp(-\delta E_b/J_0)$ if the jammer is pulsed.
   (c) Find the worse-case δ and $\max_{0<\delta<1} P_b$.

11.2 Consider a coded PSK-DS spread sprectrum signal and assume that the pulsed jammer is Gaussian with $J_0/\delta >> N_0$. The signal has a processing gain of 30 dB.
   (a) Find the maximum jamming margin for rate $\frac{1}{3}$ convolutional code of constraint length 7 with Viterbi soft-decision decoding at $P_b = 10^{-4}$, 10^{-5}, and 10^{-6}.
   (b) Find the coding gain.

11.3 Consider a coded PSK-DS spread spectrum signal and assume that the pulse jammer is not Gaussian with $J_0/\delta >> N_0$. The code used is the (23,12) Golay code with hard-decision decoding. Use the result of Prob. 11.1 to estimate the maximum tolerable jamming margin at $P_b = 10^{-6}$, assuming a processing gain of 30 dB, and the coding gain.

11.4 Consider a PSK-DS spread spectrum signal with PN sequence power spectral density $S_p(\omega)$. The signal is jammed by a continuous jammer whose power spectral density is $S_J(\omega)$. The receiver has a RF filter with transfer function $H(\omega)$. Determine the average output power caused by the jammer.

11.5 Design the small earth terminal using the system parameters as follows:
   Antenna noise temperature = 70 K
   LNA noise temperature = 90 K
   High-power amplifier = 7 dBW
   EIRP to saturate transponder = 80 dBW
   Transponder saturation EIRP = 36 dBW
   Satellite G/T = −3dB/K
   Modulation: PSK
   Information bit rate = 4.8 kbps
   PN sequence period = 255 chips
   Filtering: square root 100% cosine roll-off
   Performance without coding: $P_b < 10^{-5}$

11.6 Consider a sequence-synchronous CDMA satellite channel using PSK as a carrier modulation scheme. Assume that the interference from users 2 through $k + 1$ is uniformly distributed over the spread bandwidth B. Using the Chernoff bound show that the average probability of bit error for user 1 in upper-bounded by $P_b \leq \exp(-T_b B/k)$.

11.7 A 36-MHz satellite transponder is accessed by sequence-synchronous CDMA users with bit rate 9.6 kbps. Gold sequences with a length of $m = 10$ are used. The link carrier-to-noise plus interference ratio is $C/\mathcal{N} = -8$ dB. Assume PSK is used as carrier modulation. Find the number of users the transponder can accommodate at $P_b = 10^{-5}$.

11.8 DS-CDMA can be extended to serve a large population of users whose traffic is bursty (duty cycle is 0.01 or less). For this type of channel, data is grouped into packets, and users transmit their packet at will to a satellite channel. This form of random multiple access is termed direct sequence code division random multiple access. To analyze the packet throughput of a DS-CDRMA channel we assume an infinite population and that the channel can tolerate at most $k + 1$ overlapping packets. Let G be the channel traffic.

(a) Find the probability of a successful transmission if no retransmission attempt is made by a user after the first failure.

(b) Find the channel throughput in a no-retransmission mode.

(c) Find the probability of a successful transmission with a finite number of retransmissions after the first failure.

(d) Find the channel throughput in a retransmission mode.

11.9 Consider a FSK-FH satellite channel under partial band jamming. The channel employs rate $\frac{1}{2}$ convolutional code of constraint length $K = 5$ with Viterbi soft-decision decoding. Find the maximum tolerable jamming margin at $P_b = 10^{-6}$, assuming a processing gain of 30 dB. What is the coding gain?

11.10 Consider a fast-hop FSK-FH-CDMA satellite channel where M hops are performed per bit. The number of frequency slots in the channel is n. Find the probability of intercepting k users in one bit interval.

11.11 Consider a slow-hop noncoherent FSK spread spectrum signal using (23,12) Golay code with hard-decision decoding. Find the worst-case partial band jamming margin at $P_b = 10^{-5}$, assuming a processing gain of 33 dB. What is the coding gain?

11.12 Consider a slow-hop noncoherent FSK spread spectrum signal using rate $\frac{1}{2}$ convolutional code with Viterbi soft-decision decoding as shown in Fig. 11.7. The signal must achieve a bit error rate of 10^{-5} in a worst-case partial band jamming of δ percent of its slots. What is the required constraint length of the code if $\delta = 5$, 10, and 15 percent?

11.13 Consider a FSK-FHMA channel with 20,000 slots and 500 users using (23,12) Golay code. The link coded $E_c/N_0 = 9$ dB. Find the link average probability of bit error.

11.14 Consider the PSK-DS spread spectrum signal in Prob. 11.2. Find the integration time needed to obtain a probability of detection of 0.98 and a probability of a false alarm of 10^{-4}. The link bit error probability is 10^{-5}.

11.15 Let $x_1, x_2, \ldots, x_n$ be statistically independent Gaussian variables with zero mean and unit variance.

(a) Find the probability density function of $y_i = x_i^2$.

(b) Find the probability density function of $s = x_1^2 + x_2^2 + \cdots + x_n^2$.

(c) Find the probability density function of $s' = x'^2_1 + x'^2_2 + \cdots x'^2_n$, where x_i is a statistically independent Gaussian variable with zero mean and variance σ^2.

(d) Express the probability density functions of s and s' in terms of the gamma function $\Gamma(n/2)$, where

$$\Gamma(v) = \int_0^\infty z^{v-1} \exp(-z)\, dz \qquad v > 0$$

$$\Gamma(v) = (v - 1)! \qquad v \text{ an integer}, v > 0$$

$$\Gamma\left(\frac{1}{2}\right) = \sqrt{\pi} \qquad \Gamma\left(\frac{3}{2}\right) = \frac{\sqrt{\pi}}{2}$$

The probability density functions of s and s' are central chi-squared with n degrees of freedom.

11.16 Let $x_1, x_2, \ldots, x_n$ be statistically independent Gaussian variables with zero mean and variance σ_i^2. Show that the probability density function of

$$q = \sum_{i=1}^{n} (A_i/\sigma_i + x_i/\sigma_i)^2$$

is

$$p(q) = \frac{1}{2}\left(\frac{q}{\lambda}\right)^{(n-2)/4} \exp\left(-\frac{q}{2} - \frac{\lambda}{2}\right) I_{n/2-1} (\sqrt{q\lambda})$$

where

$$\lambda = \sum_{i=1}^{n} A_i^2/\sigma_i^2$$

This is called the noncentral chi-squared probability density function with n degrees of freedom.

11.17 Verify (11.46) using the result in Prob. 11-16.

11.18 Verify (11.48) using the result in Prob. 11.15.

Index

Index